EDEXCEL
GCSE MATHS
FOUNDATION

Marguerite Appleton, David Bowles, Dave Capewell, Geoff Fowler, Derek Huby, Jayne Kranat, Steve Lomax, Peter Mullarkey, James Nicholson, Matt Nixon and Katherine Pate

Powered by MyMath

OXFORD
UNIVERSITY PRESS

OXFORD
UNIVERSITY PRESS

Great Clarendon Street, Oxford, OX2 6DP, United Kingdom

Oxford University Press is a department of the University of Oxford. It furthers the University's objective of excellence in research, scholarship, and education by publishing worldwide. Oxford is a registered trade mark of Oxford University Press in the UK and in certain other countries

British Library Cataloguing in Publication Data
Data available

978-0-19-835150-4

10 9 8 7 6 5 4

Paper used in the production of this book is a natural, recyclable product made from wood grown in sustainable forests. The manufacturing process conforms to the environmental regulations of the country of origin.

Printed in India by Multivista Global Pvt. Ltd

Acknowledgements

The publisher would like to thank David Bowles, John Guilfoyle and Katie Wood for their contributions to this book.

Although we have made every effort to trace and contact all copyright holders before publication this has not been possible in all cases. If notified, the publisher will rectify any errors or omissions at the earliest opportunity.

Approval message from Edexcel

In order to ensure that this resource offers high-quality support for the associated Pearson qualification, it has been through a review process by the awarding body. This process confirms that; this resource fully covers the teaching and learning content of the specification or part of a specification at which it is aimed. It also confirms that it demonstrates an appropriate balance between the development of subject skills, knowledge and understanding, in addition to preparation for assessment.

Endorsement does not cover any guidance on assessment activities or processes (e.g. practice questions or advice on how to answer assessment questions), included in the resource nor does it prescribe any particular approach to the teaching or delivery of a related course. While the publishers have made every attempt to ensure that advice on the qualification and its assessment is accurate, the official specification and associated assessment guidance materials are the only authoritative source of information and should always be referred to for definitive guidance.

Pearson examiners have not contributed to any sections in this resource relevant to examination papers for which they have responsibility. Examiners will not use endorsed resources as a source of material for any assessment set by Pearson.

Endorsement of a resource does not mean that the resource is required to achieve this Pearson qualification, nor does it mean that it is the only suitable material available to support the qualification, and any resource lists produced by the awarding body shall include this and other appropriate resources.

Contents

1 Calculations 1

Introduction 2
Place value 4
Rounding 8
Adding and subtracting 12
Multiplying and dividing 16
Summary and review 20
Assessment 1 22

2 Expressions

Introduction 24
Terms and expressions 26
Simplifying expressions 30
Indices 34
Expanding and factorising 1 38
Summary and review 42
Assessment 2 44

3 Angles and polygons

Introduction 46
Angles and lines 48
Triangles and quadrilaterals 52
Congruence and similarity 56
Angles in polygons 60
Summary and review 64
Assessment 3 66

4 Handling data 1

Introduction 68
Sampling 70
Organising data 74
Representing data 1 78
Representing data 2 82
Averages and spread 1 86
Summary and review 90
Assessment 4 92

5 Fractions, decimals and percentages

Introduction 94
Decimals and fractions 96
Fractions and percentages 100
Calculations with fractions 104
Fractions, decimals and percentages 108
Summary and review 112
Assessment 5 114
Lifeskills 1: The business plan 116

6 Formulae and functions

Introduction 118
Substituting into formulae 120
Using standard formulae 124

Equations, identities and functions 128
Expanding and factorising 2 132
Summary and review 136
Assessment 6 138
Revision exercise 1 140

7 Working in 2D

Introduction 142
Measuring lengths and angles 144
Area of a 2D shape 148
Transformations 1 152
Transformations 2 156
Summary and review 160
Assessment 7 162

8 Probability

Introduction 164
Probability experiments 166
Expected outcomes 170
Theoretical probability 174
Mutually exclusive events 178
Summary and review 182
Assessment 8 184

9 Measures and accuracy

Introduction 186
Estimation and approximation 188
Calculator methods 192
Measures and accuracy 196
Summary and review 200
Assessment 9 202

10 Equations and inequalities

Introduction 204
Solving linear equations 1 206
Solving linear equations 2 210
Quadratic equations 214
Simultaneous equations 218
Inequalities 222
Summary and review 226
Assessment 10 228
Lifeskills 2: Starting the business 230

11 Circles and constructions

Introduction 232
Circles 1 234
Circles 2 238
Constructions 242
Loci 248
Summary and review 252
Assessment 11 254

12 Ratio and proportion

Introduction 256
Proportion 258
Ratio 262
Percentage change 266
Summary and review 270
Assessment 12 272
Revision exercise 2 274

13 Factors, powers and roots

Introduction 276
Factors and multiples 278
Prime factor decomposition 282
Powers and roots 286
Summary and review 290
Assessment 13 292

14 Graphs 1

Introduction 294
Drawing straight-line graphs 296
Equation of a straight line 300
Kinematic graphs 304
Summary and review 308
Assessment 14 310

15 Working in 3D

Introduction 312
3D shapes 314
Volume of a prism 318
Volume and surface area 322
Summary and review 328
Assessment 15 330
Lifeskills 3: Getting ready 332

16 Handling data 2

Introduction 334
Frequency diagrams 336
Averages and spread 2 340
Scatter graphs and correlation 344
Time series 348
Summary and review 352
Assessment 16 354

17 Calculations 2

Introduction 356
Calculating with roots and indices 358
Exact calculations 362
Standard form 366
Summary and review 370
Assessment 17 372

18 Graphs 2

Introduction 374
Properties of quadratic functions 376
Sketching functions 380
Real-life graphs 384
Summary and review 388
Assessment 18 390
Revision exercise 3 392

19 Pythagoras and trigonometry

Introduction 394
Pythagoras' theorem 396
Trigonometry 1 400
Trigonometry 2 404
Vectors 408
Summary and review 412
Assessment 19 414

20 Combined events

Introduction 416
Sets 418
Possibility spaces 422
Tree diagrams 426
Summary and review 430
Assessment 20 432
Lifeskills 4: The launch party 434

21 Sequences

Introduction 436
Sequence rules 438
Finding the nth term 442
Special sequences 446
Summary and review 450
Assessment 21 452

22 Units and proportionality

Introduction 454
Compound units 456
Direct proportion 460
Inverse proportion 464
Growth and decay 468
Summary and review 472
Assessment 22 474
Revision exercise 4 476

Formulae 478
Key phrases and terms 479
Answers 481
Index 526

About this book

This book has been specially created for the new Edexcel GCSE Mathematics 9-1.

It has been written by an experienced team of teachers, consultants and examiners and is designed to help you obtain the best possible grade in your maths GCSE.

As well as mathematical fluency, Assessment Objective 1 (AO1), the new course places an increased emphasis on your ability to reason, AO2, and your ability to apply mathematical knowledge to problem solving, AO3. This change of emphasis is built into the way topics are covered in this book.

In each chapter the lesson are organised in pairs. The first lesson is focussed on helping you to master the basic skills required (AO1) whilst the second lesson applies these skills in questions that develop your reasoning and problem solving abilities (AO2 & 3).

Throughout the book four-digit MyMaths codes are provided allowing you to link directly, using the search bar, to related lessons on the MyMaths website: so you can see the topic from a different perspective, work independently and revise.

At the end of a chapter you will find a summary of what you should have learnt together with a review section that allows you to test your fluency with the basic skills (AO1). Depending on how well you do a *What next?* box provides suggestions on how you could improve even further. This includes links to InvisiPen worked solution videos contained on the accompanying online Kerboodle. Finally there is an Assessment section which allows you to practise exam-style questions (AO1 – 3).

At the end of the book you will find a guide to understanding exam questions and a full set of answers to all the exercises.

The GCSE maths specification identifies two types of content at foundation level. A coloured band in the top-right corner indicates what type of content is included in a lesson.

Standard — All students should develop confidence and competence with this content.

Underlined — All students will be assessed on this content; more highly attaining students should develop confidence and competence with this content.

We wish you well with your studies and hope that you enjoy this course and achieve exam success.

1 Calculations 1

Introduction

When you go shopping in a supermarket, you are presented with hundreds of products, often looking similar, as well as lots of different offers. You need to be able to do arithmetic in your head to ensure that you are keeping within your budget and also that you choose the best value offer.

What's the point?

Being able to add, subtract, multiply and divide doesn't just mean that you're good at maths at school – it means that you can confidently look after your own finances in the real world.

Objectives

By the end of this chapter you will have learned how to …

- Use place value when calculating with decimals.
- Order positive and negative integers and decimals using the symbols $=$, $\neq$, $<$, $>$, $\leqslant$, $\geqslant$.
- Round to a number of decimal places or significant figures.
- Add and subtract positive and negative integers and decimals.
- Multiply and divide positive and negative integers and decimals.
- Use BIDMAS in multi-stage calculations.

Check in

1   Write the number three hundred and four in figures.

2   Calculate

   a 7×10 **b** $5 - 17$ **c** $40 \div 8$

3   Put these numbers in order starting with the smallest.

   $-8, -1, 2, -5, -3, 4$

Chapter investigation

Computers do not use the decimal system that we are familiar with, that is the digits 0 to 9. They use the binary system, which is composed only of the digits 1 and 0.

Investigate the binary system. How would you write the decimal number 37 in binary?

1.1 Place value

- In the **decimal system**, the value of each **digit** in a number depends upon its **place value**.

In the number 4237.65

Thousands 1000	Hundreds 100	Tens 10	Units 1	•	Tenths $\frac{1}{10}$	Hundredths $\frac{1}{100}$
4	2	3	7	•	6	5

You write this number in words as four thousand two hundred and thirty-seven point six five.

- Inequality and equality signs show the relationship between two numbers.

  < means less than   ≤ means less than or equal to   = means equal to

  > means greater than   ≥ means greater than or equal to   ≠ means not equal to

EXAMPLE

Place the correct symbol < > or = between the numbers in each pair.

  a 5.07 5.7   **b** 397 379   **c** −10 5   **d** −19 −24   **e** $\frac{3}{2}$ 1.5

  a 5.07 < 5.7   **b** 397 > 379   **c** −10 < 5   **d** −19 > −24   **e** $\frac{3}{2}$ = 1.5

You can use a place value table to multiply and divide by 10 or 100.

- To multiply a number by 10 move all the digits one place to the left.

- To divide a number by 100 move all the digits two places to the right.

$37 \times 10 = 370$

Hundreds	Tens	Units
	3	7
3	7	0

× 10

The 0 holds the digits in place.

$4850 \div 100 = 48.5$

Thousands	Hundreds	Tens	Units	•	Tenths
4	8	5	0	•	
		4	8	•	5

÷ 100

- **Negative numbers** are numbers below zero.

−14 is further away from zero than −13, so it is smaller.

EXAMPLE

Place these numbers in **order**, starting with the smallest.
−13, −14, 2, −5, −3, 4

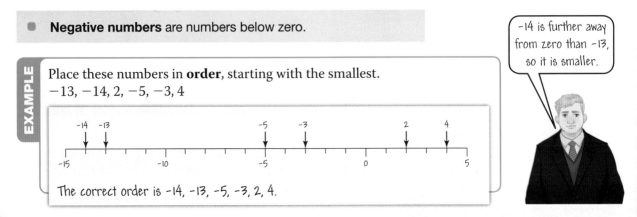

The correct order is −14, −13, −5, −3, 2, 4.

Exercise 1.1S

1 Write each of these numbers in figures.

a eighty-seven

b one hundred and forty-three

c four hundred and six

d four hundred and sixty

e two thousand and fifty-three

f eight thousand, five hundred and three

g eight thousand, five hundred and thirty

h thirty-four thousand, six hundred and forty

i thirty thousand, four hundred and sixty-four

j four hundred and seventeen point three

k five hundred and thirty-seven point four zero three

2 Write each of these numbers in words.

a 456   **b** 13 200

c 115 020   **d** 460 340

e 4 325 400   **f** 55 670 345

g 45.8   **h** 367.03

i 4503.34   **j** 2700.02

3 Put each list of numbers in order, starting with the smallest.

a 56   34   9   112   178   89   139

b 2372   1784   2386
     1990   3233   3022

c 40 500   45 045   4555
     4005   40 545   44 054

d 240 440   204 044   24 445
     245 004   42 024   242 404

4 Put these lists of numbers in order, starting with the smallest.

a 5.103   5.099   5.2   5.12   5.007

b 0.545   0.55   0.525   0.5   0.509

c 7.302   7.403   7.35   7.387   7.058

d 0.4   4.2   0.42   42   2.4

e 27.6   26.9   27.06   26.97   27.1

f 13.3   14.15   13.43   13.19   14.03

5 Place the correct symbol = or ≠ between the numbers in each pair.

a 3.6   3.60

b 450   four hundred and five

c 80.71   eighty-seven point one

d 9.50   nine point five

6 Place the correct symbol < or > between the numbers in each pair.

a 15   16   **b** 21   12

c 3.7   7.3   **d** 6.9   7

e 3.01   3.002   **f** 14.9   14.99

g −6   −5   **h** −8   −9

i −2.4   −2.7   **j** −14.75   −13.75

7 Put these lists of numbers in order from lowest to highest.

a −13   −6   0   17   −12   15

b 0   −5   −6   −8   −3   −7

c 3   8   6   −9   −1   2

d −1   −3   0   −4.5   5.5   −2.5

e −5   −5.1   −6   −5.8   −5.7   −5.4

8 Find the number that lies exactly halfway between each pair of numbers.

a 25 and 26   **b** 1.8 and 1.9

c 30 and 70   **d** 4.9 and 5

e 1.25 and 1.5   **f** 0.7 and 0.71

g −2.5 and −2.6   **h** −25 and −45

i −3.3 and −1.1   **j** −8.5 and −8.55

9 Calculate

a 12×10 **b** 4×100

c $320 \div 10$ **d** $4600 \div 100$

e 30×10 **f** 4.6×10

g $230 \div 100$ **h** $659 \div 10$

i 34×1000 **j** 3.56×100

k $23.6 \div 10$ **l** 0.345×100

m $12.4 \div 1000$ **n** 0.0814×100

 1013, 1069, 1072, 1392   SEARCH

1.1 Place value

RECAP

- The value of each digit in a number depends on its place value.
- To multiply a number by 10, move all the digits one place to the left.
- To divide a number by 10, move all the digits one place to the right.
- The inequality and equality signs are used to show the relationship between numbers.
- Negative numbers are below zero.

$<$ less than
$\leq$ less than or equal to
$>$ greater than
$\geq$ greater than or equal to
$=$ equal to
$\neq$ not equal to

HOW TO

1. Take time to understand the situation in the question.
2. Use your knowledge of place value.
3. Write the answer, include any units.

▲ Antarctica is the coldest place on earth with temperatures as low as $-94.7°C$.

EXAMPLE

Write the reading shown on the scale.

0   10   20   30 mm

1. The length of the pencil is between 20 and 30 mm.
2. There are 5 spaces between 20 and 30 mm.

   So 5 spaces represent 10 mm.

   Each space represents $10 ÷ 5 = 2$ mm.
3. The pencil is 22 mm long.

EXAMPLE

These are the top results from the Men's 400 m hurdles final at the London 2012 Olympics.
The times are given in seconds.
Who won the gold, silver and bronze medals?

Culson	48.10	Sanchez	47.63
Gordon	48.86	Taylor	48.25
Greene	48.24	Tinsley	47.91

1. The winner has the shortest time.
2. Find the three fastest times.
3. Gold: Sanchez   47.63s
   Silver: Tinsley   47.91s
   Bronze: Culson   48.10s

EXAMPLE

Use $14 \times 35 = 490$ to help solve this problem, without using a calculator.

Maria buys 35 bottles of water.
Each bottle costs £1.40.

How much does Maria spend?

1. Compare the cost of the water with the calculation you have been given.
   Cost of water $= 1.40 \times 35$
2. $14 = 1.40 \times 10$. The answer in the calculation, 490, is 10 times too big.
3. Divide 490 by 10. Give your answer in pounds.
   $490 ÷ 10 = 49$
   Maria spends £49.

Exercise 1.1A

1 Write the number each of the arrows is pointing to.

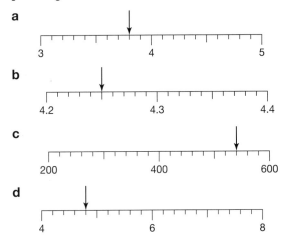

a

b

c

d

2 Write the reading shown on each scale.

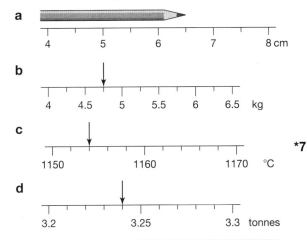

a

b

c

d

3 These are the top results from the Women's long jump final at the London 2012 Olympics.

The distances are given in metres.

DeLoach	6.89
Kolchanova	6.76
Nazarova	6.77
Sokolova	7.07
Radevica	6.88
Reese	7.12

a Who won the gold, silver and bronze medals?

b Radevica's personal best is 6.92 m. If she had jumped 6.92 m at the Olympics, would she have won a medal?

4 Dan has given a customer a bill for £356.28. He realises he has mixed up the 6 and the 2. How much does he have to pay back to the customer?

5 You are given that $\boxed{15 \times 14 = 210}$

Use this to help solve these problems, without using a calculator.

a The exchange rate between US dollars and British pounds is £1 = $1.40. Serena changes £15 into dollars. How many dollars will she receive?

b Malik wants to tile an area of 21 m². Each pack of tiles covers an area of 1.5 m². How many packs of tiles will Malik need?

6 Sonjia has five numbered cards.

| 3 | 5 | 9 | 7 | 1 |

a Write down the largest number that Sonjia can make with her cards.

b Use the cards to make a the smallest possible number that is greater than 20 000.

c Use the cards to make the largest possible number that is less than 60 000.

*7 Jessica has five symbol cards.

| < | ⩾ | = | ⩽ | > |

Jessica can choose one pair of numbers.

① 5.61   5.72

② 4.5 $4\frac{1}{2}$

Jessica scores one point for each card that she places between her chosen pair of numbers.

Jessica says that she will score more points if she chooses pair ①.

Explain why Jessica is wrong.

8 a List all of the numbers that can be made using the digits 1, 2 and 3. List the numbers in order: 1, 2, 3, 12, …

b List all of the numbers that can be made using the digits 4, 7 and 8. List the numbers in order: 4, 7, 8, 47, …

c How many four-digit numbers can be made using the numbers 3, 5, 6 and 9?

 Q 1013, 1069, 1072, 1392   SEARCH

1.2 Rounding

You can **round** a number to the nearest 10, 100, 1000 and so on.

You can also round a number to the nearest whole number or to a given number of **decimal places**.

Round 16.473

a to the nearest whole number   b to 1 decimal place

c to 2 decimal places   d to the nearest 10.

a 16.473 ≈ 16. Look at the tenths digit.

b 16.473 ≈ 16.5. Look at the hundredths digit.

c 16.473 ≈ 16.47. Look at the thousandths digit.

d 16.473 ≈ 20. Look at the units digit.

Look at the next digit. If it is 5 or more then the number is rounded up. Otherwise it is rounded down.

● The first digit that is not zero in a number is called the **first significant figure**. It has the highest value in the number.

You can round a number to one significant figure.

Round the numbers to one significant figure.

a 7560   b 52.3   c 1.5

	Thousands	Hundreds	Tens	Units	•	Tenths
a	7	5	6	0	•	
b			5	2	•	3
c				1	•	5

a First significant figure is 7, so round up to 8000.

b First significant figure is 5, so round down to 50.

c First significant figure is 1, so round up to 2.

Round 54.76 to 2 significant figures.

Look at the 3rd significant figure.

Tens	Units	•	Tenths	Hundredths
5	4	•	7	6

The 3rd significant figure is 7, so the number is rounded up to 55.

54.76 ≈ 55 (to 2 significant figures).

dp and **sf** are abbreviations for 'decimal places' and 'significant figures'.

Exercise 1.2S

1 Round each of these numbers to the nearest 10.

   a 48   **b** 89

   c 483   **d** 792

   e 2638   **f** 6193

2 Round each of these numbers to the nearest 100.

> You could use a number line sketch to help you.

   a 343   **b** 484

   c 882   **d** 2732

   e 5678   **f** 16491

3 Round each of these numbers to the nearest 1000.

   a 3448   **b** 2895

   c 4683   **d** 36927

   e 62532   **f** 261932

4 Round each of these numbers to the nearest
   i 1000   **ii** 100   **iii** 10.

   a 3472   **b** 81382

   c 1236.4   **d** 283.4

   e 13998   **f** 9999

5 Round each of these numbers to the nearest whole number.

   a 4.8   **b** 3.9

   c 11.6   **d** 25.074

   e 16.286   **f** 435.972

6 Round each of these numbers to the nearest whole number.

   a 3.7   **b** 8.7

   c 18.63   **d** 69.49

   e 109.9   **f** 6.899

7 Round each of these numbers to 1 decimal place.

   a 0.27   **b** 2.89

   c 3.82   **d** 12.48

   e 0.327   **f** 2.869

   g 3.802   **h** 14.458

   i 3.738   **j** 28.77

   k 468.63   **l** 369.29

8 Round each of these numbers to the nearest
   i 3 dp   **ii** 2 dp   **iii** 1 dp.

   a 3.4472   **b** 8.9482

   c 0.1284   **d** 28.3872

   e 17.9989   **f** 9.9999

   g 0.003987   **h** 2785.5555

9 Round each of these numbers to one significant figure.

   a 3487   **b** 3389

   c 14853   **d** 57792

   e 92638   **f** 86193

   g 3438.9   **h** 74899.36

10 Round these whole numbers to two significant figures.

   a 483   **b** 1206

   c 488   **d** 13562

   e 533   **f** 14511

11 Round these numbers to two significant figures.

   a 0.355   **b** 0.421

   c 0.0566   **d** 0.004673

   e 1.357   **f** 0.000004152

12 Round each of these numbers to the nearest
   i 3 sf   **ii** 2 sf   **iii** 1 sf.

   a 8.3728   **b** 18.82

   c 35.84   **d** 278.72

   e 1.3949   **f** 3894.79

   g 0.008372   **h** 2399.9

   i 8.9858   **j** 14.0306

   k 1403.06   **l** 140306

13 Round each number to the accuracy in brackets.

   a 9.732 (3 sf)   **b** 0.36218 (2 dp)

   c 147.49 (1 dp)   **d** 28.613 (2 sf)

   e 0.5252 (2 sf)   **f** 4.1983 (2 dp)

   g 1245.4 (3 dp)   **h** 0.00425 (3 dp)

   i 273.6 (2 sf)   **j** 459.97314 (1 dp)

Q 1001, 1004, 1005   SEARCH

1.2 Rounding

- Numbers round up if the next digit is a 5 or more and round down if the next digit is a 4 or less.
- When rounding to a given number of significant figures, start counting at the first non-zero digit.

HOW TO

1. Take time to understand the situation in the question.
2. Use your knowledge of rounding.
3. Write the answer, include any units.

EXAMPLE

The diagram shows a plan of Carli's garden.
Find the area of the garden.
Give your answer to

a 2 dp   **b** 2 sf.

12.45 m

9.94 m

1. Area = length × width

   Area = 12.45 × 9.94   Use your calculator.

   = 123.753

2. Round to 2 dp and 2 sf.

   a   123.753 = 123.75 (2 dp)   **b**   123.753 = 120 (2 sf)

3. Add units to the final answer.

   123.75 m²   120 m²

▲ The Miracle Garden in Dubai contains over 45 million flowers and is the world's largest natural flower garden.

EXAMPLE

Three friends share a prize of £40 equally.
How much do they get each?

1. Divide the prize money by the number of people.

   40 ÷ 3 = 13.333333

2. Money is given to 2 decimal places.

3. Give your answer in pounds and pence.

   They each get £13.33.

EXAMPLE

A shelf 2 m long is filled with files 42 cm wide.
How many files are there?

1. Convert 2 m to cm and divide by 42.

   2 m = 200 cm

   200 ÷ 42 = 4.7619...

2. 3. You round down here as you will not be able to fit in a 5th file.

   4 files.

▲ On a tax return you give all figures in pounds only – no pence. You round your earnings down, and your expenses up. So you pay less tax (well slightly) than if you followed the maths rules!

Exercise 1.2A

1 Here is a table showing the populations of five cities.

City	Population
London	7 172 091
Paris	2 142 800
New York	8 085 742
Mumbai	16 368 084
Beijing	7 441 000

Round the population of each city to the nearest 10 000 and then place the cities in order of size from smallest to largest.

2 Votes for four politicians were declared.

CON 25 958   LIB 2705

LAB 26 057   UKIP 5651

The local newspaper decides to round these to the nearest 1000 in its report.

a What would each result be reported as?

b Why would there be a problem if the election officials also decided to round to the nearest 1000?

c Would there be the same problem if they rounded to the nearest 100?

3 The diagram shows a plan of Phil's living room.

5.65 m
3.15 m

Find the area of Phil's living room.

Give your answer to

a 2 dp   **b** 2 sf.

c Phil wants to estimate the cost of a new living room carpet.
Should he use the answer from part **a** or **b**? Give reasons for your answer.

d A carpet fitter wants to know how much carpet to order.
Should he or she use the answer from part **a** or **b**? Give reasons for your answer.

4 **a** £175 is shared equally between eight people. How much money does each person receive?

b Elisa is moving house. She packs her belongings into 32 boxes.
Elisa's car can carry five boxes.
How many car journeys will Elisa need to make?

c A school is planning a coach trip.
There are 80 people on the trip.
Each coach can carry 24 people.
How many coaches are needed?

d A lorry can carry a maximum of 28 000 kg.
A crate weighs 800 kg.
How many crates can the lorry carry?

5 Andrea chooses two numbers from the list.

4.37	4.44	4.48	4.53
4.55	4.63	4.67	4.71

When she rounds the two numbers to 1 decimal place they are equal.
When she rounds the two numbers to 1 significant figure, they are not equal.
Find Andrea's numbers.

6 A pencil is measured to be 11.6 cm long to the nearest mm. What are the minimum and maximum possible lengths of the pencil?

7 A student has to work out $\sqrt{3} \times 25^3$ and leave her answer correct to 3 sf.

She makes a mistake and rounds each number to 3 sf before multiplying them.

a What answer does she get?

b She then tries again, this time rounding at the end. What is the difference between her two answers?

 1001, 1004, 1005   SEARCH

1.3 Adding and subtracting

There are two rules for adding and subtracting negative numbers.

- Adding a negative number is the same as subtracting a positive number.

- Subtracting a negative number is the same as adding a positive number.

EXAMPLE

Calculate

a $-5 + -6$

b $+4 - -2$

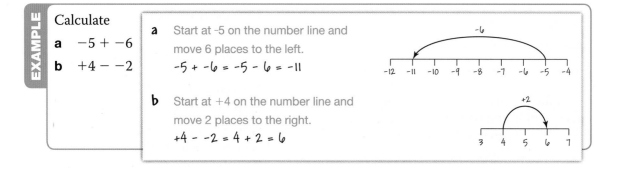

a Start at -5 on the number line and move 6 places to the left.
$$-5 + -6 = -5 - 6 = -11$$

b Start at +4 on the number line and move 2 places to the right.
$$+4 - -2 = 4 + 2 = 6$$

You can use mental methods to add and subtract whole numbers and decimals.

- Use **partitioning** to split the numbers you are adding or subtracting into parts.

- Use **compensation** when the number you are adding or subtracting is nearly a whole number, a multiple of 10 or a multiple of 100.

EXAMPLE

Calculate   **a** $19.5 - 7.2$ **b** $5.8 + 4.8$

a $19.5 - 7.2 = 19.5 - 7 - 0.2$ Split the smaller number into parts.
      $= 12.5 - 0.2$ Subtract the units from the highest number.
      $= 12.3$ Subtract the tenths.

b $5.8 + 4.8 = 5.8 + 5 - 0.2$ Rewrite **add 4.8** as **add 5 − 0.2**.
      $= 10.8 - 0.2$ Add the 5 to the highest number.
      $= 10.6$ Subtract 0.2.

> Part a uses partitioning and part b uses compensation.

To use a written method, set out the calculation in columns and line up the decimal points.

- When adding, you may need to carry digits into the next column.

- When subtracting, you may need to borrow from the next column.

- Always estimate the answer before you start.

EXAMPLE

Calculate these using a written method.

a $102.773 + 28.47$ **b** $26.44 - 1.105$

> If the numbers have different numbers of decimal digits, add extra zeros to the number with fewer digits.

a Initial estimate: $100 + 30 = 130$

```
  1 0 2 . 7 7 3
+   2 8 . 4 7 0
  1 3 1 . 2 4 3
```

b Initial estimate: $30 - 1 = 29$

```
      ³ 10
  2 6 . 4 4̶ 0
-   1 . 1 0 5
  2 5 . 3 3 5
```

Exercise 1.3S

1 Write the answer to each of these calculations.

   a $150 + 120$ **b** $170 - 90$

   c $160 + 170$ **d** $130 - 90$

   e $1900 + 1900$ **f** $210 - 140$

   g $320 + 110$ **h** $510 - 120$

2 Calculate

   a $13 + -5$ **b** $6 + -8$

   c $-5 + -8$ **d** $-3 + -11$

   e $4 - -8$ **f** $-2 - -5$

   g $-12 - -7$ **h** $-14 - -8$

   i $-12 + 7 - 4$ **j** $-12 + -7 - 4$

   k $-20 + -5 - -2$ **l** $-12 - 7 + -4$

3 Use a mental method for each of these calculations. Write the method you have used.

   a $257 + 98$ **b** $448 + 112$

   c $427 + 523$ **d** $256 + 552$

   e $354 + 213$ **f** $561 + 328$

   g $16.2 - 1.9$ **h** $5.8 + 14.9$

4 Use an appropriate method for each of these calculations.

   a $62 - 47$ **b** $83 - 68$

   c $487 - 356$ **d** $852 - 728$

   e $548 - 387$ **f** $589 - 387$

5 Use an appropriate method for each of these calculations.

   a $25 + 38 + 68$ **b** $123 + 76 - 58$

   c $173 - 27 + 56$ **d** $327 + 176 - 255$

6 Use a mental or written method to work out these calculations.

   a $33.4 + 15.2$ **b** $34.6 + 13.7$

   c $19.8 + 8.8$ **d** $18.7 + 26.5$

7 Use a mental or written method to work out these calculations.

   a $8.7 - 2.5$ **b** $15.8 - 8.4$

   c $26.3 - 7.9$ **d** $53.6 - 27.8$

   e $63.9 - 41.7$ **f** $81.2 - 58.6$

8 Use a written method to work out these additions.

   a $3.52 + 4.6$ **b** $13.62 + 2.9$

   c $8.5 + 14.81$ **d** $75.8 + 28.39$

9 Use a written method to work out these subtractions.

   a $17.3 - 4.22$ **b** $16.6 - 3.47$

   c $37.7 - 18.86$ **d** $57.28 - 38.4$

10 Use a written method for each of these calculations.

   a $16.4 + 9.87$ **b** $49.2 + 7.72$

   c $9.42 - 5.9$ **d** $26.9 + 9.82$

   e $36.57 - 8.59$ **f** $36.28 - 17.4$

11 Use a mental or written method to work out these calculations.

   a $23.4 + 13.4$ **b** $24.6 + 53.7$

   c $19.7 + 7.4$ **d** $27.8 + 14.3$

12 Use a mental or written method to work out these calculations.

   a $9.6 - 3.4$ **b** $16.7 - 9.6$

   c $16.3 - 7.8$ **d** $61.7 - 33.8$

13 Use a written method to work out these additions.

   a $4.32 + 6.4$ **b** $16.32 + 3.4$

   c $4.5 + 13.61$ **d** $73.2 + 68.79$

14 Use a written method to work out these subtractions.

   a $16.3 - 8.25$ **b** $12.6 - 7.87$

   c $67.3 - 28.56$ **d** $47.38 - 28.7$

15 Use a written method for each of these calculations.

   a $25.3 + 8.76$ **b** $38.1 + 6.61$

   c $8.31 - 4.8$ **d** $15.8 + 8.79$

   e $25.46 - 7.48$ **f** $47.39 - 18.5$

16 Calculate these using a mental or written method.

   a $12.3 + 2.7 + 7.08$

   b $38.76 + 16.9 - 8.32$

   c $61.3 + 14.85 + 7.02$

1007, 1020, 1028, 1068   SEARCH

1.3 Adding and subtracting

- You can use mental or written methods to add and subtract whole numbers and decimals.
- When you use a written method for adding or subtracting decimals, you should estimate first.
- When you do decimal addition or subtraction, you must take care with the position of the decimal points.

Line up the decimal points, and add or subtract the columns from right to left.

$$\begin{array}{r} 135.23 \\ +\ 27.8 \\ \hline 163.03 \\ {\scriptstyle 1\ 1} \end{array}$$

$$\begin{array}{r} {\scriptstyle 2\ 13\quad 1\ 5\ 10} \\ 3\cancel{4}.5\cancel{6} \\ -18.729 \\ \hline 15.831 \end{array}$$

HOW TO

1. Take time to understand the situation in the question. Estimate the answer when working with decimals.
2. Add or subtract using a mental or written method. Check that your answer agrees with your estimate.

You may need to carry digits into the next column.

You may need to borrow from the next column.

EXAMPLE

Hera the baby gorilla weighs 21.62 kg, and Horace her brother weighs 46.34 kg. The total weight of Henna the mother gorilla and her two children is 254.66 kg.

How much does Henna weigh?

1. Estimate the answer first. Round to the nearest 10.

$$20 + 50 = 70$$
$$250 - 70 = 180$$

2. Set out the calculation in columns, making sure you line up the decimal points.

$$\begin{array}{r} 21.62 \\ +46.34 \\ \hline 67.96 \end{array}$$

$$\begin{array}{r} {\scriptstyle 2\ 14\ 13} \\ 2\cancel{5}\cancel{4}.\cancel{6}6 \\ -\ 67.96 \\ \hline 186.70 \end{array}$$

Borrow digits from the next column. Try this subtraction for yourself!

Henna weighs 186.7 kg.

EXAMPLE

One of these calculations is correct. Decide which calculation has been carried out correctly without working out the answers. Explain what went wrong in the incorrect answers.

a
$$\begin{array}{r} {\scriptstyle 1\ 0\ 9\ 1} \\ \cancel{1}\cancel{1}.\cancel{0}3 \\ -2.55 \\ \hline 8.48 \end{array}$$

b
$$\begin{array}{r} 38.53 \\ +2.474 \\ \hline 6.327 \\ {\scriptstyle 1\ 1} \end{array}$$

c
$$\begin{array}{r} 100.773 \\ -28.782 \\ \hline 128.011 \end{array}$$

1. Estimate the answer to each calculation to decide which is correct.

a $11 - 3 = 8$ b $39 + 2 = 41$ c $100 - 30 = 70$

2. Explain what went wrong.

This one is correct.

Here the decimal points, and therefore all of the columns, were not properly aligned.

Here the smaller digit in each column has been subtracted from the bigger one.

Exercise 1.3A

1 Use a mental method of calculation to solve each of these problems.

   a Charlie has to travel 435 km. After 2 hours he has travelled 187 km. How much further does he have to travel?

   b In a test, Alex scores 93 marks and Sophie scores 75 marks. Alex, Sophie and Louise score 265 marks altogether. How many marks did Louise score?

   c A recycling box is full of things to be recycled. The empty box weighs 1.073 kg.

Bottles	12.45 kg
Cans	1.675 kg
Paper	8.7 kg
Plastic objects	? kg

    The total weight of the box and all the objects to be recycled is exactly 25 kg. What is the weight of the plastic objects?

2 Copy and complete this addition pyramid.

Each number is the sum of the two numbers beneath it.

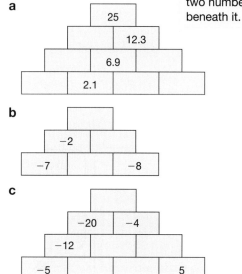

   a 25 / 12.3 / 6.9 / 2.1

   b −2 / −7, −8

   c −20, −4 / −12 / −5, 5

3 Find the missing digits.

   a
$$
\begin{array}{r}
2\,\square . 7 \\
\square 1. 5 \\
+\square 4. \square \\
\hline
2\,\square 8. 1
\end{array}
$$

   b
$$
\begin{array}{r}
3\,\square . 5\,8 \\
-\square 7 .7\,\square \\
\hline
8. \square 9
\end{array}
$$

4 Find the balance in these bank accounts after the transactions shown.

   a Opening balance £133.45. Deposits of £45.55 and £63.99, followed by withdrawals of £17.50 and £220.

     A negative number represents an overdraft.

   b Opening balance − £459.77. Deposit of £650, followed by a withdrawal of £17.85.

5 Each calculation is incorrect. Explain the mistake in each answer and find the correct answer.

   a
$$
\begin{array}{r}
3\,6 . 4\,1 \\
+ 6\,4 . 5\,6 \\
\hline
9\,0 . 9\,7
\end{array}
$$

   b
$$
\begin{array}{r}
1\,4\,6 . 5 \\
+ 9\,8 . 6 \\
\hline
2\,3\,5 . 1 \\
\scriptstyle 1\ \ 1\ \ 1
\end{array}
$$

   c
$$
\begin{array}{r}
3\,8\,5 . {}^8\!\!\not{9}\,{}^1 5 \\
- 2\,2\,4 . 6\,7 \\
\hline
1\,6\,9 . 2\,8
\end{array}
$$

   d
$$
\begin{array}{r}
{}^0\!\!\not{1}\ {}^{12}\!\!\not{3}\ {}^1 1 . {}^1 0\ 6 \\
- \qquad\quad 4\,3 . 2\,4 \\
\hline
8\,8 . 8\,2
\end{array}
$$

6 Fit the digits 1, 2, 3, 4, 5, 6, 7, 8 into the eight spaces to make this subtraction correct.

$$
\begin{array}{r}
\square\ \square\square \\
- \square\ \square\square \\
\hline
\square\square
\end{array}
$$

*7 **a** Find a pair of numbers in this table where their total is twice their difference.

   b Can you find a pair where the total is three times the difference?

   c Can you find four numbers where one pair adds up to twice the total of the other pair?

1026	432
724	342
1448	522

8 Lucia is making a bean casserole. Here is her recipe.

Bean casserole	
Kidney beans	1.6 kg
Red onions	0.375 kg
Celery	0.15 kg
French beans	▬
Tomatoes	1.2 kg

The total weight of ingredients is 3.525 kg. Find the weight of French beans.

Q 1007, 1020, 1028, 1068 SEARCH

1.4 Multiplying and dividing

Remember these rules:

- Multiplying or dividing a positive number by a negative number gives a negative number.
- Multiplying or dividing a negative number by a negative number gives a positive number.

If the signs are different the answer will be negative. If the signs are the same the answer will be positive.

For standard methods of multiplication and division, work with the significant digits from the numbers in the question.

Use an **estimate** to adjust the **place value** correctly.

EXAMPLE

Calculate

a 18.5×7.9

b $47.52 \div 1.8$

a   Estimate $20 \times 8 = 160$

```
      1 8 5
    ×   7 9
    1 6 6 5
  1 2 9 5 0
  1 4 6 1 5
```

b   Estimate $50 \div 2 = 25$

```
        2 6 4
  18)4 7 5 2
      3 6
      1 1 5
      1 0 8
          7 2
          7 2
            0
```

Use your estimate to adjust the place value.

$18.5 \times 7.9 = 146.15$ $47.52 \div 1.8 = 26.4$

- The order in which operations are carried out is **BIDMAS**

  Brackets, **I**ndices or powers, **D**ivision, **M**ultiplication, **A**ddition, **S**ubtraction

EXAMPLE

Calculate

a $11 + 13(8^2 - 47)$ **b** $30 \div (15 - (12 - 7))$

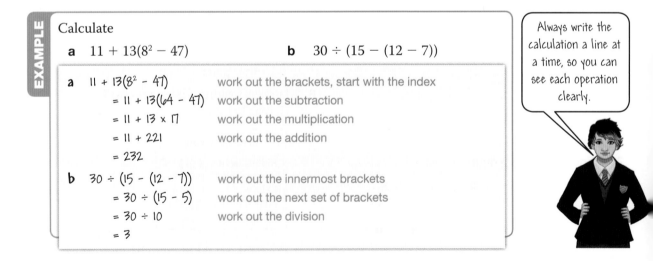

a $11 + 13(8^2 - 47)$ work out the brackets, start with the index

  $= 11 + 13(64 - 47)$ work out the subtraction

  $= 11 + 13 \times 17$ work out the multiplication

  $= 11 + 221$ work out the addition

  $= 232$

b $30 \div (15 - (12 - 7))$ work out the innermost brackets

  $= 30 \div (15 - 5)$ work out the next set of brackets

  $= 30 \div 10$ work out the division

  $= 3$

Always write the calculation a line at a time, so you can see each operation clearly.

Exercise 1.4S

1 Calculate

  a $+5 \times -5$ **b** $+4 \times 8$

  c $-8 \times +9$ **d** $-4 \times +5$

  e -3×-10 **f** -7×-7

  g $+8 \times +2$ **h** $+5 \times -4$

  i $-2 \times +9$ **j** -13×-2

  k $-7 \times +6$ **l** $+12 \times -4$

2 Calculate

  a $-18 \div +9$ **b** $-20 \div +4$

  c $-30 \div -6$ **d** $-12 \div -3$

  e $-66 \div +3$ **f** $+47 \div -47$

  g $-80 \div -2$ **h** $+24 \div +6$

  i $-45 \div -9$ **j** $-51 \div +3$

  k $+57 \div -19$ **l** $-81 \div -3$

3 Use a mental method to work out

  a 14×7 **b** 19×8

  c 21×13 **d** 17×19

4 Now use a written method to work out the answers for question **3**. Check that you get the same answers with both methods.

5 Use a written method to work out

  a 4.7×5.3 **b** 1.53×2.8

  c 21.6×4.9 **d** 33.65×3.89

  e 21.58×1.99 **f** 42.77×8.64

6 Use a written method to work out

  a $34.83 \div 9$ **b** $5.425 \div 7$

  c $7.328 \div 8$ **d** $451.8 \div 60$

  e $54.39 \div 3$ **f** $58.65 \div 17$

  g $66.4 \div 16$ **h** $185.76 \div 24$

  i $7.752 \div 1.9$ **j** $3.055 \div 1.3$

7 Use a written method to work these out, giving your answers correct to two decimal places.

  a $14.73 \div 2.8$ **b** $51.99 \div 1.8$

  c $193.8 \div 0.14$ **d** $1013 \div 5.77$

  e $23.78 \div 0.83$ **f** $65.79 \div 0.59$

8 Use a calculator to check your answers to questions **5–7**.

9 Given that $43 \times 67 = 2881$, find

  a 4.3×6.7 **b** 430×0.067

  c $2881 \div 670$ **d** $28.81 \div 430$

  e $2.881 \div (0.43 \times 0.67)$

10 Calculate these using the order of operations.

  a $2 + 8 \times 3$ **b** $4 \times 11 - 7$

  c $4 \times 3 + 5 \times 8$ **d** $5 + 12 \div 6 + 3$

  e $(2 + 9) \times 3$ **f** $(1.5 + 18.5) \div 4$

  g $(12 + 3) \times (14 - 2)$

  h $5 + (3 \times 8) \div 6$

11 Calculate these using the order of operations.

  a $(4 + 3) \times 2^2$ **b** $3^2 \times (15 - 7)$

  c $(6^2 - 16) \div 4$ **d** $2^4 \times (3^2 - 2 \times 4)$

  e $128 \div (2 + 2 \times 3)^2$

  f $(8^2 - 7^2) \times 5$

12 Calculate

  a $\dfrac{7^2 - 9}{5 \times 8}$ **b** $\dfrac{4 \times 8}{4^2}$

  c $\dfrac{15 \times 4}{6 \times 5}$ **d** $\dfrac{2 \times (3 + 4)^2}{7}$

  e $\dfrac{(6 + 4)^2}{20} + 7 \times 5$

  f $\dfrac{6 + (2 \times 4)^2 + 7}{11}$

13 Calculate

  a $12 + 6 - 4$ **b** $5 \times 4 \div 2$

  c $(40 \div 10) \div 2$ **d** $28 - 12 - 4$

  e $13(2 + 5)$ **f** $14(2 + 6)$

  g $5^2 + 9(8 - 3)$ **h** $4^2 + 3(16 - 9)$

  i $4 + (12 - (3 + 12))$

  j $120 \div (8 \times (7 - 2))$

14 Solve each of these calculations.

  a $(15.7 + 1.3) \times (8.7 + 1.3)$

  b $\dfrac{7^2}{(2.3 \times 4)^2}$ **c** $\dfrac{(7 + 5)^2}{(25 + 7 \times 8)}$

> Decide whether to use a mental, written or calculator method. Where appropriate give your answer to 2 decimal places.

 Q 1167, 1393, 1916, 1917   SEARCH

1.4 Multiplying and dividing

- You can make an **estimate** before starting a written calculation. You can use it to check your answer or to adjust place value.

- In a multi-stage calculation you need to remember the **order of operations, BIDMAS**.

BIDMAS

Brackets
Indices (or powers)
Division or
Multiplication
Addition or
Subtraction

1. Take time to understand the situation in the question.
   Estimate the answer when working with decimals.

2. Multiply or divide using a mental or written method.
   Remember to use BIDMAS.
   Check that your answer agrees with your estimate.

A firework display costs £28 800.
The display lasts for 15 minutes.
On average eight fireworks are set off every second.
What is the average cost of a firework?

1. Divide the cost by the total number of fireworks.

2. Start by finding the number of fireworks set off in the display.
   Use a mental method to convert 15 minutes to seconds.

15 minutes = 15 × 60 seconds = 900 seconds   15 × 6 × 10 = 90 × 10

Total fireworks = 900 × 8 = 7200   9 × 8 × 100 = 72 × 100

Find the average cost.

Average cost = 28800 ÷ 7200

= 288 ÷ 72

= 4

The average cost is £4 per firework.

Adam explained how he would calculate $\dfrac{5 + \sqrt{9}}{4}$
What is the problem here?
Find the correct answer.

> There is a root, an addition and a division. I'll do the root first, then divide, and then add.

1. Explain how Adam applied BIDMAS incorrectly.

Even though there are no brackets in this expression, the whole of the 'top line' is divided by 4, so you need to find the square root, then add and then divide.

2. Find the correct answer. The expression could be written as $(5 + \sqrt{9}) \div 4$.

$(5 + \sqrt{9}) \div 4 = (5 + 3) \div 4$

$= 8 \div 4$

$= 2$

Exercise 1.4A

1 Skye buys 18 packets of rice.
Each packet of rice costs £1.17.
Skye also buys 12 packets of pasta.
She spends £37.50 in total.
How much does one packet of pasta cost?

2 Nancy says, 'I am working out $36 \div (2 + 3)$.
I'll get the same result if I do $36 \div 2$ and
$36 \div 3$, and then add the answers together.'
Explain why Nancy is wrong.

3 Copy each of these calculations.

Insert brackets where necessary to make
each of the calculations correct.

a $5 \times 2 + 1 = 15$

b $5 \times 3 - 1 \times 4 = 40$

c $20 + 8 \div 2 - 7 = 17$

d $2 + 3^2 \times 4 + 3 = 65$

e $2 \times 6^2 \div 3 + 9 = 33$

f $4 \times 5 + 5 \times 6 = 150$

4 Fill in the missing numbers to make the
multiplications complete.

The carry digits are not included in parts **b**
and **c**.

a

```
        □ □ □
  ×       □ □
  ─────────────
    3  0₂ 5₄ 9
+ 3  4₂ 9₅ 6  0
─────────────
  3  8₁ 0₁ 1  9
```

b

```
        □ □ □
  ×       □ 9
  ─────────────
    7  6  5  9
+ 4  2  5  5  0
─────────────
  5  0  2  0  9
```

c

```
        □ □ □
  ×       □ 9
  ─────────────
    3  8  6  1
+ 1  □  8  7  0
─────────────
  □  6  □  3  1
```

5 Fill in the missing numbers to make the
divisions complete.

a

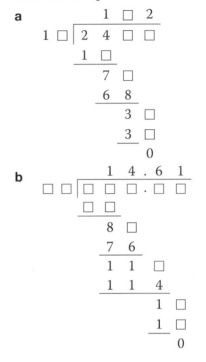

b

6 There are 18 questions in a quiz.

A correct answer scores 4 points.
An incorrect answer scores -3 points.

A question not answered scores nothing.
It is possible to have a negative total.

a What are the maximum and minimum
points that you could score on the quiz?

b Anna answers seven of the questions.
Three are correct. Explain why her
score is zero.

c Write down three different ways in
which a student could have a total score
of 12 points.

*7 Think of a number between 100 and 999.
Repeat it to make it a 6-digit number.
Enter this number on your calculator.
Now divide it by 13.
And then by 11.
And then by 7.
Do you end up with what you first thought
of? Does it work for any 3-digit number?

> Hint: Multiply your number by 1001.
> What do you notice?

Summary

Checkout

You should now be able to...

	Test it Questions
✔ Use place value when calculating with decimals	**1, 2**
✔ Order positive and negative integers and decimals using the symbols =, ≠, <, >, ≤, ≥.	**3, 4**
✔ Round to a number of decimal places or significant figures.	**5, 6**
✔ Add and subtract positive and negative integers and decimals.	**7 – 9**
✔ Multiply and divide positive and negative integers and decimals.	**10 – 13**
✔ Use BIDMAS in multi-stage calculations.	**14, 15**

Language   Meaning   Example

Language	Meaning	Example
Decimal system	A number system using a base of 10.	Thousands Hundreds Tens Units • tenths hundredths
Digit	The individual symbols 0, 1, 2, 3, 4, 5, 6, 7, 8, 9 that are used on their own or put together to make numbers.	$1000 \quad 100 \quad 10 \quad 1 \quad \frac{1}{10} \quad \frac{1}{100}$ $5 \quad 6 \quad 3 \quad 8 \quad • \quad 2 \quad 7$ 5, 6, 3, 8, 2 and 7 are digits. In the number 5638.27 the 6 has a value of 6 hundreds.
Place value	The value of a digit depends on its position in the number.	
Negative numbers	A negative number is less than zero.	$-3, -2, -1$
Rounding	Making a number less accurate but easier to estimate with.	552.631 = 553 (nearest whole number)
Decimal places (dp)	The number of digits after the decimal point. A number can be rounded to a given number of decimal places.	552.631 = 552.63 (2 dp)
Significant figures (sf)	Describe the relative importance of digits in a number. A number can be rounded to a given number of significant figures.	In the number 0.00487 the 4 is the 1st significant figure, 8 the 2nd and 7 the 3rd.
First significant figure	The first digit from the left that is not zero.	
Partitioning	Splitting a number into smaller numbers which add up to the original number.	127 = 100 + 20 + 7 152 = 80 + 40 + 32 5.1 + 12.7 = 5.1 + 10 + 2 + 0.7
Compensation	One number is rounded to simplify the calculation, then the answer is adjusted to compensate for the original change.	142 − 39 = (142 − 40) + 1 = 102 + 1 = 103 158 − 18.9 = (158 − 20) + 1.1 = 139.1

Significant figures example table:

4 sf	3 sf	2 sf	1 sf
45920	45900	46000	50000
78.02	78.0	78	80
0.003256	0.00326	0.0033	0.003

Review

1 Work out the value of these expressions.

   a 67×100 **b** 8.52×10

   c 0.24×1000 **d** 0.05×100

2 Work out the value of these expressions.

   a $450 \div 10$ **b** $6210 \div 1000$

   c $7.9 \div 100$ **d** $0.06 \div 10$

3 Copy the numbers and write $>$ or $<$ between them to show which is larger.

   a $905 \square 961$ **b** $14.7 \square 14.9$

   c $0.7 \square 0.09$ **d** $0.214 \square 0.22$

4 Put each list of numbers in order, starting with the smallest.

   a 53909 503099 530909 503909 53099

   b 4.3 4.289 4.32 4.09 4.29

   c -8 9 -4 0 -14

5 Round

   a 845 to the nearest 10

   b 25.3 to the nearest whole number

   c 0.846 to 1 decimal place

   d 62.938 to 2 decimal places.

6 Round to the stated level of accuracy

   a 351 1 sf   **b** 5070 2 sf

   c 45.72 3 sf   **d** 0.0845 1 sf

   e 0.0902 2 sf   **f** 0.99 1 sf.

7 Calculate.

   a $4 + -5$ **b** $-3 + -2$

   c $5 - -1$ **d** $-3 - -5$

   e $-7 - -4 + 3$ **f** $6 - -3 + -12$

8 Calculate using mental or written methods.

   a $467 + 891$ **b** $14.9 + 23.5$

   c $905.4 + 8.67$ **d** $0.58 + 6.821$

9 Calculate using mental or written methods.

   a $965 - 45$ **b** $657 - 389$

   c $257.4 - 38.2$ **d** $9.57 - 5.9$

10 Calculate

   a $-3 \times +7$ **b** -8×-4

   c $-35 \div +7$ **d** $-52 \div -4.$

11 Calculate using mental methods.

   a 12×20 **b** 25×16

   c $240 \div 6$ **d** $960 \div 120$

12 Calculate using written methods.

   a 47×63 **b** 192×78

   c $224 \div 8$ **d** $312 \div 12$

13 Calculate using written methods.

   a 3.2×5.6 **b** 4.31×2.7

   c $19.6 \div 8$ **d** $118.65 \div 21$

14 Evaluate these without using a calculator.

   a $13 + 5 \times 4$ **b** $20 - 12 \div 4$

   c $5 \times (2 + 9)$ **d** 5×3^2

   e $12 + \sqrt{36}$ **f** $7(9 - 1) \div 2^2$

15 Use your calculator to work these out.

   a $4.8 + 5.2 \times 6$

   b $\dfrac{39 - \sqrt{25} + 24}{4^2}$

What next?

<table>
<tr><td rowspan="3">Score</td><td>0 – 5</td><td></td><td>Your knowledge of this topic is still developing.
To improve look at MyMaths: 1001, 1004, 1005, 1007, 1013, 1020, 1028, 1068, 1069, 1072, 1167, 1392, 1393, 1916, 1917</td></tr>
<tr><td>6 – 12</td><td></td><td>You are gaining a secure knowledge of this topic.
To improve look at InvisiPens: 01Sa – p</td></tr>
<tr><td>13 – 15</td><td></td><td>You have mastered these skills. Well done you are ready to progress!
To develop your exam technique look at InvisiPens: 01Aa – f</td></tr>
</table>

Assessment 1

1 Carli tries to order each set of numbers from smallest to largest.
   One pair of numbers in each list is in the wrong order.
   Explain where Carli has made a mistake and put each list of numbers in order,
   starting with the smallest.

   a   8, 19, −33, 44, 303, 576   [3]   b   −19, −576, 8, 33, 44, 303   [3]

2 Ben tries to put these numbers in ascending order.

   $42 \div 100, 0.3 \times 10, 4236 \div 1000, 516 \div 10, 42 \times 100, 216 \times 1000$

   Has he ordered the numbers correctly? Give reasons for your answer.   [4]

3 In the number grids shown the number in each cell is the sum of the two adjacent
   cells above it. Copy and complete the grids shown.

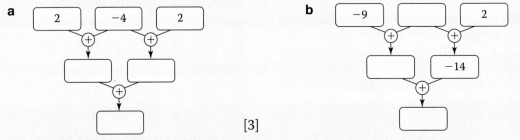

   a   b

   [3]   [3]

4 A magic square is a square grid of numbers where each number is different.
   The sum of the numbers in each row, each column and each diagonal is the same.

   a   What number must be the sum of each row, column and diagonal?   [1]

   b   Fill in the missing values in the magic square.   [5]

7		
6	8	10

5 The world's tallest man, Robert Wadlow, was 271.78 cm (2 dp) tall when measured to 2 dp.
   The world's tallest woman, Yao Defen, is 233.34 cm (2 dp) tall when measured to 2 dp.

   a   A challenger to the world's tallest man record measured his height as 271.8 cm
       to 1 dp. Has the challenger definitely beaten the world record?
       Give reasons for your answer.   [1]

   b   A challenger to the world's tallest woman record measured her height as
       233.341 cm. Has the challenger definitely beaten the world record?
       Give reasons for your answer.   [1]

6 Dave is 36 and Jane is 44. Jane says that she and Dave are the same age to 1 sf.

   a   Is Jane correct?   [1]

   b   Will Dave and Jane be the same age to 1 sf in one years time?
       Give reasons for your answer.   [2]

   c   How old will Dave and Jane be the next time their ages are the same to 1 sf?   [2]

7 As the Earth spins on its axis, everything on the Earth's surface moves with it.

The distance travelled in one day due to the Earth's rotation is $3.142x$, where x is the diameter of the circular path.

Abena lives on the equator and Edward lives in the UK.

$x = 12\,756$ for Abena and $x = 8134$ for Edward

  a Write both values of x to 2 sf.   [2]

  b Use your answer to part **a** to estimate how much further Abena travels than Edward in one day.   [3]

  c Explain how using values of x to correct to 1 sf would affect the estimate in part **b**.   [3]

(Labels near globe: Edward's path, Abena's path)

8 Maria writes some calculations.
Which calculation gives a different answer in each group?

  a   **A** 3.45×8 **B** 4.6×6 **C** 3.95×7 [1]

  b   **A** 3.68×9 **B** 4.74×7 **C** 4.14×8 [1]

  c   **A** $36.6 \div 8$ **B** $55.2 \div 12$ **C** $45.75 \div 10$ [1]

  d   **A** $222.2 \div 20$ **B** $88.8 \div 8$ **C** $199.8 \div 18$ [1]

9 Amelia has eight cards labelled **A** to **H**.
The bottom half of each card has a question and the top half has an answer.
Choose any card and answer the question.
Find the card with the matching answer and answer the question at the bottom.
Now find the card with the answer. Keep going until you have answered all the questions and write down the sequence of your answers.
The question on your final card should match the answer on your first card.   [10]

A	B	C	D
20.16	6.8	29.6	17.71
$7.4 \times 4 = ?$	$88.55 \div 5 = ?$	$4.88 \times 2 = ?$	$134.2 \div 11$

E	F	G	H
46.27	9.76	12.2	5.4
$48.6 \div 9 = ?$	$6.61 \times 7 = ?$	$6.72 \times 3 = ?$	$27.2 \div 4 = ?$

10 Jasmine's bike has wheels of circumference 2.5 m.
When Jasmine cycles to school, the wheels go round 850 times.
How far does Jasmine cycle to school?   [3]

11 Work out the following without using a calculator. Include units in your answer.

  a A bag of sweets weighing 113 g includes wrappings of 0.5 g. Each sweet weighs 4.5 g. How many sweets are in the bag?   [2]

  b A football stadium has 135 400 m² of seating for its fans. Each fan is allowed 5.4 m² of space. How many fans, to the nearest 1000, is the stadium capable of holding?   [2]

12 Write one pair of brackets in each calculation to make the answer correct.

  a $3 + 4 \times 5 + 2 = 37$ [1]   **b** $60 \div 5 + 7 + 5 = 10$ [1]

2 Expressions

Introduction

The real world is messy and complicated, and always in motion. Algebra is a vital part of maths because it attempts to describe aspects of the world, such as fluid flow or the forces acting on a suspension bridge. Through equations and formulae, algebra provides a mathematical model to describe the real-world situation, from which understanding can be gleaned and predictions made. It is only able to do this if it makes assumptions that simplify the situation, meaning that the model is only ever an inaccurate reflection of the real world. However simplifying a situation helps us to understand the forces that lie behind it.

What's the point?

Without algebra, we would not be able to work with large mechanical forces – so there would be no skyscrapers or suspension bridges; we would also not be able to understand electronics, so there would be no tablets or mobile phones.

Objectives

By the end of this chapter you will have learned how to …

- Use algebraic notation.
- Substitute numbers into formulae and expressions.
- Use and understand the words expressions, equations, formulae, terms and factors.
- Collect like terms and simplify expressions involving sums, products, powers and surds.
- Use the laws of indices.
- Multiply a single term over a bracket.
- Take out common factors in an expression.

Check in

1 Work out these powers.

   a 3^2 **b** 2^2 **c** 4^2

2 There are 5 CDs in a packet.
   How many CDs are there in 3 packets?

3 Work out

   a $4 - -3$ **b** $2 + -3$ **c** $-3 + 5$ **d** $-4 - -1$

   e -3×2 **f** 4×-2 **g** $6 \div -3$ **h** $-8 \div -2$

4 **a** Write all the factors of these numbers.

     i 18   **ii** 12   **iii** 24

   b Write all the common factors of 18, 12 and 24.

   c What is the highest common factor of 18, 12 and 24?

Chapter investigation

Think of a number between 1 and 10.

- Double it.
- Add 4.
- Halve your answer.
- Take away the number you first thought of.

What do you notice? Investigate why this is the case.

Can you invent different instructions that give similar results?

2.1 Terms and expressions

- An **expression** is a collection of letters and numbers with no = sign, for example $3x + 1$

- An **equation** contains an = sign, and an unknown letter to be solved, for example $3x + 1 = 10$

- A **formula** is a relationship between two or more letters, and it contains an = sign, for example $P = IV$

- **Terms** are groups of symbols in an expression separated by + and − signs. $3x$ and 1 are the terms in $3x + 1$.

> **EXAMPLE**
>
> One apple costs 20 pence.
>
> **a** Work out the cost of 3 apples.
> **b** Write an expression for the cost of y apples.
>
>
>
> **a** 3 apples cost $3 \times 20 = 60$ pence **b** y apples cost $y \times 20 = 20y$ pence

There are several conventions used in algebra.

- Multiplication signs aren't written.

- Terms involving letters are written in alphabetical order.

- Terms involving letters and numbers are written in alphabetical order with the number first.

- If there is a bracket write the term outside the bracket first.

- x squared is written using the power two.

$4y$	✓	$4 \times y$	✗
ab	✓	ba	✗
$4ab$	✓	$ab4$	✗
$4(x + 2)$	✓	$(x + 2)4$	✗
x^2	✓	xx	✗

> **EXAMPLE**
>
> Write an algebraic expression for these descriptions.
>
> **a** a number, n, add 4
> **b** a number, n, take away 6
> **c** a number, n, multiplied by 7
> **d** a number, n, divided by 2
> **e** a number, n, multiplied by itself
>
> **a** $n + 4$
> **b** $n - 6$
> **c** $7n$
> **d** $\dfrac{n}{2}$ or $n \div 2$
> **e** n^2

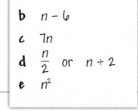

The conventions make it easier to tell when two expressions are equal.

- You can substitute values into expressions and formulae.

Take care when substituting more than one variable into an expression.

> **EXAMPLE**
>
> Work out the value of each expression when $a = 3$ and $b = -2$.
>
> **a** $\dfrac{4a}{6}$ **b** $\dfrac{a - b}{5}$
>
> **a** $\dfrac{4a}{6} = \dfrac{4 \times a}{6}$ **b** $\dfrac{a - b}{5} = \dfrac{3 - -2}{5}$
>
> $\quad = \dfrac{4 \times 3}{6} = \dfrac{12}{6} = 2$ $\quad = \dfrac{3 + 2}{5} = \dfrac{5}{5} = 1$

Exercise 2.1S

1 Stamps cost 30 pence each.
  a How much do 8 stamps cost?
  b How much do n stamps cost?

2 Chews cost 20 pence each.
How much do x chews cost?

3 Match the algebraic expression with the correct description.

$n + 2$	a number subtract 2
$n - 2$	a number divided by 2
$2n$	2 take away a number
$\dfrac{n}{2}$	a number multiplied by 2
$2 - n$	a number add 2

4 Match the algebraic expression with the correct description.

$4x + 2$	x add 4 multiplied by 2
$2x + 4$	x add 2 multiplied by 4
$2(x + 4)$	x multiplied by 4 add 2
$4(x + 2)$	x multiplied by 2 add 4

5 Liam thinks that the value of $7a$ is 73 when $a = 3$.
Do you agree with Liam? Explain your answer.

6 If $a = 3$, work out the value of these expressions.
  a $2a$ **b** $a + 1$ **c** $4a$
  d $a - 2$ **e** $2a + 3$ **f** $3a - 4$

7 Find the value of each expression when $m = 2$ and $n = 5$.
  a $m + n$ **b** $n - m$
  c $2m + n$ **d** $m - n$
  e m^2 **f** $n - 4m$
  g $3m + n - 5$ **h** $2m - 3n + 1$

8 Work out the value of each expression when $x = 3$, $y = 5$ and $z = 2$.
  a $4x + 3$ **b** $2z + y$
  c $3y - z$ **d** $x + y + z$
  e $2x - y + 3z$ **f** $3z - 2x + y$
  g $4x^2$ **h** $2(x + y - z)^2$

9 Work out the value of these when $a = 3$, $b = 5$, $c = 4$ and $d = 6$.
  a a^2 **b** $2a^2$ **c** b^2
  d $2b^2$ **e** c^2 **f** $2c^2$
  g d^2 **h** $2d^2$ **i** $abcd$

10 If $x = -3$, $y = 2$ and $z = 4$, work out the value of these expressions.
  a $x + y$ **b** $y^2 - 5$
  c $x^2 + 2$ **d** $2y + z$
  e $3y + 2x$ **f** $z^2 - 2y$
  g $3z + 2x$ **h** $3x + 2y - z$

11 Work out the value of each expression when $e = 4$, $f = 2$, $g = 5$ and $h = 23$.
  a $\dfrac{3e}{g}$ **b** $\dfrac{8g}{10}$
  c $\dfrac{6f}{3}$ **d** $\dfrac{2g + 3}{7}$
  e $\dfrac{6g}{f}$ **f** $\dfrac{3h + 4g}{8}$
  g $\dfrac{5e}{fg}$ **h** $\dfrac{(2g + 6)}{(e - f)}$

12 Calculate the value of each expression when $r = 2$, $s = 4$ and $t = -3$.
  a $\dfrac{s}{2}$ **b** $\dfrac{6r}{3}$ **c** $\dfrac{t}{3}$
  d $\dfrac{s + r}{3}$ **e** $\dfrac{t - 5}{2}$ **f** $\dfrac{s \times r}{3}$
  g $\dfrac{3r}{t}$ **h** $\dfrac{st}{r}$ **i** $\dfrac{r + s}{t}$

13 For each item, state whether it is an expression, equation or formula.
  a $2x - 3$ **b** $8 - 2y = 12$
  c $V = IR$ **d** $3a + 2b$
  e $v = u + at$ **f** $5p + 1 = 16$

14 A **term** is part of an expression.
In the expression $3x - 2y$, the terms are $3x$ and $-2y$.
List the terms in each of these expressions.
  a $2p + 4q$
  b $3x + y$
  c $2a - 5b + 3c$

> Hint: In part **c**, don't forget the minus sign.

2.1 Terms and expressions

- In algebra, you use letters to represent unknown numbers.
- You can replace letters with number values. This is called **substituting**.
- An expression is a collection of letters and numbers with no = sign.

HOW TO

1. Give every '**unknown**' a letter and write it down.
2. Translate the words into letters and symbols.
3. Test your expression using real values.

EXAMPLE

Write an expression for the number of pens in 3 boxes plus 5 extra pens.

1. Give every '**unknown**' a letter.
2. Translate into letters and symbols.

number of pens in one box = m

3 boxes + 5 pens

$3m + 5$

3. Test your expression using real values.

If $m = 10$, then there are $3 \times 10 + 5 = 35$ pens.

EXAMPLE

In a fruit shop, apples cost 20p each and oranges cost 15p each. Write an expression for the cost of x apples and y oranges.

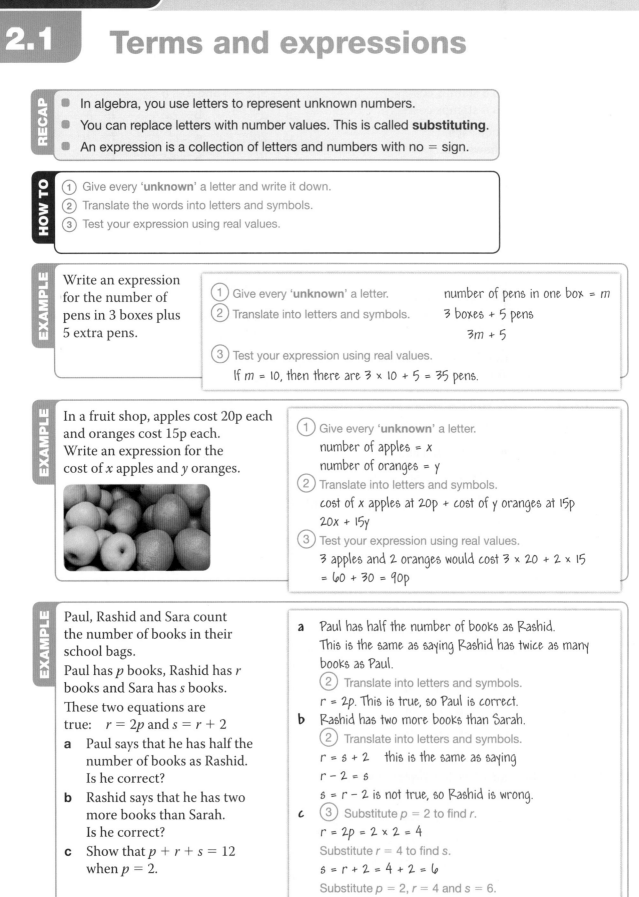

1. Give every '**unknown**' a letter.

number of apples = x
number of oranges = y

2. Translate into letters and symbols.

cost of x apples at 20p + cost of y oranges at 15p
$20x + 15y$

3. Test your expression using real values.

3 apples and 2 oranges would cost $3 \times 20 + 2 \times 15$
$= 60 + 30 = 90p$

EXAMPLE

Paul, Rashid and Sara count the number of books in their school bags.
Paul has p books, Rashid has r books and Sara has s books.
These two equations are true: $r = 2p$ and $s = r + 2$

a Paul says that he has half the number of books as Rashid. Is he correct?

b Rashid says that he has two more books than Sarah. Is he correct?

c Show that $p + r + s = 12$ when $p = 2$.

a Paul has half the number of books as Rashid. This is the same as saying Rashid has twice as many books as Paul.

2. Translate into letters and symbols.

$r = 2p$. This is true, so Paul is correct.

b Rashid has two more books than Sarah.

2. Translate into letters and symbols.

$r = s + 2$ this is the same as saying
$r - 2 = s$
$s = r - 2$ is not true, so Rashid is wrong.

c 3. Substitute $p = 2$ to find r.

$r = 2p = 2 \times 2 = 4$

Substitute $r = 4$ to find s.

$s = r + 2 = 4 + 2 = 6$

Substitute $p = 2$, $r = 4$ and $s = 6$.

$p + r + s = 2 + 4 + 6 = 12$

Algebra Expressions

Exercise 2.1A

1 Match each expression in box **A** with an expression in box **B**.

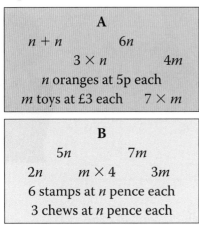

A

$n + n$ $6n$

$3 \times n$ $4m$

n oranges at 5p each

m toys at £3 each $7 \times m$

B

$5n$ $7m$

$2n$ $m \times 4$ $3m$

6 stamps at n pence each

3 chews at n pence each

2 Tony's Taxis calculates its fares using the formula $F = 2 + 4n$

where F is the fare in pounds and n is the number of miles.

Carla's Cabs calculates its fares using the formula $F = 6 + 3n$

Jessica wants to travel 7 miles.

Which company should she use?

3 A rent-a-car company charges for its cars using the formula $C = 30 + 20n$

where C is the cost in £s and n is the number of days hired.

a Bobbi says that it costs an extra £30 a day for each extra day that you rent a car. Do you agree with Bobbi?

b What is the charge for 10 days' hire?

c The company also charges a cleaning fee when the car is returned.
Bobbi is charged a total of £260 for car rental and cleaning.
What is the maximum number of days that Bobbi hired the car for?

4 Work out the value of each capital letter and then read the coded word.

$m = 3$ $n = 8$ $p = 5$ $q = -2$

H $= 4n + 2q$ L $= 6m + 3p$
O $= m^2 + 7$ A $= 2q + 10$
T $= pq$ C $= 2n - 3p$ E $= np + 2m$

| 28 | 16 | | 16 | 33 | 6 | -10 | 46 |

5 Daniel has x DVDs. Lisa has twice as many DVDs as Daniel.
Sareeta has 3 fewer DVDs than Lisa.
Write an expression in terms of x for the number of DVDs Sareeta has.

6 In one month, Dan sends x texts.
Alix sends 4 times as many texts as Dan.
Kris sends 8 more texts than Alix.
How many texts does Kris send?

7 In a pizza takeaway

● a medium pizza has 6 slices of tomato

● a large pizza has 10 slices of tomato.

Write an expression for the total number of slices of tomato needed for c medium and d large pizzas.

8 Write algebraic expressions for the cost of

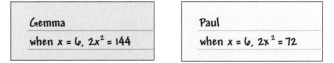

Café price list

Tea   50p
Fruit juice   80p
Milk   60p
Scone   30p
Flapjack   40p

a f teas and g scones

b j fruit juices and k flapjacks

c x teas, y milks and z scones

d p milks, q fruit juices and r flapjacks.

9 Gemma and Paul evaluated $2x^2$ when $x = 6$.
Who was right? Explain why.

Gemma
when $x = 6$, $2x^2 = 144$

Paul
when $x = 6$, $2x^2 = 72$

10 Audrey, Billie and Cerys count the amount of money they each have. Audrey has £a, Billie has £b and Cerys has £c.
These two equations are true:
$b = 3a$ $c = a + b$

a Audrey says that she has three times as much money as Billie. Cerys says that she has four times as much money as Audrey. Are they both correct?
Explain your answer fully.

b Audrey has £5. How much money do the three friends have in total?

Q 1158, 1186, 1187   SEARCH

2.2 Simplifying expressions

● Terms with the same letter are called like terms.
You can simplify expressions by collecting like terms.

$y = 1y$

EXAMPLE

Simplify
a $4m + 2p + 3m$
b $2x + 5y + 3x + y$

Rearrange the terms to collect like terms together.
a $4m + 2p + 3m = 4m + 3m + 2p$
       $= 7m + 2p$
b $2x + 5y + 3x + y = 2x + 3x + y + 5y$
           $= 5x + 6y$

EXAMPLE

Simplify these expressions.
a $2e + 5f + 6e - 2f$ b $6u - v - 3u + 4v$ c $3m + 2n - m - 4n$

a $2e + 5f + 6e - 2f$
   $= 2e + 6e + 5f - 2f$
   $= 8e + 3f$

b $6u - v - 3u + 4v$
   $= 6u - 3u + 4v - v$
   $= 3u + 3v$

c $3m + 2n - m - 4n$
   $= 3m - m + 2n - 4n$
   $= 2m + -2n$
   $= 2m - 2n$

Keep each term with its sign.

There are conventions for writing expressions in algebra.
● Do not include the multiplication sign. $3 \times p \to 3p$
● Write divisions as fractions. $3 \div p \to \frac{3}{p}$
● Write numbers first as products. $q \times 4 \to 4q$
● Write letters in products in alphabetical order. $4 \times q \times r \times p \to 4pqr$

You can add, subtract, multiply or divide algebraic terms.

$3n + 5n + 8n = 16n$ $4p - p = 3p$ $2 \times 6p = 12p$ $8r \div 4 = 2r$

● To simplify an expression, you follow the same order of operations as in arithmetic.

Brackets → Indices → Division or Multiplication → Addition or Subtraction

EXAMPLE

Simplify
a $4n + 2 \times 5n$
b $3r \times 2s$
c $4t^2 - 3 \times t^2 + t$

a $4n + 2 \times 5n = 4n + (2 \times 5n)$
             $= 4n + 10n$
             $= 14n$
b $3r \times 2s = 3 \times r \times 2 \times s$
   Rearrange: numbers first, then letters.
             $= 3 \times 2 \times r \times s$
             $= 6rs$
c $4t^2 - 3 \times t^2 + t = 4t^2 - (3 \times t^2) + t$
             $= 4t^2 - 3t^2 + t$
   Collect like terms.
             $= t^2 + t$

You can use the acronym BIDMAS to remember the order.

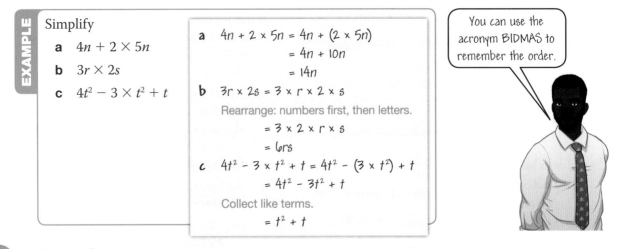

Exercise 2.2S

1 Simplify these expressions.

   a $m + m + m$

   b $n + n + n + n + n + n + n$

   c $y + y + y + y + y + y$

   d $z + z + z - z + z$

2 Collect the like terms and simplify each expression.

   a $6a + 7a$ **b** $5n + 12n$

   c $6t + 9t$ **d** $3x + 16x$

   e $9r - 5r$ **f** $6f - 2f$

   g $12g - 5g$ **h** $15r - 7r$

   i $16n + 4n + 3n$ **j** $6c + 8c + 3c$

3 Write each of these as a single term.

   a $4 \times m$ **b** $p \times 3$

   c $r \times s$ **d** $12 \times q$

   e $5 \times g$ **f** $c \times 4$

   g $d \times 8$ **h** $j \times k$

   i $n \times n$ **j** $e \times 15$

   k $t \times t$ **l** $6 \times m \times n$

4 Simplify each expression.

   a $8n + 3n + 4n$ **b** $3m + 2m + 7m$

   c $8p + 6p + 3p$ **d** $5q + 7q + 6q$

   e $12x - 7x - 4x$ **f** $8w - w - 3w$

   g $4a + 6a - 3a$ **h** $12b - 3b - 4b$

   i $3j - 4j + 2j$ **j** $k - 5k + 6k$

5 Simplify these divisions. The first one has been done for you.

   a $d \div 4 = \dfrac{d}{4}$ **b** $x \div 3$

   c $y \div 7$ **d** $t \div 9$

   e $2a \div 3$ **f** $3n \div 4$

   g $5p \div 7$ **h** $2v \div 4$

6 Simplify each expression by collecting like terms together.

   a $a + b + a + b$ **b** $c + d + c + c + c$

   c $3e + 2f + 4e + 3f$ **d** $5g + 7h + 2g + h$

   e $3i + j + 4i + 5j$ **f** $3u + 5v + 2v + u$

7 Simplify these expressions.

   a $4a + 2b - 3a + b$

   b $6x + 4y - 2x + 2y$

   c $5m + 3n + 2m - n$

   d $8s + 3t + s - 2t$

   e $3p + 2q - p + 3q$

   f $2c + 3d + 3c - 2d$

8 Simplify these by collecting like terms.

   a $4a - 6b - 3a + 8b$

   b $3c - 4d - 2c + 5d$

   c $7u - 4v - 2u + 5v$

   d $6x - 5y - 5x + 11y$

   e $12m - 5n - 3m + 10n$

   f $9p - 4q - 3p + 6q$

9 Simplify these expressions.

   a $5e - 3f - 2e + 2f$

   b $4g - 6h - 3g + 4h$

   c $j - 3k + 4j + 2k$

   d $2r - 4s - r + s$

   e $7t + 3u - 4t - 5u$

   f $9v - 5w + 3v - 2w$

10 Simplify these expressions.

   a $2p + 5q + 3p + q$

   b $6x + 2y + 3x + 5y$

   c $4m + 2n - 2m + 6n$

   d $5x + 3y - 4x + 2y$

   e $7r - 4s + r - 2s$

   f $2f - 3g + 5g - 6f$

   g $3a + 2b + 5c - a + 4b$

   h $7u - 5v + 3w + 3v - 2u$

   i $5x - 3y - 2x + 4z - y + z$

   j $4r + 6s - 3t + 2r + 5t - s$

11 Simplify these expressions.

   a $3r + 3 \times 2r$ **b** $2m^2 + 2m \times m$

   c $2t \times 4v$ **d** $5m \times 2n$

   e $3x \times 2y^2$ **f** $x^2 + x^2 + x$

   g $3 \times 3w - 2 \times 4$ **h** $z \times z^2 + 3z + 1$

Q 1178, 1179   SEARCH

2.2 Simplifying expressions

- Terms with the same letter are called like terms.
- You can simplify expressions by collecting like terms.

$2a$ and $6a$ are like terms.
$2a$ and $4b$ are not like terms.
a and a^2 are not like terms.

HOW TO

$2a + 6a = 8a$

(1) Use the information in the question to write an algebraic expression.

(2) Collect like terms to simplify the expression.

EXAMPLE

Bags of sweets come in two sizes.

There are n sweets in a small bag.

There are p sweets in a large bag.

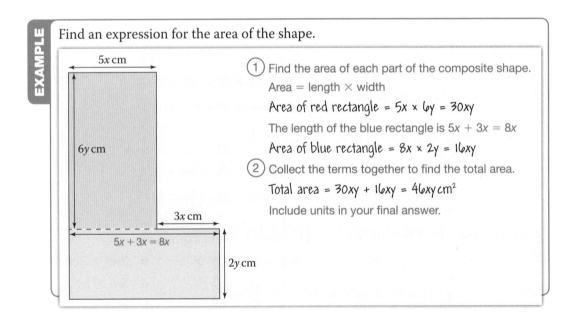

Rupal buys 1 large bag and 1 small bag.

Maisie buys 2 small bags and 3 large bags.

Val buys 2 small bags and 1 large bag.

How many sweets do they have altogether?

(1) Write the words as a number sentence.

Rupal buys 1 small bag + 1 large bag
$$= n \text{ sweets} + p \text{ sweets}$$
$$= n + p \text{ sweets}.$$

Maisie buys 2 small bags + 3 large bags
$$= 2 \times n \text{ sweets} + 3 \times p \text{ sweets}$$
$$= 2n + 3p \text{ sweets}.$$

Val buys 2 small bags + 1 large bag $= 2n + p$ sweets.

Add all the sweets together.

They have $n + p + 2n + 3p + 2n + p$ sweets.

(2) Collect the terms.

They have $5n + 5p$ sweets.

EXAMPLE

Find an expression for the area of the shape.

5x cm

6y cm

3x cm

5x + 3x = 8x

2y cm

(1) Find the area of each part of the composite shape.
Area = length × width
Area of red rectangle = $5x \times 6y = 30xy$
The length of the blue rectangle is $5x + 3x = 8x$
Area of blue rectangle = $8x \times 2y = 16xy$

(2) Collect the terms together to find the total area.
Total area = $30xy + 16xy = 46xy \text{ cm}^2$
Include units in your final answer.

Exercise 2.2A

1 Bags of peanuts come in two sizes.
There are x peanuts in a small bag.
There are y peanuts in a large bag.
Sebastian buys 1 small bag and 1 large bag.
Gabi buys 2 small bags and 1 large bag.
Kofi buys 3 small and 2 large bags.

   a Sebastian and Gabi combine their bags of peanuts.
Do they have more peanuts than Kofi?

   b Write an expression for the total number of peanuts that all three friends buy.

2 Make sets of three matching expressions, using one expression from each box in each set.

A
$3x + 5y - x + 2y$
$2x - 4y + 3x + 2y$
$2x + 4y - x$
$2y + 3x - x + 3y$

B
$2x + 5y$
$3y + 7x + 4y - 5x$
$5x - 2y$
$3x + 6y - 2x - 2y$

C
$7y - 3x + 5x - 2y$
$4x + 4y - 3x$
$7y + 2x$
$2x - 4y + 2y + 3x$

3 Three students tried to simplify $3m + 5$.
Which of them did it correctly?

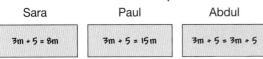

   Sara   Paul   Abdul

    $3m + 5 = 8m$ $3m + 5 = 15m$ $3m + 5 = 3m + 5$

4 Rearrange each set of cards to make a correct statement.

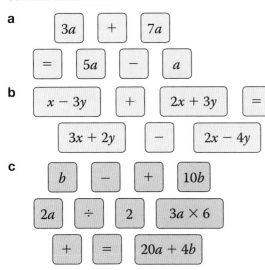

   a $3a$ $+$ $7a$ $=$ $5a$ $-$ a

   b $x - 3y$ $+$ $2x + 3y$ $=$ $3x + 2y$ $-$ $2x - 4y$

   c b $-$ $+$ $10b$ $2a$ $\div$ 2 $3a \times 6$ $+$ $=$ $20a + 4b$

5 **a** Write a simplified expression for the
     i perimeter
     ii area of this rectangle.

      $4p$
      8

   b Two of these rectangles are joined to make a composite shape.
     i Write an expression for the area of the composite shape.

     ii The perimeter of the composite shape is $32 + 8p$.
Draw a possibility for the composite shape.

6 What are the measurements of a rectangle with perimeter $6x + 4y$ and area $6xy$?

7 Find the area of the shape.

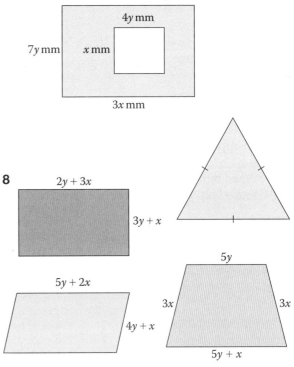

    $4y$ mm
    $7y$ mm x mm
    $3x$ mm

8 $2y + 3x$ $3y + x$

    $5y + 2x$
    $4y + x$
    $5y$ $3x$ $3x$ $5y + x$

   a The triangle and the parallelogram have the same perimeter. Find the length of each side of the triangle.

   b Does the rectangle or the trapezium have the greater perimeter?
Explain how you decide.

2.3 Indices

- You can use **index** notation to write repeated multiplication.

Indices is the plural of index.

$5 \times 5 \times 5 \times 5 = 5^4 \quad m \times m \times m \times m = m^4$

5 is the **base** and 4 is the index.   You say 'm to the **power** of 4'.

You can simplify expressions with **indices** and numbers.

EXAMPLE

Simplify

a $3 \times p \times p \times p \times q \times q$

b $2 \times s \times s \times 3 \times t \times t \times t$

a $3 \times p \times p \times p \times q \times q$
$= 3 \times p^3 \times q^2$
$= 3p^3q^2$

b $2 \times s \times s \times 3 \times t \times t \times t$
$= 2 \times s^2 \times 3 \times t^3$
$= 2 \times 3 \times s^2 \times t^3$
$= 6s^2t^3$

You can simplify expressions with powers of the same base.

$$n^2 \times n^2$$
$$= n \times n \times n \times n = n^4$$

$$t^5 \div t^2 = \frac{t^5}{t^2} = \frac{{}^1t \times {}^1t \times t \times t \times t}{{}^1t \times {}^1t} = t^3$$

$$(v^2)^3 = v^2 \times v^2 \times v^2$$
$$= v^{2 \times 3} = v^6$$

- To multiply powers of the same base, add the indices.

$x^a \times x^b = x^{(a + b)}$

- To divide powers of the same base, subtract the indices.

$x^a \div x^b = x^{(a - b)}$

- When finding the 'power of a power', multiply the indices.

$(x^a)^b = x^{a \times b}$

You can apply the index laws to positive and negative indices.

- When terms have numerical **coefficients**, deal with these first.

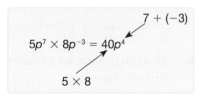
$$5p^7 \times 8p^{-3} = 40p^4$$
$7 + (-3)$
5×8

EXAMPLE

Simplify each of these using the index laws.

a $\dfrac{(k^3 \times k^2)^7}{k}$

b $(5p^2)^3 \times 2p^{-7}$

a $\dfrac{(k^3 \times k^2)^7}{k} = \dfrac{(k^5)^7}{k}$ $3 + 2 = 5$
                                       $5 \times 7 = 35$

        $= \dfrac{k^{35}}{k}$ Remember k is really k^1.

        $= k^{34}$ $35 - 1 = 34$

b $(5p^2)^3 \times 2p^{-7}$ $2 \times 3 = 6$
   $= 125p^6 \times 2p^{-7}$ $125 \times 2 = 250$
   $= 250p^{-1}$ $6 + -7 = -1$

EXAMPLE

Expand $(2p^2q)^3$

$2^3 = 8$, $(p^2)^3 = p^6$, $(q^1)^3 = q^3$.

$(2p^2q)^3 = 8p^6q^3$

Everything inside the bracket is cubed.

Algebra Expressions

Exercise 2.3S

1 Simplify these expressions.

   a $y \times y \times y \times y$

   b $m \times m \times m \times m \times m \times m$

2 Simplify

   a $3 \times t \times t$

   b $4 \times p \times q \times q$

   c $6 \times v \times v \times w \times w \times w$

   d $2 \times r \times r \times r \times r \times s$

   e $2 \times m \times m \times 3 \times n$

   f $4 \times y \times y \times y \times 2 \times z \times z$

3 Simplify

   a $3m^2 \times 2$ **b** $3 \times 4p^3$

   c $2x \times 3y^2$ **d** $5r^2 \times 2s^2$

4 Kyle thinks that $a^5 \times a^2 = a^{10}$.
Do you agree with Kyle? Explain your
answer.

5 Simplify

   a $n^2 \times n^3$ **b** $s^3 \times s^4$

   c $p^3 \times p$ **d** $t \times t^3$

6 Write each of these as a single power in the
form x^n.

   a $x^2 \times x^2 \times x^3$ **b** $x \times x^5 \times x^2$

   c $x^3 \times x^2 \times x^4$ **d** $x^5 \times x \times x$

7 Write each of these as a single power in the
form r^n.

   a $r^4 \div r^2$ **b** $r^5 \div r^4$

   c $r^7 \div r^2$ **d** $r^8 \div r^5$

8 Simplify

   a $\dfrac{m^6}{m^2}$ **b** $\dfrac{x^4}{x^3}$ **c** $\dfrac{t^7}{t^5}$ **d** $\dfrac{y^4}{y}$

9 Simplify

   a $\dfrac{x^2 \times x^3}{x^4}$ **b** $\dfrac{m^3 \times m}{m^2}$

   c $\dfrac{s^2 \times s^4}{s^3}$ **d** $\dfrac{v \times v^3 \times v^3}{v^4}$

   e $\dfrac{q^2 \times q^3 \times q^2}{q^4}$ **f** $\dfrac{t^3 \times t \times t^2}{t^2}$

   g $\dfrac{p^4 \times p^2 \times p^2}{p^7}$ **h** $\dfrac{y^2 \times y^4 \times y}{y^3 \times y^2}$

10

> The index rules say that you add the powers
> when two terms are multiplying each other.

Tracey thinks that $4y^5 \times 2y^2 = 6y^7$.
Do you agree with Tracey?

11 Simplify these expressions, leaving your
answers in index form.

   a $3x^5 \times x^2$ **b** $5y^2 \times y^5$ **c** $4b^2 \times 3b^6$

   d $2p^4 \times 5p^7$ **e** $5h^5 \times 6h^6$ **f** $4s^3 \times 3t^4$

Explain your answers.

12

> The index rules say that you subtract the
> powers when two terms are dividing each other.

Andy thinks that $12p^{12} \div 3p^4 = 9p^8$.
Do you agree with Andy?

13 Simplify these expressions, leaving your
answers in index form.

   a $10y^6 \div 5y^2$ **b** $6a^9 \div 3a^3$

   c $20k^7 \div 4k^3$ **d** $18p^8 \div 6p^3$

   e $35x^{10} \div 7x^4$ **f** $4x^8 \div 8y^4$

14 Simplify these expressions, leaving your
answers in index form.

   a $(a^3)^2$ **b** $(y^2)^6$ **c** $(k^3)^5$

   d $(p^7)^8$ **e** $(a^3)^7$ **f** $(a^7)^3$

15 Simplify these expressions, leaving your
answers in index form.

   a $(2a^3)^2$ **b** $(3y^2)^6$ **c** $(5k^3)^2$

   d $(6p^7)^3$ **e** $(2a^3)^7$ **f** $(4a^4)^4$

16 Simplify these expressions, leaving your
answers in index form.

   a $2a^3 \times 4a^4 \times a^5$ **b** $\dfrac{m^{11}}{m^4}$

   c $\dfrac{12y^9}{3y^3}$ **d** $\dfrac{3y^3 \times 6y^5}{2y^4}$

17 Simplify

   a $g^8 \times g^{-5}$ **b** $\dfrac{h^{-2}}{h^4}$ **c** $(b^{-4})^3$

   d $j^{-4} \times j^{-2}$ **e** $(t^{-5})^{-2}$ **f** $n^{-8} \div n^{-6}$

18 Simplify fully

   a $(2p^8)^{-2}$ **b** $10r^3 \times 6r^{-4}$

   c $(3h^{-3})^3$ **d** $9b^3 \div 3b^{-5}$

2.3 Indices

- An **index** is a power. The **base** is the number which is raised to this power.

- You can simplify expressions with the same base using the three index laws.

  - When multiplying, add the indices.

  - When dividing, subtract the indices.

  - To find the 'power of a power', multiply the indices.

$$3^4 = 3 \times 3 \times 3 \times 3 = 81$$

base   index

$$x^2 \times x^5 = x^7$$
$$y^8 \div y^4 = y^4$$
$$(z^3)^2 = z^6$$

① Read the question carefully. You may need to apply your knowledge of other topics.

② Use the index laws to calculate or simplify. Deal with numbers first, then powers.

③ Answer the question. Give an explanation if the question asks for one.

EXAMPLE

A rectangle has area $28x^2y^6$.

The width of the rectangle is $4xy^3$.

Find the perimeter of the rectangle.

> The perimeter is the distance around the edge of the rectangle.

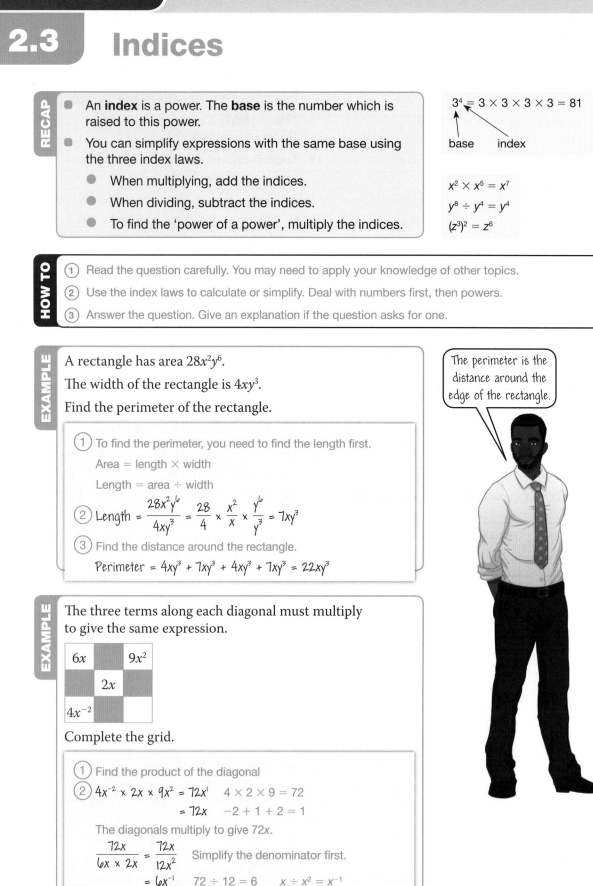

① To find the perimeter, you need to find the length first.

Area = length × width

Length = area ÷ width

② Length $= \dfrac{28x^2y^6}{4xy^3} = \dfrac{28}{4} \times \dfrac{x^2}{x} \times \dfrac{y^6}{y^3} = 7xy^3$

③ Find the distance around the rectangle.

Perimeter $= 4xy^3 + 7xy^3 + 4xy^3 + 7xy^3 = 22xy^3$

EXAMPLE

The three terms along each diagonal must multiply to give the same expression.

$6x$		$9x^2$
	$2x$	
$4x^{-2}$		

Complete the grid.

① Find the product of the diagonal

② $4x^{-2} \times 2x \times 9x^2 = 72x^1$ $4 \times 2 \times 9 = 72$

$\qquad\qquad\qquad\quad = 72x$ $-2 + 1 + 2 = 1$

The diagonals multiply to give $72x$.

$\dfrac{72x}{6x \times 2x} = \dfrac{72x}{12x^2}$ Simplify the denominator first.

$\qquad\qquad = 6x^{-1}$ $72 \div 12 = 6$ $x \div x^2 = x^{-1}$

③ The missing entry is $6x^{-1}$.

Exercise 2.3A

1 Match each of the pairs.

2 A cube has sides of length $2xy^2$ cm.
Find the volume of the cube.

3 Find an expression for the area of each shape.

a

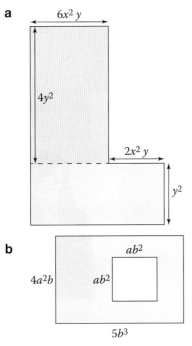

$6x^2 y$

$4y^2$

$2x^2 y$

y^2

b

ab^2

$4a^2b$ ab^2

$5b^3$

4 A rectangle has area $20p^4q^2$.
The width of the rectangle is $2p^2q$.
Find the perimeter of the rectangle.

5 A square has area $16a^2b^6$.
Find the length of the sides of the square.

6 Rearrange each set of cards to make a
correct statement.

a

| (| p^2q | × | = |

| pq | $)^3$ | p^5q^4 |

b

| (| xy | × | xy | $)^2$ |

| + | $4x^3y^3$ | = | xy |

7 Simplify $((x^4)^2)^5$

8 Is this statement true or false?

$(xy^2)^3 = (xy^3)^2$

Explain your answer.

9 Aria, Emily, Hanna and Spencer have all
made mistakes in their indices homework.
Explain their mistakes and write down the
correct answer.

> **Aria**
> $6a^3b^2 \times 2a^4b^3 = 12a^7b^6$

> **Hanna**
> $10p^8q^5 \div 2p^4q^2 = 5p^2q^3$

> **Emily**
> $5x^2y^3 + 4x^2y^3 = 9x^4y^6$

> **Spencer**
> $(4m^2n^3)^2 = 16m^4n^9$

10 Match each of the pairs.

| $x^{-2} \div x^{-6}$ | | $x^6 \div x^5$ |

| $(x^2)^2$ | | $(x^{-4} \times x^4)^2$ |

| $x^3 \times x^{-3}$ | | $x^{-2} \times x^3$ |

11 Write a simplified
expression for
the area of this
triangle.

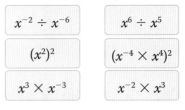

$15w^3$

$4w^6$

12 In a multiplication pyramid, you multiply
the numbers directly below to get the
number above.

| 10 |
| 2 | 5 |

Complete this multiplication pyramid.

x

x^2

13 The three terms on each
edge of this grid must
multiply to give 60.

Complete the grid.

		$2x^{-1}$
$3x^{-4}$		$10x^4$
	$4x$	

2.4 Expanding and factorising 1

You can use **brackets** in algebraic expressions.

- You can multiply out brackets.
  - You multiply each term inside the bracket by the term outside.

$2(x + 4) = 2 \times x + 2 \times 4 = 2x + 8$

Expand means 'multiply out'.

EXAMPLE

Expand

a $2(3x + 1)$ **b** $n(n + 5)$

a $2(3x + 1) = 2 \times 3x + 2 \times 1$
$= 6x + 2$

b $n(n + 5) = n \times n + 5 \times n$
$= n^2 + 5n$ $n \times n = n^2$

To **simplify** expressions with brackets, expand the brackets and collect like terms.

Like terms have the same power of the same letter.

EXAMPLE

Simplify

a $m(m + 2) + m$ **b** $3(x + 1) + 2(4x + 2)$

c $5y(2y - 3)$ **d** $-4(x - 6)$

a $m(m + 2) + m = m^2 + 2m + m$
$= m^2 + 3m$

b $3(x + 1) + 2(4x + 2)$
$= 3x + 3 + 8x + 4$
$= 11x + 7$

c $5y(2y - 3) = 5y \times 2y + 5y \times -3$
$= 10y^2 - 15y$

d $-4(x - 6) = -4 \times x + -4 \times -6$
$= -4x + 24$

- To factorise an expression, look for a **common factor** for all the terms.

To factorise completely, use the highest common factor.

EXAMPLE

a Find the common factors of $12x$ and 8.

b Factorise $12x + 8$.

a 2 and 4 are common factors of $12x$ and 8.

b The highest common factor of $12x$ and 8 is 4.
$12x + 8 = 4 \times 3x + 4 \times 2 = 4(3x + 2)$

EXAMPLE

Factorise fully

a $12pq - 4pw$ **b** $5x + 10x^2 - 25xy$

a $12pq - 4pw$
Deal with numbers first, then letters.
HCF of $12pq$ and $4pw$ is $4p$.
So, $12pq - 4pw = 4p(3q - w)$ $4p(3q - w) = 12pq - 4pw$ ✓

b The HCF of $5x$, $10x^2$ and $25xy$ is $5x$.
$5x + 10x^2 - 25xy = 5x(1 + 2x - 5y)$ $5x(1 + 2x - 5y) = 5x + 10x^2 - 25xy$ ✓

Exercise 2.4S

1 Expand the brackets in these expressions.

  a $3(m + 2)$ **b** $4(p + 6)$

  c $2(x + 4)$ **d** $5(q + 1)$

  e $2(6 + n)$ **f** $3(2 + t)$

  g $4(3 + s)$ **h** $2(4 + v)$

2 Expand these expressions.

  a $3(2q + 1)$ **b** $2(4m + 2)$

  c $3(4x + 3)$ **d** $2(3k + 1)$

  e $5(2 + 2n)$ **f** $3(4 + 2p)$

  g $4(1 + 3y)$ **h** $2(5 + 4z)$

3 Expand and simplify each of these expressions.

  a $3(p + 3) + 2p$ **b** $2(m + 4) + 5m$

  c $4(x + 1) - 2x$ **d** $2(5 + k) + 3k$

  e $4(2t + 3) + t - 2$ **f** $3(2r + 1) - 2r + 4$

4 Find all the common factors of

  a $2x$ and 6

  b $4y$ and 12

  c 10 and $20j$

  d 6 and $12p$

  e 9 and $6q$

  f $6t$ and 4

  g $4x$ and 10

  h $24t$ and 8

> Hint for **4a**: 2 and x are factors of $2x$. 1, 2, 3 and 6 are factors of 6. 2 is the common factor of $2x$ and 6.

5 Find the highest common factor of

  a $3x$ and 9 **b** $12r$ and 10

  c $6m$ and 8 **d** 4 and $4z$

6 Find the highest common factor of

  a y^2 and y **b** $4s^2$ and s

  c $7m$ and m^3 **d** $2y^2$ and $2y$

7 Factorise these expressions.

  a $2x + 10$ **b** $3y + 15$

  c $8p - 4$ **d** $6 + 3m$

  e $5n + 5$ **f** $12 - 6t$

  g $14 + 4k$ **h** $9z - 3$

  i $12m + 15$ **j** $28 - 6y$

  k $40z - 24$ **l** $18 - 30b$

8 Expand and simplify each of these expressions.

  a $3(r + 2) + 2(r - 1)$

  b $4(s + 1) - 2(s + 2)$

  c $3(2j + 3) - 2(j + 2)$

  d $3(4t - 2) + 3(t - 1)$

9 Expand and simplify each of these expressions.

  a $2(n + 3) + 3(n + 2)$

  b $4(p + 1) + 2(3 + p)$

  c $4(2x + 1) + 2(x + 3)$

  d $2(3n + 2) + 3(4n + 1)$

10 Expand

  a $x(4x + 1)$ **b** $m(m^2 + 2)$

  c $2t(t^2 + 4)$ **d** $3p(p^2 + 1)$

11 Expand and simplify

  a $4m(m - 3)$ **b** $2p(p - 6)$

  c $-3(x + 2)$ **d** $-5(2m - 4)$

12 Factorise

  a $w^2 + w$ **b** $z - z^2$

  c $4y + y^2$ **d** $2m^2 - 3m$

  e $4p^2 + 5p$ **f** $7k - 2k^2$

  g $3n^3 - 2n$ **h** $5r + 3r^2$

13 Factorise

  a $4y - 12$ **b** $2x^2 + 3x$

  c $3y^2 - 1$ **d** $15 + 5t^2$

  e $3m + 9m^2$ **f** $2r^2 - 2r$

  g $4v^3 + v$ **h** $3w^2 + 3w$

> Check your answers by expanding.

14 The cards show expansions and factorisations. Match the cards in pairs.

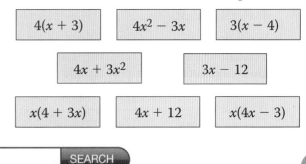

$4(x + 3)$ $4x^2 - 3x$ $3(x - 4)$

$4x + 3x^2$ $3x - 12$

$x(4 + 3x)$ $4x + 12$ $x(4x - 3)$

2.4 Expanding and factorising 1

RECAP

- Expand means multiply each term inside the bracket by the term outside.
- Factorising is the 'opposite' of expanding brackets.

$$2(x + 4) \xrightarrow{\text{expand}} 2x + 8$$
$$\underset{\text{factorise}}{\longleftarrow}$$

HOW TO

1. Read the question carefully. Give any unknown values a letter.
2. Expand the brackets …or… find the HCF of all of the terms and collect like terms   and factorise.
3. Answer the question fully.

Look for a **common factor** for all the terms.

EXAMPLE

Fill in the missing values.
$3(2x + 11) + \Box(2x − 5) = 8(2x + \Box)$

1. Label the empty boxes so that you can tell them apart.
$3(2x + 11) + ⊛(2x − 5) = 8(2x + •)$
2. Expand the brackets on both sides.
$6x + 33 + ⊛ × 2x − 5 × ⊛ = 16x + 8 × •$
Look at the x terms. $6x + ⊛ × 2x = 16x$
$6 + 2 × ⊛ = 16$
$2 × ⊛ = 10$ so $⊛ = 5$
Now look at the constant terms. $33 − 5 × ⊛ = 8 × •$
Substitute $⊛ = 5$ $33 − 5 × 5 = 8 × •$
$33 − 25 = 8 × •$
$8 = 8 × •$ so $• = 1$
3. $3(2x + 11) + 5(2x − 5) = 8(2x + 1)$

EXAMPLE

A rectangle of width x has length 1 cm more than its width. Its area is 200 cm².

Show that $x^2 + x = 200$. Explain why x must be between 13 cm and 14 cm.

1. The length is $(x + 1)$ cm
Area of rectangle = length × width
$200 = x(x + 1)$
2. $200 = x^2 + x$
3. Substitute $x = 13$ and $x = 14$. $13^2 + 13 = 169 + 13 = 182$ smaller than 200
x is between 13 cm and 14 cm. $14^2 + 14 = 196 + 14 = 210$ larger than 200

Sketch a diagram to help:
x ↕   ← $x + 1$ →

EXAMPLE

A cuboid has volume $48x^2 + 16xy$ cm³.

Write down a possibility for the dimensions of the cuboid.

2. Factorise the expression. The HCF of $48x^2$ and $16xy$ is $16x$.
$48x^2 + 16xy = 16x(3x + y)$
3. Volume of a cuboid = length × width × height.
$16x(3x + y) = 16 × x × (3x + y)$
Possible dimensions are 16 cm × x cm × $(3x + y)$ cm
There are many other possibilities including 8 cm × 2x cm × $(3x + y)$ cm
and 4 cm × 4x cm × $(3x + y)$ cm.

Exercise 2.4A

1 At a pick-your-own farm, Lucy picks n apples. Mary picks 5 more apples than Lucy.

Nat picks 3 times as many apples as Mary.

Write down, in terms of n, the number of apples Nat picks.

2 Jake is n years old.
Jake's sister is 4 years older than Jake.
Jake's mother is 3 times older than his sister.
Jake's father is 4 times older than Jake.
Jake's uncle is 2 years younger than Jake's father.
Jake's grandmother is twice as old as Jake's uncle.

a Copy the table and write each person's age in terms of n.

Jake	Sister	Mother	Father	Uncle	Grandmother
n					

b Find, in terms of n, how much older Jake's grandmother is than his mother. Give your answer in its simplest form.

3 Debbie, Kate and Bryn factorise $16x^2 + 4x$. Here are their answers.

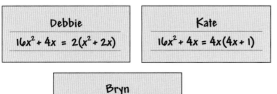

Debbie
$16x^2 + 4x = 2(x^2 + 2x)$

Kate
$16x^2 + 4x = 4x(4x + 1)$

Bryn
$16x^2 + 4x = 2x(8x^2 + 2)$

a Who is correct?

b Explain where the other two have gone wrong.

4 Copy each equation and fill in the missing values.

a $3(x + \square) = 3x + 12$

b $2(\square x + 5) = 12x + \square$

c $2(x + 4) + \square = 2x + 11$

d $\square(2x - 1) = 8x - \square$

e $6(x - \square) + \square(2x + 1) = 16x - 19$

f $2(x + 1) + \square(3x - 1) = 20x - \square$

g $\square(4x + 3) + 2(x - 1) = 7(2x + \square)$

h $5(5x + 6) - 7(\square x + 2) = \square(x + 4)$

5 **a** Using brackets, write a formula for the area of this rectangle.

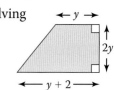

3

$2x - 1$

b Expand the brackets.

c The area of the rectangle is $15\,\text{cm}^2$. Show that $6x - 18 = 0$.

6 An expression expands to give $24x + 16$.

a What could the expression have been if it involved one pair of brackets?

b What could the expression have been if it involved adding two single brackets?

7 Write an expression involving brackets for the area of this trapezium.

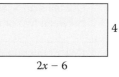

$\leftarrow y \rightarrow$

$2y$

$\leftarrow y + 2 \rightarrow$

Expand the brackets and simplify your expression.

8 Write a factorised expression for

a The perimeter of this rectangle.

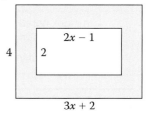

4

$2x - 6$

b The perimeter of a square with sides $5b + 10$.

9 Show that the shaded area of this rectangle is $2(4x + 5)$.

4

2

$2x - 1$

$3x + 2$

10 A cuboid has volume $20xy + 50x^2y\,\text{cm}^3$.

Write down three possibilities for the dimensions of the cuboid.

11 A cuboid has volume $16x^2y - 54x^2\,\text{cm}^3$.

a Write down three possibilities for the dimensions of the cuboid.

b Explain why y must be greater than $3.375\,\text{cm}$.

Q 1155, 1247   SEARCH

Summary

Checkout

You should now be able to...	Questions
✔ Use algebraic notation.	1
✔ Substitute numbers into formulae and expressions.	2 – 5
✔ Use and understand the words expressions, equations, formulae, terms and factors.	6, 7
✔ Collect like terms and simplify expressions involving sums, products, powers and surds.	8 – 10
✔ Use the laws of indices.	11
✔ Multiply a single term over a bracket.	12
✔ Take out common factors in an expression.	13

Language Meaning / Example

Language	Meaning	Example
Expression	A collection of letters and numbers without an $=$ sign.	$5x - 2$
Equation	Contains an $=$ sign and an unknown letter to be solved.	$6x + 2 = 14 \qquad x = 2$ $x^2 = 5 + 4x \qquad x = 5$ or -1
Formula **Formulae**	Contains an $=$ sign and describes a relationship between two or more letters.	$V = IR$
Term	One of the quantities in an expression. Terms are linked with addition or subtraction signs.	In the expression $4x^3 + 3x^2 - 7y + 9$ $4x^3$, $3x^2$, $7y$ and 9 are all terms.
Substituting	Replacing a letter with a number and working out the value.	Substituting $x = 2$ in $4x^2 + 3x$ gives $4 \times 2^2 + 3 \times 2 = 22$
Unknown	An unknown quantity represented by a letter.	$3x + 4 = 16$ The unknown value of x can be found by solving the equation.
Index **Base** **Power**	In index notation, the index or power shows how many times the base has to be multiplied. The plural of index is **indices**.	Power or index $5^3 = 5 \times 5 \times 5$ Base
Index laws	A set of rules for calculating with numbers written in index notation.	$3^2 \times 3^5 = 3^7 \qquad a^m \times a^n = a^{m+n}$ $5^6 \div 5^2 = 5^4 \qquad a^m \div a^n = a^{m-n}$ $(2^3)^4 \quad = 2^{12} \qquad (a^m)^n \quad = a^{mn}$
Coefficient	A number in front of a letter that shows how many of that letter are required.	In $6x + 1$, 6 is a coefficient.
Brackets	Used to show part of an expression that has to be evaluated before the rest of the expression.	In $3(x + 9)$, 9 is added to x before multiplying by 3.
Simplify	Expand brackets, collect like terms or factorise to make an expression easier to use.	$2(6x + 3) - 3x + 2y = 12x + 6 - 3x + 2y$ $= 9x + 2y + 6$
Highest common factor	The longest expression that divides exactly into two or more expressions.	HCF of 15 and 35 is 5. HCF of $15x$ and $3xy$ is $3x$.

Review

1 Simplify these expressions.

   a $y \times 13$ b $y \times 7 \times x$

   c $x + x + x$ d $y \times y \times y$

   e $2 \times x \div 4$ f $4yx + yx$

2 Pens cost 15 p each.

   a How much do 9 pens cost?

   b How much do y pens cost?

3 Substitute $t = 7$ to find the value of these expressions.

   a $5t$ b $t + 8$

   c $10 - 2t$ d t^2

4 Substitute $x = 3$ and $y = -4$ to find the value of these expressions.

   a $7y$ b xy c $2y^2$

   d $2x - 3y$ e $x^2 + 2y$ f $2x^2 + y^2$

5 Calculate the value of these expressions when $m = 4$, $n = 2$ and $p = -1$.

   a $\dfrac{3m}{n}$ b $\dfrac{m}{p}$ c $\dfrac{2p - 8}{5}$

   d $\dfrac{5m - 3n}{7}$ e $\dfrac{mp}{n}$ f $\dfrac{(5m + 2p)}{(n + 4)}$

6

   | $5x - y$ $S = \dfrac{D}{T}$ $3x - 2 = 7$ |

   Give an example of each of these from the box.

   a equation   b formula

   c expression

7 List the terms in each of these expressions.

   a $5f + 6g - 2h$ b $5p - 6q + q^2$

8 Write each of these expressions in the simplest way possible.

   a $5a \times b$ b $c + c + c + c$

   c $d \times d \times d$ d $2f \times 5g$

   e $4 \times 2e - 5 \times 3$ f $a^2 \times a + a^2 + 3$

9 Simplify these expressions.

   a $7r - 5r$

   b $2a + 5b + 3a + b$

   c $7d - 4e - 3e - 2d$

   d $2x + 3x^2 - x$

10 Four friends have simplified the expression $a + \sqrt{a} + \sqrt{a} + a$.
   Which answers are correct?

   a $2a + \sqrt{2a}$ b $2(a + \sqrt{a})$

   c $2a + 2\sqrt{a}$ d $3a$

11 Simplify these expressions involving indices.

   a $c^2 \times c^4$ b $d^8 \div d^3$

   c $(r^3)^4$ d $t^9 \times t^{-3}$

   e $2u^3 \times 3u^5$ f $\dfrac{12v^7}{4v^6}$

12 Expand the brackets in these expressions.

   a $2(a + 1)$ b $8(4b - 2c)$

   c $-5(3d - 4)$ d $h(h + 2)$

13 Fully factorise these expressions.

   a $14b + 7$ b $8 - 4c$

   c $3x^2 + 6x$ d $4ab - 12b$

What next?

Score			
	0 – 5		Your knowledge of this topic is still developing.
			To improve look at MyMaths: 1033, 1155, 1158, 1178, 1179, 1186, 1187, 1247
	6 – 10		You are gaining a secure knowledge of this topic.
			To improve look at InvisiPens: 02Sa – f
	11 – 13		You have mastered these skills. Well done you are ready to progress!
			To develop your exam technique look at InvisiPens: 02Aa – f

Assessment 2

1 Debs has 4 packets of crisps, each worth c pence. She says it cost her $c + c + c + c$ pence.
Her friend Gill says the crisps cost Debs $4c$ pence.
Who is right? Give reasons for your answer.   [2]

2 Vic is V years old.

   a How old was Vic 4 years ago?   [1]

   b How old is Vic's daughter if she is half of Vic's age?   [1]

   c How old will Vic's daughter be in five years time?   [1]

   d How old will Vic be when his daughter is twice her current age?   [2]

3 The book of a film costs £b, a DVD of the same film costs £d and the Blu-ray version costs £r.
What do the following statements mean?

   a $d = 8$ [1]   **b** $r - d = 7$ [1]   **c** $b = \frac{1}{2}d$ [1]

   d $r + d = 23$ [1]   **e** $b + d + r = 27$ [1]

4 A rectangle has a length of l m and a width of w m.

   a What expression represents the perimeter of the rectangle?   [1]

   A second rectangle is 5 m longer and 5 m narrower.

   b What expression represents the perimeter of this rectangle?   [1]

   c Do the two rectangles have the same perimeter? Give reasons for your answer.   [2]

5 The formula for the curved surface area of a cone is $A = \pi r l$, where r is the radius of the
base and l is the slant height.

   a Find A for a cone with base radius 5 cm and slant height 10 cm.   [2]

   b Find r when $A = 45\,\text{m}^2$ and $l = 4\,\text{m}$.   [3]

   c Find l when $A = 126.4\,\text{in}^2$ and $r = 12.3\,\text{in}$.   [3]

6 Amanda, Bengt, Colin, Davide, Erik and Fiona try to put the expression $9x - 4x - 2$
into its simplest form. Here are their results

     Amanda $13x + 2$ Bengt $5(x - 2)$ Colin $5x + 2$

     Davide $9 - 6x$ Erik $3x$ Fiona $5x - 2$.

   Who is correct? Give reasons for your answer.   [1]

7 There are two different charges to post letters of two different sizes. Small letters must
be smaller than 24 cm in length and 16.5 cm wide. Large letters have maximum dimensions
of 35.3 cm by 25 cm. Small letters cost 62 p to post and large letters 93 p.

   a Using the letter S to represent the number of small letters posted and L
the number of large letters, write down a formula to represent the cost C
of sending some letters of both sizes.   [2]

   b Navinder posts 4 small letters and 2 large letters.
Use your formula to confirm that Navinder paid £4.34.   [2]

   c Julie posts 5 small letters and 3 large letters.
Use your formula to calculate how much they cost to post.   [2]

   d Amelia gets 31 birthday cards through the post. 26 are small letter size and the
rest are large letter size. How much did her friends and family spend to send them?   [2]

   e Eilidh has to post 7 items. 3 are small letter size, two measure 25 cm by 15 cm and the
other two are large letter size. How much does she have to pay to post the items?   [3]

8 The area of a trapezium is given by the formula $A = \dfrac{(a + b) \times h}{2}$
where a and b are the parallel sides and h is the perpendicular height.
Find the area of the trapezium when $a = 2z^2$, $b = 3z^2$ and $h = 4z$. [3]

9 Romeo buys Juliet a present. It is a cuboid with a square base of side y cm and height 20 cm.

   a Calculate, in terms of y, the total surface area of the cuboid in its simplest form. [4]

   b The volume is 200 cm³. Romeo says that the exact value of $y = 10$.
Is he correct? Give reasons for your answer. [4]

10 Wanda is W years old.

   a Wanda has twin brothers who are 5 years less than twice Wanda's age.
The total of the ages of the three children can be written as $\square(W - \square)$.
Find the missing values. [4]

   b Wanda's dad's age is three years more than four times Wanda's.
Her mum is 2 years younger than her dad.
The sum of Wanda's parents' ages can be written as $\square(\square W + \square)$.
Find the missing values. [5]

11 A room is L-shaped with a width w. Greta draws a plan of the room with these dimensions:
Can Greta's diagram be correct? Give reasons for your answer. [6]

12 A rectangular pyramid has a base with width $2p$ and length $5p$. Two of the four triangular
sides have an area of $2p(3p + 2)$ each. The other two have an area of $5p(p - 3)$ each.

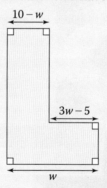

Find, simplify and factorise the expression for the total surface area of the pyramid. [5]

13 Mark draws a rectangle $ABCD$ as shown. He joins AB and CD with
the line EF which is parallel to BC and AD.

$AD = a$ cm
$AE = (a + 4)$ cm
$FC = 9$ cm

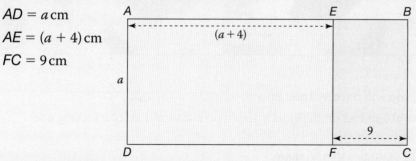

   a He says that in its simplest form the length of $DC = 9a + 4$.
Is he correct? If not, give the correct expression. [2]

   b Find, simplify and factorise the expression for the perimeter of the rectangle $ABCD$. [3]

   c Write down the expressions for the areas of $AEFD$ and $EBCF$.
Give a geometric reason for your answers. [3]

3 Angles and polygons

Introduction

Tiling is a fascinating topic that is highly mathematical, involving angles and shapes. There are some wonderful tiling patterns to be found in architecture, particularly in Islamic floors to be found in palaces and mosques.

The tiling shown in this picture is from the Alhambra Palace in Granada, Spain.

What's the point?

An understanding of angles and shapes allows us to create beautiful things.

Objectives

By the end of this chapter you will have learned how to …

- Describe and apply the properties of angles at a point, on a line and at intersecting and parallel lines.
- Derive and use the sum of angles in a triangle.
- Derive and apply the properties and definitions of special types of quadrilaterals.
- Solve geometrical problems on coordinate axes.
- Identify and use congruence and similarity.
- Deduce and use the angle sum in any polygon and derive properties of regular polygons.

Chapter investigation

A quadrilateral has been drawn on a 3 × 3 square dotty grid.

How many different quadrilaterals can you find?

3.1 Angles and lines

● An **angle** is a measure of turn. You measure angles in **degrees**.

An **acute** angle is less than 90°.

A **right angle** is exactly 90°.

An **obtuse** angle is between 90° and 180°.

A **reflex** angle is between 180° and 360°.

● You should know these facts:

There are 360° at a point.

There are 180° on a straight line.

Vertically opposite angles are equal.

EXAMPLE

Calculate the values of p, q and r.

a [260°, p]

b [q, 76°, 65°]

c [85°, r]

a $360° - 260° = 100°$

$p = 100°$

Angles at a point add to 360°.

b $76° + 65° = 141°$

$180° - 141° = 39°$

$q = 39°$

Angles on a straight line add to 180°.

c $r = 85°$

Vertically opposite angles are equal.

● **Parallel** lines are always the same distance apart.

Parallel lines are shown by sets of arrows.

● **Perpendicular** lines meet at a right angle.

Parallel lines never **intersect** (cross) each other.

When a line crosses **parallel** lines, eight angles are formed.

● **Alternate angles** are equal.

● **Corresponding angles** are equal.

Alternate angles – look for a Z shape.
Corresponding angles – look for a F shape.

EXAMPLE

Find the angles marked by letters. State whether each answer is acute, obtuse or reflex

[115°, b, c, a]

$a = 115°$ Alternate angles are equal.

$b = 115°$ Corresponding angles are equal.

$c = 180° - 115° = 65°$

Angles on a straight line add to 180°.

$115°$ is an obtuse angle.

$65°$ is an acute angle.

Exercise 3.1S

1 Choose one of these to describe each angle.

| acute | right angle | obtuse | reflex |

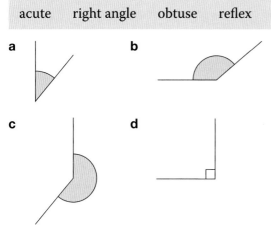

a

b

c

d

2 Choose one of these to describe each angle.

| acute | right angle | obtuse | reflex |

a 90°   **b** 40°   **c** 140°

d 200°   **e** 270°   **f** 36°

g 137°   **h** 248°   **i** 302°

j 33°   **k** 96°   **l** 239°

3 a Draw two lines that are parallel. Label them with >.

b Draw two lines that are perpendicular. Label them with ⌐.

4 Give the values in degrees of the coloured angles.

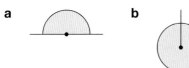

a

b

5 Calculate the size of the angle marked by a letter in each diagram.
Give a reason for each answer. The diagrams are not accurately drawn.

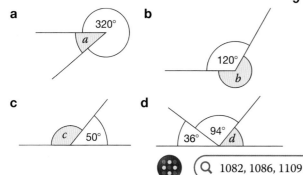

a   320°   *a*

b   120°   *b*

c   *c*   50°

d   36°   94°   *d*

6 Find the value of each angle marked with a letter.
Give a reason for each answer.

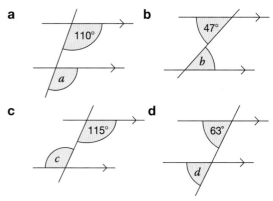

a   110°   *a*

b   47°   *b*

c   115°   *c*

d   63°   *d*

7 Calculate the size of the angles marked by letters in each diagram.
Give a reason for each answer. The diagrams are not accurately drawn.

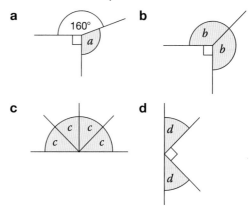

a   160°   *a*

b   *b*   *b*

c   *c*   *c*   *c*

d   *d*   *d*

8 Calculate the sizes of the angles marked by letters in each diagram.
Give a reason for each of your answers.

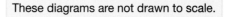

166°   *h*   *g*   *i*   37°   *l*   *j*   *k*

These diagrams are not drawn to scale.

9 Find the value of each angle marked with a letter. Give a reason for each answer.

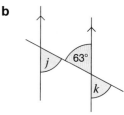

a   *i*   *h*   130°

b   *j*   63°   *k*

3.1 Angles and lines

- Angles at a point add up to 360°.
- Angles at a point on a straight line add up to 180°.
- Vertically opposite angles are equal.
- When two lines are parallel, alternate angles are equal and corresponding angles are equal.

Vertically opposite angles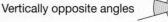

Alternate angles   Corresponding angles

North, east, south or west are often not enough to give an accurate direction.

- A **bearing** is an angle measured clockwise from north.

You use a 360° **scale** or a **bearing** to give a direction accurately.

- To give a bearing accurately you measure from north, measure clockwise and use three figures.

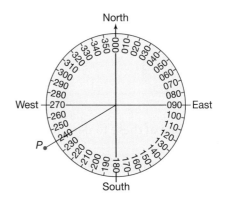

The bearing of *P* is 240°.

HOW TO

1. You may need to draw a sketch and label the angles that you know.
2. Look for parallel lines or places where angles meet at a point.
3. Write down each rule that you use during each stage of your calculation.

000° = North
090° = East
180° = South
270° = West

EXAMPLE

Find the unknown angle in this diagram. Give reasons for your answer.

70°
130°
y

①
y 130°
70°
130°

② Corresponding angles

y = 180° − 130°
  = 50°

③ Angles on straight line add to 180°.

EXAMPLE

The bearing of *G* from *B* is 028°.

Find the bearing of *B* from *G*.

N
N
28°
G
28°
B

① Draw a sketch.

② Corresponding angle

Bearing of *B* from *G* is the angle at *G* measured clockwise from north to *B*.

③ Bearing of *B* from *G* = 28° + 180° = 208°

Geometry Angles and polygons

Exercise 3.1A

1 This diagram is wrong.
Explain why.

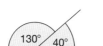

2 Find the missing angles in each diagram.
Write down which angle fact you are using
each time.

a   **b**

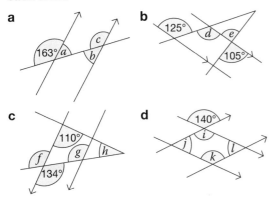

c   **d**

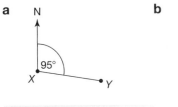 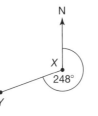

3 These diagrams have not been drawn
accurately.
Find the bearing of *X* from *Y* in each case.

a   N   **b**   N

95°
X   *Y*   *X*
                  248°
               Y

> Draw a sketch to help you.

4 Find these bearings.

   a The bearing of *A* from *B* is 104°.
      Work out the bearing of *B* from *A*.

   b The bearing of *E* from *F* is 083°.
      Work out the bearing of *F* from *E*.

   c The bearing of *J* from *K* is 297°.
      Work out the bearing of *K* from *J*.

5 A plane takes off from a runway in a
north-west direction.
It then turns through an angle of 75°
to its right.
A helicopter, flying on a bearing of 232°,
needs to turn to fly in the same direction.
What turn must the helicopter make?

6 Jennifer draws this diagram.

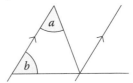

Use Jennifer's diagram to prove these
statements.

   a The exterior angle of a triangle is equal
      to the sum of the interior angles of the
      other two vertices.

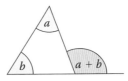

> An exterior angle
> is formed by
> extending a side.

   b The interior angles of a
      triangle add up to 180°.

$$a + b + c = 180°$$

7 Prove that the opposite angles in a
parallelogram are equal.

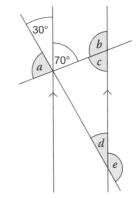

8 Work out the missing angles.

Triangles and quadrilaterals

There are 180° on a straight line.

You can draw any triangle ... tear off the corners ... and put them together to make a straight line.

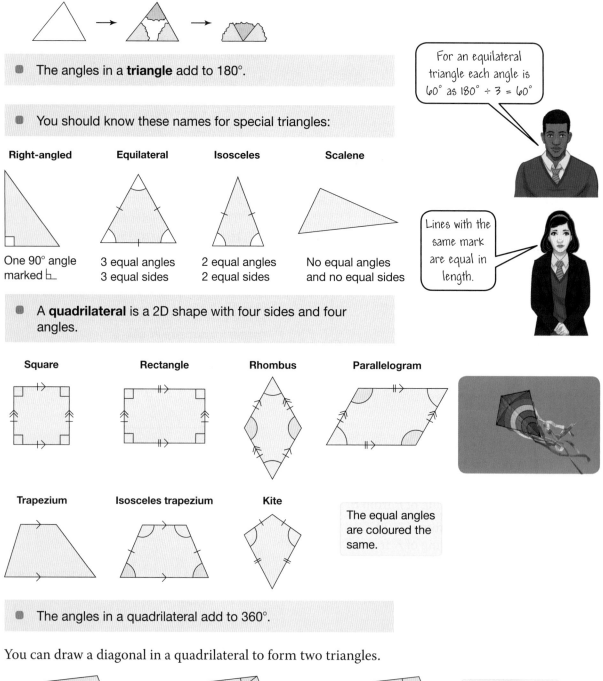

- The angles in a **triangle** add to 180°.

For an equilateral triangle each angle is 60° as 180° ÷ 3 = 60°

- You should know these names for special triangles:

Right-angled	Equilateral	Isosceles	Scalene
One 90° angle marked ∟	3 equal angles 3 equal sides	2 equal angles 2 equal sides	No equal angles and no equal sides

Lines with the same mark are equal in length.

- A **quadrilateral** is a 2D shape with four sides and four angles.

Square   **Rectangle**   **Rhombus**   **Parallelogram**

Trapezium   **Isosceles trapezium**   **Kite**

The equal angles are coloured the same.

- The angles in a quadrilateral add to 360°.

You can draw a diagonal in a quadrilateral to form two triangles.

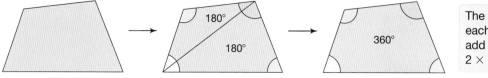

The angles in each triangle add to 180°. 2 × 180° = 360°

Geometry Angles and polygons

Exercise 3.2S

1 Give the mathematical name of each coloured shape in the regular hexagon.

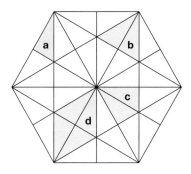

2 Calculate the size of the unknown angle in each diagram.
The diagrams are not drawn to scale.

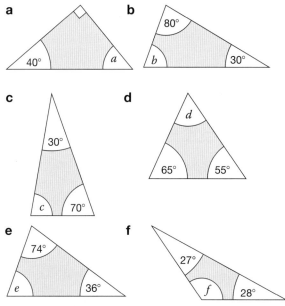

3 List any triangles in question **2** that are
  a right-angled
  b isosceles.

4 State the type of each triangle, if the sides of the triangle are
  a 8 cm, 8 cm, 8 cm
  b 6 cm, 7 cm, 8 cm
  c 3 cm, 5 cm, 5 cm
  d 25 mm, 10 mm, 2.5 cm.

5 Calculate the third angle of the triangle and state the type of each of these triangles.
  a 30°, 60°   **b** 70°, 40°
  c 60°, 60°   **d** 35°, 65°
  e 45°, 45°   **f** 20°, 30°

6 State the value of each unknown angle.

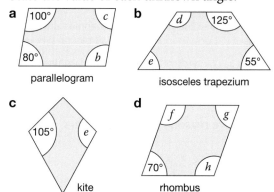

7 Find the unknown angle in each quadrilateral and state the type of quadrilateral.

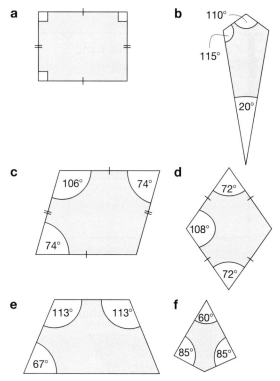

3.2 Triangles and quadrilaterals

- A triangle has three sides and its angles add to 180°.
- A quadrilateral has four sides and its angles add to 360°.

▲ Osaka Castle in Japan

HOW TO

① Sketch a diagram (unless one is given).
Mark (or look for) known angles and equal or parallel sides.

② Use the properties of triangles and quadrilaterals.
Look for parallel lines or places where angles meet at a point.

③ Write down each rule that you use during each stage of your calculation.

EXAMPLE

Calculate the values of x and y.

Give reasons for your answers.

① As the triangle is isosceles, two of the angles are equal.

② $180° - 129° = 51°$

$x = 51°$

③ Angles on a straight line add to 180°.

② $51° + 51° = 102°$

$180° - 102° = 78°$

$y = 78°$

③ Angles in a triangle add to 180°.

EXAMPLE

Calculate the values of y and z.
Give a reason for each of
your answers.

① The quadrilateral is a rhombus.

② $y = 36°$

③ Opposite angles of a rhombus are equal.

② $36° + 36° = 72°$

$360° - 72° = 288°$

$288° \div 2 = 144°$

$z = 144°$

③ Opposite angles of a rhombus are equal and angles in a
quadrilateral add to 360°.

EXAMPLE

Are these statements true or false?
Give reasons for your answer.

a All squares are rhombuses.

①

b Parallelograms are rectangles.

①

a ② A **square** is a special type of rhombus with all
angles equal.

③ True: all properties of a rhombus are also
properties of a square.

b ② A **rectangle** is a special type of parallelogram with
all angles equal.

③ False: a parallelogram does not necessary have all
its angles equal.

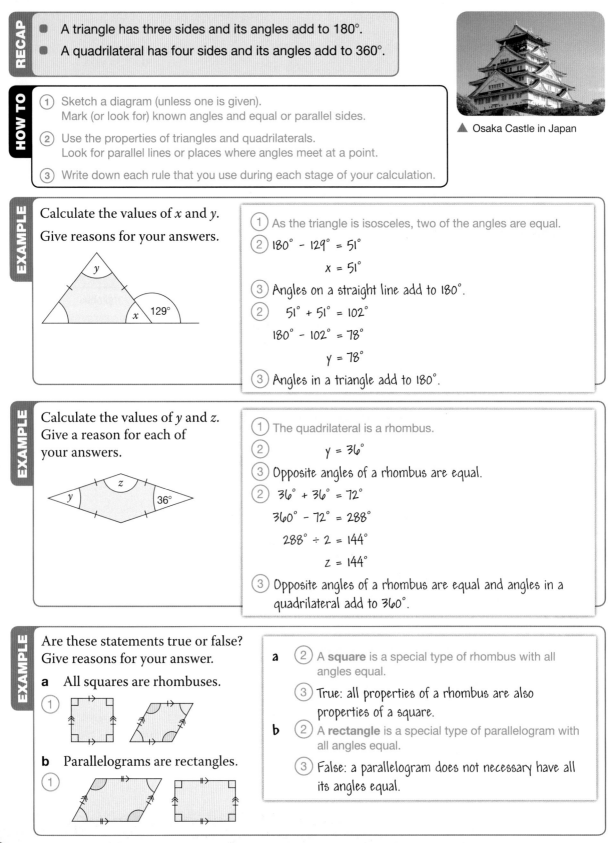

Geometry Angles and polygons

Exercise 3.2A

1 Choose three of these angles that could be put together to make the angles in a triangle.

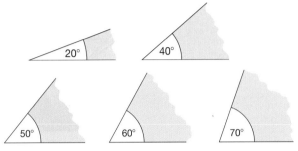

2 Calculate the size of the unknown angles in each diagram.
The diagrams are not drawn to scale.

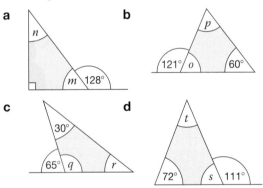

3 The points A $(-1, -2)$ and B $(3, -2)$ are shown. Give the coordinates of a point C, so that triangle ABC

a is isosceles

b is right-angled but scalene

c is right-angled and isosceles

d is scalene

e is equilateral (only an approximate value of y is needed)

f has an area of $4 \, \text{cm}^2$.

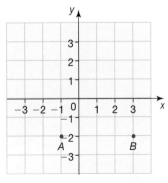

4 Calculate the value of x for each quadrilateral.

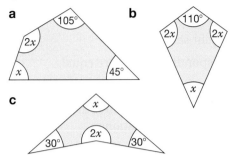

5 Find the value of each angle marked with a letter. Give a reason.

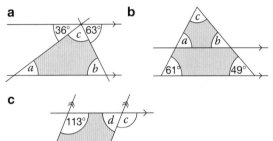

6 Find the values of the angles marked by letters.
Give reasons for each step in your answer.

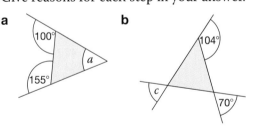

7 Are these statements true or false?
Give reasons for your answers.

a All squares are rectangles.

b All kites are rhombuses.

c All rhombuses are rectangles.

8 Prove that the pink triangle is isosceles.

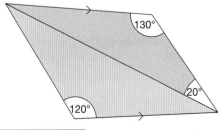

1082, 1102, 1130, 1141 SEARCH

3.3 Congruence and similarity

● **Congruent** shapes are exactly the same shape and size.

> Congruent shapes fit exactly on top of each other.

In congruent shapes

● corresponding angles are equal

● corresponding sides are equal.

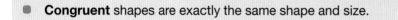

Two triangles are congruent if they satisfy one of four sets of conditions.

SSS: three sides the same

SAS: two sides and the included angle the same

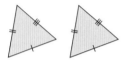

> You should learn the four conditions for congruency.

ASA: two angles and the included side the same

RHS: right-angled triangles with **hypotenuse** and one other side the same.

EXAMPLE

This is triangle A.

Are triangles B, C or D congruent to triangle A?

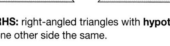

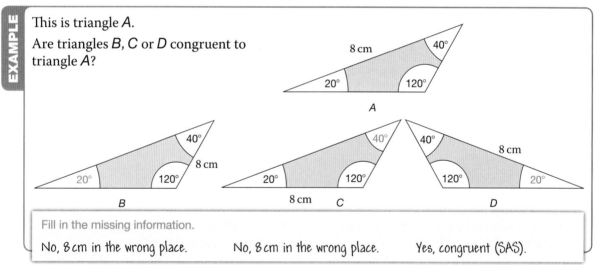

Fill in the missing information.

No, 8 cm in the wrong place.   No, 8 cm in the wrong place.   Yes, congruent (SAS).

In an **enlargement**, the object and the image are **similar**:

● the angles stay the same

● the lengths increase in proportion.

You use corresponding lengths to find the **scale factor**.

● Scale factor = $\dfrac{\text{length of image}}{\text{length of object}}$

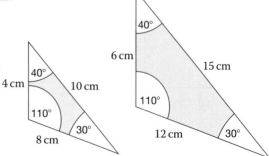

Exercise 3.3S

1 a Find **all** the pairs of quadrilaterals that are congruent.

b Give the name of the shape for each pair.

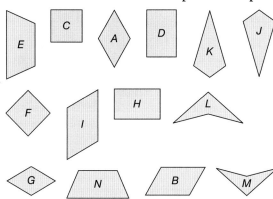

2 Here is triangle A.

Which of these triangles are congruent to triangle A?

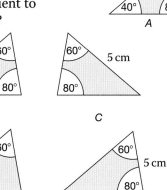

3 Which of these triangles are congruent to triangle A?

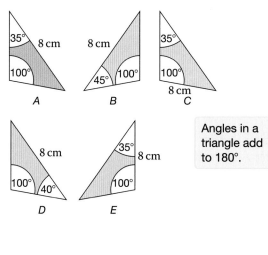

> Angles in a triangle add to 180°.

4 In each question, the two triangles are similar. Find the value of the unknown angles.

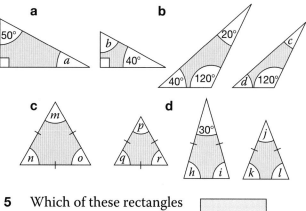

5 Which of these rectangles are similar to the green rectangle?

For the ones that are similar, give the scale factor of the enlargement.

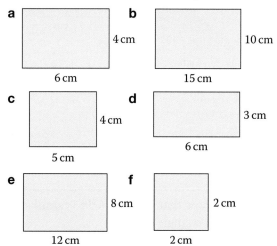

6 In each question, the two triangles are similar. Calculate the scale factor of the enlargement and the unknown length.

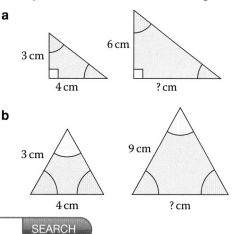

Q 1119, 1148　SEARCH

3.3 Congruence and similarity

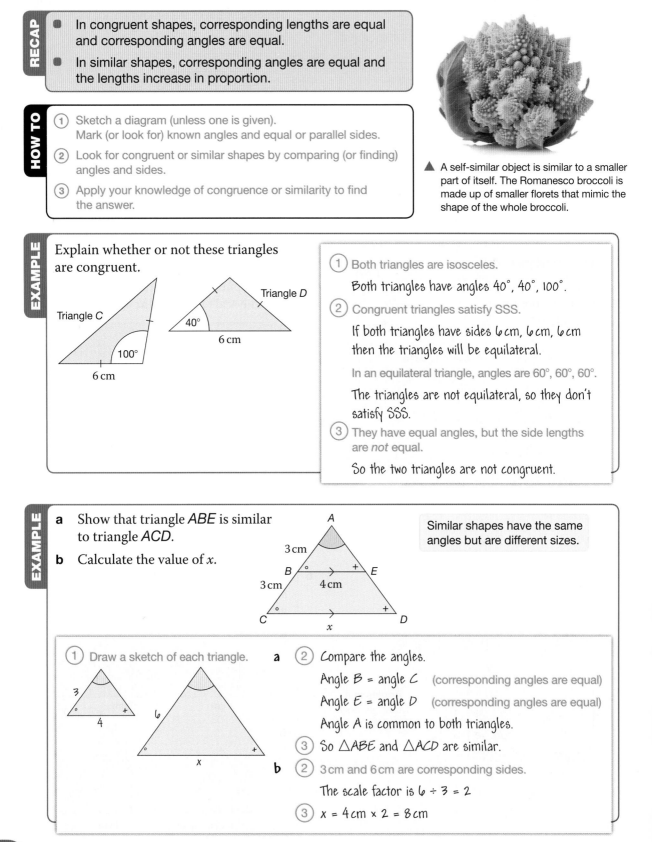

- In congruent shapes, corresponding lengths are equal and corresponding angles are equal.
- In similar shapes, corresponding angles are equal and the lengths increase in proportion.

HOW TO

(1) Sketch a diagram (unless one is given).
Mark (or look for) known angles and equal or parallel sides.

(2) Look for congruent or similar shapes by comparing (or finding) angles and sides.

(3) Apply your knowledge of congruence or similarity to find the answer.

▲ A self-similar object is similar to a smaller part of itself. The Romanesco broccoli is made up of smaller florets that mimic the shape of the whole broccoli.

EXAMPLE

Explain whether or not these triangles are congruent.

Triangle C

Triangle D

40°

6 cm

100°

6 cm

(1) Both triangles are isosceles.

Both triangles have angles 40°, 40°, 100°.

(2) Congruent triangles satisfy SSS.

If both triangles have sides 6 cm, 6 cm, 6 cm then the triangles will be equilateral.

In an equilateral triangle, angles are 60°, 60°, 60°.

The triangles are not equilateral, so they don't satisfy SSS.

(3) They have equal angles, but the side lengths are *not* equal.

So the two triangles are not congruent.

EXAMPLE

a Show that triangle *ABE* is similar to triangle *ACD*.

b Calculate the value of *x*.

A

3 cm

B ○ → + E

3 cm   4 cm

○ → +

C   x   D

Similar shapes have the same angles but are different sizes.

(1) Draw a sketch of each triangle.

3

○ → +

4

6

○ → +

x

a (2) Compare the angles.

Angle B = angle C   (corresponding angles are equal)

Angle E = angle D   (corresponding angles are equal)

Angle A is common to both triangles.

(3) So △ABE and △ACD are similar.

b (2) 3 cm and 6 cm are corresponding sides.

The scale factor is 6 ÷ 3 = 2

(3) x = 4 cm × 2 = 8 cm

Geometry   Angles and polygons

Exercise 3.3A

1 How many pairs of congruent triangles are there in this kite?

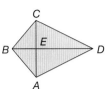

2 Explain why triangles *PQR* and *WXY* are congruent.

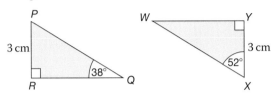

3 *ABCD* is a rectangle.

Prove that triangles *ABD* and *CDB* are congruent.

4 *KLMN* is a kite.

a Explain why triangles *KLN* and *MLN* are not congruent.

b Explain why triangles *KLM* and *KNM* are congruent.

5 The dotted lines show two different ways of splitting a rectangle into two congruent shapes.

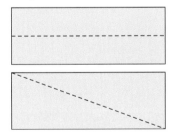

a Draw two copies of the rectangle.

Draw dotted lines to show two more ways of splitting the rectangle into congruent shapes.

b For each, state whether or not the dotted line is also a line of symmetry.

6 Calculate the value of each unknown length.

a

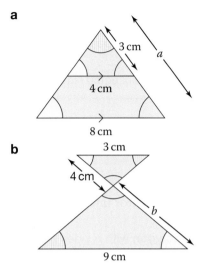

b

7 A diagram is 10 cm wide and 12 cm high. A photocopier is used to reduce the size of the diagram.

a Find the new height of the diagram when the new width is 8 cm.

b Find the new width of the diagram when the new height is 8 cm.

8 A shop sells picture frames in the following sizes.

Size	Width	Height
Mini	24 cm	30 cm
Small	30 cm	40 cm
Medium	40 cm	50 cm
Large	50 cm	70 cm
Extra Large	60 cm	80 cm

Which of these sizes are similar in shape? Explain your answer.

9 Which of these types of shapes will **always** be similar, no matter what their size?

Give reasons for your answers.

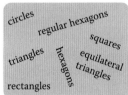

3.4 Angles in polygons

A **polygon** is a 2D shape with three or more straight sides.

- A **regular** shape has equal sides and equal angles.

Sides	Name	Sides	Name
3	triangle	7	heptagon
4	quadrilateral	8	octagon
5	pentagon	9	nonagon
6	hexagon	10	decagon

A regular hexagon has six equal sides and six equal angles.

- You should know the names of the polygons in the table.

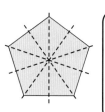

A line of symmetry divides the shape into two identical halves, each of which is the mirror image of the other.

- A **regular polygon** with *n* sides has *n* lines of symmetry.

The **order of rotational symmetry** is the number of times a shape looks exactly like itself in a complete turn.

- A **regular polygon** with *n* sides has rotational symmetry of order *n*.

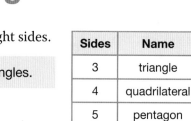

A regular pentagon has 5 lines of symmetry and rotational symmetry of order 5.

The **interior angles** are inside the polygon.

The **exterior angles** are made by extending each side in the same direction.
Exterior angles are outside the polygon.

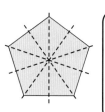

exterior

interior

- The exterior angles of any polygon add to 360°.

- Interior angle + Exterior angle = 180°
  (angles on a straight line add to 180°).

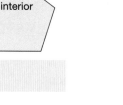

int ext

You can divide any polygon into triangles by drawing diagonals from a **vertex** (corner). The number of triangles is always two less than the number of sides.

- The sum of the interior angles of any polygon
  = (number of sides − 2) × 180°

Calculate the sum of the interior angles for a pentagon.

Draw in the two diagonals.
Three triangles formed:

3 × 180° = 540°
Sum of interior
angles = 540°

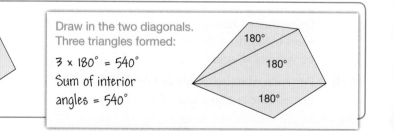

180°

180°

180°

Geometry Angles and polygons

Exercise 3.4S

1 State the number of lines of symmetry for each of these regular polygons.

a   **b**

c   **d**

e   **f**

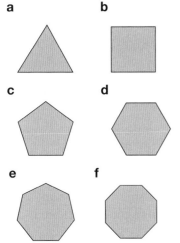

2 State the order of rotational symmetry for each polygon in question **1**.

3 **a** Calculate the value of one interior angle of an equilateral triangle.

Two equilateral triangles are placed together to form a rhombus.

b Calculate the value of each interior angle of this rhombus.

c Calculate the sum of the interior angles of a rhombus.

4 Copy and complete this table.

Regular polygon	No. of sides	Exterior angle	Interior angle
Triangle	3		
Quadrilateral	4		
Pentagon	5		
Hexagon	6		
Heptagon	7		
Octagon	8		
Nonagon	9		
Decagon	10		

5 A regular polygon has 15 sides.

a Use the angle sum of this polygon to work out the size of an interior angle.

b Use the sum of the exterior angles to check your answer to part **a**.

6 A regular polygon has 18 sides.

a Calculate the size of an interior angle of this polygon.

b Use another method to check your answer to part **a**.

7 Explain how to find the sum of all the interior angles in an octagon.

8 Find the angle x.

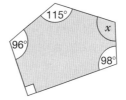

9 Two of the angles of a hexagon are right angles and three angles are equal to 132°. Find the other angle.

10 A dodecagon is a polygon with 12 sides. Eleven of the angles of a dodecagon are equal to 154°.
Calculate the other angle.

*11 The diagram shows a regular nonagon divided into congruent triangles.

a Find the angles marked x and y.

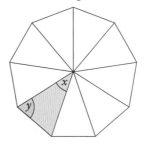

b Use the value of y to check your answers in question **4** for a regular nonagon.

 1100, 1320   SEARCH

3.4 Angles in polygons

- The exterior angles of any polygon add to 360°.
- Interior angle + Exterior angle = 180°
- The sum of the interior angles of any polygon = (number of sides − 2) × 180°

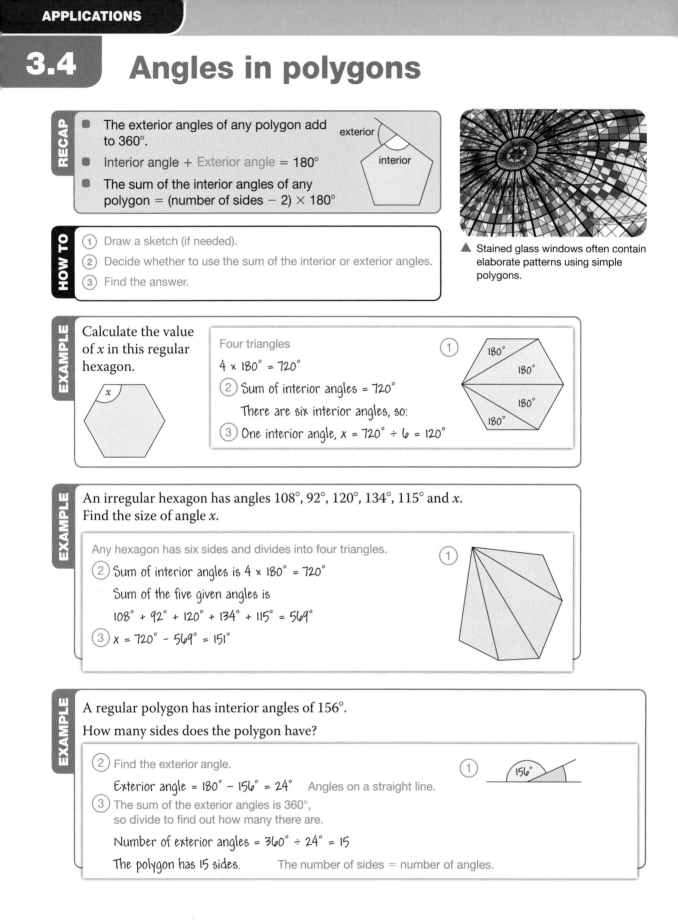

▲ Stained glass windows often contain elaborate patterns using simple polygons.

HOW TO

① Draw a sketch (if needed).

② Decide whether to use the sum of the interior or exterior angles.

③ Find the answer.

EXAMPLE

Calculate the value of x in this regular hexagon.

Four triangles

$4 \times 180° = 720°$

① ② Sum of interior angles = 720°

There are six interior angles, so:

③ One interior angle, $x = 720° \div 6 = 120°$

180°
180°
180°
180°

EXAMPLE

An irregular hexagon has angles 108°, 92°, 120°, 134°, 115° and x.
Find the size of angle x.

Any hexagon has six sides and divides into four triangles.

② Sum of interior angles is $4 \times 180° = 720°$

Sum of the five given angles is

$108° + 92° + 120° + 134° + 115° = 569°$

③ $x = 720° - 569° = 151°$

①

EXAMPLE

A regular polygon has interior angles of 156°.

How many sides does the polygon have?

② Find the exterior angle.

① 156°

Exterior angle = 180° − 156° = 24°   Angles on a straight line.

③ The sum of the exterior angles is 360°,
so divide to find out how many there are.

Number of exterior angles = 360° ÷ 24° = 15

The polygon has 15 sides.   The number of sides = number of angles.

Exercise 3.4A

1 The interior angle of a regular polygon is 162°.

   a Calculate the value of an exterior angle.

   b State the sum of the exterior angles of the polygon.

   c Calculate the number of exterior angles in the polygon.

   d State the number of sides of the regular polygon.

2 A regular polygon has 15 sides.

   a Calculate the value of an exterior angle.

   b Calculate the value of an interior angle.

3 Calculate the size of the angles marked with letters in these polygons.

   a

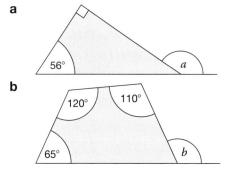

   b

4 An interior angle of a regular polygon is three times the exterior angle.

   a Calculate the value of each exterior angle.

   b Calculate the value of each interior angle.

   c Give the name of the regular polygon.

5 Find the missing angles in these irregular polygons.

   a   **b**

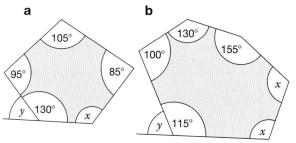

6 Find the size of smallest angle of this pentagon.

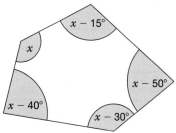

7 How many sides does a regular polygon have if

   a it has exterior angle 30°?

   b it has interior angle 135°?

8 A tessellation is a tiling pattern with no gaps or overlaps. Explain why a regular hexagon tessellates.

9 Does a regular pentagon tessellate? Explain your answer.

*10 The diagram shows a regular hexagon attached to a square.

   a Calculate angle z.

   b Name one or more regular polygons that would fit exactly into this space.

11 A regular polygon can be made by fitting together squares and equilateral triangles. Find angle x.

12 *ABCDEF* is a regular hexagon.

   Show that angle *ACD* is a right angle.

Summary

Checkout
You should now be able to...

Test it
Questions

✔ Describe and apply the properties of angles at a point, on a line and at intersecting and parallel lines.	1 – 2
✔ Derive and use the sum of angles in a triangle.	3 – 4
✔ Derive and apply the properties and definitions of special types of quadrilaterals.	5 – 7
✔ Solve geometrical problems on coordinate axes.	8
✔ Identify and use congruence and similarity.	9 – 10
✔ Deduce and use the angle sum in any polygon and derive properties of regular polygons.	11

Language	Meaning	Example
Acute angle	An angle smaller than a right angle	
Right angle	90° or one-quarter turn.	
Obtuse angle	Greater than 90° but smaller than 180°.	
Reflex angle	Greater than 180° but smaller than 360°.	
Alternate angles	When referring to parallel lines: angles in the corners of a Z shape.	Alternate angles Corresponding angles
Corresponding angles	When referring to parallel lines: angles under the arms of an F shape.	
Three-figure bearing	A direction defined by a three-figure angle measured clockwise from north.	East is 090°. South-west is 225°.
Polygon	A 2D shape with straight edges.	Pentagon (5), Hexagon (6), Octagon (8), Decagon (10).
Triangle	A three sided polygon.	Right-angled, equilateral, isosceles, scalene.
Quadrilateral	A four sided polygon.	Square, rectangle, rhombus, trapezium, parallelogram, kite.
Congruent	Exactly the same shape and size.	
Similar	The same shape but different in size.	
Scale Factor	The ratio of corresponding lengths in two similar shapes.	A and B are similar; the scale factor is 2. A and C are congruent.
Interior angle	The angle between two sides inside a polygon.	
Exterior angle	The angle between one side of a polygon and the next side extended.	

Review

1 Calculate the size of angles *a*, *b*, *c* and *d* and give a reason for each answer.

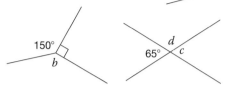
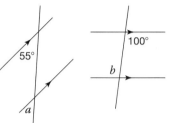

2 Calculate the size of angles *a* and *b* and give a reason for each answer.

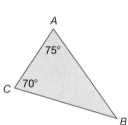

3 Calculate the size of angle *ABC* in this triangle.

A
75°
70°
C
B

4 The triangle *DEF* is not drawn accurately. Which side is the same length as side *DE*?

D
F   E

5 Use your knowledge about triangles to prove that the angles in a quadrilateral add up to 360°.

6 Calculate the size of angles *a* and *b* in the rhombus.

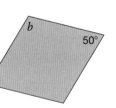

7 Which quadrilateral is being described below?

a Two pairs of parallel and equal sides.

b One pair of parallel sides and no equal sides.

c Two pairs of equal sides but no parallel sides.

8 Draw coordinate axes with *x* and *y* from 0 to 6. Now plot these points A (3, 5), B (6, 2) and C (1, 0).

Join up the dots to form a triangle, what type of triangle is this?

9 Are these two triangles congruent?

Give a reason for your answer.

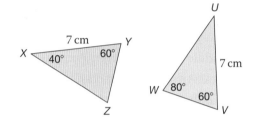

10 These triangles are similar, what is the length of *DE*?

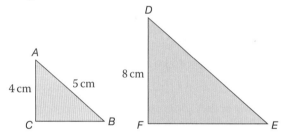

11 What do the interior angles of a pentagon add up to?

What next?

Score			
	0 – 4		Your knowledge of this topic is still developing. To improve look at MyMaths: 1082, 1086, 1100, 1102, 1109, 1119, 1130, 1320, 1141, 1148
	5 – 9		You are gaining a secure knowledge of this topic. To improve look at InvisiPens: 03Sa – k
	10 – 11		You have mastered these skills. Well done you are ready to progress! To develop your exam technique look at InvisiPens: 03Aa – g

Assessment 3

1 How many turns do these actions require?

   a The hour hand of a clock moving for 8 hours. [2]

   b The Earth rotating about the Sun for 9 months. Ignore the Earth spinning on its own axis. [2]

   c An athlete running 3 laps of a track. [1]

2 The time on a clock is ten minutes to four.
Yolanda says that the angle between the hour hand and the minute hand is 175°.
Explain why Yolanda is correct. Show all your workings and give reasons for your answer.

   a 5 o'clock [1]   **b** half past 8 [2]

3 The cruise ship 'Black Watch' sails on a bearing of 070°. To avoid a storm it changes
course to a bearing of 130°. What angle has it turned through? [2]

4 Three villages, East Anywhere, Middle Anywhere and
West Anywhere, are the vertices of the triangle shown.
The bearing of East Anywhere from West Anywhere
is 075°.

Write down the bearing of

   a Middle Anywhere from West Anywhere [1]

   b West Anywhere from East Anywhere [2]

   c Middle Anywhere from East Anywhere. [2]

5 Rafa and Sunita were standing at the top of a hill. Rafa was facing due north and Sunita
was facing due south. Explain why they could see each other. [1]

6 Siobhan fits three triangles together at one point.

Will they make a straight line? Give reasons for your answer. [2]

7 Sienna says that the missing angles have these values

$a = 56°$ $b = 75°$ $c = 46°$ $d = 148°$ $e = 121°$ $f = 121°$.

   a [2]   **b** [2]   **c** [2]

   d [2]   **e** [4]

Decide if her value for each angle is correct or incorrect.
Give reasons for your answers.

8 Manuel made the following statements about triangles. Write down if each of his statements is true
or false. If the statement is true draw an example. If the statement is false explain why it is false.

   a Some triangles have two obtuse angles. [2]

   b Some triangles have one obtuse angle and two acute angles. [2]

   c Some triangles have one right angle, one acute angle and one obtuse angle. [2]

   d Some triangles have two acute angles and one right angle. [2]

9 Mark draws triangle *XZY* so that it is congruent to triangle *ABC*.
Write down the sizes of all the angles in triangle *XZY*.   [2]

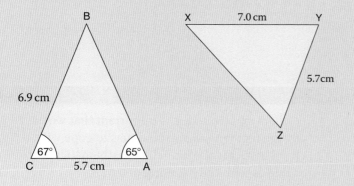

10 Sunita is standing in the shade of a tree, as shown.

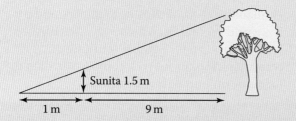

How high is the tree?   [2]

11 a Natasha draws a pentagon that has angles of 125°, 155° and 74°.
She wants to draw the other two angles so that they are equal.
What size should Natasha draw these two angles?   [3]

b Shivani draws an octagon that has five angles each of 114°.
She wants to draw the other three angles so that they are equal.
What size should Shivani draw these three angles?   [3]

c Tyler draws a hexagon that has three angles of 137° each. He wants to draw
the remaining three angles so that they have the values $w°$, $w° + 120°$, and $w° − 30°$.
What value should Tyler use for w?   [4]

d Dean draws a regular polygon with 60 sides.
What size should he use for the interior angles?   [2]

12 Stewart wants to make a patio with seven identical
hexagonal paving slabs, as shown.

a He says that UVWY is a trapezium.
Is he correct?
Give reasons for your answer.   [2]

b What angle does he need to use for
i   the external angle   [2]
ii the internal angle of the regular hexagons?   [2]

c What size does he need to make angle VXY?   [4]

d He says that triangle UWY is isosceles.
Is he correct?
If not, what type of triangle is it?   [1]

4 Handling data 1

Introduction

In the modern world, there is an ever-increasing volume of data being continually collected, analysed, interpreted and stored. It was estimated that in 2007, there was 295 exabytes (or 295 billion gigabytes) of data being stored around the world. It has increased significantly since then. To put this into perspective, if all that data were recorded in books, it would cover the area of China in 13 layers of books. With all this information being generated, it is important that we have the mathematical techniques to cope with it. Statistics is the branch of mathematics that deals with the handling of data.

What's the point?

Availability of data helps us to understand the world around us, whether it is scientists looking for trends in global warming or consumers looking at cost comparison data to help inform purchasing decisions.

Objectives

By the end of this chapter you will have learned how to …

- Identify when a sample may be biased.
- Construct and interpret frequency tables and two-way tables.
- Construct and interpret pictograms, bar-line charts and bar charts.
- Interpret and construct pie charts and know their appropriate use.
- Compare distributions using median, mean, mode and range and identify outliers.

Check in

1   Put these numbers in order of size, smallest first.

   a   37, 42, 17, 6, 30, 19, 26, 29

   b   118, 135, 106, 121, 130, 115

   c   156, 145, 154, 165, 166, 155, 144

2   Calculate

   a $63 + 58$ **b** $48 + 69$ **c** $73 + 95 + 84$ **d** $138 + 275$ **e** $63 + 5$

   f $38 - 15$ **g** $96 - 47$ **h** $136 - 54$ **i** $258 - 69$ **j** $432 - 166$

3   Calculate the value of the angle x.

   a   **b**

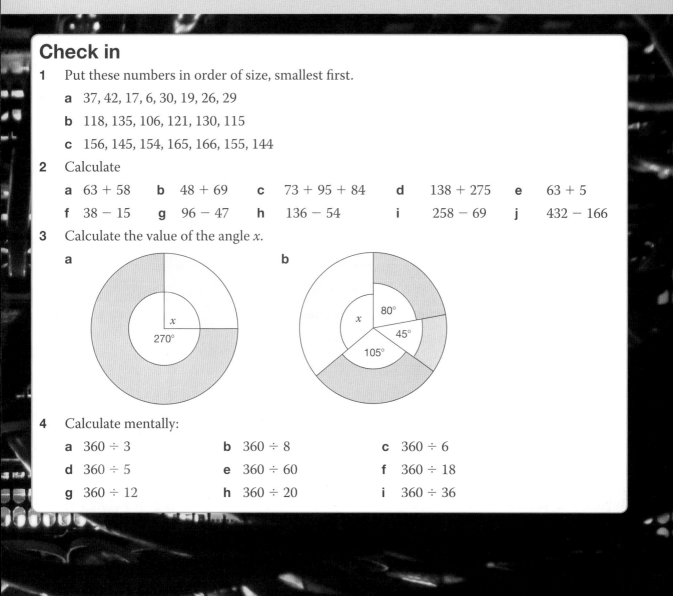

4   Calculate mentally:

   a $360 \div 3$ **b** $360 \div 8$ **c** $360 \div 6$

   d $360 \div 5$ **e** $360 \div 60$ **f** $360 \div 18$

   g $360 \div 12$ **h** $360 \div 20$ **i** $360 \div 36$

Chapter investigation

What is the average age of students in your class?

4.1 Sampling

To find out information about a **population**, you can collect **data** using a **survey**. A survey of the population is called a **census**.

For example, if you want to know what proportion of students in a class walk to school, you can ask each student, complete a **data collection sheet** and work out the answer.

> A population is the group of people or items that you want to find out information about.

EXAMPLE

A class has 30 students. Lisa asks each student whether they walk to school (Y) or not (N).

Y	N	N	Y	N	Y	N	N	Y	N
Y	N	Y	Y	N	N	N	Y	N	N
N	N	N	N	N	Y	Y	Y	N	Y

a   Complete a data collection sheet (a **tally chart**).

b   Find the proportion of students in the class who walk to school.

a

Do you walk to school?	Tally	Number
Yes	IIII IIII II	12
No	IIII IIII IIII III	18

b   The proportion of students who walk to school is 12 out of 30.

Did you know...

The National Census is carries out every 10 years in the UK. The information is used to plan services like schools and hospitals. The next Census will be in 2021.

What if you want to find out what proportion of all the students in the school walk to school?

Sometimes it is impossible to collect data from all the population and so a **random sample** is used.

Instead of asking every student in your school, you could choose a random sample of 50 students.

- The larger the sample, the more accurate the data will be.

A good sample must represent the whole group.
It must not be **biased**.

- In a random sample, each person or item has the same chance of being included.

> A sample is biased if individuals or groups from the population are not represented in the sample.

> Taking every 4th student from an alphabetical or form list is biased.

EXAMPLE

Describe a method to choose a random sample of 30 students for a year group of 120 students.

- Number each student from 1 to 120.
- Put a different number on 120 pieces of paper.
- Place the pieces of paper in a bag.
- Pick out 30 numbers from the bag.

Exercise 4.1S

1 Chris counts the number of passengers in passing cars.

```
1  0  2  0  1  0  1  3  1  0  3  2
0  1  0  0  2  1  0  0  0  1  2
```

Number of passengers	Tally	Number of cars
0		
1		
2		
3		

a Copy and complete the data collection sheet to show this information.

b Calculate the total number of cars that passed.

2 James uses a data collection sheet for a survey to find his class' favourite soup. He limits the choice to Tomato (T), Vegetable (V), Fish (F) or Other (O).

```
T  T  T  V  F  V  V  T  T  V
F  O  T  V  O  O  T  V  T  F
O  T  V  O  T  V  O  O  O  O
```

a Copy and complete the tally chart to show the data.

Type of soup	Tally	Number of students

b Calculate the number of students in James's class.

c State the class' favourite soup.

3 Felicity counts the number of occupants in passing cars.

```
1  2  1  4  3  1  2  1  5  4
1  1  1  2  1  3  1  4  1  1
1  2  1  1  1  2  1  1  4  2
2  2  1  2  1  3  2  1  2  1
```

a Construct a data collection sheet to show this data.

b Calculate the total number of cars that passed.

c Calculate the total number of occupants of the cars that passed.

4 A class of 30 students decide to elect a class representative by a random process. State whether these methods of selection are random or biased.
Give a reason for each answer.

a Arrange the class list into alphabetical order and select the first name on the list.

b Arrange the class list into alphabetical order and select the last name on the list.

c Put the names of the students on cards of equal size. Put the cards into a bag and pick out one card.

d Arrange the students in order of height and select the smallest student.

e Hide a gold star in the classroom, and select the student who finds the star.

f Number the students from 1 to 30. Roll a dice and select the student with that number.

5 Megan made this table to show the advantages and disadvantages of using a sample or carrying out a census of the whole population.

Census	**Advantages**	
	• Everyone is represented	
	Disadvantages	
	• Takes a long time	
	• Expensive	
	• Produces a lot of data	
Sample	**Advantages**	
	• Quick to do	
	• Cheap to carry out	
	Disadvantages	
	• Could be biased depending on the sample chosen	

Copy the table. Add these statements to the correct region of Megan's table.

● Difficult to make sure that the whole population is included

● Less data to analyse

● Unbiased

● Not everyone is represented

Q 1212, 1248, 1249   SEARCH

4.1 Sampling

- To find out information about a population you can collect data using a survey and a data collection sheet.

- Sometimes it is too difficult or time-consuming to collect data from everyone. In these cases you should survey a sample.

- A good sample should represent the whole population. It must not be biased.

- In a random sample, each person or item has the same chance of being included.

> You can use the results from a sample to draw conclusions for the whole population. The larger the sample, the more reliable the results.

HOW TO

① Describe the population and the sample.

② Decide whether each person in the whole population has an equal chance of being chosen (random sample) or not (biased sample).

③ Give a reason for your answer.

EXAMPLE

James carries out a survey to find out if people in his town enjoy sport.

He stands outside a football ground and surveys people's opinions as they go in to watch a match.

   a   Is the sample biased? Give your reasons.

   b   Give two possible ways that the sample could be unrepresentative of the views of everyone in the town.

> **a** ① Population: all the people in the town. Sample: people going to see a football match
>
>   ② If someone is not at the football match, then they can't be included in the sample. The sample is biased.
>
> **b** ③ People who watch football are more likely to enjoy sport.
> More men than women go to watch football so the survey could be gender biased.

EXAMPLE

A train company carried out a survey about a local rail service.

They telephoned 100 people from a page of the telephone directory to answer a questionnaire on the rail service.

   a   Explain why the sample is biased.

   b   Write three reasons why this sample could be unrepresentative.

> **a** ① Population: all the people who live in the local area.
> Sample: people listed in one page of the local telephone directory
>
>   ② Sample is biased as it is impossible for some people in the town to be included in the survey.
>
> **b** ③ Only people who have a land-line telephone can be included in the sample.
>
>   Only people on one page of the directory can be included in the sample.
>
>   Some people may not be at home when they are phoned, so the sample may be incomplete.

Exercise 4.1A

1 Which of the following statements are true? Give reasons for your answers.

   a A sample will always give the same results if the data is collected using an unbiased method.

   b A sample uses half the population.

   c A sample is quicker to do than a census.

   d A smaller sample gives better results than a larger sample.

2 Energising! make batteries. The company wants to investigate how long their batteries last. Should Energising! use a census or a sample? Give reasons your answer.

3 Katy is doing a survey to find out how often people go to the cinema and how much they spend. She stands outside a cinema and asks people as they go in.

   Why could this sample be biased?

4 Sally wants to find out how often people play sport. Sally belongs to an athletics club. She asked members in her athletics club.

   How could this sample be biased?

5 James wants to know which flavour crisps he should stock in the school tuck shop.

   a He asks his mum, dad, auntie and uncle. Say why this is not a good sample to use.

   b He asks only Year 11 at his school. Say why this sample could be biased.

   c Describe how James could take a sample of 50 people. (There are 1000 people in his school.)

6 Merlin wants to find out how far people would travel to see their favourite band perform.

   a He asks all his friends. Write a reason why this sample could be biased.

   b He goes into town one Saturday morning and asks anyone listening to music on a MP3 player. How could this sample be biased?

7 Jenny carries out a survey to find out the most popular band. She asks 10 of her friends – all girls. How could this sample be biased?

8 Wayne carries out a survey to find out the most popular car colour. He stands on a street corner and notes the colour of the first 15 cars that pass by. Write a reason why this sample could be biased.

9 Lisa wants to find out how people travel to work.

   a She asks people at a bus stop one morning. How could this sample be biased?

   b She opens the telephone directory at a random page and phones everyone on that page. How could this sample be biased?

10 Dervla asked people at a netball club

> Do you agree that tennis is the most exciting sport?

Write two reasons why this is not a good way to find out which sport people find most exciting.

11 Sanjit is collecting data on TV viewing habits among 15 year-olds. He asks a sample of 15 year-olds from his media studies class. How could this sample be biased?

12 Billie is interested in how many times adults go swimming in one year. She asks five parents of her friends

> How many times did you go swimming last year?

Billie records her findings the next day.

Find three causes of bias in Billie's method.

Organising data

- You can organise data in a table, such as a **frequency table**, using **rows** and **columns**.

Method of travel	Number of students
Car	14
Bus	11
Walk	5

14 students travelled by car.

- A **two-way table** shows more detail and links two types of information, for example, method of travel and gender.

	Boys	Girls
Car	4	10
Bus	6	5
Walk	2	3

5 girls travelled by bus.

- You can use a **stem-and-leaf diagram** to display numerical data.

A stem-and-leaf diagram shows

- the shape of the distribution
- each individual value of the data.

This stem-and-leaf diagram is **ordered**, as the data is in numerical order.

13	5
12	0 6 9
11	2 2 6 8
10	3 7

This means 135.

This means 120.

Key: 11 | 2 means 112

stem   leaf

Always give a key.

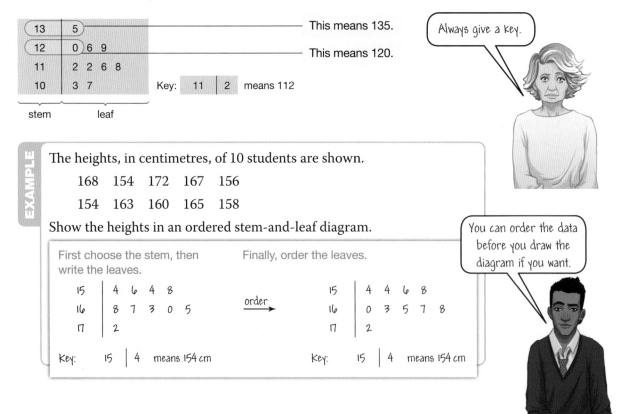

EXAMPLE

The heights, in centimetres, of 10 students are shown.

168   154   172   167   156

154   163   160   165   158

Show the heights in an ordered stem-and-leaf diagram.

You can order the data before you draw the diagram if you want.

First choose the stem, then write the leaves.

15	4 6 4 8
16	8 7 3 0 5
17	2

Key: 15 | 4 means 154 cm

order →

Finally, order the leaves.

15	4 4 6 8
16	0 3 5 7 8
17	2

Key: 15 | 4 means 154 cm

Exercise 4.2S

1 Tickets to see an Irish band cost £5, £10, £15 or £20. Tickets with the following prices are sold one morning.

These values are all in pounds (£).

15	10	10	5	5	5	5	5
10	20	20	5	5	5	5	10
5	10	5	15	15	20	20	5
5	10	10	10	10	5	5	5
10	5	10	15	20	20	20	20

Price	Tally	Number of tickets
£5		
£10		
£15		
£20		

a Copy and complete the tally chart.

b How many £5 tickets were sold?

c How many tickets were sold altogether?

2 A tetrahedron dice is numbered 1, 2, 3, 4.

a Make a data collection sheet to show these scores.

4	2	1	3	1	3	4	2
3	3	2	2	2	3	4	4
2	4	1	1	2	4	1	2
1	2	1	3	2	1	4	3
2	3	1	4	4	2	3	1

b State the number of times a 3 was rolled.

3 There are two cinemas in a Cinecomplex. The number of people in each cinema is shown in the two-way table.

| | Cinema | |
	1	2
Adult	31	47
Child	12	8

Calculate the number of

a adults in the Cinecomplex

b children in the Cinecomplex

c people in Cinema 1

d people in Cinema 2

e people in the whole Cinecomplex.

4 The numbers of houses on each street of a town are shown.

12   5   25   27   7   35   15   10   22   18
34   30   14   19   20   28   32   4   25   36

a Copy and complete the stem-and-leaf diagram.

```
0 | 5 7 4
1 |
2 |
3 |
```

Key: 1 | 8 means 18 houses

b Redraw the table to give an ordered stem-and-leaf diagram.

5 The times taken, in seconds, for 25 athletes to run 400 metres are given.

44.3   44.4   43.3   43.2   44.0   45.2
45.0   44.5   45.6   43.9   46.5   46.3
46.0   46.5   44.7   46.9   44.1   43.8
45.0   46.9   43.0   46.1   45.1   43.8

Draw an ordered stem-and-leaf diagram, using stems of 43, 44, 45 and 46.

Use the key: 43 | 3 means 43.3.

6 The weights, in kilograms, of 30 students are shown.

48   47   48   53   61   70
45   56   57   60   42   46
44   55   63   65   49   50
55   65   70   53   64   61
46   47   56   40   41   54

Draw an ordered stem-and-leaf diagram. Remember to give the key.

7 The attempted heights, in centimetres, during a high jump event are shown.

214   204   225   230   244
210   207   209   240   232
230   216   242   233   238
206   217   236   216   211
208   230   209   237   241

Draw an ordered stem-and-leaf diagram.

4.2 Organising data

- You can organise data in a frequency table.
- A two-way table shows more detail and links two types of information.
- You can also use a stem-and-leaf diagram to show numerical data.

Frequency tables and two-way tables make it easy to spot trends within the data set. They are also useful for organising large data sets and non-numerical data.

Stem-and-leaf diagrams are useful because they display all the raw data.

It can be hard to spot trends from stem-and-leaf diagrams for large data sets because there is so much information!

HOW TO

1. Construct or complete the table or diagram.
2. Interpret information from the table.
3. Add entries in columns or rows if necessary to find the answer.

EXAMPLE

A class of 30 students study either History or Geography.

There are 13 girls in the class, with 8 girls and 9 boys studying Geography.

How many boys study History?

1. Draw a two-way table to show the information that you know.
2. Use the totals to fill in the other entries.

	History	Geography	Total
Boys		9	30 − 13 = 17
Girls	13 − 8 = 5	8	13
Total		9 + 8 = 17	30

↓

	History	Geography	Total
Boys	17 − 9 = 8	9	17
Girls	5	8	13
Total	30 − 17 = 13	17	30

3. Read the answer from the table.

8 boys study History.

EXAMPLE

The number of televisions in each house in Fern's street is shown in the frequency table.

a Calculate the number of houses in Fern's street.

b Calculate the total number of televisions in Fern's street.

1. Complete the table to show the total number of televisions.

a 2. Total number of houses
= 1 + 5 + 12 + 9 + 1 = 28

b 3. Multiply the number of TVs by the number of houses for each row in the table. Find the total.

No. of TVs	No. of houses	No. of TVs × no. of houses
0	1	0 × 1 = 0
1	5	1 × 5 = 5
2	12	2 × 12 = 24
3	9	3 × 9 = 27
4	1	4 × 1 = 4

Total number of televisions = 0 + 5 + 24 + 27 + 4 = 60

Exercise 4.2A

1 The table shows some information about the eye colour of children in a nursery.

	Brown	Blue	Other	Total
Boys	5		6	36
Girls				
Total	23	28		64

Are there more blue-eyed girls or brown-eyed boys?

2 A group of 40 students study either French or Spanish.
There are 17 boys in the group, with 15 girls and 10 boys studying French.
How many girls study Spanish?

3 The table shows the number of students in a school who take part in the Duke of Edinburgh scheme.

Year 8	13
Year 9	12
Year 10	11

There are

- seven year 8 boys
- five Year 10 girls
- equal numbers of boys and girls in year 9.

Emily says that there are the same number of boys and girls taking part.
Do you agree with Emily? Say why.

4 Sophie did a survey to find the number of DVDs owned by the students in her class.
The results are shown in the frequency table.

Number of DVDs	Number of students
0	1
1	8
2	6
3	8
4	2
5	3

a Calculate the number of students in Sophie's class.

b Calculate the total number of DVDs owned by the whole class.

5 In a traffic survey, the colour and speed of 100 cars are recorded. The results are summarised in the two-way table.

	Not speeding	Over the speed limit
Red	5	55
Not red	10	30

Maria claims that drivers of red cars tend to break speed limits. Use the two-way table to decide whether you agree with Maria.

6 Kyra and Ravi collect data on the number of minutes per day students in their year play computer games.

Kyra collects this data from a sample of 14 students.

40	26	75	55	33	39	28
66	67	71	64	37	52	47

a Kyra starts to make a frequency table to display the data.

Minutes	Frequency
26	1
27	0
28	1
29	0

Give two advantages of using a stem-and-leaf diagram instead of Kyra's table.

b Ravi suggest that Kyra uses this grouped frequency table instead.

Minutes	Frequency
20 to 29	2
30 to 39	3
40 to 49	2
50 to 59	2
60 to 69	3
70 to 79	2

Give one advantage and one disadvantage of using a grouped frequency table to display the data.

c Ravi takes a census all of all 152 students in the year. Give one disadvantage of using a stem-and-leaf diagram to display the data.

 1193, 1214, 1215   SEARCH

4.3 Representing data 1

You can use a **pictogram** to display data.

- Pictograms use symbols to show the size of each category.

You may have to use part of a symbol to **represent** some quantities.

EXAMPLE

The number of cars that a car salesman sells is given in the table.

Week	Cars sold
1	4
2	6
3	12
4	10

Draw a pictogram to illustrate this information.

Use 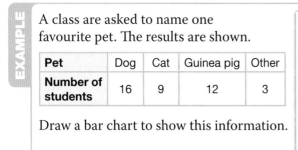 to represent 4 cars.

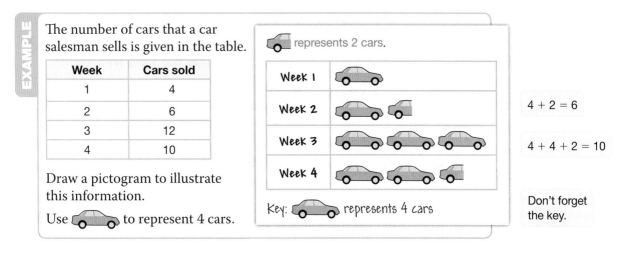

represents 2 cars.

Key: represents 4 cars

$4 + 2 = 6$

$4 + 4 + 2 = 10$

Don't forget the key.

You can use a **bar chart** to display data.

Bar charts give a visual picture of the size of each category.

- A bar chart should have
  - labels on each axis
  - bars with equal width
  - equal gaps between the bars
  - values on the vertical axis evenly spaced starting at zero.

Lines are drawn instead of bars in a bar-line chart.

The bars can be horizontal or vertical.

- **Bar-line charts** are a good way to display (discrete) numerical data.

EXAMPLE

A class are asked to name one favourite pet. The results are shown.

Pet	Dog	Cat	Guinea pig	Other
Number of students	16	9	12	3

Draw a bar chart to show this information.

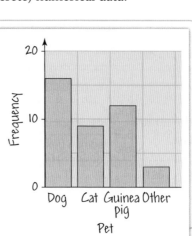

Use a sensible scale on the vertical axis. Try to make the graph as large as possible.

Exercise 4.3S

1 The pictogram shows where people use the internet the most.
Copy the pictogram and represent this information on your diagram.

Library 6   School 9   Work 17

Key:

 represents
2 people

Home	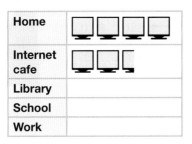
Internet cafe	
Library	
School	
Work	

2 The results of a survey about people's favourite hot drink are given in the table.

Letting ⌣ represent 20 people, draw a pictogram for this data.

Tea	80
Coffee	60
Hot chocolate	50
Soup	30
Other	20

3 The number of foreign language teachers at a school is shown in the pictogram.

   a Which subject has only one teacher?

   b How many French teachers teach at the school?

   c How many more German teachers are there than Russian teachers?

   d Calculate the total number of foreign language teachers at the school.

French	𝕏 𝕏 𝕏 ⁇
Spanish	𝕏 𝕏 𝕏
Russian	𝕏
German	𝕏 𝕏 𝕏 ⁇
Italian	⁇

Key: 𝕏 represents 2 teachers

4 The number of concerts held at various venues is given.

Venue	Number of concerts
NEC	10
Arena	15
MEN	20
NIA	18
Wembley	9

Draw a bar chart to show this information.

5 The number of Bank Holidays in different countries is shown.

Draw a bar chart to show this information.

Country	Number of Bank Holidays
UK	8
Italy	16
Iceland	15
Spain	14

6 The size of donations to a charity are shown in the frequency table.
Draw a bar-line chart to show this information.

Donation	Frequency
£1	90
£2	100
£5	80
£10	70
£20	110

7 The numbers of buses that stop at a village through the week are shown on the bar chart.

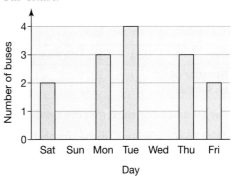

   a How many buses stop at the village on Thursday?

   b On which days is there no bus service?

   c Calculate the total number of buses that stop at the village throughout the week.

1193, 1205   SEARCH

4.3 Representing data 1

- You can use a pictogram or a bar chart to display data.
- A pictogram uses symbols to show the size of each category.
- A bar chart uses a bar to show the size of each category.
- Pictograms and bar charts show all the data, but in categories. They are good at comparing the frequencies of different categories.

> It can be time consuming to draw lots of symbols when using pictograms for large data sets.

HOW TO

① Read information from the diagram. Make sure that you read the scale or key carefully.

② Use the information from the diagram to answer the question.

EXAMPLE

The pictogram shows the number of students attending school in a week.

Four students were absent on Monday.

Calculate the number of students that were absent on Thursday.

Monday	○○○○
Tuesday	○○○
Wednesday	○○○
Thursday	○◔
Friday	○○◁

Key: ○ represents 4 students

① Each circle represents 4 students, so three-quarters of a circle represents 3 students.

Number of students attending on Thursday = 4 + 3 = 7

② Number of students attending on Monday = 4 + 4 + 4 + 4 = 16

Total number of students = 16 + 4 = 20   There were 4 students absent on Monday.

Subtract the number of students who attended on Thursday from the total.

Number of students absent on Thursday = 20 – 7 = 13

EXAMPLE

A class are asked to name one favourite pet. The results are shown in the table.

Pet	Dog	Cat	Guinea pig	Other
Number of students	16	9	12	3

Jess uses her computer to create this graph.

Make two criticisms of the graph.

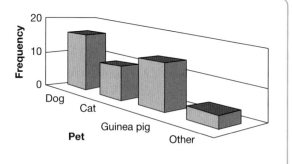

① Compare the graph with the frequency table and graph in 4.3 Skills. Look at the scale of the graph.

The scale chosen makes it is difficult to read information from the graph.

② Bar charts should let you compare the different categories.

The perspective makes it difficult to compare the heights of the bars.

Exercise 4.3A

1 The pictogram shows the number of people who eat different take-away food. 240 people were surveyed in total.

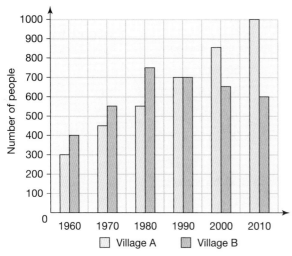

How many people ate Chinese food?

2 The dual bar chart shows the population of two villages.

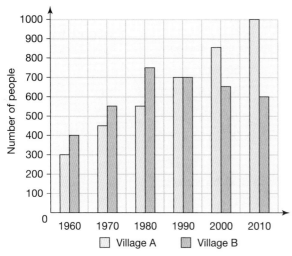

Village A ☐   Village B ▨

a During what year were the populations of the two villages the same?

b Compare the changes in population of the two villages over time.

3 Compare and contrast a stem-and-leaf diagram with a bar chart which has been 'turned on its side'.

4 A newspaper headline says

Team USA won 28 medals at the 2014 Winter Olympics in Sochi.

● Team USA won more bronze medals than gold medals.

● A quarter of the medals were silver medals.

Draw a bar chart to show the number of bronze, silver and gold medals that could have been won by the team.

5 The number of students in different year groups is given in the frequency table.

Department	Frequency
Year 7	120
Year 8	126
Year 9	130
Year 10	128
Year 11	125

a Give one disadvantage of using a pictogram to display the data. Marnie draws this bar chart for the data.

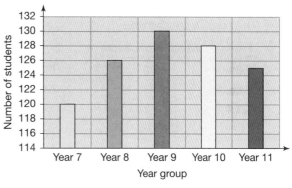

b Give one reason why Marnie's graph is misleading.

6 Travel insurance prices are shown on the bar chart.

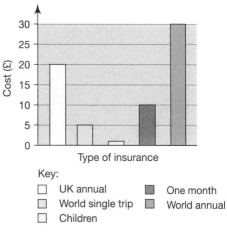

Key:
☐ UK annual   ▨ One month
▨ World single trip   ▨ World annual
☐ Children

Suzie plans to make four trips abroad during the year. She can either buy World Single Trip insurance each time or World Annual insurance.
Which is her cheaper option? Give reasons.

4.4 Representing data 2

You can use a **pie chart** to display data.

Pie charts use a circle to give a quick visual picture of all the data.

The size of each **angle** shows the size of each **category**.

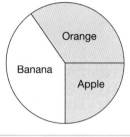

Apple is a quarter of all the data.

Banana is the biggest **sector**.

All the data must be included.

- A pie chart shows
  - the **proportion** or fraction of each category compared to the whole circle
  - all the data, but in categories.

240 people are asked to name their favourite fruit. The results are shown.

Fruit	Number of people
Apple	50
Banana	80
Orange	72
Other	38

Draw a pie chart to illustrate the information.

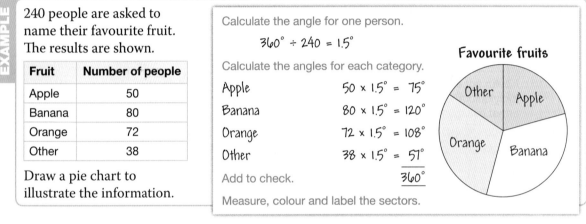

Calculate the angle for one person.

$360° ÷ 240 = 1.5°$

Calculate the angles for each category.

Apple $50 × 1.5° = 75°$
Banana $80 × 1.5° = 120°$
Orange $72 × 1.5° = 108°$
Other $38 × 1.5° = 57°$

Add to check. $\underline{360°}$

Measure, colour and label the sectors.

Favourite fruits

- You can interpret information from a pie chart by using the angles for each category to find the frequencies.

60 vehicles are shown on the pie chart.

Calculate the numbers of cars, vans, buses and lorries.

Vehicles parked in the High St.

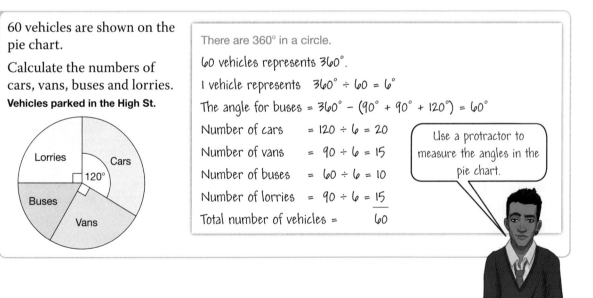

There are 360° in a circle.

60 vehicles represents 360°.

1 vehicle represents $360° ÷ 60 = 6°$

The angle for buses = $360° − (90° + 90° + 120°) = 60°$

Number of cars   = $120 ÷ 6 = 20$
Number of vans   = $90 ÷ 6 = 15$
Number of buses   = $60 ÷ 6 = 10$
Number of lorries = $90 ÷ 6 = 15$
Total number of vehicles = $\underline{60}$

Use a protractor to measure the angles in the pie chart.

Exercise 4.4S

1 Alvin makes eight sandwiches.

1 tuna

2 cheese and tomato

3 chicken

2 corned beef

Draw a pie chart to show this information.

2 Seven boys and five girls attend an after-school homework club.

    a Calculate the total number of students.

    b Calculate the angle one student represents in a pie chart.

    c Calculate the angles to represent boys and girls.

    d Draw a pie chart to show the information.

3 The weather record for 60 days is shown in the frequency table. This gives the predominant weather for that particular day.

    a Calculate the angle that one day represents in a pie chart.

    b Calculate the angle of each category in the pie chart.

Weather	Number of days
Sunny	15
Cloudy	18
Rainy	14
Snowy	3
Windy	10

    c Draw a pie chart to show the data.

4 A school fete is open from 10 am to 4 pm.

    a Calculate the number of minutes the school fete is open.

A teacher has offered to help. She spends these times on each stall.

    b Draw a pie chart to show this information.

Stall	Time
Bat the Rat	30 mins
Hook a Duck	25 mins
Smash a Plate	35 mins
Roll a Coin	80 mins
Tombola	70 mins
Break 1	60 mins
Break 2	60 mins

5 The pie chart shows the survey results for the favourite band of 100 people.

    a What fraction of the pie chart represents

        i Kasabian   **ii** The Prodigy

        iii Glasvegas?

    b Calculate the number of people who voted for

        i Kasabian   **ii** The Prodigy

        iii Glasvegas.

6 A shop sells 12 loaves of three different types: organic, wholegrain and white.

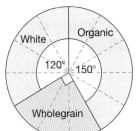

    a Copy and complete   12 loaves = 360°
                            1 loaf = ___°

    b Calculate the number of loaves that are

        i organic   **ii** wholegrain

        iii white.

7 A car dealer sells 18 cars in one week of three different types: diesel, petrol and electric.

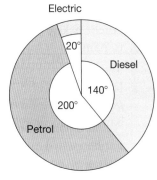

    a Calculate the angle that represents one car.

    b Calculate the number of cars sold that are

        i diesel   **ii** petrol

        iii electric.

🔍 1206, 1207   SEARCH

4.4 Representing data 2

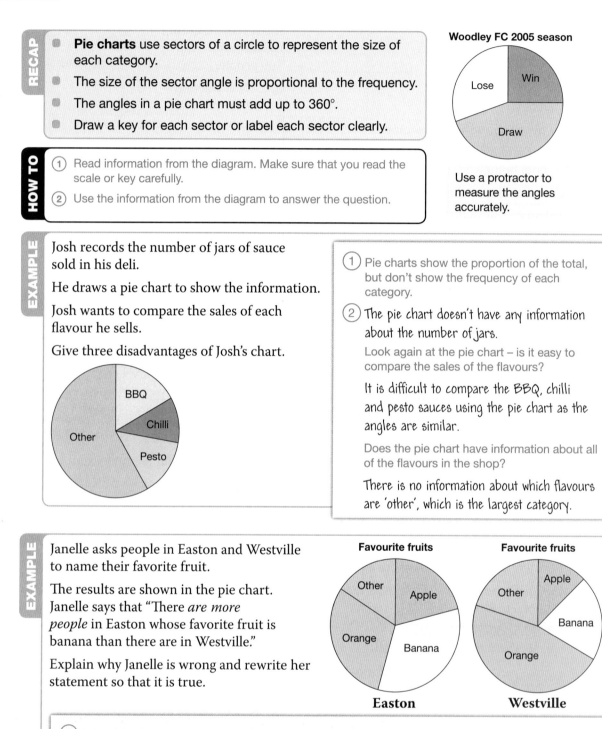

- **Pie charts** use sectors of a circle to represent the size of each category.
- The size of the sector angle is proportional to the frequency.
- The angles in a pie chart must add up to 360°.
- Draw a key for each sector or label each sector clearly.

Woodley FC 2005 season

Use a protractor to measure the angles accurately.

HOW TO

(1) Read information from the diagram. Make sure that you read the scale or key carefully.

(2) Use the information from the diagram to answer the question.

EXAMPLE

Josh records the number of jars of sauce sold in his deli.

He draws a pie chart to show the information.

Josh wants to compare the sales of each flavour he sells.

Give three disadvantages of Josh's chart.

(1) Pie charts show the proportion of the total, but don't show the frequency of each category.

(2) The pie chart doesn't have any information about the number of jars.

Look again at the pie chart – is it easy to compare the sales of the flavours?

It is difficult to compare the BBQ, chilli and pesto sauces using the pie chart as the angles are similar.

Does the pie chart have information about all of the flavours in the shop?

There is no information about which flavours are 'other', which is the largest category.

EXAMPLE

Janelle asks people in Easton and Westville to name their favorite fruit.

The results are shown in the pie chart. Janelle says that "There *are more people* in Easton whose favorite fruit is banana than there are in Westville."

Explain why Janelle is wrong and rewrite her statement so that it is true.

Favourite fruits — Easton

Favourite fruits — Westville

(1) Pie charts don't show the frequency of each category.

(2) Janelle doesn't know how many people in Westville are represented in the pie chart.
Although the sector is larger for Easton, it could represent fewer people.
Pie charts show the proportion of each category.

There is a higher proportion of people in Easton whose favorite fruit is banana than there are in Westville.

Exercise 4.4A

1 This pie chart shows the number of tourists visiting some regions in England over a certain period.

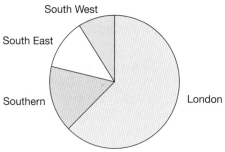

The pie chart is not drawn to scale.

Complete this table.

Region	Number of tourists	Angle
London	225000	
Southern		60°
South East	35000	
South West		40°
	360000	

2 The number of drinks sold during lunchtime in a shop are shown in the table.

Drink	No. sold		Drink	No. sold
Apple	6		Lemonade	3
Blackcurrant	5		Mocha	2
Cappuccino	8		Orange	7
Cola	12		Pineapple	4
Espresso	3		Tea	9
Latte	5		Water	9

Explain one disadvantages of using a pie chart to display the data.

3 The test scores of five students are shown.

Name	Mark
Abi	55
Carlos	52
Daniel	58
Emily	51
Parminder	57

Explain one disadvantages of using a pie chart to display the data.

4 The pie charts show the number of year 11 students studying different languages at two nearby schools.

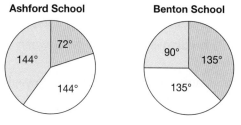

☑ French ☐ Spanish ☐ German

a Jermaine says "twice as many students at Ashford school study German compared to those studying Spanish". Do you agree with Jermaine? Give a reason for your answer.

b Lydia says it must be the case that more students at Ashford schools study French than at Benton school. Is she correct? You must give reasons for your answer.

5 The bar chart shows the number of letters delivered in one week to No. 10 and No. 12.

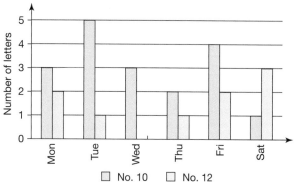

☐ No. 10 ☐ No. 12

a Draw an accurate pie chart to show the number of letters delivered each day to No. 10.

b Draw an accurate pie chart for No. 12.

c Mark wants to compare the proportion of letters received on each day for the two houses. Should he use the bar chart or the pie charts? Give your reasons.

d Peter want to compare the number of letters received each day for the two houses. Should he use the bar chart or the pie charts? Give your reasons.

 🔍 1206, 1207   SEARCH

Averages and spread 1

- The **mean** of a set of data is the total of all the values divided by the number of values.
- The **mode** is the value that occurs most often.
- The **median** is the middle value when the data is arranged in order.
- The **range** is the highest value – lowest value.

Outliers are values that 'lie outside' most of the other values of a set of data.

In this data set:

1, 1, 2, 2, 3, 4, 4, 4, 16

16 is an outlier.

The mean and range are both affected by outliers.

Ten people took part in a golf competition. Their scores are shown in the frequency table.

Calculate the mean, mode and median of the scores.

Score	Frequency
67	1
68	4
69	3
70	1
71	1

4 people scored 68.

1 person scored 71.

The results can be written in numerical order.

67, 68, 68, 68, 68, 69, 69, 69, 70, 71   Mean = 687 ÷ 10 = 68.7   Mode = 68 (occurs 4 times)

Median = (68 + 69) ÷ 2 = 68.5

There are 10 values so the median is the mean of the 5th and 6th value.

Alternatively, you can calculate the mean directly from the frequency table.

Score	Frequency	Score × Frequency
67	1	67
68	4	272
69	3	207
70	1	70
71	1	71
	10	687

68 + 68 + 68 + 68 or 68 × 4.

The total of all the scores of the 10 golfers.

Mean = 687 ÷ 10 = 68.7

Median = (5th value + 6th value) ÷ 2 = (68 + 69) ÷ 2 = 68.5

Mode = 68 as 68 occurs the most often.   The mode is 68, not 4.

There are advantages and disadvantages of the three types of averages.

	Mode	Median	Mean
Advantages	• Can be used with all types of data (numbers, colours …) • Not affected by outliers • Mode is an actual value in the data set	• Not affected by outliers • Easy to calculate	• Uses all of the values
Disadvantages	• Not always a mode • Can be more than one mode	• Not always an actual value in the data set	• Not always an actual value in the data set • Can be affected by outliers

Exercise 4.5S

1 Calculate the mean for each set of numbers.

   a 4, 9, 7, 12   **b** 8, 11, 8

   c 3, 2, 2, 3, 0   **d** 1, 9, 8, 6

   e 2, 2, 4, 1, 0, 3 **f** 23, 22, 25, 26

   g 17, 19, 19, 20, 18, 17, 16

   h 103, 104, 105 **i** 14, 10, 24, 12

   j 4, 6, 7, 6, 4, 4, 3, 7, 3, 6

2 a Write these numbers in order, smallest first.

     i 7, 16, 8, 5, 4, 3, 3

     ii 11, 12, 10, 9, 9

     iii 38, 35, 24, 37, 34

     iv 101, 98, 103, 97, 99, 97, 95

     v 3, 2, 0, 0, 1, 2, 1, 2, 3

   b Use your answers to find the median of each set of numbers.

   c Which data sets contain outliers? Have the outliers affected the median?

3 Calculate the mode and range of each set of numbers.

   a −6, 0, 1, 1, 1, 1, 2, 2, 2, 3, 3, 4, 4

   b 5, 5, 6, 6, 6, 7, 7, 7, 7, 8, 8, 8, 8, 8

   c 10, 11, 11, 11, 12, 12, 13, 14

   d 21, 22, 23, 24, 24, 25, 25, 25, 36

   e 8, 8, 8, 9, 9, 10, 11

   f 4, 3, 5, 5, 6, 6, 4, 3, 4, 5, 6, 5, 3

   Which data sets contain outliers? Have the outliers affected the mode or range?

4 The numbers of flowers on eight rose plants are shown in the frequency table.

Number of flowers	Tally	Frequency
3	||||	4
4	||	2
5	||	2

   a List the eight numbers in order of size, smallest first.

   b Calculate the mean, mode, median and range of the eight numbers.

5 The number of days that 25 students were present at school in a week are shown in the frequency table.

   a List the 25 numbers in order of size, smallest first.

   b How many students were present for 5 days of the week?

   c Calculate the mean, mode, median and range of the 25 numbers.

Number of days	Tally	Frequency
0		0
1	||||	4
2	⊬⊦ |	6
3	||	2
4	⊬⊦	5
5	⊬⊦ |||	8

6 For these sets of numbers work out the

    i range   **ii** mode

    iii mean   **iv** median.

   a 5, 9, 7, 8, 2, 3, 6, 6, 7, 6, 5

   b 45, 63, 72, 63, 63, 24, 54, 73, 99, 65, 63, 72, 39, 44, 63

   c 97, 95, 96, 98, 92, 95, 96, 97, 99, 91, 96

   d 13, 76, 22, 54, 37, 22, 21, 19, 59, 37, 84

   e 89, 87, 64, 88, 82, 88, 85, 83, 81, 89, 90

   f 53, 74, 29, 32, 67, 53, 99, 62, 34, 28, 27, 27, 27, 64, 27

   g 101, 106, 108, 102, 108, 105, 106, 109, 103, 105, 107, 104, 104, 105, 105

7 a Subtract 100 from each of the numbers in question **6 g** and write down the set of numbers you get.

   b For your set of numbers in **a**, work out the

    i range   **ii** mode

    iii mean   **iv** median

   c Compare your answer for the range with **6 g**.

   What do you notice?

 1192, 1202, 1254   SEARCH

4.5 Averages and spread 1

- To calculate the mean of a set of data, add all the values and divide by the number of values.
- To find the median, arrange the data in order and choose the middle value.
- To find the mode, choose the value that occurs most often.
- To find the range, subtract the smallest value from the largest value.

> The mean, median and mode are averages. The range is a measure of spread.

HOW TO

To compare data sets

① Calculate the mean, median, mode, range or interquartile range for each set of data.

② To compare the averages, look at the mean, median or mode.

③ To compare the spread look at the range.

EXAMPLE

A team of 7 girls and a team of 8 boys did a sponsored run for charity.
The distances the girls and boys ran are shown.

Girls

Distance (km)	Frequency
1	3
2	2
3	1
4	1
5	0

Boys

3, 5, 5, 3, 4, 4, 5, 5
all distances in kilometres

By calculating the mean, median and range, compare each set of data.

① Calculate the mean, median and range of each set of data.

Girls

$\quad$ Mean $= (3 + 4 + 3 + 4 + 0) \div 7 = 14 \div 7 = 2$

$\quad$ Median $= 2$ $\qquad$ The 4th distance is the middle value.

$\quad$ Range $= 4 - 1 = 3$ $\qquad$ Highest value – lowest value.

Boys

$\quad$ Mean $= 34 \div 8 = 4.25$

$\quad$ Median $= (4 + 5) \div 2 = 9 \div 2 = 4.5$ $\qquad$ The mean of the 4th and 5th distances.

$\quad$ Range $= 5 - 3 = 2$ $\qquad$ Highest value – lowest value.

② Compare the mean and median.

$\quad$ The mean and median show the boys ran further on average than the girls.

③ Compare the range.

$\quad$ The range shows that the girls' distances were more spread out than the boys' distances.

Exercise 4.5A

1 The number of bottles of milk delivered to two houses is shown in the table.

	Sat	Sun	Mon	Tues	Wed	Thur	Fri
Number 45	2	0	1	1	1	1	1
Number 47	4	0	2	2	2	2	2

    a Calculate the range for Number 45 and Number 47.

    b Use your answers for the range to compare the number of bottles of milk delivered to each house.

2 The number of cars at each house on Ullswater Drive are

    2 4 1 0 1 2 1 2 3 2

    a Copy and complete the frequency table.

Number of cars	Tally	Number of houses
0		
1		
2		
3		
4		

    b Calculate the mean, mode and median number of cars for Ullswater Drive. The mean, mode and median number of cars at each house on Ambleside Close are

Mean	Mode	Median
0.7	0	1

    c Use the mean, mode and median to compare the number of cars on Ullswater Drive and Ambleside Close.

3 The range of these numbers is 17.

    Find two possible values for the unknown number.

4 Monica has five numbered cards. One of the cards is numbered −2.6. Monica's cards have

    ● range = 7.2

    ● median = 3.5

    ● mode = 4.1

    Write down the five numbers on Monica's cards.

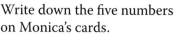

5 The bar chart shows the test results for a class of 20 students.

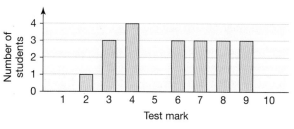

    a Copy and complete the frequency table to illustrate these results.

Mark			1	2	3	4	5	6	7	8	9	10
Number of students			0	1								

    b Calculate the mean, mode, median and range for the 20 students.

    c If a new student joined the class and got a mark of 10, how would this affect your three answers to part **b**?

6 There are nine passengers on a bus.
The mean age of the passengers is 44 years old.
Jasmine gets on the bus.
Jasmine is 14 years old.
Find the new mean age of the passengers on the bus.

7 Reuben counted the raisins in 21 boxes.
The mean number of raisins per box was 14.095 (3 dp).
Reuben records the information for 20 boxes in the table.

Number of raisins	Number of boxes
13	5
14	8
15	7

    Find number of raisins in the last box.

8 The mean mark in a statistics test for a class was 84%.
There are 32 students in the class, 12 of whom are girls.
The mean mark in the test for these girls was 93%.
Work out the mean mark in the statistics test for the boys.

Summary

Checkout

You should now be able to...

You should now be able to...	Test it Questions
✓ Identify when a sample may be biased.	1
✓ Construct and interpret frequency tables and two-way tables.	2 – 3
✓ Construct and interpret pictograms, bar-line charts and bar charts.	4 – 5
✓ Interpret and construct pie charts and know their appropriate use.	6 – 7
✓ Compare distributions using median, mean, mode and range and identify outliers.	8 – 9

Language   Meaning   Example

Language	Meaning	Example
Population	The whole group of people or items that are to be investigated.	To find out how long students spend each week doing homework at one school with a population of 1000, survey a group of 50 students from that school.
Sample	A set chosen to represent a population.	
Survey	Gather data from a sample to find out the characteristics of a population.	
Data collection sheet	A sheet on which data can be recorded and organised.	
Tally chart	A data collection sheet for data collected by counting.	
Frequency table	A table that records the frequency of each piece of data.	
Frequency	The number of times each piece of data occurs.	2 cars are red
Pictogram	A frequency diagram using a symbol to represent a number of units of data.	See lesson 4.3.
Bar-chart	The height of each bar represents the frequency.	See lesson 4.3.
Bar-line chart	The length of each line represents the frequency.	See lesson 4.3.
Pie chart	A circular chart divided into sectors. The size of the angle of each sector is proportional to the frequency.	See lesson 4.4.
Mean	An average found by adding all the values together and dividing by the number of values.	Data: 1, 1, 2, 2, 4, 5, 7, 8, 8, 8, 9, 22 $\\qquad$ Mean $= 77 \div 11$ $\\qquad\\qquad = 7$
Mode	The value that occurs most often.	Mode $= 8$
Median	The middle value when the data is arranged in order of size. If there is an even number of data, the median is the mean of the middle two values.	1, 1, 2, 2, 4, 5, 7, 8, 8, 8, 9, 22 $\\qquad$ Median $= 6$
Range	The difference between the largest value and the smallest value.	Range $= 22 - 1 = 21$
Outlier	A value that lies outside most of the other values in a set of data.	22 is an outlier.

The frequency table example:

Colour of car	Tally	Frequency
Red	II	2
Silver	IIII III	8
Black	IIII IIII	10
Other	IIII	4

Review

1 A sample is taken of people visiting a garden. This is done by recording all the people that arrive between 10:00 am and 10:15 am.

Their ages are recorded as

45, 37, 24, 8, 38, 40, 59, 68, 41, 80

  a Use this sample to estimate the mean age of all people visiting the garden.

  b What are the possible issues with estimating the mean in this way?

2 Organise this data into a frequency table.

3, 4, 1, 3, 4, 3, 4, 3, 2, 3, 1, 2, 3, 3, 4, 4, 2, 2, 3

3 The results of a handedness survey are shown. Use the two-way table to find the number of

  a people surveyed

  b boys

  c left-handed girls

  d right-handed people.

	Left	Right
Girls	2	15
Boys	1	12

50, 43, 21, 24, 78, 80, 85, 5, 11, 14, 48, 62, 19

4 The pictogram shows the number of children playing football each lunchtime at a school.

Monday	⚽ ⚽ ⚽ ⚽
Tuesday	⚽ ⚽ ⚽ ◖
Wednesday	⚽ ⚽ ◗
Thursday	
Friday	

⚽ represents 4 children

  a How many children played football on a

    i Monday   **ii** Wednesday?

  b Ten children played on a Thursday and 15 on a Friday, complete a copy of the pictogram to show this information.

5 Draw a bar-line chart to represent the data in question **2**.

6 The pie chart shows the favourite fruit of 80 people.

How many people prefer

  a apples

  b bananas?

7 Suzy spends 60 p of her pocket money on a snack, £1.80 on a magazine and saves £3.60.

Show this information in a pie chart.

8 Jacob counts the number of sweets in a selection of packets, his results are

5, 8, 4, 3, 7, 9, 3, 6, 7, 8, 7, 5

Calculate.

  a the mean   **b** the median

  c the mode   **d** the range.

9 A zookeeper shows the number of birds in different aviaries at a zoo.

13, 11, 2, 5, 24, 34, 12, 11, 24, 30 , 1, 35, 11

Write down the

  a highest number

  b range

  c modal number

  d median number of birds in an aviary.

In a different zoo the median number of birds per aviary is 15 and the range is 20.

  e Compare the two zoos.

What next?

Score			
	0 – 3		Your knowledge of this topic is still developing. To improve look at MyMaths: 1192, 1193, 1202, 1205, 1206, 1207, 1212, 1214, 1215, 1248, 1249, 1254
	4 – 7		You are gaining a secure knowledge of this topic. To improve look at InvisiPens: 04Sa – n
	8 – 9		You have mastered these skills. Well done you are ready to progress! To develop your exam technique look at InvisiPens: 05Aa – g

Assessment 4

1 These figures show the number of one pint cartons of milk bought in a small corner shop over 50 days.

5	2	1	1	2	3	4	1	2	4
2	2	2	3	1	4	0	1	4	2
3	2	4	6	1	2	0	2	3	3
0	1	3	2	1	1	3	4	1	5
4	0	2	2	0	1	1	1	1	0

   a Draw a frequency table for this data. [3]

   b Draw a bar chart for this data. [3]

2 For each of the five investigations answer these questions.

   i Would you investigate the whole population or take a sample?

   ii How would you collect your data?

   iii How would you avoid bias?

      a What is the most common make of the car for teachers in your school? [3]

      b How tall are the children in your class? [3]

      c Are winters in the UK getting warmer? [3]

      d Can boys estimate lengths better than girls? [3]

      e Which mobile phone provider is the most popular in your school? [3]

3 Ms Connell collected information about the way that the 35 students in her class access information electronically.

	Smart phone	Tablet/Laptop	Broadband on a home computer	Internet café
Girls	15	9	12	0
Boys	20	16	10	3

   a How many students use a smart phone? [1]

   b How many boys have a tablet/laptop? [1]

   c How many more boys than girls use an internet café? [1]

   d Explain why the sum of the entries in the table does not add up to the total number of students in the class. [1]

4 In a group of 35 business executives, 15 wear glasses and 5 are left-handed. 4 executives are left-handed but don't wear glasses. Display this information in a two-way table. [4]

5 A lorry driver recorded the distances driven in 10 journeys (in miles).

56   113 88 67 163 90 88 109 135 121

   a Calculate the

      i mean [2]   **ii** median [2]   **iii** mode. [1]

   b Does he average 100 miles per day? Give reasons for your answer. [2]

   c Calculate the range. [2]

d He realises that one distance recorded as 88 should have been 98. Without performing any additional calculations, decide what effect this will have on the mean, median, mode and range of the distances. Give reasons for your answers. [4]

6 The bar chart shows the different types of coins in Harry's till.

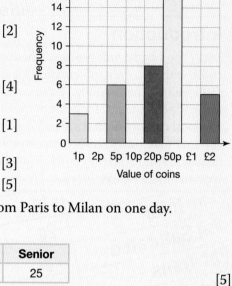

a There are also seven 2p coins, seventeen 10p coins and six £1 coins in his till. Add this data to a copy of the bar chart. [2]

b Draw a pictogram representing the data shown. Use one circle to represent two coins. [4]

c Which coin is found most often in the till? [1]

d What is the value of the median coin in the till? [3]

e Find the mean value of the coins in the till. [5]

7 Mia records all the tickets she sells on the train service from Paris to Milan on one day. Draw a pie chart to show this information.

Ticket class	Business	Standard	First	Child	Senior
Frequency	17	18	14	6	25

[5]

8 This pie chart shows how 1200 hospital staff members get to work in summer.

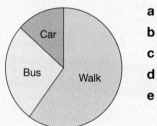

a What is the modal form of travel? [1]

b Which is the least favourite form of travel? [1]

c What percentage of staff take the bus? [2]

d How many staff drive to work? [3]

e How many staff cycle to work? [1]

9 The table shows information about the number of children and pets in different families.

a How many families have 1 pet? [2]

b How many 2 child families have 1 pet? [1]

c How many families are there where the number of pets is the same as the number of children? [2]

d What is the modal number of pets? [1]

e Calculate the mean number of pets per family. [4]

		Number of pets					
		0	1	2	3	4	5
	1	3	5	4	1	0	1
Number of children in family	2	4	2	2	0	0	0
	3	1	0	2	1	0	0
	4	0	1	0	0	0	0
	5	2	1	0	0	0	0

5 Fractions, decimals and percentages

Introduction

Many food and drink products that you buy come with nutrition information clearly displayed on the labelling. This is usually in the form of percentages, for example 'saturated fat 5%, total carbohydrate 12%, calcium 9%, etc'. Product manufacturers are expected by law to display this information and the labels often tend to have a similar format.

What's the point?

Awareness of what you are eating and drinking is important in achieving a healthy, balanced diet and percentages allow you to monitor this.

Fat Total	1g
Fat Saturated	2g
Fat Trans	2g
Carbohydrate	0g
Sugar	23g
Dietary Fibre	15g
Sodium	0g
	45mg

Recommended Daily Allowances

| Vitamin A | 0% | Vitamin C | |
| Calcium | 15% | Iron | |

	per Serving	Intake (per Serving)	
Energy	607 kJ (145Cal)	7%	243kJ (58Cal)
Protein	2.5g	5%	1.0g
Fat, Total†	0.6g	0.9%	0.2g
- saturated	0.2g	0.9%	0.1g
Carbohydrate	31.2g	10%	12.5g
- sugars	8.0g	9%	3.2g
Sodium	815mg	36%	325mg

Objectives

By the end of this chapter you will have learned how to …

- Convert between terminating decimals and their corresponding fractions.
- Compare decimals and fractions using the symbols $>$ and $<$.
- Find fractions and percentages of amounts.
- Add and subtract simple fractions and mixed numbers.
- Multiply and divide simple fractions and mixed numbers.
- Convert between fractions, decimals and percentages.

Check in

1 Calculate

   a $13 \div 10$ **b** 0.03×10

2 Round each of these numbers to 2 decimal places.

   a 3.5624   **b** 8.0392   **c** 0.0551

3 Put these decimals in order from smallest to largest.

   0.75   0.8   0.7   0.875

	Average Quantity Per Serving	Average Quantity Per 100ml
	146 kJ	86 kJ
	1.9 g	1.1 g
	0.6 g	0.4 g
	0.1 g	0.1 g
otal	5.5 g	3.2 g
urated	4.0 g	2.4 g
bohydrate	620 mg	365 mg
ugars		
odium		

		Total	6.3g	7.4g
PROTEIN				
FAT	Saturated		1.4g	1.7g
	Trans		Less than 0.1g	
	Polyunsaturated		2.4g	2.8g
	Omega-3		0.5g	0.6g
	EPA		210mg	250mg
	DHA		290mg	350mg
	Monounsaturated		2.5g	2.9g
CARBOHYDRATE	Total		0.3g	0.4g
	Sugars		0.2g	0.2
SODIUM			391mg	460m

Nutrition Facts

Serving Size 1 Rounded Scoop (32g)
Servings Per Container 73

Amount Per Serving

Calories 132   Calories from Fat 10

% Daily Value*

2%
3%

0%
3%
2%

Sugars 3g

Protein 24g

Chapter investigation

You can divide a rectangle into two halves by drawing one straight line.

To divide the rectangle into thirds requires two lines.

How many lines do you need to divide the rectangle into quarters?

Investigate the number of lines needed for different fractions of the rectangle.

Describe your findings.

5.1 Decimals and fractions

● You can use a **fraction** to describe a part of a whole. To use a fraction the whole must be divided into **equal**-sized parts.

Numerator: the top number shows how many parts you have.

Denominator: the bottom number shows how many equal-sized parts the whole has been divided into.

EXAMPLE

Here is a fuel gauge from a car. How full is the petrol tank? Give your answer as a fraction.

Empty   Full

3 out of 5 sections are coloured.

The car is $\frac{3}{5}$ full.

● You can find equivalent fractions by multiplying or dividing the numerator and denominator by the same number.

$$\overset{\times 3}{\underset{\times 3}{\frac{4}{7} = \frac{12}{21}}}$$

● You can simplify a fraction by dividing the numerator and denominator by a common factor.

$$\overset{\div 5}{\underset{\div 5}{\frac{15}{20} = \frac{3}{4}}}$$

Mixed numbers like $1\frac{2}{3}$ and **improper fractions** like $\frac{5}{3}$ describe numbers greater than 1.

EXAMPLE

a Change $\frac{13}{8}$ into a mixed number.   **b** Change $1\frac{3}{5}$ into an improper fraction.

a $\frac{13}{8} = \frac{8}{8} + \frac{5}{8} = 1 + \frac{5}{8} = 1\frac{5}{8}$

b $1\frac{3}{5} = 1 + \frac{3}{5} = \frac{5}{5} + \frac{3}{5} = \frac{8}{5}$

A **decimal** is another way of writing a **fraction**.

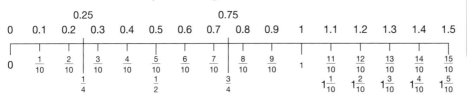

You should learn some common fractions and their decimal **equivalents**.

EXAMPLE

a Convert 0.8 decimal to a fraction in its simplest form.

b Convert $\frac{5}{4}$ to a decimal.

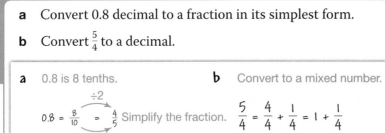

a 0.8 is 8 tenths.

$$\overset{\div 2}{\underset{\div 2}{0.8 = \frac{8}{10} = \frac{4}{5}}}$$ Simplify the fraction.

b Convert to a mixed number.

$$\frac{5}{4} = \frac{4}{4} + \frac{1}{4} = 1 + \frac{1}{4}$$
$$= 1 + 0.25 = 1.25$$

Number Fractions, decimals and percentages

Exercise 5.1S

1 Write the fraction of each of these shapes that is shaded.

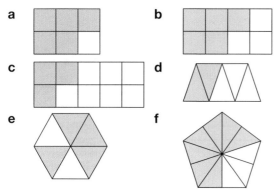

a   b

c   d

e   f

2 Write the fraction indicated by each of the pointers.

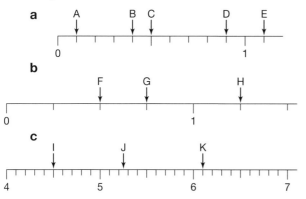

a

b

c

3 **i** Write the fraction of each of these shapes that is shaded.

ii Write your fraction in its simplest form.

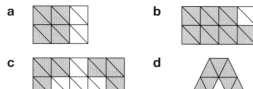

a   b

c   d

4 Cancel down each of these fractions into its simplest form.

a $\frac{4}{12}$ b $\frac{21}{28}$ c $\frac{24}{40}$ d $\frac{28}{63}$

e $\frac{45}{72}$ f $\frac{42}{126}$ g $\frac{64}{144}$ h $\frac{23}{93}$

5 Change each of these fractions to an improper fraction.

a $1\frac{1}{2}$ b $3\frac{2}{3}$ c $4\frac{3}{8}$ d $2\frac{2}{9}$

e $5\frac{6}{7}$ f $7\frac{4}{5}$ g $8\frac{8}{11}$ h $12\frac{4}{7}$

6 Change each of these fractions to a mixed number.

a $\frac{5}{4}$ b $\frac{8}{5}$ c $\frac{11}{7}$ d $\frac{9}{4}$

e $\frac{11}{5}$ f $\frac{20}{7}$ g $\frac{23}{5}$ h $\frac{28}{9}$

7 Find the missing number in each of these pairs of equivalent fractions.

a $\frac{2}{3} = \frac{?}{12}$ b $\frac{3}{4} = \frac{?}{36}$

c $\frac{5}{7} = \frac{40}{?}$ d $\frac{7}{8} = \frac{?}{64}$

e $\frac{12}{30} = \frac{?}{5}$ f $\frac{6}{7} = \frac{?}{105}$

g $\frac{5}{4} = \frac{?}{68}$ h $\frac{?}{10} = \frac{154}{220}$

8 Put these lists of numbers in order, starting with the smallest.

a   2.13   2.09   2.2   2.12   2.07

b   0.345   0.35   0.325   0.3   0.309

c   1.32   1.4   1.35   1.387   1.058

d   5.306   5.288   5.308   5.29   5.3

9 Write the number each of the arrows is pointing to.

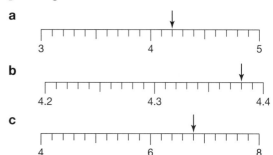

a

b

c

10 a Copy this decimal number line.

b Mark these fractions.

i $\frac{8}{10}$ **ii** $\frac{3}{5}$ **iii** $\frac{13}{10}$

11 Convert each decimal to a fraction in its simplest form.

a   0.1   b   0.6   c   0.75

d   1.25   e   1.5   f   1.2

12 Convert these fractions to decimals.

a $\frac{3}{10}$ b $\frac{4}{5}$ c $\frac{1}{4}$

d $1\frac{1}{10}$ e $\frac{14}{10}$ f $\frac{8}{5}$

Q 1016, 1019, 1042, 1075   SEARCH

5.1 Decimals and fractions

RECAP

- The numerator (top) of a fraction shows how many parts you have.
- The denominator (bottom) shows how many equal-sized parts the whole has been divided into.
- You can find equivalent fractions by multiplying or dividing the numerator and denominator by the same number.
- You can convert between fractions and decimals.

▲ The line separating the numerator and denominator is called the vinculum. The notation was developed by the mathematician Al-Hassar who lived in Morocco during the 12th century.

HOW TO

(1) Read the problem and decide how to use your knowledge of equivalent fractions and decimals.

(2) Find equivalent fractions or decimals.

(3) Answer the question and include any units.

EXAMPLE

There are 25 students in a class. 10 students are boys.

What fraction of the class are girls?

(1) The fraction is the number of girls over the total.

Number of girls = 25 − 10 = 15

Fraction of girls = $\frac{15}{25}$

(2) Simplify the fraction.

$$\frac{15}{25} \underset{\div 5}{\overset{\div 5}{=}} \frac{3}{5}$$

(3) $\frac{3}{5}$ of the students are girls.

EXAMPLE

Each of these shapes is partly shaded.

Shape A

Shape B

Shape C

Which shape has the greatest fraction shaded?

Show how you decide.

(1) The fraction shaded is the number of shaded sections over the total.

Shape A = $\frac{4}{8}$ Shape B = $\frac{6}{10}$ Shape C = $\frac{6}{8}$

Method 1

(2) Find an equivalent decimal.

Shape A = $\frac{1}{2}$ Shape B = 0.6   Shape C = $\frac{3}{4}$

= 0.5   = 0.75

(3) Writing the fractions as decimals makes it easier to order the fractions.

Shape C has the greatest fraction shaded because 0.75 is bigger than 0.5 and 0.6.

Method 2

(2) Simplify the fractions.

Shape A = $\frac{1}{2}$ Shape B = $\frac{3}{5}$ Shape C = $\frac{3}{4}$

Find equivalent fractions with the same denominator.

= $\frac{10}{20}$ = $\frac{12}{20}$ = $\frac{15}{20}$ 20 is a multiple of 2, 4 and 5.

(3) Compare the numerators.

Shape C has the greatest fraction shaded because 15 is bigger than 10 and 12.

Exercise 5.1A

1 There are 28 students in a class. 15 are boys and 13 are girls.
What fraction of the class are

  a   boys   **b**   girls?

2 Tyrone has eight pairs of brown shoes and seven pairs of black shoes.
What fraction of his shoes are

  a   brown   **b**   black?

3 Wesley earns £300 a week. He pays £91 of his money each week in tax.
He saves £60 each week. What fraction of his weekly wage does Wesley

  a   pay in tax   **b**   save?

4 A teacher works for eight hours at school and three hours at home.
What fraction of the day does the teacher work

  a   at school   **b**   at home?

5 Rory has a collection of 13 CDs, 15 DVDs and nine computer games.
What fraction of his collection is

  a   CDs   **b**   DVDs

  c   computer games?

6 Rearrange these cards so that each row has four cards of the same value.

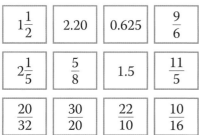

$1\frac{1}{2}$ 2.20   0.625 $\frac{9}{6}$

$2\frac{1}{5}$ $\frac{5}{8}$ 1.5 $\frac{11}{5}$

$\frac{20}{32}$ $\frac{30}{20}$ $\frac{22}{10}$ $\frac{10}{16}$

7 Here are two fractions, $\frac{1}{3}$ and $\frac{2}{5}$.
Explain which is the larger fraction.
Copy the grids to help with your explanation.

8 For each pair of fractions, write which is the larger fraction.

  a $\frac{3}{8}$ and $\frac{2}{5}$ **b** $\frac{3}{5}$ and $\frac{2}{3}$

  c $\frac{4}{7}$ and $\frac{2}{5}$ **d** $\frac{5}{6}$ and $\frac{7}{9}$

  e $\frac{5}{9}$ and $\frac{4}{7}$ **f** $\frac{7}{5}$ and $\frac{10}{7}$

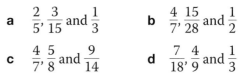

Convert each of the fractions to an equivalent fraction with the same denominator.

9 Put these fractions in order from smallest to largest.

  a $\frac{2}{5}, \frac{3}{15}$ and $\frac{1}{3}$ **b** $\frac{4}{7}, \frac{15}{28}$ and $\frac{1}{2}$

  c $\frac{4}{7}, \frac{5}{8}$ and $\frac{9}{14}$ **d** $\frac{7}{18}, \frac{4}{9}$ and $\frac{1}{3}$

10 Each of these shapes is partly shaded.

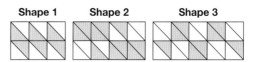

Shape 1   Shape 2   Shape 3

  a   Put the shapes in order from lowest fraction to highest fraction shaded.

  b   Put the shapes in order from most to least shaded.

11 Jamie partly shades a shape.

  a   Copy this circular shape. Shade the sectors so that the shape is shaded in the same fraction as Jamie's shape.

  b   Corrine wants to shade the squares in this shape so that the shape in shaded in the same fractions as Jamie's shape. Explain why Corrine can't shade the shape in this way.

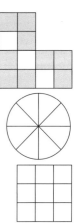

12 Match the equivalent pair of cards.

$\frac{1}{2}a + 0.4b - 0.1a + \frac{1}{4}b$	$0.4a + 0.65b$
$1.8a + 0.4b - 1\frac{2}{5}a + \frac{1}{5}b$	$0.4a + 0.6b$
$\frac{1}{10}a + \frac{7}{4}b + \frac{2}{5}a - 1.1b$	$0.5a + 0.65b$

5.2 Fractions and percentages

● You find fractions of a quantity by multiplying.

For example, two thirds of $5 = \frac{2}{3} \times 5 = \frac{2 \times 5}{3} = \frac{10}{3} = 3\frac{1}{3}$

Notice that you cannot give $3\frac{1}{3}$ as an **exact** answer in decimals because $\frac{1}{3}$ is a recurring decimal.

● When the quantity and the denominator of the fraction have a **common factor**, you can **cancel** this factor before multiplying.

For example, $\frac{2}{{}_1\cancel{3}} \times \cancel{24}^{8} = \frac{2}{1} \times 8 = 16$

EXAMPLE

Calculate **a** $\frac{3}{4}$ of 28   **b** $\frac{5}{8}$ of 6   **c** $\frac{4}{9}$ of 12   **d** $\frac{5}{9}$ of 25

a $\frac{3}{{}_1\cancel{4}} \times \cancel{28}^{7} = 3 \times 7 = 21$

b $\frac{5}{{}_4\cancel{8}} \times \cancel{6}^{3} = \frac{5 \times 3}{4} = \frac{15}{4} = 3\frac{3}{4}$

c $\frac{4}{{}_3\cancel{9}} \times \cancel{12}^{4} = \frac{4 \times 4}{3} = \frac{16}{3} = 5\frac{1}{3}$

d $\frac{5}{9} \times 25 = \frac{125}{9} = 13\frac{8}{9}$

● You can use mental methods to find percentages of amounts.

To find 50%, divide by 2.   To find 25%, divide by 4.

To find 10%, divide by 10.   To find 1%, divide by 100.

A percentage tells you how many parts per 100 there are. 78% means 78 out of 100.

● You can find a percentage of an amount by multiplying the quantity by an equivalent decimal or fraction.

To find 47% of an amount, multiply the amount by $\frac{47}{100}$ or 0.47.

EXAMPLE

Calculate   **a** 45% of 60 cm   **b** 34% of 85 kg   **c** 16% of £25

a 50% of 60 = 30   50% is a half.

5% of 60 = 3

45% of 60 cm = 30 cm – 3 cm = 27 cm

b 34% = $\frac{34}{100}$ = 0.34   0.34 is the **decimal equivalent** of 34%.

34% of 85 kg = 0.34 × 85

= 28.9 kg   By calculator.

= 29 kg to 2 sf.

c 16% of £25 = $\frac{{}^4\cancel{16}}{\cancel{100}_{{}_1}} \times \cancel{25}^{1}$ Cancel the quantity (25) and the denominator (100). Then cancel 16 and 4.

= £4

Exercise 5.2S

1 Use a mental method to calculate each of these amounts.

a $\frac{1}{2}$ of 40 sheep   **b** $\frac{1}{3}$ of 15 apples

c $\frac{1}{5}$ of 25 shops   **d** $\frac{1}{4}$ of 48 marks

2 Calculate each of these, leaving your answer in its simplest form.

a $5 \times \frac{1}{2}$ **b** $8 \times \frac{1}{4}$

c $8 \times \frac{1}{3}$ **d** $13 \times \frac{1}{7}$

e $\frac{1}{12} \times 24$ **f** $\frac{1}{3} \times 4$

3 Calculate each of these, leaving your answer in its simplest form.

a $6 \times \frac{2}{3}$ **b** $5 \times \frac{3}{4}$

c $2 \times \frac{2}{5}$ **d** $4 \times \frac{7}{6}$

e $5 \times \frac{9}{20}$ **f** $\frac{4}{5} \times 28$

g $\frac{4}{9} \times 30$ **h** $\frac{11}{18} \times 14$

4 a What fraction of this shape is shaded?

b What percentage of the shape is shaded?

c Copy the shape and shade in more squares so that 70% of the shape is shaded.

5 Calculate these percentages without using a calculator.

a 50% of £60   **b** 50% of 40

c 50% of 272 p   **d** 10% of 40

e 10% of 370 p   **f** 1% of £700

g 50% of 12 kg   **h** 50% of £31

i 1% of 200   **j** 1% of 420 m

6 Calculate these percentages using a mental method.

a 10% of £40   **b** 10% of £340

c 5% of £120   **d** 20% of 210 km

e 20% of $530   **f** 25% of £300

g 20% of £32   **h** 5% of 28 m

7 Use a mental or written method to work these out.
Leave your answers as fractions in their simplest form where appropriate.

a $\frac{3}{10}$ of €40   **b** $\frac{2}{5}$ of £70

c $\frac{3}{4}$ of 50 m   **d** $\frac{4}{7}$ of 64 km

e $\frac{3}{8}$ of £1000   **f** $\frac{5}{6}$ of 70 mm

g $\frac{11}{12}$ of 1500 m   **h** $\frac{4}{13}$ of 60 g

8 Use a suitable method to calculate each of these.
Where appropriate round your answer to 2 decimal places.

a $\frac{8}{15}$ of 495 kg   **b** $\frac{9}{10}$ of $5000

c $\frac{5}{9}$ of 8 kg   **d** $\frac{7}{9}$ of 1224 cups

e $\frac{13}{18}$ of 30 tonnes   **f** $\frac{4}{15}$ of 360°

g $\frac{12}{31}$ of 360°   **h** $\frac{13}{15}$ of 1 hour

i $\frac{17}{15}$ of £230   **j** $\frac{13}{11}$ of 200 kg

9 Calculate these percentages using a mental or written method.
Show all the steps of your working out.

a 11% of £18   **b** 60% of 7300 km

c 8% of £30   **d** 2% of €3000

e 7% of 60 m   **f** 13% of 40 cm

g 75% of 48 m   **h** 3% of £70

i 30% of 250 kg   **j** 80% of 2100 km

10 Calculate these using a mental or written method.
Show all the steps of your working out.
Give your answers to 2 decimal places as appropriate.

a 12% of £17   **b** 16% of 87 km

c 8% of £38   **d** 32% of €340

e 17% of 65 m   **f** 73% of 46 cm

g 85% of 148 m   **h** 2% of £76.40

i 25% of £85   **j** 35% of £3.25

 1018, 1030, 1031, 1962, 1963   SEARCH

5.2 Fractions and percentages

RECAP

- You find a fraction of an amount by multiplying the fraction and the quantity.
- You can find a percentage of an amount by multiplying the quantity by an equivalent decimal or fraction.

HOW TO

(1) Read the problem and decide what fraction or percentage of the amount you need.

(2) Calculate the answer mentally or with a calculator.

(3) Answer the question and include any units.

▲ The term percent is derived from the Latin per centum, meaning "by the hundred"

EXAMPLE

Jane earns £320 each week.

She spends $\frac{3}{5}$ of the money and deposits the rest into a savings account.

How much money will Jane save each year?

(1) Jane will save $\frac{2}{5}$ of the money. $1 - \frac{3}{5} = \frac{2}{5}$

(2) Find $\frac{2}{5}$ of £320. $\frac{2}{5} \times 320 = 128$

(3) There are 52 weeks in a year. $128 \times 52 = 6656$

Jane saves £6656 each year.

EXAMPLE

Tom has £42. He spends $\frac{1}{3}$ of it on Monday. On Tuesday he spends $\frac{3}{4}$ of the remainder. How much does he spend on Tuesday?

(1) First, work out how much money Tom has left on Tuesday.

(2) Remaining money $= \frac{2}{3}$ of $42 = \frac{2}{3} \times 42 = 28$

Tom spends $\frac{3}{4}$ of the remainder.

$\frac{3}{4}$ of $28 = \frac{3}{4} \times 28 = 21$

(3) Tom spends £21 on Tuesday.

EXAMPLE

Which deal gives most money off?

Was £255
Now ⅓ off!

Was £238
Now 35% off!

(1) Use a calculator to work out the discount on each deal. Round to the nearest penny.

(2) Multiply the amount by the fraction.

$\frac{1}{3}$ of £255 = £255 $\times \frac{1}{3}$ = £85   or you could divide by 3.

35% = 0.35. Multiply the amount by the decimal.

35% of £238 = £238 $\times$ 0.35 = £83.30

Write in the answer in pounds and pence.

(3) The first deal gives more money off.

Exercise 5.2A

1 Calculate each of these, leaving your answer in its simplest form.

   a What is the total mass of four packets that each weigh $\frac{1}{5}$ kg?

   b A cake weighs $\frac{7}{20}$ of a kg. What is the mass of 10 cakes?

   c What is the total capacity of 12 jugs that each have a capacity of $\frac{3}{5}$ of a litre?

   d What is the total mass of 16 bags of flour that each weigh $\frac{9}{10}$ kg?

2 Rearrange these cards to make three correct statements.

25% of	£450	= £128
30% of	£640	= £130
20% of	£520	= £135

3 Calculate these fractions of amounts without using a calculator.

   a A jacket normally costs £130. In a sale the jacket is priced at $\frac{4}{5}$ of its normal selling price. What is the new price of the jacket?

   b Hector rents out a holiday flat. His flat is available for 45 weeks of the year. He rents the flat out to tourists for $\frac{7}{9}$ of the time it is available. For how many weeks is Hector's flat occupied by tourists?

4 Calculate these percentages without using a calculator.

   a Kelvin collects models. He owns 170 models. He has painted 20% of the models. How many of the models has he painted?

   b A barrel can hold 70 litres. Water is poured into the barrel until it is 80% full. How much water is there in the barrel?

5 An empty swimming pool is to be filled with water. It takes 12 hours to fill the pool and the full pool contains 98 m³ of water. How much water will the pool contain after 5 hours? Show your working.

6 Mrs Jones has a conservatory built, which costs £12 000. She pays an initial deposit of 15%. The remainder is to be paid in 24 equal monthly instalments. How much is each of these instalments?

7 A restaurant adds a 12% service charge to the bill. What will be the total cost for a meal that is £65.80 before the service charge is added?

8 A school has 1248 pupils, and 48% of them are girls. How many boys are there in the school? Show your working.

9 Julia earns a salary of £47 800 per year. She is awarded a 2.7% pay rise. Calculate her new salary.

10 Rajin earns £18 500 per year; the first £6550 is untaxed; the remainder is taxed at 20%. How much tax does he pay per month?

11 Alfie sees the same coat in two different shops.

| Top Bloke | Merton |
| £85 with 15% off | £95 with $\frac{1}{4}$ off |

Where should Alfie buy the coat?

12 Fill in the gaps.

   a 45% of £48 = ☐% of £45

   b 2% of 400 m = $\frac{1}{5}$ of ☐ m

   c 25% of 8 kg = $\frac{1}{☐}$ of 4 kg

13 To obtain a driving licence, you must pass both the theory and the practical test. You cannot take the practical test if you fail the theory test.
200 people applied for a driving licence.
80% of them passed the theory test.
Three out of four people who took the practical test passed.
How many people obtained their driving licence?

5.3 Calculations with fractions

You can only add and subtract fractions if they have common denominators.

$$\frac{2}{8} + \frac{5}{8} = \frac{7}{8}$$

Add the numerators.

- To add or subtract fractions with different denominators, change them to equivalent fractions with the same denominator.

- To multiply fractions, multiply the numerators and then the denominators, and cancel any common factors.

- To divide by a fraction, multiply by its reciprocal.

The reciprocal of a fraction is the original fraction 'turned upside down'.

The reciprocal of $\frac{3}{5}$ is $\frac{5}{3}$.

EXAMPLE

Work out each of these calculations.

a $\frac{3}{5} + \frac{11}{16}$ **b** $\frac{2}{7} - \frac{1}{4}$ **c** $\frac{3}{5} \div 8$ **d** $\frac{4}{9} \div \frac{1}{3}$ **e** $\frac{3}{8} \times \frac{5}{9}$ **f** $2\frac{4}{5} \div 3\frac{1}{2}$

a $\frac{3}{5} + \frac{11}{16} = \frac{48}{80} + \frac{55}{80}$

The LCM of 5 and 16 is 80.

$= \frac{48 + 55}{80}$

$= \frac{103}{80}$

Write the answer as a mixed number.

$= 1\frac{23}{80}$

b $\frac{2}{7} - \frac{1}{4} = \frac{8}{28} - \frac{7}{28}$

The LCM of 7 and 4 is 28.

$= \frac{1}{28}$

c $\frac{3}{5} \div 8 = \frac{3}{5} \times \frac{1}{8}$

The reciprocal of 8 is $\frac{1}{8}$.

$= \frac{3 \times 1}{5 \times 8}$

$= \frac{3}{40}$

d $\frac{4}{9} \div \frac{1}{3} = \frac{4}{9} \times \frac{3}{1} = \frac{4 \times 3}{9 \times 1}$

The reciprocal of $\frac{1}{3}$ is $\frac{3}{1}$ (which is 3).

$= \frac{12}{9}$

$= \frac{4}{3}$

$= 1\frac{1}{3}$

e $\frac{\overset{1}{\cancel{3}}}{8} \times \frac{5}{\underset{3}{\cancel{9}}} = \frac{1}{8} \times \frac{5}{3}$

Notice how you can **cancel common factors** before multiplying.

$= \frac{5}{24}$

Cancelling the fractions before multiplying gives the same answer as simplifying after multiplying.

f $2\frac{4}{5} \div 3\frac{1}{2} = \frac{14}{5} \div \frac{7}{2}$

Change mixed numbers to improper fractions first.

$= \frac{\overset{2}{\cancel{14}}}{5} \times \frac{2}{\underset{1}{\cancel{7}}}$

$= \frac{4}{5}$

Exercise 5.3S

1 Copy and complete each of these equivalent fraction families.

a $\dfrac{1}{2} = \dfrac{2}{4} = \dfrac{?}{10} = \dfrac{?}{16}$

b $\dfrac{3}{10} = \dfrac{6}{20} = \dfrac{?}{30} = \dfrac{15}{?} = \dfrac{30}{?}$

c $\dfrac{4}{5} = \dfrac{?}{10} = \dfrac{12}{?} = \dfrac{?}{25} = \dfrac{?}{100}$

d $\dfrac{5}{8} = \dfrac{10}{?} = \dfrac{?}{24} = \dfrac{?}{80} = \dfrac{75}{?}$

2 Find an equivalent fraction for each fraction. Both of your fractions should have the same denominator.

a $\dfrac{1}{2}$ and $\dfrac{1}{3}$ b $\dfrac{1}{5}$ and $\dfrac{1}{3}$ c $\dfrac{1}{2}$ and $\dfrac{1}{5}$

d $\dfrac{2}{3}$ and $\dfrac{1}{4}$ e $\dfrac{3}{10}$ and $\dfrac{1}{3}$ f $\dfrac{4}{5}$ and $\dfrac{1}{4}$

3 Calculate each of these. Give your answer in its simplest form.

a $\dfrac{1}{3} + \dfrac{1}{3}$ b $\dfrac{2}{5} + \dfrac{1}{5}$ c $\dfrac{3}{10} + \dfrac{7}{10}$

d $\dfrac{19}{16} - \dfrac{7}{16}$ e $\dfrac{12}{35} + \dfrac{8}{35}$ f $\dfrac{23}{30} - \dfrac{7}{30}$

4 Change these improper fractions into mixed numbers.

a $\dfrac{5}{4}$ b $\dfrac{9}{5}$ c $\dfrac{13}{8}$ d $\dfrac{17}{4}$

5 Change these mixed numbers to improper fractions.

a $1\dfrac{3}{4}$ b $1\dfrac{7}{16}$ c $1\dfrac{5}{9}$ d $2\dfrac{4}{7}$

6 Work out each of these, leaving your answer in its simplest form.

a $1\dfrac{2}{3} + \dfrac{2}{3}$ b $4\dfrac{2}{7} - \dfrac{5}{7}$

7 Work out

a $\dfrac{1}{3} + \dfrac{1}{2}$ b $\dfrac{1}{4} + \dfrac{3}{5}$ c $\dfrac{3}{5} - \dfrac{1}{3}$

d $\dfrac{4}{5} - \dfrac{2}{7}$ e $\dfrac{5}{8} + \dfrac{1}{3}$ f $\dfrac{4}{9} + \dfrac{2}{5}$

8 Calculate each of these, leaving your answer in its simplest form.

a $3 \times \dfrac{1}{2}$ b $6 \times \dfrac{1}{3}$ c $10 \times \dfrac{1}{3}$

d $15 \times \dfrac{1}{7}$ e $\dfrac{1}{10} \times 25$ f $\dfrac{1}{3} \times 13$

9 Calculate each of these, leaving your answer in its simplest form.

a $3 \times \dfrac{2}{3}$ b $6 \times \dfrac{2}{3}$ c $5 \times \dfrac{2}{3}$

d $2 \times \dfrac{7}{24}$ e $4 \times \dfrac{3}{20}$ f $\dfrac{11}{8} \times 17$

10 Calculate each of these, leaving your answer in its simplest form.

a $4 \div \dfrac{1}{2}$ b $2 \div \dfrac{1}{5}$ c $2 \div \dfrac{1}{7}$

d $10 \div \dfrac{1}{2}$ e $12 \div \dfrac{1}{4}$ f $22 \div \dfrac{1}{10}$

11 Work out each of these, leaving your answer in its simplest form.

a $1\dfrac{1}{3} + 1\dfrac{1}{4}$ b $1\dfrac{2}{7} + \dfrac{3}{5}$ c $2\dfrac{2}{5} - \dfrac{1}{3}$

d $3\dfrac{3}{8} - 1\dfrac{1}{2}$ e $4\dfrac{1}{3} - 2\dfrac{3}{4}$ f $3\dfrac{4}{7} - 2\dfrac{8}{9}$

12 Calculate each of these, leaving your answer in its simplest form.

a $2 \div \dfrac{5}{6}$ b $12 \div \dfrac{6}{7}$ c $20 \div \dfrac{5}{12}$

d $5 \div \dfrac{7}{9}$ e $3 \div 1\dfrac{1}{2}$ f $3 \div 1\dfrac{2}{5}$

13 Calculate each of these, leaving your answer in its simplest form.

a $\dfrac{2}{5} \times \dfrac{3}{4}$ b $\dfrac{3}{5} \times \dfrac{3}{4}$ c $\dfrac{5}{7} \times \dfrac{3}{4}$

d $\dfrac{4}{7} \times \dfrac{3}{5}$ e $\dfrac{5}{6} \times \dfrac{4}{5}$ f $\dfrac{3}{8} \times \dfrac{7}{9}$

g $\dfrac{3}{5} \times \dfrac{10}{9}$ h $\dfrac{15}{16} \times \dfrac{12}{5}$ i $\left(\dfrac{3}{7}\right)^2$

j $1\dfrac{3}{4} \times \dfrac{2}{7}$ k $3\dfrac{2}{3} \times \dfrac{7}{11}$ l $1\dfrac{3}{8} \times 1\dfrac{2}{5}$

14 Calculate each of these, leaving your answer in its simplest form.

a $4 \div \dfrac{2}{5}$ b $\dfrac{2}{3} \div \dfrac{4}{5}$ c $\dfrac{4}{5} \div \dfrac{3}{4}$

d $\dfrac{4}{7} \div \dfrac{2}{3}$ e $\dfrac{3}{7} \div \dfrac{4}{9}$ f $\dfrac{3}{5} \div \dfrac{1}{3}$

g $\dfrac{3}{4} \div 3$ h $\dfrac{4}{7} \div 5$ i $\dfrac{4}{11} \div 5$

j $\dfrac{7}{4} \div \dfrac{2}{3}$ k $\dfrac{7}{4} \div \dfrac{3}{2}$ l $\dfrac{9}{5} \div \dfrac{5}{3}$

m $1\dfrac{1}{2} \div \dfrac{3}{4}$ n $2\dfrac{1}{4} \div \dfrac{2}{3}$ o $2\dfrac{2}{5} \div \dfrac{9}{7}$

1017, 1040, 1046, 1047, 1074 SEARCH

5.3 Calculations with fractions

- To add or subtract fractions, convert to equivalent fractions with the same denominator, and then add or subtract the numerators.
- To multiply fractions, multiply the numerators and then the denominators, and cancel any common factors.
- Dividing by a fraction is the same as multiplying by its reciprocal.

> Convert mixed numbers to improper fractions before calculating.

HOW TO

① Read the problem and decide what calculation you must do.

② Calculate using your knowledge of fractions. Simplify fully and include any units.

EXAMPLE

Jodi wrote

$$\frac{2}{3} + \frac{3}{4} = \frac{2+3}{3+4} = \frac{5}{7} \ \text{✗}$$

Without calculating the correct answer, explain how you know that Jodi's answer is incorrect.

① Explain how you know that the answer is incorrect.

The answer $\frac{5}{7}$ cannot be correct. Both $\frac{2}{3}$ and $\frac{3}{4}$ are bigger than $\frac{1}{2}$, so the answer must be greater than 1.

② Explain the mistake that Jodi made in her calculation.

Jodi added the numerators and denominators instead of finding the common denominator first.

EXAMPLE

Paul cuts 75 cm from a piece of rope $2\frac{2}{3}$ m long. Find the exact length of the remaining rope.

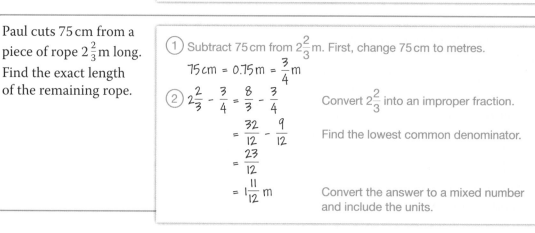

① Subtract 75 cm from $2\frac{2}{3}$ m. First, change 75 cm to metres.

$$75\,cm = 0.75\,m = \frac{3}{4}\,m$$

② $2\frac{2}{3} - \frac{3}{4} = \frac{8}{3} - \frac{3}{4}$ Convert $2\frac{2}{3}$ into an improper fraction.

$$= \frac{32}{12} - \frac{9}{12}$$ Find the lowest common denominator.

$$= \frac{23}{12}$$

$$= 1\frac{11}{12}\,m$$ Convert the answer to a mixed number and include the units.

EXAMPLE

Write in ascending order $\frac{7}{8}$ $\frac{5}{6}$ $\frac{3}{4}$

> Ascending – going up
> Descending – going down

① Convert the fractions to fractions with a common denominator.

LCM of 8, 6 and 4 = 24

$$\frac{7}{8} = \frac{21}{24} \qquad \frac{5}{6} = \frac{20}{24} \qquad \frac{3}{4} = \frac{18}{24}$$

② Order the fractions by comparing the numerators.

$$\frac{18}{24} < \frac{20}{24} < \frac{21}{24}$$

In ascending order the fractions are: $\frac{3}{4}, \frac{5}{6}, \frac{7}{8}$

Exercise 5.3A

1 Write whether each of these statements is true or false.

> < means 'less than', > means 'more than'.

a $\frac{9}{2} < 3$ **b** $\frac{17}{24} > \frac{5}{8}$ **c** $\frac{11}{12} > \frac{8}{9}$

d $\frac{2}{3} < \frac{5}{7}$ **e** $\frac{3}{5} > \frac{4}{7}$ **f** $\frac{26}{25} > \frac{16}{15}$

g $\frac{5}{4} < \frac{12}{7}$ **h** $\frac{7}{4} > \frac{12}{10}$

Work out each of these problems, leaving your answer in its simplest form.

2 Pete walked $3\frac{2}{3}$ miles before lunch and then a further $2\frac{1}{4}$ miles after lunch. How far did he walk altogether?

3 A bag weighs $2\frac{3}{16}$ lb when it is full. When empty the bag weighs $\frac{3}{8}$ lb. What is the weight of the contents of the bag?

4 Henry and Paula are eating peanuts. Henry has a full bag weighing $1\frac{3}{16}$ kg. Paula has a bag that weighs $\frac{4}{5}$ kg. What is the total mass of their two bags of peanuts?

5 Simon spent $\frac{2}{3}$ of his pocket money on a computer game. He spent $\frac{1}{5}$ of his pocket money on a ticket to the cinema. Work out the fraction of his pocket money that he had left.

6 Calculate each of these, leaving your answer in its simplest form.

a What is the total weight of seven boxes that each weigh $\frac{2}{5}$ kg?

b What is the total length of five pieces of wood that are each $\frac{3}{7}$ of a metre long?

c Find the total capacity of eight bottles that each hold a capacity of $\frac{7}{10}$ litre.

7 Calculate each of these, rounding your answer as appropriate.

a A plank of wood is $4\frac{3}{5}$ metres long. How many pieces of wood of length $1\frac{1}{4}$ metres can be cut from the plank?

b A paint pot can hold $2\frac{3}{4}$ litres of paint. Hector buys $11\frac{1}{5}$ litres of emulsion paint. How many times can Hector fill the paint pot with emulsion paint?

8 Calculate the perimeter of each of these swimming pools.

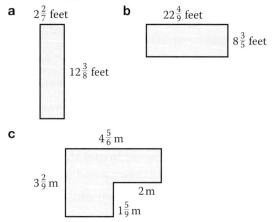

a $2\frac{2}{7}$ feet $12\frac{3}{8}$ feet

b $22\frac{4}{9}$ feet $8\frac{3}{5}$ feet

c $4\frac{5}{6}$ m $3\frac{2}{9}$ m 2 m $1\frac{5}{9}$ m

9 Copy the equation grid and fill in the empty squares to make all of the equations correct both horizontally and vertically.

$\frac{1}{2}$	$+$	$\frac{1}{6}$	$=$	
$\times$		$+$		$\bigstar$
	$\times$		$=$	
$=$		$=$		$=$
$\frac{3}{20}$	$\times$		$=$	$\frac{1}{10}$

Enter each fraction in its simplest form. Which of the four operations ($+$, $-$, $\times$ or $\div$) must replace the '$\bigstar$'?

10 The diagram shows three identical pentagons A, B and C.

$\frac{5}{7}$ of shape A is shaded.

$\frac{4}{5}$ of shape C is shaded.

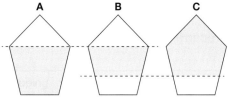

A   B   C

What fraction of shape B is shaded?

11 A unit fraction has 1 as a numerator and an integer (whole number) as a denominator. $\frac{1}{2}$ and $\frac{1}{3}$ are both unit fractions. Find three unit fractions that add up to $\frac{1}{2}$.

 Q 1017, 1040, 1046, 1047, 1074   SEARCH

5.4 Fractions, decimals and percentages

- To convert a **terminating** decimal to a fraction:
  - Write the decimal as a fraction with **denominator** 10, 100, 1000, ... according to the number of decimal places.
  - Simplify the fraction.
- To convert a percentage to a fraction, divide the percentage by 100.

EXAMPLE

Convert these to fractions   **a**   0.306   **b**   45%   **c**   32.5%   **d**   0.52

3 decimal places so denominator is 1000.

a $0.306 = \dfrac{306}{1000} = \dfrac{153}{500}$

b $45\% = \dfrac{45}{100} = \dfrac{9}{20}$

c $32.5\% = 0.325 = \dfrac{325}{1000} = \dfrac{65}{200} = \dfrac{13}{40}$

d $0.52 = \dfrac{52}{100} = \dfrac{26}{50} = \dfrac{13}{25}$

- To convert a fraction to a decimal divide the numerator by the denominator.

EXAMPLE

Write these fractions as decimals.   **a** $\dfrac{5}{8}$ **b** $\dfrac{5}{9}$

a $\dfrac{5}{8} = 5 \div 8 = 8\overline{)5.000}^{\,0.625}$

b $\dfrac{5}{9} = 5 \div 9 = 9\overline{)5.000...}^{\,0.555...} = 0.\dot{5}$

0.625 is a **terminating** decimal.

$0.555 ... = 0.\dot{5}$ is a **recurring** decimal. The dot over the 5 shows the recurring digit.

To convert a fraction to a percentage

- write it as a decimal
- multiply the decimal by 100%.

$100\% = \dfrac{100}{100} = 1$

so you are multiplying by 1 which does not change the value of the decimal.

EXAMPLE

Write as percentages   **a** $\dfrac{5}{8}$ **b** $\dfrac{5}{9}$

a $\dfrac{5}{8} = 0.625 = 0.625 \times 100\% = 62.5\%$

b $\dfrac{5}{9} = 0.\dot{5} = 0.\dot{5} \times 100\% = 55.5\dot{5}\% = 55.6\%$ to 1 dp

- A **recurring decimal** has digits that keep repeating.

The dots show which digits repeat.

$\dfrac{1}{3} = 1 \div 3 = 0.333333 ... = 0.\dot{3}$ $\dfrac{4}{33} = 4 \div 33 = 0.121212 ... = 0.\dot{1}\dot{2}$

$\dfrac{1}{7} = 0.142857142857 ... = 0.\dot{1}4285\dot{7}$

- All recurring decimals are exact fractions.
- Not all exact fractions are recurring decimals.

$0.\dot{1} = \dfrac{1}{9}$

$\dfrac{1}{5} = 0.2$

Use short division

$9\overline{)1.^{1}0\;^{1}0\;^{1}0}^{\;0.1\;1\;1...}$

Exercise 5.4S

1 Write these percentages as decimals.

a	67%	**b**	78%	**c**	99%
d	70%	**e**	39%	**f**	88%
g	150%	**h**	125%	**i**	99.9%
j	110%	**k**	75%	**l**	37.6%

2 Write these decimals as percentages.

a	0.32	**b**	0.22	**c**	0.85
d	0.03	**e**	0.54	**f**	0.63
g	0.38	**h**	0.375	**i**	0.333
j	1.25	**k**	0.0015	**l**	0.995

3 Write these decimals as fractions in their simplest forms.

a	0.8	**b**	0.28	**c**	0.325
d	0.05	**e**	0.12	**f**	0.375

4 Change these fractions to decimals. Give your answers to 2 decimal places as appropriate.

a $\frac{3}{10}$ **b** $\frac{7}{25}$ **c** $\frac{7}{12}$

d $\frac{9}{15}$ **e** $\frac{15}{7}$ **f** $\frac{19}{9}$

5 Write these percentages as fractions in their simplest forms.

a	25%	**b**	40%	**c**	65%
d	15%	**e**	145%		

6 Write each of these decimals as a fraction in its simplest form.

a	0.3	**b**	0.6	**c**	0.64
d	0.45	**e**	0.375	**f**	1.08
g	3.2375	**h**	3.0625	**i**	4.25

7 Change these fractions to decimals without using a calculator.

a $\frac{3}{10}$ **b** $\frac{11}{25}$ **c** $\frac{26}{25}$

d $\frac{124}{200}$ **e** $\frac{27}{60}$ **f** $\frac{39}{75}$

g $\frac{42}{150}$ **h** $3\frac{21}{60}$ **i** $\frac{89}{25}$

8 Change these fractions into decimals using an appropriate method. Give your answers to 2 decimal places where necessary.

a $\frac{22}{50}$ **b** $\frac{2}{3}$ **c** $\frac{27}{20}$

d $\frac{11}{15}$ **e** $\frac{8}{7}$ **f** $1\frac{2}{5}$

g $2\frac{11}{66}$ **h** $\frac{11}{13}$

9 Write each of these percentages as a fraction in its simplest form.

a	40%	**b**	90%	**c**	35%
d	65%	**e**	1%	**f**	362%
g	15.25%	**h**	2.125%		

10 Write each of these fractions as a percentage without using a calculator.

a $\frac{27}{50}$ **b** $\frac{2}{5}$ **c** $\frac{17}{20}$

d $\frac{13}{25}$ **e** $\frac{2}{3}$ **f** $\frac{48}{200}$

g $1\frac{3}{15}$ **h** $\frac{33}{75}$

11 Write these percentages as decimals.

a	37%	**b**	7%
c	189%	**d**	45%

12 Write these decimals as percentages.

a	0.72	**b**	0.2
c	1.25	**d**	0.03

13 Write these fractions as percentages. Give your answers to 1 decimal place as appropriate.

a $\frac{48}{70}$ **b** $\frac{16}{25}$ **c** $\frac{17}{19}$

d $1\frac{11}{12}$ **e** $\frac{5}{19}$

14 Write these fractions as recurring decimals, using the correct notation. Try to use a written method of division.

a $\frac{2}{3}$ **b** $\frac{3}{11}$

c $\frac{2}{9}$ **d** $\frac{3}{7}$

1015, 1016, 1029   SEARCH

5.4 Fractions, decimals and percentage:

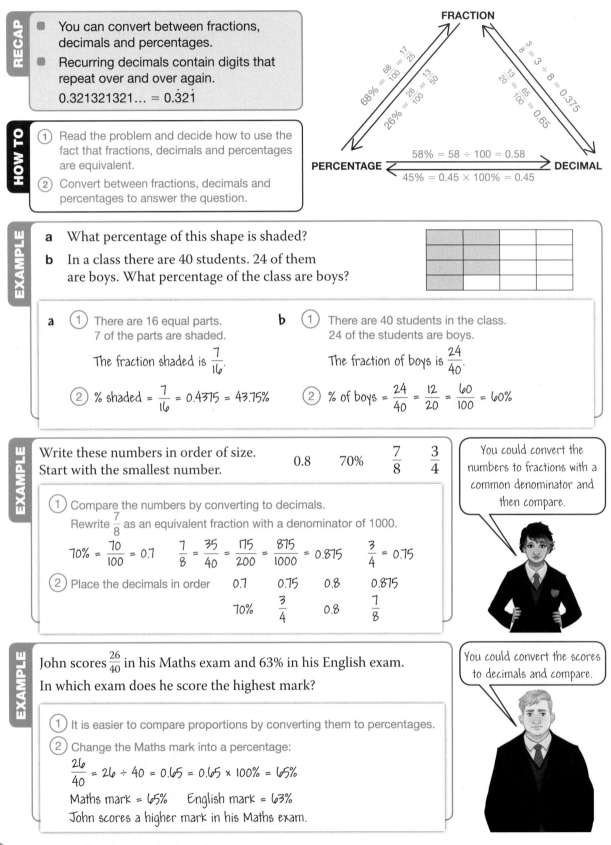

RECAP

- You can convert between fractions, decimals and percentages.
- Recurring decimals contain digits that repeat over and over again.
  $0.321321321... = 0.\dot{3}2\dot{1}$

FRACTION

$68\% = \frac{68}{100} = \frac{17}{25}$

$26\% = \frac{26}{100} = \frac{13}{50}$

$\frac{3}{8} = 3 \div 8 = 0.375$

$\frac{13}{20} = \frac{65}{100} = 0.65$

PERCENTAGE $\rightleftarrows$ **DECIMAL**

$58\% = 58 \div 100 = 0.58$

$45\% = 0.45 \times 100\% = 0.45$

HOW TO

① Read the problem and decide how to use the fact that fractions, decimals and percentages are equivalent.

② Convert between fractions, decimals and percentages to answer the question.

EXAMPLE

a What percentage of this shape is shaded?

b In a class there are 40 students. 24 of them are boys. What percentage of the class are boys?

a ① There are 16 equal parts. 7 of the parts are shaded.

The fraction shaded is $\frac{7}{16}$.

② % shaded $= \frac{7}{16} = 0.4375 = 43.75\%$

b ① There are 40 students in the class. 24 of the students are boys.

The fraction of boys is $\frac{24}{40}$.

② % of boys $= \frac{24}{40} = \frac{12}{20} = \frac{60}{100} = 60\%$

EXAMPLE

Write these numbers in order of size. Start with the smallest number.

$0.8 \quad 70\% \quad \frac{7}{8} \quad \frac{3}{4}$

> You could convert the numbers to fractions with a common denominator and then compare.

① Compare the numbers by converting to decimals.

Rewrite $\frac{7}{8}$ as an equivalent fraction with a denominator of 1000.

$70\% = \frac{70}{100} = 0.7 \qquad \frac{7}{8} = \frac{35}{40} = \frac{175}{200} = \frac{875}{1000} = 0.875 \qquad \frac{3}{4} = 0.75$

② Place the decimals in order

$0.7 \qquad 0.75 \qquad 0.8 \qquad 0.875$

$70\% \qquad \frac{3}{4} \qquad 0.8 \qquad \frac{7}{8}$

EXAMPLE

John scores $\frac{26}{40}$ in his Maths exam and 63% in his English exam.

In which exam does he score the highest mark?

> You could convert the scores to decimals and compare.

① It is easier to compare proportions by converting them to percentages.

② Change the Maths mark into a percentage:

$\frac{26}{40} = 26 \div 40 = 0.65 = 0.65 \times 100\% = 65\%$

Maths mark = 65%   English mark = 63%

John scores a higher mark in his Maths exam.

Exercise 5.4A

1 What percentage of each of these shapes is shaded?

a

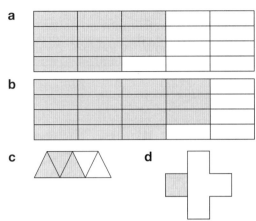

b

c   **d**

2 Put these fractions, decimals and percentages in order from smallest to largest. Show your working.

a 47%, $\frac{12}{25}$ and 0.49   **b** $\frac{4}{5}$, 78% and 0.81

c $\frac{5}{8}$, 66% and $\frac{7}{12}$ **d** $\frac{5}{16}$, 0.3, 29% and $\frac{7}{22}$

3 a Jo scores 68% in his French exam and gets $\frac{37}{54}$ in his German exam. In which subject did he do the best? Explain your answer.

b In a school survey 23% of the students said they did not like eating meat. In Sarah's class $\frac{7}{31}$ students said they did not like eating meat. How do the results of Sarah's class compare with the rest of the school?

4 Put these fractions, decimals and percentages in order from lowest to highest.
Order these quantities without a calculator.

a $\frac{3}{5}$, 61%, 0.63   **b** $\frac{7}{10}$, $\frac{17}{25}$, 69%, 0.71

c $\frac{7}{20}$, $\frac{2}{5}$, $\frac{3}{8}$, 0.36, 34%

Order these quantities with a calculator.

d $\frac{3}{7}$, 42%, $\frac{2}{5}$ **e** 15%, $\frac{3}{19}$, 0.14, $\frac{1}{5}$

f 81%, $\frac{8}{9}$, 0.93, 0.9, $\frac{19}{20}$

5 Solve each of these problems without using a calculator. Express each of your answers

i as a fraction in its lowest form

ii as a percentage.

a Brian's mark in a Geography test was 42 out of 70. What proportion of the test did he answer correctly?

b Hannah collects dolls. She has 25 dolls altogether. 13 of the 25 dolls are from Russia. What proportion of Hannah's dolls are from Russia?

c A restaurant makes a service charge of 3p in every 20p. Work out 3p as a proportion of 20p.

d Class 11X2 has 30 students. 21 of these students are boys. What proportion of the class are boys?

*6 All fractions can be turned into a decimal by dividing the numerator by the denominator. Some produce recurring decimals.

For example $\frac{1}{3} = 1 \div 3 = 0.333\,333\,333\ldots$

a Convert each of the fractions less than 1 with a denominator of 7 into a decimal using your calculator. Write down all the decimal places in your answer. For example

$\frac{1}{7} = 1 \div 7 = 0.142\,857\,142\ldots$

$\frac{2}{7} = 2 \div 7 = \ldots$

b Write what you notice about each of your answers.

c Repeat for all the fractions less than 1 with a denominator of 13.

*7 Rewrite these sets of numbers in ascending order. Show your working.

a 33.3%, 0.33, 33, $33\frac{1}{3}$ %

b 0.45, 44.5%, 0.454, 0.4̇

c 0.2̇3̇, 0.232, 22.3%, 23.22%, 0.233

d $\frac{2}{3}$, 0.66, 0.6̇5̇, 66.6%, 0.6666

e $\frac{1}{7}$, 14%, 0.142, $\frac{51}{350}$, 14.1̇%

f 86%, $\frac{5}{6}$, 0.86̇, 0.866, $\frac{6}{7}$

 1015, 1016, 1029   SEARCH

Summary

Checkout

You should now be able to...

Test it

Questions

	Questions
✓ Convert between terminating decimals and their corresponding fractions.	1 – 4
✓ Compare decimals and fractions using the symbols $>$ and $<$.	5
✓ Find fractions and percentages of amounts.	6, 7
✓ Add and subtract simple fractions and mixed numbers.	8
✓ Multiply and divide simple fractions and mixed numbers.	9, 10
✓ Convert between fractions, decimals and percentages.	11

Language   Meaning   Example

Language	Meaning	Example
Fraction	A fraction compares the size of a part with the size of a whole. All the parts that make up the whole have to be the same size.	$\frac{5}{8}$ is shaded
Equal	Exactly the same quantity or size.	$\frac{1}{2} = \frac{2}{4}$
Numerator	The part of a fraction above the line.	$\frac{2}{5}$ ← numerator
Denominator	The part of a fraction below the line.	← denominator
Mixed number	A number containing a whole number and a fraction.	$1\frac{1}{3}$ and $3\frac{5}{8}$ are mixed numbers.
Improper fraction	A fraction with the numerator larger than the denominator.	$\frac{5}{2}$ and $\frac{102}{51}$ are improper fractions.
Decimal	A way of expressing values of fractions less than 1.	$0.376 = \frac{3}{10} + \frac{7}{100} + \frac{6}{1000} = \frac{376}{1000}$
Decimal equivalent	A number written as a decimal that has the same value as a fraction or percentage.	$\frac{5}{8} = 0.625$
Common factor	A factor that is shared by two or more numbers.	$15 = 3 \times 5 \qquad 35 = 5 \times 7$ 5 is a common factor of 15 and 35.
Cancel	Common factors in the numerator and denominator of a fraction can be 'cancelled'.	$\frac{2}{3}$ of $12 = \frac{2}{3} \times 12^4 = 2 \times 4 = 8$ 3 is a common factor of 12 and 3.
Lowest common denominator	The smallest number into which the denominators of two or more fractions will divide.	The lowest common denominator of $\frac{1}{6}$ and $\frac{1}{8}$ is 24. $\frac{1}{6} = \frac{4}{24}, \frac{1}{8} = \frac{3}{24}$
Ascending	Going up.	1, 2, 3, 4, 5, 6 are in ascending order.
Descending	Going down.	6, 5, 4, 3, 2, 1 are in descending order.
Terminating	A decimal with a definite number of digits.	$\frac{1}{8} = 0.125$
Recurring decimal	A decimal with a repeating pattern that goes on forever.	$\frac{1}{3} = 0.333... = 0.\dot{3}$ $\frac{9}{11} = 0.818181... = 0.\dot{8}\dot{1}$

Review

1 Convert these mixed numbers to improper fractions.

   a $1\frac{2}{5}$

   b $3\frac{4}{7}$

2 Convert these improper fractions to mixed numbers.

   a $\frac{9}{4}$

   b $\frac{11}{6}$

3 Write these decimal numbers as fractions in their simplest form.

   a 0.3

   b 0.25

   c 0.88

   d 0.05

4 Convert these fractions to decimals.

   a $\frac{1}{2}$

   b $\frac{7}{10}$

   c $\frac{2}{100}$

   d $\frac{25}{40}$

5 For each pair of fractions, write which is the larger fraction.

   a $\frac{2}{7}$ and $\frac{1}{5}$

   b $\frac{8}{3}$ and $\frac{21}{8}$

6 Calculate the following fractions of amounts.

   a $\frac{1}{5}$ of 45

   b $\frac{3}{7}$ of 28

   c $1\frac{1}{3}$ of 6

   d $\frac{9}{8}$ of 32

7 Calculate the following percentages of amounts.

   a 25% of 64

   b 40% of 120

   c 15% of 80

   d 110% of 90

8 Calculate

   a $\frac{3}{11} + \frac{4}{11}$

   b $\frac{4}{5} - \frac{1}{10}$

   c $\frac{3}{8} + \frac{2}{3}$

   d $1\frac{2}{5} - \frac{1}{4}$

   e $5\frac{2}{3} - 2\frac{3}{4}$

   f $2\frac{5}{6} + 1\frac{1}{4}$

9 Calculate

   a $4 \times \frac{5}{8}$

   b $\frac{2}{7} \times \frac{1}{5}$

   c $\frac{4}{5} \times \frac{3}{8}$

   d $1\frac{2}{3} \times \frac{5}{6}$

10 Calculate

   a $\frac{4}{5} \div 10$

   b $4\frac{2}{3} \div 7$

   c $5 \div \frac{1}{4}$

   d $4 \div 1\frac{3}{7}$

   e $\frac{5}{6} \div \frac{2}{3}$

   f $5\frac{1}{4} \div \frac{3}{7}$

11 Copy and complete the table to show the conversions between fractions, decimals and percentages. Ensure your fractions are fully simplified.

Fraction	Decimal	Percentage
$\frac{3}{5}$		
	0.01	
		65%
	1.2	

What next?

Score			
	0 – 4		Your knowledge of this topic is still developing. To improve look at MyMaths: 1015, 1016, 1017, 1018, 1019, 1029, 1030, 1031, 1040, 1042, 1046, 1047, 1074, 1075
	5 – 9		You are gaining a secure knowledge of this topic. To improve look at InvisiPens: 05Sa – h
	10 – 11		You have mastered these skills. Well done you are ready to progress! To develop your exam technique look at InvisiPens: 05Aa – e

Assessment 5

Do not use a calculator, with the exception of questions **14** and **16**.

1 Jasmin ordered the following fractions as shown. Reorder them correctly. [4]

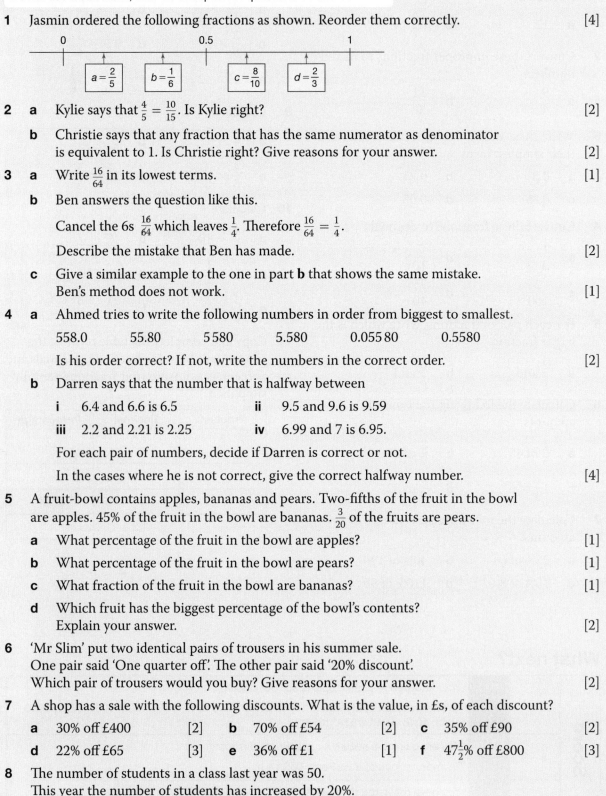

2 a Kylie says that $\frac{4}{5} = \frac{10}{15}$. Is Kylie right? [2]

   b Christie says that any fraction that has the same numerator as denominator
   is equivalent to 1. Is Christie right? Give reasons for your answer. [2]

3 a Write $\frac{16}{64}$ in its lowest terms. [1]

   b Ben answers the question like this.

   Cancel the 6s $\frac{16}{64}$ which leaves $\frac{1}{4}$. Therefore $\frac{16}{64} = \frac{1}{4}$.

   Describe the mistake that Ben has made. [2]

   c Give a similar example to the one in part **b** that shows the same mistake.
   Ben's method does not work. [1]

4 a Ahmed tries to write the following numbers in order from biggest to smallest.

   558.0   55.80   5 580   5.580   0.055 80   0.5580

   Is his order correct? If not, write the numbers in the correct order. [2]

   b Darren says that the number that is halfway between

   i   6.4 and 6.6 is 6.5   **ii**   9.5 and 9.6 is 9.59

   iii 2.2 and 2.21 is 2.25   **iv**   6.99 and 7 is 6.95.

   For each pair of numbers, decide if Darren is correct or not.

   In the cases where he is not correct, give the correct halfway number. [4]

5 A fruit-bowl contains apples, bananas and pears. Two-fifths of the fruit in the bowl
are apples. 45% of the fruit in the bowl are bananas. $\frac{3}{20}$ of the fruits are pears.

   a What percentage of the fruit in the bowl are apples? [1]

   b What percentage of the fruit in the bowl are pears? [1]

   c What fraction of the fruit in the bowl are bananas? [1]

   d Which fruit has the biggest percentage of the bowl's contents?
   Explain your answer. [2]

6 'Mr Slim' put two identical pairs of trousers in his summer sale.
One pair said 'One quarter off'. The other pair said '20% discount'.
Which pair of trousers would you buy? Give reasons for your answer. [2]

7 A shop has a sale with the following discounts. What is the value, in £s, of each discount?

   a 30% off £400   [2]   **b** 70% off £54   [2]   **c** 35% off £90   [2]

   d 22% off £65   [3]   **e** 36% off £1   [1]   **f** $47\frac{1}{2}$% off £800   [3]

8 The number of students in a class last year was 50.
This year the number of students has increased by 20%.
How many students are there this year? [2]

9 Branson asked some friends which e-mail provider they use.

Provider	Number of boys	Number of girls
Air	5	0
Minus-net	1	2
Chat-Chat	3	6
PT	8	8
Redyonder	8	4
	Total 25	Total 20

  a Which provider do 20% of the girls use? [1]

  b Which provider do 20% of the boys use? [1]

  c Which provider do 20% of the total number of pupils use? [2]

  d Branson said 'In my survey, PT was equally popular with both boys and girls.'
Was Branson correct or incorrect?
Give your reasons. [2]

10 **a** What percentage of the word MISSISSIPPI is made up by the letter S? [2]

  b Old MacDougal had a farm.

    He had 21 pigs, 8 lambs, 150 sheep, 8 calves, 27 cows, 35 chickens and 1 bull.

    What percentage of his livestock were chickens? [2]

11 In the 4×100 m relay, the first three runners took the following proportion of the total race time.

First runner 25%   Second runner $\frac{7}{25}$ Third runner 0.35

  a What decimal of the time was taken by the 4th member? [2]

  b Which team member ran the fastest leg? [1]

  c Which team member ran the slowest leg? [1]

12 S. Crumpy has an orchard. The orchard contains $3\frac{3}{4}$ hectares of apple trees.
He needs to treat $\frac{2}{5}$ of the area for disease prevention.
What area does he need to treat? [3]

13 In a football match the goalkeeper kicked the ball from the goal line for $\frac{5}{9}$ of the length of the pitch. Another player then kicked it a further $\frac{7}{20}$ of the way.
If the length of the pitch is 90 yards, how many yards further is the opposite goal line? [4]

14 In her garden, busy Lizzie planted flowers in 27% of the total area.

  a The total area was 540 m². What area did she use for flowers? [2]

  b 36% of the *remainder* of her garden was taken up by the patio.
Find the area of the patio, to the nearest m². [3]

15 In 'Topmarks College', $\frac{8}{11}$ of the students are girls.

Of these girls $\frac{3}{4}$ are brunette and of these brunettes $\frac{5}{9}$ wear earrings.
What fraction of the school students are brunette girls who wear earrings? [2]

16 'FALSEPRINT' film laboratories sell prints in sizes 12.5 cm by 7.5 cm and 15 cm by 10 cm. Their adverts say that their 15 cm by 10 cm prints are more than 50% bigger than the 12.5 cm by 7.5 cm size. Are they correct? [4]

17 Convert the following fractions into recurring decimals.

  a $\frac{1}{3}$ [2]   **b** $\frac{5}{9}$ [2]   **c** $\frac{6}{7}$ [2]   **d** $\frac{7}{11}$ [2]

18 Put the following list in ascending order of size.

  0.34 $\frac{3}{8}$ $33\frac{1}{3}\%$ $\frac{5}{14}$ 0.334   33.3% [5]

Life skills 1: The business plan

Four friends – Abigail, Mike, Juliet and Raheem – are planning to open a new restaurant in their home town of Newton-Maxwell. They have a lot to think about and organise!

They start by creating a business plan. This plan needs to include: market research to understand their potential customers, what their costs and revenues are expected be, how big a loan they could afford to borrow and how any profits should be shared.

Task 1 – Market research

The friends decide to investigate how much people are prepared to pay for a three course meal. They carry out a small pilot survey. This involves stopping people in the street and asking them a few questions.

a Draw a back-to-back stem and leaf diagram to show the ages of the women and men interviewed. Describe what this shows.

b Abigail and Mike think that men will be prepared to pay more for a good meal than women. Do the results back up this theory? Calculate averages to justify your conclusions.

Pilot survey results (15 men and 15 women)

M 24, £33	F 20, £23	F 22, £25
M 37, £36	M 62, £33	F 47, £36
M 47, £35	M 42, £32	F 19, £16
F 52, £32	M 31, £22	M 66, £25
M 26, £24	M 55, £40	F 38, £35
F 18, £20	M 39, £35	M 40, £30
M 21, £21	F 23, £21	M 20, £30
F 58, £40	F 35, £32	F 32, £28
F 22, £30	F 61, £37	F 28, £20
M 23, £27	M 51, £27	F 44, £34

Key   Gender (M/F) Age (years), amount prepared to spend

Task 2 – Projected revenue

The friends are estimating the revenue (money coming in) for their restaurant for the first year. To do this they make some assumptions. Use their assumptions to answer the following questions.

a What would the mean amount paid for a meal be after VAT is added?

b Use the amount paid for a meal excluding VAT to estimate the revenue for the first year.

c Write down a formula for the profit (money left after costs), P, in terms of the other variables listed to the right.

d Find P if G = £40 000, S = £80 000 and C = £50 000.

e Make S the subject of the formula you found in part **c**.

The cost of a meal includes Value Added Tax (VAT) charged at 20%.

Assumptions

- The mean amount paid for a meal, excluding VAT, is £24.42.
- The restaurant is open 364 days a year.
- The mean number of meals sold a day is 25.

Variables

G = cost of food bought by restaurant

S = salary costs C = other costs

R = revenue P = profit

Broad age band	Percentage
16–24	19
25–49	47
50–64	18
65 and over	16

▲ Percentage of the population in Newton-Maxwell in different age bands.

Ownership shares

Abigail $\frac{2}{5}$ Raheem $\frac{1}{4}$

Mike and Juliet both own an equal share of the remainder.

Task 3 – The survey

Following from the pilot survey, the friends do a larger survey of 200 people. They decide their sample should be stratified by age, so that the proportion of people from each age group in their sample is the same as the percentage in the total population of Newton-Maxwell.

How many people from each of the age groups should they include in their sample?

Task 4 – Shares in the business

The friends invested different capital (initial amounts of money) into the business. Based on this, they each own shares that determine the fraction of profit they are entitled to.

a   What fraction of the business do Mike and Juliet each have?

b   Draw a pie chart to show how much of the business each person owns.

c   How much profit would each owner get from a yearly profit of £50 000?

Three year repayment formula

C = amount borrowed A = amount repaid each year

i = annual interest rate (AIR), expressed as a decimal

$$C = A\left(\frac{1}{1 + i} + \frac{1}{(1 + i)^2} + \frac{1}{(1 + i)^3}\right)$$

Task 5 – The business loan

The friends decide to take out a business loan in order to equip the restaurant.

They borrow £C at an annual interest rate (AIR) of i (expressed as a decimal), and repay an amount £A at the end of each year for three years, as shown by the three year repayment formula.

a   What is the value of i if the annual interest rate is 6%?

b   If they decide that the maximum they can afford to repay each year is £5000, how much can they borrow at an AIR of

   i   6%   ii   8%?

c   If they decide that they need to borrow £30 000, how much will each yearly repayment be at an AIR of

   i   6%   ii   8%?

6 Formulae and functions

Introduction

Nurses often use mathematical formulae when they are administering drugs, for example, in converting from one unit to another, calculating the amount of a drug based on somebody's weight, or working out concentrations from solutions.

Working with formulae is a topic within algebra, and is a good example of the practical use of mathematics.

What's the point?

The ability to apply a mathematical formula accurately when calculating a patient's dose of a particular medicine is vitally important.

Objectives

By the end of this chapter you will have learned how to …

- Substitute numerical values into formulae and expressions.
- Rearrange formulae to change the subject.
- Identify inequalities, equations, formulae and identities.
- Expand double brackets.
- Factorise quadratic expressions of the form $x^2 + bx + c$ and the difference of two squares.

Chapter investigation

This grid of numbers uses each of the numbers from 1 to 9.

4	9	2
3	5	7
8	1	6

Every row, column and diagonal adds up to 15. It is called a magic square.

Create your own magic square.

6.1 Substituting into formulae

You can write a **formula** to represent an everyday situation.

- Write the formula in words and then using letters.
- Explain what the letters represent.

The formula to work out the area of a rectangle is

Area = length × width

length

width

You can write this using letter symbols as

$$A = l \times w$$

or $$A = lw$$

In algebra you do not write the × sign.

where A is the area, l is the length and w is the width.

You can use this formula to calculate the area of any rectangle.

l, w and A are called **variables**.

l and w can take **any** values.

The value of A is determined by the values of l and w.

- You can simplify a formula by collecting **like terms**.

Like terms have exactly the same letter.

EXAMPLE

Write a formula for the perimeter of this rectangle.
Simplify your formula as much as possible.

l

w w

l

Perimeter $P = l + w + l + w$

$\quad = l + l + w + w$

$P = 2l + 2w$

If you know the values of l and w, you can **substitute** the values of l and w into the formula.

This means you write the formula replacing l and w with their number values.

EXAMPLE

Use the formula $P = 2l + 2w$ to work out

a the perimeter of a rectangular field with length 20 m and width 8 m

b the perimeter of a table mat with length 25 cm and width 30 cm.

a $l = 20$ m and $w = 8$ m

$P = 2l + 2w$

$P = 2 \times 20 + 2 \times 8$

$\quad = 40 + 16$

$P = 56$ m

b $l = 25$ cm and $w = 30$ cm

$P = 2l + 2w$

$P = 2 \times 25 + 2 \times 30$

$\quad = 50 + 60$

$\quad = 110$ cm

Exercise 6.1S

1 Plain ribbon costs 30p per metre.

Tartan ribbon costs 42p per metre.

a How much does 2 m of plain ribbon cost?

b How much does 3 m of tartan ribbon cost?

c Write a formula connecting:

| cost of ribbon | length of ribbon |

price per metre

2 Employees in a factory are paid by the hour.

Under 18s and adults are paid different hourly rates.

Write a formula for any employee, connecting:

number of hours worked

hourly rate

pay

3 To fix a car, Tony charges £20 per hour for labour. He also charges the cost of any parts used.

Write a formula connecting:

cost of the repair

cost of parts

number of hours worked

hourly rate for labour

4 Write these formulae using letter symbols.

Explain what each letter symbol represents.

a Repair cost = labour cost + parts cost

b Cost of electric cable = price per metre × length in metres

c Monthly cost = annual cost ÷ 12

d Cost of apples = price per kg × number of kg

e Distance in metres = distance in km × 1000

f Length in metres = length in centimetres ÷ 100

5 **a** Write a formula for the perimeter of this equilateral triangle.

Simplify your formula as much as possible.

b Find the perimeter when $x = 4$.

6 **a** Write a formula for the perimeter of this square.

Write your formula in its simplest form.

b Find the perimeter when $y = 7$.

7 Here is a formula: $p = 5m$

Work out the value of p when

a $m = 4$ **b** $m = 3$

c $m = 10$ **d** $m = 2.5$

8 Here is a formula: $y = ax$

Work out the value of y when

a $x = 3$ and $a = 2$

b $x = 7$ and $a = 4$

c $x = 4.5$ and $a = 3$

Work out the value of x when

d $y = 10$ and $a = 2$

e $y = 25$ and $a = 10$

9 For each formula, work out the value of y when $x = 4$ and $c = 6$.

a $y = 3x + c$ **b** $y = 4x - c$

c $y = \dfrac{4c}{x}$ **d** $y = cx + 10$

e $y = x^2$ **f** $y = c^2$

g $y = x^3$ **h** $y = 2x^2 + c$

10 Use the formula $s = \dfrac{d}{t}$ — Be careful with the units.

where s = average speed, d = distance and t = time, to work out the average speed when

a $d = 150$ miles, $t = 3$ hours

b $d = 190$ km, $t = 2$ hours

c $d = 500$ metres, $t = 10$ seconds

d $d = 60$ km, $t = 0.5$ hours

 Q 1158, 1167, 1186, 1187 SEARCH

6.1 Substituting into formulae

- Write the formula in words and then using letters. Explain what the letters represent.
- You can substitute numbers into a formula written using letters.

① Define what letters you will use and write the formula using letters.

② Use your knowledge of substitution.

③ Give your answer in the context of the question.

EXAMPLE

a Write a formula to show your total amount of pocket money P (£s), if you receive £3 per month with an extra £2 for every job (j) you do at home.

b Explain why you can't earn exactly £20 of pocket money in one month.

① Write a formula for the amount of pocket money. Try the situation with various numbers, before putting it into algebra.

a If I don't do any jobs, I get £3.

If I do 4 jobs I get £3 plus 4 × £2, which is £11.

If I do 10 jobs I get £3 plus 10 × £2, which is £23.

If I do j jobs I get £3 plus j × £2.

$P = 3 + 2j$

② Try substituting values for j that will give values close to £20.

b $j = 8, P = 3 + 2 \times 8 = 3 + 16 = £19$

$j = 9, P = 3 + 2 \times 9 = 3 + 18 = £21$

③ The formula only gives odd values for P.

You can't earn exactly £20.

EXAMPLE

Write a formula for the total area of this shape. Let the total area be A.

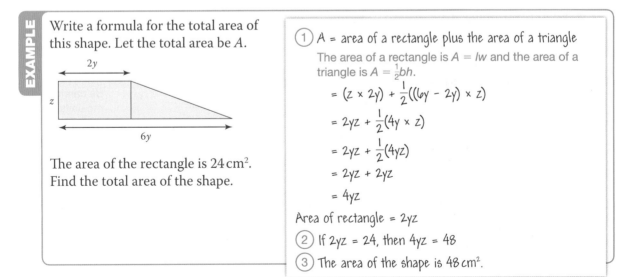

The area of the rectangle is 24 cm^2. Find the total area of the shape.

① A = area of a rectangle plus the area of a triangle

The area of a rectangle is $A = lw$ and the area of a triangle is $A = \frac{1}{2}bh$.

$= (z \times 2y) + \frac{1}{2}((6y - 2y) \times z)$

$= 2yz + \frac{1}{2}(4y \times z)$

$= 2yz + \frac{1}{2}(4yz)$

$= 2yz + 2yz$

$= 4yz$

Area of rectangle = $2yz$

② If $2yz = 24$, then $4yz = 48$

③ The area of the shape is 48 cm^2.

Exercise 6.1A

1 Sareeta is making a row of coins in the High Street for charity.

She collects coins from the public and arranges them like this on the pavement:

 ...

...

...

a What is the value of each vertical column?

b A 10 cm length includes four 10p pieces.

What is the total value of a 10 cm length?

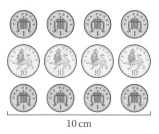

10 cm

c What is the total value of a 1 m length? Give your answer in pounds (£).

d Copy and complete:
Total value of row = _____ × length of row in metres

2 To make a cover for a square cushion, you need two squares of fabric.

Write a formula for the area of fabric (in square metres) needed for a square cushion of side k metres.

3 An electrician charges £35 for each job + £20 per hour.

a Write a formula for the electrician's charge in pounds.

b Use your formula to find the charge for a job that takes 3 hours.

4 The cost of a taxi is £2 for a callout + 60p for each mile.

a Write a formula for the cost of a taxi in pounds.

b Work out the cost for a journey of
  i 5 miles **ii** 15 miles.

5 In Spain a hire car costs €75 plus €35 a day.

a Write a formula for the cost of hiring a car in euros.

b How much does it cost to hire a car for 7 days?

c Louise has a budget of €400. How many days can she hire a car for?

6 Pencils are arranged in rectangles.

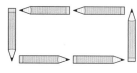

The number of pencils needed to make a rectangle is

2 × number of pencils along the bottom + 2.

a Write this formula using letters.

b Check that your formula gives the correct answer for a rectangle of length 5.

c Work out how many pencils are needed for a rectangle of length 8.

d Ella uses the formula
2 × (number of pencils along the bottom + 1).
Explain why Ella's formula is also correct.

7 Jan uses the formula

$$s = ut + \frac{1}{2}at^2$$

to calculate s when $u = 0$, $t = 2$ and $a = 3$.

Her working is shown on the right.

Jan's answer is **wrong**.

Work out the correct value of s.

Explain which **three** mistakes Jan has made in her working.

$s = 0 \times 3 + \frac{1}{2} \times 2 \times 3^2$
$= 3 + \frac{1}{2} \times 6^2$
$= 3 + \frac{1}{2} \times 36$
$= 3 + 18$
$s = 21$

A function machine has

an **input** → an **operation** → an **output**

the value you put in

what you do to the input

the end value

EXAMPLE

Work out the outputs for these function machines.

a 2 → +5 →

b 12 → ÷3 →

a 2 → +5 → 7

b 12 → ÷3 → 4

> The inverse of 'add' is 'subtract'.
> The inverse of 'multiply' is 'divide'.

● Every operation has an **inverse** operation.
  The inverse operation 'undoes' the operation.

input   output

10 → −4 → 6

10 ← +4 ← 6

● You can work backwards through a function machine using inverse operations.

EXAMPLE

Draw the inverse machines for these function machines.

a 3 → ×2 → 6   **b** 5 → +13 → 18   **c** 10 → ÷2 → 5

a 3 ← ÷2 ← 6   **b** 5 ← −13 ← 18   **c** 10 ← ×2 ← 5

● The **subject** of a formula is the variable before the equals sign.

For example, in $y = mx + c$, y is the subject of the formula.

● You can **rearrange** a formula in order to change its subject.

EXAMPLE

Rearrange these formulae to make x the subject.

a $y = mx + c$

b $y = \dfrac{x}{2} - b$

a $y = mx + c$ Subtract c from both sides.
        $y - c = mx$ Divide both sides by m.
        $\dfrac{y - c}{m} = x \rightarrow x = \dfrac{y - c}{m}$

b $y = \dfrac{x}{2} - b$ Add b to both sides.
        $y + b = \dfrac{x}{2}$ Multiply both sides by 2.
        $2(y + b) = x \rightarrow x = 2(y + b)$

Exercise 6.2S

1 Work out the outputs for these function machines.

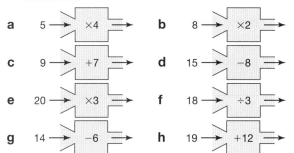

a $5 \rightarrow \boxed{\times 4} \rightarrow$ **b** $8 \rightarrow \boxed{\times 2} \rightarrow$

c $9 \rightarrow \boxed{+7} \rightarrow$ **d** $15 \rightarrow \boxed{-8} \rightarrow$

e $20 \rightarrow \boxed{\times 3} \rightarrow$ **f** $18 \rightarrow \boxed{\div 3} \rightarrow$

g $14 \rightarrow \boxed{-6} \rightarrow$ **h** $19 \rightarrow \boxed{+12} \rightarrow$

2 Copy and complete these function machines.

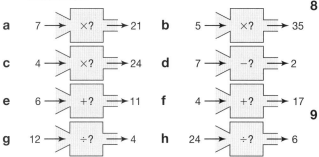

a $7 \rightarrow \boxed{\times ?} \rightarrow 21$ **b** $5 \rightarrow \boxed{\times ?} \rightarrow 35$

c $4 \rightarrow \boxed{\times ?} \rightarrow 24$ **d** $7 \rightarrow \boxed{- ?} \rightarrow 2$

e $6 \rightarrow \boxed{+ ?} \rightarrow 11$ **f** $4 \rightarrow \boxed{+ ?} \rightarrow 17$

g $12 \rightarrow \boxed{\div ?} \rightarrow 4$ **h** $24 \rightarrow \boxed{\div ?} \rightarrow 6$

3 Write the inverse operation for each of these operations.

a $\times 2$ **b** $+4$

c -3 **d** $\div 6$

e -7 **f** $\times 5$

g $\div 2$ **h** $+11$

4 Rearrange these formulae to make a the subject.

a $H = a + 2$ **b** $H = a - 2$

c $H = 2a$ **d** $H = \dfrac{a}{2}$

5 Rearrange each of these formulae to make x the subject.

a $y = x + 3$ **b** $y = x + c$

c $y = x - 4$ **d** $y = x - c$

e $y = 5x$ **f** $y = xc$

g $y = \dfrac{x}{6}$ **h** $y = \dfrac{x}{c}$

6 Rearrange each formula.

a $y = mx + c$, make x the subject

b $v = u + at$, make t the subject

c $y = \dfrac{x}{2} + d$, make x the subject

d $x + 3y = 4$, make y the subject

7 For each of these formulae make t the subject.

a $s = 3t - 6$

b $2x = 5t + 9$

c $12 + 2t = 3x$

8 Rearrange each formula to make y the subject.

a $4x + 6y = 2$

b $3y - 2x = 6$

c $3x - 5y = z$

9 Rearrange each of these formulae to make b the subject.

a $H = b + a^2$ **b** $A = b - a^2$

c $Q = 2b + a^2$ **d** $F = 2b - a^2$

e $M = p^2 + b$ **f** $L = bp^2$

g $T = a^2b$ **h** $W = \dfrac{b}{a^2}$

10 Mo is rearranging the formula $y = 10 - x$ to make x the subject of the formula.

He thinks the answer is $x = y - 10$.

Do you agree with Mo? Explain your answer.

11 Rearrange each of these formulae to make x the subject.

a $y = 5 - x$ **b** $y = 20 - x$

c $y = m - x$ **d** $y = 2b - x$

e $y = s^2 - x$ **f** $y = \sqrt{p} - x$

12 Rearrange this formula to make T the subject.

$$S = \dfrac{D}{T}$$

6.2 Using standard formulae

- You can substitute numbers into a formula written using letters.
- The subject of a formula is the letter on its own on one side of the equals sign.
- You can change the subject of a formula using the balance method.

Inverse operations

$+ \longleftrightarrow -$

$\times \longleftrightarrow \div$

$^2 \longleftrightarrow \sqrt{}$

$y = 2x + 3$

$y - 3 = 2x$

$\dfrac{y - 3}{2} = 2$

HOW TO

1. Define what letters you will use and write the formula using letters.
2. Use your knowledge of substitution or changing the subject of a formula.
3. Give your answer in the context of the question.

EXAMPLE

A plumber charges £25 for a callout and £30 per hour of work.

Write a formula for the number of hours worked.

Use your formula to work out how many hours it would take for the plumber to earn £475.

1. It's easier to find a formula for the total charge first.

Charge in pounds = 25 + 30 × number of hours of work

$$C = 25 + 30h$$

where C = charge in pounds, h = number of hours of work.

2. Change the subject of the formula to make h the subject.

$$C - 25 = 30h$$

$$\frac{c - 25}{30} = h$$

Substitute C = £475 into the formula.

$$h = \frac{475 - 25}{30} = \frac{450}{30} = 15$$

3. The plumber would need to work for 15 hours.

EXAMPLE

Parminder and Quinn are rearranging the formula $M = 4(x + a)$ in order to make x the subject. Their solutions are different but are both correct.

Parminder's solution   Quinn's solution

$$x = \frac{M - 4a}{4} \qquad x = \frac{M}{4} - a$$

If you divide Parminder's answer by 4, you get Quinn's solution.

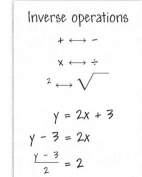

Here are the steps of each of their workings. The steps have been jumbled up.

$M = 4(x + a)$ $M - 4a = 4x$ $x = \frac{M}{4} - a$ $x = \frac{M - 4a}{4}$

$\frac{M}{4} = x + a$ $M = 4x + 4a$ $M = 4(x + a)$

Sort the steps to form Parminder's and Quinn's solutions.

2. Start with the original formula. You could expand the brackets and then change the subject.

$M = 4(x + a)$ $M = 4x + 4a$ $M - 4a = 4x$ $x = \frac{M - 4a}{4}$

3. Parminder's solution.

2. You could divide both sides by 4 and then change the subject.

$M = 4(x + a)$ $\frac{M}{4} = x + a$ $x = \frac{M}{4} - a$

3. Quinn's solution.

Exercise 6.2A

1 Use function machines to solve these 'think of a number' problems.

   a I think of a number and add 12. The answer is 25. What number did I think of?

   b I think of a number and divide it by 4. The answer is 5. What number did I think of?

2 A two-step function machine has input 4.

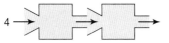

Use any **two** of these function machines

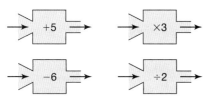

to make a two-step machine that gives the output

   a 3   **b** 7   **c** 17

Make as many different outputs as you can. Draw the function machines you use each time.

3 These patterns are made with pencils.

The formula to work out the number of pencils (P) in a row of n huts is

$$P = 4n + 1$$

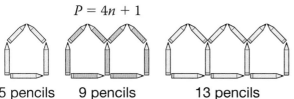

5 pencils   9 pencils   13 pencils

   a Rearrange this formula to make n the subject.

   b Use your formula to work out the number of huts you can make with

      i 37 pencils   **ii** 53 pencils

      iii 229 pencils

4 The formula for calculating the cost in pounds of an electricity bill is

$$C = 17.5 + 0.1u$$

where u is the number of units used.

By rearranging the formula, work out the numbers of units used when the cost is

   a £37.50   **b** £32.50   **c** £74.50

5 An electrician charges £45 for each job plus £15 per hour.

   a Write a formula for the electricians charge in pounds.

   b Use your formula to find the charge of a job that takes 9 hours.

6 The cost of a taxi is £2 for a callout plus 160p for each mile.

   a Write a formula for the cost of a taxi in pounds.

   b Work out the cost for a journey of

      i 5 miles   **ii** 15 miles.

7 In Spain a hire car costs €75 plus €35 a day.

   a Write a formula for the cost of hiring a car in euros.

   b How much does it cost to hire a car for 7 days?

   c Louise paid €495 to hire a car. How many days did she hire it for?

8 James and Sebastian are rearranging the formula $C = a(x - b)$ in order to make x the subject. They both come up with solutions that look different but are, in fact, correct. Can you explain why?

James' solution	Sebastian's solution
$\frac{C}{a} + b = x$	$\frac{C + ab}{a} = x$

9 These are the stages in changing the subject of the formula $c = \dfrac{8(D + k)}{ab}$. Put them in order.

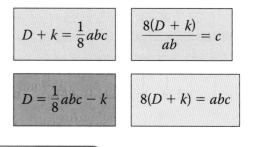

6.3 Equations, identities and functions

In algebra you use letters to represent numbers.

Here are some words you need to know:

- An **expression** is made up of algebraic **terms**. It has no equals sign.

  $2x + 3b$ and $2(l + w)$ are expressions.

- A **formula** is a rule linking two or more variables.

  $P = 2l + 2w$ is a formula for the perimeter of a rectangle.

- An **equation** is only true for particular values of a variable. You can solve the equation to find the solution.

  $x + 4 = 10$ is an equation. Its solution is $x = 6$.

Here are two other useful words:

- An **identity** is true for any values of the variables.

  $a + a + a + a \equiv 4a$ is an identity.

- A **function** links two variables. When you know one, you can work out the other.

  $x \rightarrow 3x + 2$ or $y = 3x + 2$ is a function.

The sign $\equiv$ means identically equal to.

EXAMPLE

Decide if each of these statements is an identity, an equation or a formula.

a $x^3 - 2x = x(x^2 - 2)$ **b** $5x - 1 = 2x + 3$ **c** $A = bh$

a Expand the right-hand side: $x(x^2 - 2) = x^3 - 2x$
    $x^3 - 2x$ = left-hand side so the statement is an identity.

b $5x - 3 = 2x + 3$

    $3x - 3 = 3$ This statement is only true for one value of x.

      $3x = 6$

       $x = 2$

    $5x - 3 = 2x + 3$ is an equation.

c $A = bh$ is a formula showing the relationship between the length of a pair of parallel sides, the distance between them and the area of a parallelogram.

If you know the values of b and h, you can find A from the formula or b given A and h or h given b and A.

Exercise 6.3S

1 Multiply out the brackets and write each of these as an identity.
The first one has been done for you.

a $7(x + 4) \equiv 7 \times x + 7 \times 4 \equiv 7x + 28$

b $3(x - 2)$

c $2(3 + x)$

d $5(2 - x)$

2 Copy and complete these identities by factorising.

a $8m + 4 \equiv 4 (\quad)$

b $12n - 9 \equiv \square (\quad)$

c $15p + 55 \equiv \square (\quad)$

d $q^2 + 2q \equiv q (\quad)$

e $16r - 28 \equiv \square (\quad)$

f $4pq - 10q \equiv \square (\quad)$

3 Copy the table.

Expressions	Equations	Functions	Formulae	Identities

Write these under the correct heading in your table.

a $a + bc = bc + a$ **b** $y = 3x + 2$

c $a + bc = d$ **d** $3a + 5 = -4$

e $4xy + 3x - z$ **f** $E = mc^2$

g $s = ut$ **h** $x - 1 = y$

4 Copy these statements and say whether they are identities, equations or formulae.

a $c = 2\pi r$	**b** $3x(x + 1) = 3x^2 + 3x$
c $3x + 1 = 10$	**d** $y \times y = y^2$
e $2x + 5 = 3 - 7x$	**f** $A = \frac{1}{2}(a + b)h$
g $a^2 + b^2 = c^2$	**h** $20 - x = -(x - 20)$
i $2x^2 = 50$	

5 Prove that these are identities.

a $4(a + 2) + 2(a + 1) \equiv 6a + 10$

b $3(x + 2) + 4(x - 1) \equiv 7x + 2$

c $5(y - 2) + 3(y - 3) \equiv 8y - 19$

d $y(y + 3) + 2(y + 3) \equiv y^2 + 5y + 6$

e $x(x - 4) + x(x + 2) \equiv 2x^2 - 2x$

6 Find the values of a and b such that

a $2(x + 2) + 5(x + 1) \equiv ax + b$

b $3(x - 2) + 4(x + 3) \equiv ax + b$

c $5(y + a) + 3(y - b) \equiv 8y - 19$

d $y(y + a) + 2(y + b) \equiv y^2 + 3y + 4$

e $x(x - 4) + 2x(x - 3) \equiv ax^2 - bx$

7 Are these identities are true or false? Explain your answer.

a $(a + 2)(a + 5) \equiv a^2 + 7a + 7$

b $(x + 3)(x + 4) \equiv x^2 + 7x + 7$

c $(b + 2)(b + 6) \equiv b^2 + 8b + 12$

d $(y - 2)(y + 3) \equiv y^2 + y - 6$

e $(y + 2)(y - 3) \equiv y^2 + y - 6$

f $(p - 2)(p - 3) \equiv p^2 - 5p - 6$

8 Write an identity for each of these expressions.

a $(x + y)^2$ **b** $(x - y)^2$

9 Prove that these are identities.

a $(x - 3)(x + 3) \equiv x^2 - 9$

b $(y + 4)(y - 4) \equiv y^2 - 16$

c $(5 - a)(5 + a) \equiv 25 - a^2$

d $(a + b)(a - b) \equiv a^2 - b^2$

*10 Find the values of a and b such that $x^2 + 6x - 2 \equiv (x + a)^2 - b$

11 The perimeter of a rectangle can be found using the formula $P = 2a + 2b$ or $P = 2(a + b)$. Prove that these formulae are identical.

Q 1155, 1247, 1942   SEARCH

6.3 Equations, identities and functions

- An expression is a collection of algebraic terms.
- A formula is a rule linking two or more variables.
- An equation is only true for particular values (the solution).
- A function links two variables.
- An identity is true for any values of the variables.

You can use algebra to prove identities.

- To prove a statement is true, you need to generalise it to all possible examples.

For example, show that the sum of two consecutive integers is always odd.

Let the consecutive integers be n and $n + 1$: Sum $= n + n + 1 = 2n + 1$.

$2n + 1$ is always odd since it is one more ($+1$) than a multiple of 2 ($2n$).

> To disprove the statement 'All cube numbers are even', you could use the counter-example 125 ($=5^3$) is odd.

HOW TO

① To demonstrate that a statement is true, you find examples that fit the statement.

② To **prove** that a statement is true you can use algebra to generalise it to all possible examples.

Some useful algebraic generalisations:	
Even numbers	$2n$
Odd numbers	$2n + 1$
Consecutive even	$2n, 2n + 2$
Consecutive odd	$2n + 1, 2n + 3$

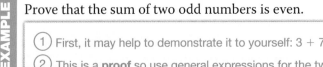

EXAMPLE

Prove that the sum of two odd numbers is even.

① First, it may help to demonstrate it to yourself: $3 + 7 = 10, 9 + 11 = 20, 1 + 51 = 52$ etc.

② This is a **proof** so use general expressions for the two odd numbers.

$2m + 1$ and $2n + 1$ are odd numbers. (They are each 1 more than a multiple of 2.)

$(2m + 1) + (2n + 1) = 2m + 2n + 2$ or $2(m + n + 1)$ by factorisation.

$2(m + n + 1)$ is even because it is a number multiplied by 2.

- To show that a statement is false, you can use a **counter-example**.

EXAMPLE

$y = 2x^2 + 11$ The value of y is prime when $x = 0, 1, 2$ or 3. The following statement is *not* true:

'$y = 2x^2 + 11$ is *always* a prime number when x is an integer'.

Show that the statement is not true.

Try different values of x until you find a counter-example.

For $x = 11$, $y = 2x^2 + 11 = 2 \times 11^2 + 11 = 253$

253 is not prime, since $253 = 23 \times 11$.

Therefore the statement is not true.

Algebra   Formulae and functions

Exercise 6.3A

1 For each of the following diagrams

   i create an identity

   ii write an equation.

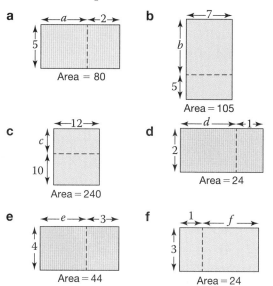

a Area = 80

b Area = 105

c Area = 240

d Area = 24

e Area = 44

f Area = 24

2 Copy and complete these expressions to create identities.

   a $3(x + 4) \equiv 3x + \square$

   b $9(x - 6) \equiv \square x - 54$

   c $6(x - 12) \equiv \square x - \square$

   d $\square(x + 4) \equiv \square x + 16$

   e $3(x + \square) \equiv \square x + 33$

   f $\square(x + \square) \equiv 5x + 35$

3 Copy and complete these expressions to create identities.

   a $x(2x + \square) \equiv \square x^2 + 6x$

   b $\square x(3x + 2) \equiv 12x^2 + \square x$

   c $\square x(7x - \square) \equiv 14x^2 - 16x$

   d $\square x(\square x - 5) \equiv 16x^2 - 20x$

   e $6(2x - \square) + 3(\square x - 4)$

       $\equiv 15x - 24$

   f $7(7x - \square) + 2x(\square x + 9)$

       $\equiv 6x^2 + \square x - 56$

4 Find a counter-example to show that each of these statements is untrue.

   a When you subtract 7 from a number, the answer is always odd.

   b When you square a number, the answer is always even.

   c When you treble a prime number, the answer is always odd.

   d When you find the product of two consecutive numbers, the answer is always odd.

5 The sum of any five consecutive integers is always a multiple of 5.

   a Demonstrate, with a few examples of your own, that this statement is true.

   b By letting the numbers be $n, n + 1$, $n + 2, n + 3$ and $n + 4$, prove the statement is true.

6 Repeat question **5** for these statements.

   a The sum of two even numbers is even.

   b Squaring an even number gives a number in the four times table.

   c The sum of an odd number and an even number is always odd.

   d If two consecutive numbers are multiplied, and the smaller number is subtracted from the result, you always get a square number.

7 a Can you find a counter-example to this statement?

> Squaring a number will always give you a value greater than the number you started with.

   b What range of values does not support this statement?

8 Decide whether each of these statements is always true, sometimes true or never true.

   a An equation is a formula.

   b A formula is an equation.

   c A formula is a function.

   d A function is a formula.

   e An inequality is an equation.

   f An inequality is a function.

   g An equation uses the letter 'x'.

   h A formula is written using algebra.

   Q 1155, 1247, 1942   SEARCH

6.4 Expanding and factorising 2

You can **expand** a double bracket in algebra by multiplying pairs of terms.

Each term in the first bracket multiplies each term in the second bracket:

F... Firsts
O ... Outers
I ... Inners
L ... Lasts

$(p+7)$ $(p+3)$ $\longrightarrow$ $p^2+3p+7p+21$ $\longrightarrow$ $p^2+10p+21$

$p^2 + 10p + 21$ is the product of $(p + 7)$ and $(p + 3)$.

EXAMPLE

Expand and simplify
$(x + 3)(x + 4)$

$(x + 3)(x + 4) = x^2 + 4x + 3x + 12$
$= x^2 + 7x + 12$

F: $x \times x = x^2$
O: $x \times 4 = 4x$
I: $3 \times x = 3x$
L: $3 \times 4 = 12$

EXAMPLE

Expand and simplify
a $(x + 3)(2x - 2)$
b $(3x + 2)^2$

a $(x + 3)(2x - 2) = 2x^2 - 2x + 6x - 6$
$= 2x^2 + 4x - 6$

b $(3x + 2)^2 = (3x + 2)(3x + 2)$
$= 9x^2 + 6x + 6x + 4$
$= 9x^2 + 12x + 4$

Use the rules for multiplying negative terms:
O: $x \times -2 = -2x$
L: $3 \times -2 = -6$

• The two numbers in the brackets **multiply** to give the number at the end and **add** to give the number of xs.

You can use this pattern to help you factorise a quadratic expression.
You can check your answer by expanding the brackets.

EXAMPLE

a Factorise $x^2 + 8x + 15$.　**b** Factorise $x^2 - 7x - 18$.
c Factorise $x^2 - 16$.　***d** Factorise $9x^2 - 16$.

It can help to write out all the factor pairs. Consider the factor pairs of −18:
−1 and 18　−18 and 1
−3 and 6　−6 and 3
−2 and 9　−9 and 2

a Look for two numbers that multiply to give +15 and add to give +8. These are +3 and +5.
$x^2 + 8x + 15 = (x + 3)(x + 5)$

b Look for two numbers that multiply to give −18 and add to give −7. The two numbers are −9 and +2.
$x^2 - 7x - 18 = (x - 9)(x + 2)$

c Look for two numbers that multiply to give −16 and add to give 0. These are +4 and −4.
$x^2 - 16 = x^2 + 0x - 16 = (x + 4)(x - 4)$

*d $9x^2 - 16 = 9x^2 + 0x - 16 = (3x + 4)(3x - 4)$

Expanding the brackets gives $9x^2 + 16 + 12x - 12x = 9x^2 - 16$

Algebra　Formulae and functions

Exercise 6.4S

1 Expand and simplify

a $(x + 2)(x + 3)$

b $(p + 5)(p + 6)$

c $(w + 1)(w + 4)$

d $(c + 5)^2$

e $(x + 4)(x - 2)$

f $(y - 2)(y + 7)$

g $(t + 6)(t - 2)$

h $(x - 2)(x - 5)$

i $(y - 4)(y - 10)$

j $(w - 1)(w - 2)$

k $(p - 5)^2$

l $(q - 12)^2$

2 Expand and simplify

a $(2x + 1)(3x + 7)$

b $(5p + 2)(2p + 3)$

c $(3y + 4)(2y + 1)$

d $(2y + 6)^2$

e $(5t - 4)(2t + 4)$

f $(5w - 1)(3w + 9)$

g $(2x + 2y)(3x - 3y)$

h $(3m - 4)^2$

i $(2p + 5q)(3p - 8q)$

j $(2m - 3n)^2$

3 Expand and simplify

a $(x + 1)(x - 1)$

b $(5x - 1)(5x + 1)$

c $(2x + 3)(2x - 3)$

d $(x + y)(x - y)$

4 Expand and simplify

a $(2x - 8)(x + 3)$

b $(3p + 2)(4p + 5)$

c $(3m - 7)(2m - 6)$

d $(5y - 9)(2y + 7)$

e $(3t - 2)^2$

f $(x + 4)(x - 6) + (x + 3)^2$

g $(4 + 8b)(2 - 3b) - (3 - b)^2$

5 Factorise each of these into double brackets.

a $x^2 + 6x + 8$

b $x^2 + 10x + 21$

c $x^2 + 11x + 28$

d $x^2 + 11x + 24$

e $x^2 - 8x + 12$

f $x^2 - 9x + 18$

g $x^2 - 13x + 36$

h $x^2 + x - 12$

i $x^2 - 2x - 35$

j $x^2 + 6x - 27$

k $x^2 - 14x - 32$

l $x^2 + 18x - 40$

6 Factorise each of these expressions.

a $x^2 + 6x - 72$

b $x^2 - 10x - 24$

c $x^2 - 20x + 75$

d $x^2 + 12x - 64$

e $x^2 - 64$

f $x^2 - 29x + 100$

7 Factorise fully each of these expressions.

a $x^2 - 4$

b $4x^2 - 1$

c $16x^2 - 9$

d $a^2 - b^2$

e $100x^2 - 25$

f $p^4 - q^4$

> These are examples of the Difference Of Two Squares (DOTS).

8 Decide if each of these are single or double bracket factorisations. Factorise each fully.

a $x^2 + 21x + 38$

b $5x^2 + 5x + xy$

c $x^2 + 22x + 121$

d $x^2 + 7x - 18$

e $33 + p^2 + 14p$

f $2x^2 + 3xy$

Q 1150, 1151, 1157   SEARCH

6.4 Expanding and factorising 2

RECAP

To expand double brackets, you multiply each term in the second bracket by each term in the first bracket.

F ...	Firsts
O ...	Outers
I ...	Inners
L ...	Lasts

$$(2x+7) \quad (3x-4) = 6x^2 - 8x + 21x - 28$$
$$= 6x^2 + 13x - 28$$

To factorise into double brackets, look for two numbers that add to give the coefficient of x and multiply to give the constant.

EXPAND

$$(x + 4)(x + 7) \quad \Longrightarrow \quad x^2 + 11x + 28$$
$$4 + 7 \quad 4 \times 7$$

FACTORISE

> *Quadratic* expressions often factorise into double brackets.

HOW TO

① Read the question carefully. Give any unknown values a letter.
Decide whether to
expand   ...or...   factorise into double brackets.

② Collect like terms   ...or...   check your answer by expanding.

③ Use your expansion or factorisation to answer the question.

EXAMPLE

Marina arranges a group of square shaped tiles to make a rectangle.
The rectangle is $x + 5$ tiles wide and $x - 1$ tiles long.
Show that adding 9 more tiles will let Marina make a square.

① Sketch a diagram:

x+5

x-1

Area of rectangle = length × width
$$A = (x + 5)(x - 1)$$
$$= x^2 - x + 5x - 5 \quad ② \text{ Expand and collect like terms.}$$
$$A = x^2 + 4x - 5$$

③ Adding 9 tiles: $x^2 + 4x - 5 + 9 = x^2 + 4x + 4$
$$= (x + 2)(x + 2)$$
$$= (x + 2)^2$$

Factorise by finding two numbers that add to 4 and multiply to give 4.

Adding 9 tiles will make a square.

EXAMPLE

a Expand $(x + 6)(x - 6)$.

b Hence calculate 106×94 without using a calculator.

a ① Expand using FOIL. $(x + 6)(x - 6) = x^2 - 6x + 6x - 36$
② Collect like terms. $= x^2 - 36$

b $106 \times 94 = (100 + 6)(100 - 6)$
$$= 100^2 - 36 \quad ③ \text{ Use } (x + 6)(x - 6) = x^2 - 36.$$
$$= 10\,000 - 36$$
$$= 9964$$

Algebra Formulae and functions

Exercise 6.4A

1 Write an expression for the areas of these shapes.

a
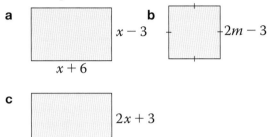
$x + 6$

b $2m - 3$

c
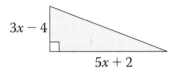
$2x + 3$

$3x - 1$

d Find an expression for the area of the triangle.

$3x - 4$

$5x + 2$

2 a Write an expression for the perimeter of this triangle.

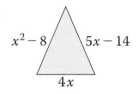
$x^2 - 8$ $5x - 14$

$4x$

b Factorise your expression.

c Explain why x cannot equal 2.

3 Charlie arranges a group of square shaped tiles to make a rectangle.
The rectangle is $x + 2$ tiles long and $x + 4$ tiles wide.
Show that adding one more tile will let Charlie make a square.

$x + 2$

$x + 4$

4 Janelle is trying to factorise $x^2 + 5x + 6$.
She says

1 and 5 multiply to give 5   1 and 5 add to give 6
$x^2 + 5x + 6 = (x + 1)(x + 5)$

a Explain why Janelle is wrong.

4 b Fill in the blanks so that the statement is correct.

□ and □ multiply to give □   □ and □ add to give □
$x^2 + 5x + 6 = (x + □)(x + □)$

5 a Expand $(a + b)^2$.

b Hence, or otherwise, calculate
$1.32^2 + 2 \times 1.32 \times 2.68 + 2.68^2$

c Write another calculation that you could work out using this expansion.

6 Factorise $2.3^2 + 2 \times 2.3 \times 1.7 + 1.7^2$ and use this to show that the calculation results in 16.

7 Copy and fill in the missing values.

a $(x + □)(x + 6) = x^2 + 9x + □$

b $(x + 4)(x - □) = x^2 + □x - 8$

c $(x + 1)(x + 2) + (x + 4)(x - □)$
$= □x^2 + 3x - □$

8 Find the mean of these three quadratic expressions.

$(x - 9)^2$ $2(x + 3)^2$ $3(x + 1)^2$

9 In a multiplication pyramid, you multiply the two numbers directly below to get the number above.

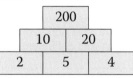

Complete this multiplication pyramid.

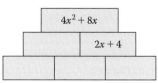

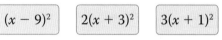

Summary

Checkout
You should now be able to...

	Test it Questions
✔ Substitute numerical values into formulae and expressions.	1 – 3
✔ Rearrange formulae to change the subject.	4
✔ Identify inequalities, equations, formulae and identities.	5, 6
✔ Expand double brackets.	7
✔ Factorise quadratic expressions of the form $x^2 + bx + c$ and the difference of two squares.	8 – 10

Language	Meaning	Example
Variable	A letter used to represent a number.	$3x + 4$ x is the variable.
Like terms	Terms that contain exactly the same variables and exactly the same powers.	$3x^2$ and $5x^2$ are like terms. $3x^2$ and $5x$ are not like terms.
Function machine	A diagram that indicates the order in which operations have to be done.	
Input	The value put in to a function.	$3 \rightarrow \boxed{\times 2} \rightarrow 6$ Input = 3 Operation = ×2 Output = 6
Output	The end value of a function after the operations have been applied.	
Operation	An operation is a rule for processing numbers.	
Inverse	The inverse operation reverses the effect of the original operation.	The inverse of $+4$ is -4. The inverse of $\times 2$ is $\div 2$.
Subject	The variable before the equals sign in a formula.	$V = IR$ V is the subject of the formula.
Rearrange	Rewrite an equation or formula as an equivalent version with a different variable as the subject.	$V = IR$ rearranged to make I the subject of the formula is $I = \dfrac{V}{R}$
Identity	An equation that is true for every possible value.	$\dfrac{a}{4} \equiv 0.25 \times a$
Function	A function is a rule that links each input value with one output value.	$f(x) = 2x$
Expand	Remove the brackets in an expression by multiplying.	$(3x +1)(2x - 3) = 6x^2 - 9x + 2x - 3$ $= 6x^2 - 7x - 3$
Factorise	Find common factors in an expression and write it using brackets; the reverse of expanding.	$x^2 + 5x = x(x + 5)$ $x^2 - 6x + 5 = (x - 1)(x - 5)$
Quadratic	A quadratic expression contains a square term such as x^2 as the highest power.	$6x^2 - 7x - 3$

Review

1 This formula gives the relationship between density, D, mass, m, and volume, v.

$$D = \frac{m}{v}$$

Calculate the density when

   a $m = 15$ g and $v = 5$ cm^3

   b $m = 25$ g and $v = 2$ cm^3.

2 Use the formula $A = \frac{1}{2}bh$ to calculate A when

   a $b = 8$ and $h = 4$

   b $b = 5$ and $h = 9$.

3 Devi earns £x per hour, Jo earns twice as much as Devi and Alice earns £y less than Devi.

   a   Write an expression for the amount earned by

      i   Jo   **ii**   Alice.

   b   If Jo earns £30 per hour, how much does Devi earn per hour?

4 Rearrange each formula to make A the subject.

   a $3 + A = b$ **b** $2A = d$

   c $5A - c = F$ **d** $\dfrac{A + h}{2} = J$

   e $A^2 - 2K = L$ **f** $b = \dfrac{2}{A}$

5

> $4z + 2$ $3b \times 4b \equiv 12b^2$
>
> $F = ma$ $2y + 3 = 7$
>
> $y = 3x + 4$

Give an example of each of these from the box.

   a   Equation   **b**   Identity

   c   Expression   **d**   Formula

   e   Function

6 Use algebra to show that this identity is true.

$$2(3x + 1) - 4 \equiv 6x - 2$$

7 Expand the brackets and simplify these expressions.

   a $(x + 3)(x + 5)$ **b** $(x - 6)(x - 2)$

   c $(x - 7)(x + 4)$ **d** $(2x + 5)(3x - 1)$

8 Factorise these quadratic expressions.

   a $x^2 + 5x$ **b** $12x^2 - 3x$

9 Factorise these quadratic expressions.

   a $x^2 + 5x + 4$ **b** $x^2 - 7x + 6$

   c $x^2 - 2x - 8$ **d** $x^2 + 3x - 10$

10 Factorise these quadratic expressions.

   a $x^2 - 36$ **b** $4x^2 - 25$

What next?

Score			
	0 – 4		Your knowledge of this topic is still developing.
			To improve look at MyMaths: 1150, 1151, 1155, 1157, 1158, 1159, 1167, 1171, 1186, 1187, 1247
	5 – 8		You are gaining a secure knowledge of this topic.
			To improve look at InvisiPens: 06Sa – l
	9 – 10		You have mastered these skills. Well done you are ready to progress!
			To develop your exam technique look at InvisiPens: 06Aa – d

Assessment 6

1 **a** Amy uses this rule to work out how far away a thunderstorm is.
Count the number of seconds, t, between the time you see the lightning and the time you hear the thunder. Divide the number by 5 to work out the distance, m, in miles.
Write down a formula connecting m and t. [1]

  b Use your formula to find

   i how far away a thunderstorm is when $t = 15$ seconds [1]

   ii how far away a thunderstorm is when $t = 2\frac{1}{2}$ seconds [1]

   iii how long the time would be for a storm $2\frac{1}{2}$ miles away. [1]

2 Jamie uses 450 g raspberries and 550 g sugar to make 1 kg raspberry jam.

  a Write a formula connecting J, the mass of jam, R, the mass of raspberries and S, the mass of sugar. [1]

  b How much fruit does Jamie need to make 4 kg of jam? [2]

  c How much sugar does Jamie need to make 6 kg of jam? [2]

  d Jamie made 9 kg jam. How much fruit and sugar did he use? [2]

  e Jamie picked 2.25 kg raspberries from his garden and had enough sugar.
How much jam could he make? [2]

3 Mr and Mrs Perfectparents use this formula to work out how many hours of sleep, s their children need depending on their age a (years): $s = 15 - 0.75a$.

  a Find s when **i** $a = 2$ [2] **ii** $a = 10$. [1]

  b Find a when **i** $s = 6$ [3] **ii** $s = 9$. [1]

  c Use the formula to find out how much sleep an 18 year old needs.
Is this a sensible formula? Give reasons for your answer. [2]

4 David uses the formula $a^2 = b$. He works out

  a the value of b when $a = 7$. He says the answer is $b = 49$. [1]

  b the value of a when $b = 121$. He says the answer is $a = 11$. [2]

  Is David correct? If he is incorrect, work out the correct answer.

5 **a** Carlo tried to expand and simplify these expressions.
Is Carlo correct? Show your workings.

   i $(p - 4)(p - 7) = p^2 + 28$ [2] **ii** $(v + 9)(v - 7) + (4 - 5v)^2 = 6v^2 - 47$ [5]

  b Carlo then tried to factorise these expressions.
Is Carlo correct? Show your workings.

   i $z^2 + 13z + 36 = (z + 4)(z + 9)$ [2] **ii** $v^2 - 100 = (v - 10)^2$ [2]

6 **a** Write down a formula to find the number of days, D, there are in W weeks. [1]

  b Bob the builder's digger is broken and he needs to hire a new one.
The hire firm charges a fixed charge of £50 and £250 for each day Bob uses it.
Write a formula to show how much it costs Bob to hire a digger for D days. [1]

7 'TextUnending' has a pay as you go phone that charges

- 15p per minute between peak times of 09:00 and 20:00
- 12p per minute at other times, called off-peak
- 14p per text at all times.

'TextUnending' rounds all extra seconds up to the next minute.

   a Using the letters P for the cost (in pence) and m for the time in minutes, write down a formula to work out the cost of

      i a call at peak times   [1]   **ii** a call at off-peak times   [1]

      iii making n texts.   [1]

   b **i** Hugo made a call for 3 min 20 s at 13:15 and sent 8 text messages.

      How much was he charged?   [4]

     ii Albert made one call for 6 min 10 s at 08:01 and another for 14 min 1 s at 09:01.

      He sent 17 text messages in total. How much was he charged?   [5]

   c Cheng made 4 off-peak calls, lasting 2 min 43 s, 7 min 17 s, 23 min 42 s and 4 min 55 s.

     She also sent 25 text messages. Work out how much she paid.   [3]

8 The formula $V = IR$ is used in electricity to work out the voltage, V volts, when an electrical current, I amps, flows through a wire with resistance R ohms. Find V when $I = (2x - 1)$ amps and $R = (3x + 5)$ ohms.   [3]

9 When a body starts with a speed u, and accelerates at a rate of 6 m/s² for t seconds, the distance it has travelled, s metres, is given by the formula $s = ut + 3t^2$. Find s when $u = (4 + 3z)$ and $t = (2 - 5z)$.   [6]

10 Selina makes the following statements. For each statement either show that it is always true or find an example to show that it is false.

   a An odd number times an even number is always odd.   [1]

   b Prime numbers have one factor.   [1]

   c The sum of two even numbers is always even.   [2]

   d Any number squared is more than 0.   [1]

   e Two prime numbers multiplied together are always odd.   [1]

   f The sum of three consecutive even numbers is always divisible by 6.   [2]

   g The square of any number is never a prime number.   [2]

11 Work out the rule that turns each input in these function machines into its corresponding output.

   a 6, 10, 12 → → → 25, 41, 49   [2]   **b** 6, 15, 36 → → → 0, 3, 10   [2]

12 The formula for the length of skid, S m, for a vehicle travelling at v km/h, is $S = \dfrac{v^2}{170}$.

   a A car skids while travelling at 100 km/h. How long is the skid?   [2]

   b A car and a lorry are travelling head-on towards each other on a wet narrow minor road. They see the danger looming and start to skid at the same instant. The car is travelling at 70 km/h and the lorry at 52 km/h. Calculate the minimum distance they were apart if they just stop in time.   [4]

   c Rearrange the formula to make v the subject. Hence find the speed of a vehicle which skidded for 60 m. Give your answer to the nearest km/h.   [2]

Revision 1

1 A pile of 80 identical boxes weigh 10 kg. How much do 45 of these boxes weigh? Give your answer in kilograms. [3]

2 Jenni gets the bus to work. The fare is £3.50 for a return journey. She works Monday to Wednesday, Friday and Saturday. Would Jenni save money if she bought a weekly ticket costing £15? Give reasons for your answer. [3]

3 Oliver is finding the 'COOL' value of numbers. To find the 'COOL' value of a number

   1 Square the number

   2 If the squared number has more than 1 digit, add the digits together

   3 Repeat step **2** until you get a single digit. This is the 'COOL' value of the number.

  a Find the 'COOL' values for the numbers 9 and 28. [3]

  b There are only 4 'COOL' values for all the numbers. Find all of them. [3]

  c Which numbers are the same as their 'COOL' values? [2]

  d There are two consecutive numbers between 10 and 15 with the same 'COOL' value. What are they? [3]

  e 16 has a 'COOL' value of 4. What is the next number after 16 to have a 'COOL' value of 4? [2]

4 The numbers 0 to 99 are in this grid.

0	1	2	3	4	5	6	7	8	9
10	11	12	13	14	15	16	17	18	19
20	21	22	23	24	25	26	....	....	....
30	31	32	33	34	....	....	....	....	....
40	41	42	43	....	....	....	....	....	....

4 Five cells are coloured to form a T shape. The one shown is T_2.

$T_2 = 1 + 2 + 3 + 12 + 22 = 40$.

  a Work out T_{13}. [2]

  b Work out T_{65}. [1]

  c Write down an expression for T_x. Simplify your answer. [3]

  d Find x when $T_x = 105$. [2]

  e Is the value of T_{20} possible? Give reasons for your answer. [1]

  f Find the value of x that gives the largest value of T_x. Show your workings. [3]

5 *PQ* and *QR* are a pair of stepladders standing on a horizontal floor. *PQ* = *PR*.

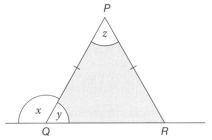

  a Angle x is 130°. Find the values of y and z. Give reasons for your answers. [4]

  b The ladders slip on the floor until angle z becomes 116°. Find the new values of x and y. [3]

  c What sort of triangle would *PQR* be if x was 120°? Give reasons for your answer. [2]

6 The data shows the number of nights and number of guests staying in a hotel.

Number of nights (*x*)	1	2	3	4	5	6	7	8	9
Number of guests (*f*)	4	9	3	6	8	11	8	4	2

  a Calculate the total number of guests that stayed in the hotel. [1]

  b Calculate the mean and median number of nights that guests stayed. [5]

  c Find the range. [1]

7 a A rectangle has sides $3y - 2$ cm and $2y + 6$ cm. Work out its perimeter. [3]

b A cuboid has sides $3y - 2$ cm, 6 cm and 8 cm. Work out its surface area. [4]

c Two sides of an equilateral triangle are labelled $3x + 2$. The other side is 35 cm. Write down an equation in x and solve it to find x. [3]

8 *ABCDEFGH* is a regular octagon.

a Sara makes the following statements. For each statement decide if Sara is correct or not. Fully describe the shape in each case.

   i *HBD* is an isosceles triangle. [2]

   ii *HBDF* is a trapezium. [2]

   iii *HBCG* is a parallelogram. [3]

b Name any triangle

   i congruent to *HOP* [1]

   ii similar to *HOP* [1]

   iii with 1 line of symmetry [1]

   iv with 1 line parallel to *AC*. [1]

c Find the value of

   i angle *ABC* [3]

   ii angle *BDF* [1]

   iii angle *OAB* [1]

   iv angle *FHD* [1]

   v angle *POF* [1]

   vi angle *HOC*. [1]

9 Sunita has £7.50. She spends $\frac{1}{5}$ on a coffee and $\frac{5}{12}$ of what she has left on a sandwich.

a How much did Sunita spend on her coffee? [1]

b How much did she spend on her sandwich? [3]

c How much did she have left at the end? [1]

10 There are 450 campers on a campsite, 24% are teenagers, 81 campers are over 65.

a How many are teenagers? [2]

b What percentage of the total number of campers are over 65? [2]

c How many campers were neither teenagers nor over 65? [1]

11 Ally the chemist is diluting sulphuric acid. Flask A contains 80 ml of sulphuric acid and flask B contains 100 ml of water. Ally transfers 20 ml of water from flask B into flask A.

a What fraction of flask A is water? [1]

b What fraction of flask A is acid? [1]

The contents of flask A is mixed together and a 20 ml spoonful is transferred to flask B.

c How many ml of acid and water does the spoon hold? [2]

d Fill in the table of volumes for each flask. [4]

e Use the table to write the ratio of acid to water in

Flask A	Acid	____ ml
	Water	____ ml
Flask B	Acid	____ ml
	Water	____ ml

   i flask A [1]

   ii flask B. [1]

12 The formula for the radius, r, of a sphere given its surface area, A, is given by

$$r = \sqrt{\frac{A}{4\pi}}.$$

a Find r when $A = 65 \text{ cm}^2$. [2]

b Write down the value of A which gives a radius of 1. Show your working. [3]

c Rearrange the formula to make A the subject. [3]

d Use this formula to find the value of A when $r = 15.4$ cm. [2]

13 The bar and pie chart show the same information. Complete both charts. [4]

7 Working in 2D

Introduction

Self-similarity is the property whereby an entire shape is mathematically similar to a part of itself. What this means is that if you 'zoom in' on a small corner of the shape, you get an exact replica of the original shape itself. Self-similarity is used in fractal images, like the one you can see here, and it has real-world use in describing the structure of coastlines, as well as the natural growth of plants such as ferns, and the formation of crystals and snowflakes.

What's the point?

The real world, being mathematically untidy, is never exactly self-similar. However self-similarity provides a highly useful model in understanding the complex geometries seen in nature, which can't usually be reduced to simple rectangles and circles.

Objectives

By the end of this chapter you will have learned how to ...

- Use standard units of measure for length. For example, mm, cm, m, km.
- Measure line segments and angles.
- Use bearings.
- Interpret maps and scale drawings.
- Know and apply formulae to calculate the area of triangles, parallelograms and trapezia.
- Identify, describe and construct reflections, rotations, translations and enlargements.

Check in

1 Evaluate

   a $6 \times \frac{8}{5}$ **b** $9 \times 2\frac{1}{2}$ **c** $6 \times 1\frac{3}{4}$

2 Evaluate

   a $5.8 + 2$ **b** $14.8 + 0.7$ **c** $6.4 + 2.6$

3 Measure this line

   a in millimetres   **b** in centimetres.

Chapter investigation

Create a snowflake!

Step 1   Draw an equilateral triangle.

Step 2   Draw equilateral triangles on each of the three sides
           (carefully – you'll have to divide each side into three equal parts).

Step 3   You now have 12 sides.
           Draw equilateral triangles on each of these.

If it's still not snowflaky enough, try once more – but you'll find it
starts getting very fiddly!

7.1 Measuring lengths and angles

You measure lengths in **millimetres** (mm), **centimetres** (cm), **metres** (m) or **kilometres** (km).

$10\,mm = 1\,cm$ $100\,cm = 1\,m$ $1000\,m = 1\,km$

A ruler **measures length** in **millimetres** (mm) or **centimetres** (cm).

This line measures 2.5 cm or 25 mm.

This line measures 2.7 cm or 27 mm.

To measure a line, line up the ruler so that the zero mark is at the start of the line.

You can **measure** and draw an **angle** in **degrees** with a **protractor**.
A protractor measures angles up to 180°.
There are 180° in a half turn.

180°   180°

● You can measure a reflex angle by measuring the associated acute or obtuse angle.

A full turn is 360°.

Angles at a point add to 360°.

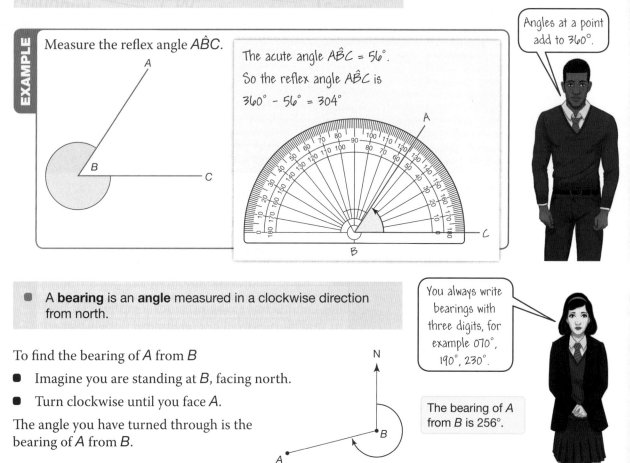

EXAMPLE

Measure the reflex angle $A\hat{B}C$.

The acute angle $A\hat{B}C = 56°$.
So the reflex angle $A\hat{B}C$ is
$360° - 56° = 304°$

● A **bearing** is an **angle** measured in a clockwise direction from north.

You always write bearings with three digits, for example 070°, 190°, 230°.

To find the bearing of A from B

● Imagine you are standing at B, facing north.

● Turn clockwise until you face A.

The angle you have turned through is the bearing of A from B.

The bearing of A from B is 256°.

Geometry Working in 2D

Exercise 7.1S

1 Convert these metric measurements of length.

   a 180 cm to mm   **b** 45 mm to cm

   c 350 cm to m   **d** 2000 m to km

   e 3500 m to km   **f** 4500 mm to m

2 Measure the lengths of these lines in

   a centimetres   **b** millimetres.

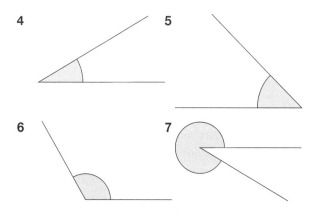

3   **a** Draw a line AB, so that $AB = 9$ cm.

   b Find the midpoint of AB and mark it with a cross.

In questions **4–7**, for each angle state

   a the type of angle – acute, right angle, obtuse or reflex

   b your estimate in degrees

   c the measurement in degrees.

Set out your answers like this:

Question	Type of angle	Estimate	Measurement
4	acute	40°	30°
5			

4

5

6

7

8 Draw and label these angles using a protractor.
State whether each angle is acute, obtuse, reflex or a right angle.

   a 40°   **b** 140°   **c** 90°

   d 36°   **e** 144°   **f** 56°

   g 124°   **h** 38°   **i** 142°

   j 85°   **k** 300°   **l** 200°

   m 320°   **n** 245°   **o** 265°

9 These diagrams are drawn accurately.
Measure the bearing of T from S in each.

   a

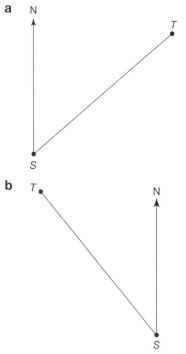

   b

10 P and Q are points 2 cm apart. Draw diagrams to show the position of points P and Q where the bearing of Q from P is

   a 070°   **b** 155°

   c 340°   **d** 260°

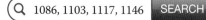

Q 1086, 1103, 1117, 1146   SEARCH

7.1 Measuring lengths and angles

RECAP

- A ruler measures lengths in cm and mm.
- You measure and draw angles using a protractor.
- A bearing is an angle measured in a clockwise direction from north. You always write a bearing with three digits.

In **scale drawings**, lines and shapes are **reduced** or **enlarged**.

Corresponding lengths are multiplied by the same **scale factor**. You can write the scale factor as a ratio.

You can write the scale factor 1 cm represents 100 cm as 1 : 100.

real length = 100 × length on the map

map length = real length ÷ 100

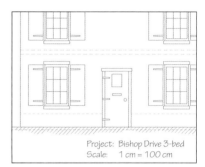

Project: Bishop Drive 3-bed
Scale: 1 cm = 100 cm

HOW TO

① Choose a scale (unless one is given).

② Work out or measure the lengths or angles required.

③ Draw a diagram (unless one is given).

④ Give the answers including units.

EXAMPLE

A church, C, is 10 km due west of a school, S.
Joe is 6 km from the school on a bearing of 320°.
He wants to walk directly to the church.

Draw a diagram to show the positions of Joe, the church and the school, and use it to find the bearing Joe should take.

① Use a scale of 1 cm to 2 km.

② Represent 10 km by a line 5 cm long.
Represent 6 km by a line 3 cm long.

Draw the line JC.

Draw the north line at J.

Measure the clockwise angle between the north line and JC.

Draw the north line at S.

Measure and draw the 320° bearing from S and, 3 cm from S, mark a point, J, to show Joe's position.

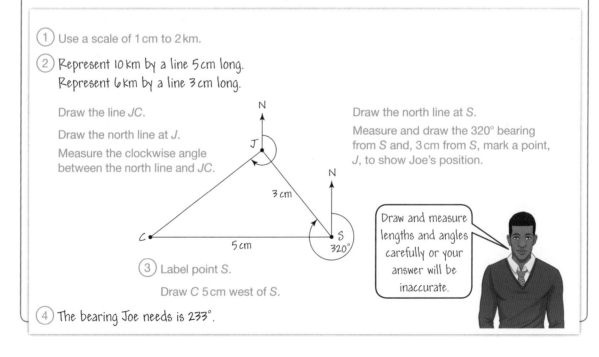

3 cm

5 cm

N

N

J

C

S
320°

Draw and measure lengths and angles carefully or your answer will be inaccurate.

③ Label point S.

Draw C 5 cm west of S.

④ The bearing Joe needs is 233°.

Exercise 7.1A

1 Measure and write the bearing of

   a Leeds from Manchester

   b Sheffield from Leeds

   c Manchester from Leeds

   d Manchester from Sheffield

   e Leeds from Sheffield.

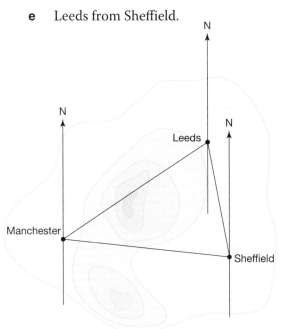

2 Liam sees a lifeboat on a bearing of 122°
from Mevagissey. Kim sees the same
lifeboat on a bearing of 225° from the
Rame Head chapel.
Kim says the lifeboat is nearer to her than
to Liam.
Use tracing paper to copy the coastline and
work out if Kim is correct.

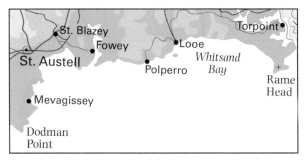

3 Copy the diagram.
The distance from
Truro to Falmouth
is 14 km. The bearing
of St. Mawes from
Falmouth is 080°.
The bearing of
St. Mawes from
Truro is 170°.

   a Mark the
position of
St. Mawes on your
diagram.

   b Calculate the
distance from
Falmouth to St.
Mawes.

   c Calculate the
distance from Truro to St. Mawes.

Scale: 1 cm represents 2 km

4 A youth club (*Y*) is 4 km due east of
a school (*S*).
Hazel leaves school and walks 5 km on
a bearing of 042° to her house (*H*).

   a Make a scale drawing to show the
positions of *Y*, *S* and *H*.
Use a scale of 1 cm to 1 km.

   b Hazel walks directly from her house to
the youth club.
What bearing does she take?

5 A lighthouse, *L*, is 6 km on a bearing of 160°
from a point *H* at the harbour. A boat, *B*, is
3 km from *L* on a bearing of 125°.

   a   **i** Make a scale drawing to show the
positions of *L*, *H* and *B*.

      ii On what bearing should *B* travel to
go directly to *H*?

     iii Estimate the distance between *B*
and *H*.

   b The boat moves 4 km due west.

      i Mark on your drawing the new
position of *B*.

     ii On what bearing should *B* now
travel to go directly to *H*?

    iii Estimate the new distance between
B and *H*.

 1086, 1103, 1117, 1146 SEARCH

7.2 Area of a 2D shape

- The **perimeter** of a shape is the distance round it.
- The **area** of a shape is the amount of space it covers.

You can use formulae to find the areas of **rectangles** and **triangles**.

Rectangle

width

length

Triangle

height

base

> The height of a triangle is always at right angles to the base.
>
> height
>
> base

- Area = length × width

- Area = $\frac{1}{2}$ × base × height

You can find the formula for the **area** of any **parallelogram**.

For this parallelogram ...

cut off one triangle ...

and fit it on the other end ... to make a rectangle.

height

base

- Area of parallelogram = **base** × **perpendicular height**

height

base

> The height must be perpendicular to the base.

You can find the formula for the area of any **trapezium**.

You can fit two **congruent** (identical) trapeziums together to make a parallelogram.

The base of the parallelogram is $a + b$ and the height is h.

Area of parallelogram = $(a + b) \times h$

Area of trapezium = half area of parallelogram.

h

b a

a b

- Area of trapezium = $\frac{1}{2} \times (a + b) \times h$

a

height = h

b

> The height is the perpendicular distance between the parallel sides.

Calculate the area of each shape.

a

3 cm

5 cm

b

3 cm

4 cm

7 cm

a Area of parallelogram = 5 × 3

= 15 cm²

b Area of trapezium = $\frac{1}{2}$ (3 + 7) × 4

= 5 × 4

= 20 cm²

Exercise 7.2S

1 Calculate the areas of these rectangles. Remember to give the units of your answers.

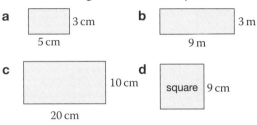

a 5 cm × 3 cm   **b** 9 m × 3 m   **c** 20 cm × 10 cm   **d** square 9 cm

2 Calculate the area of each of these right-angled triangles.
State the units of your answers.

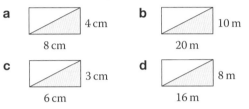

a 8 cm, 4 cm   **b** 20 m, 10 m   **c** 6 cm, 3 cm   **d** 16 m, 8 m

3 Calculate the area of each of these triangles. State the units of your answers.

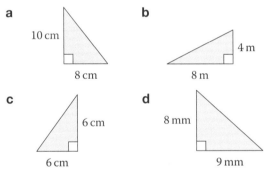

a 10 cm, 8 cm   **b** 4 m, 8 m   **c** 6 cm, 6 cm   **d** 8 mm, 9 mm

4 Calculate the area of each parallelogram.

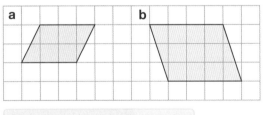

Give your answers in square units.

5 Calculate the area of each trapezium.

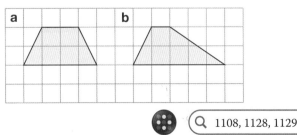

6 Calculate the area of each parallelogram. State the units of your answers.

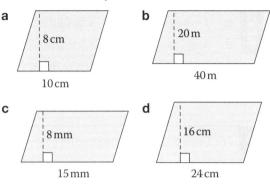

a 8 cm, 10 cm   **b** 20 m, 40 m   **c** 8 mm, 15 mm   **d** 16 cm, 24 cm

7 Calculate the area of each trapezium. State the units of your answers.

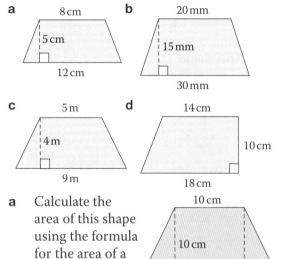

a 8 cm, 5 cm, 12 cm   **b** 20 mm, 15 mm, 30 mm   **c** 5 m, 4 m, 9 m   **d** 14 cm, 10 cm, 18 cm

8 **a** Calculate the area of this shape using the formula for the area of a trapezium.

b Calculate the area by adding the areas of the triangles and the square.

10 cm, 10 cm, 5 cm, 5 cm

9 The areas of these shapes are given. Calculate the unknown lengths.

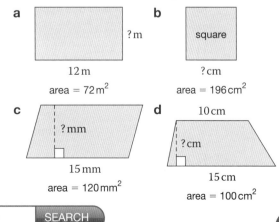

a 12 m, ? m, area = 72 m²

b square, ? cm, area = 196 cm²

c ? mm, 15 mm, area = 120 mm²

d 10 cm, ? cm, 15 cm, area = 100 cm²

Q 1108, 1128, 1129   SEARCH

7.2 Area of a 2D shape

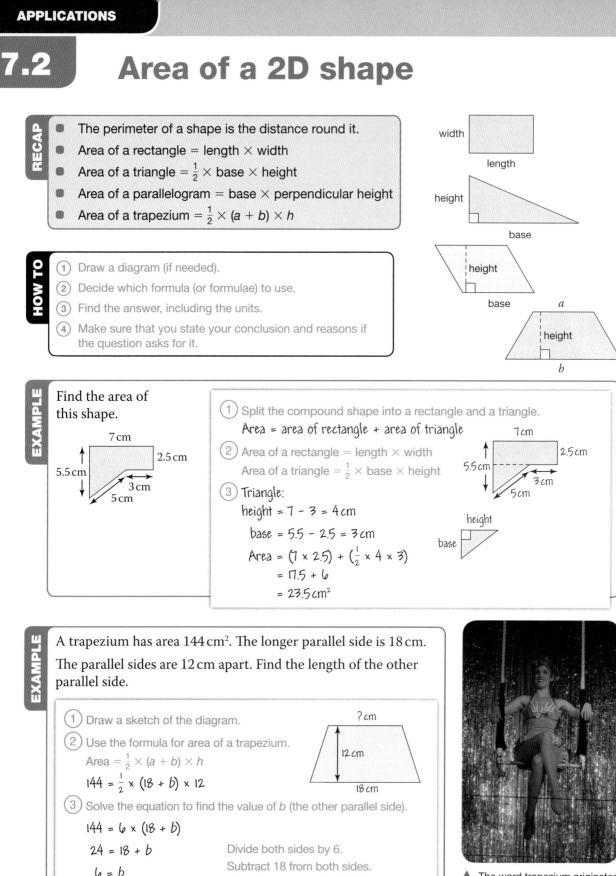

RECAP

- The perimeter of a shape is the distance round it.
- Area of a rectangle = length × width
- Area of a triangle = $\frac{1}{2}$ × base × height
- Area of a parallelogram = base × perpendicular height
- Area of a trapezium = $\frac{1}{2}$ × (a + b) × h

HOW TO

1. Draw a diagram (if needed).
2. Decide which formula (or formulae) to use.
3. Find the answer, including the units.
4. Make sure that you state your conclusion and reasons if the question asks for it.

EXAMPLE

Find the area of this shape.

7 cm
2.5 cm
5.5 cm
3 cm
5 cm

1. Split the compound shape into a rectangle and a triangle.
   Area = area of rectangle + area of triangle
2. Area of a rectangle = length × width
   Area of a triangle = $\frac{1}{2}$ × base × height
3. Triangle:
   height = 7 – 3 = 4 cm
   base = 5.5 – 2.5 = 3 cm
   Area = (7 × 2.5) + ($\frac{1}{2}$ × 4 × 3)
        = 17.5 + 6
        = 23.5 cm²

EXAMPLE

A trapezium has area 144 cm². The longer parallel side is 18 cm. The parallel sides are 12 cm apart. Find the length of the other parallel side.

1. Draw a sketch of the diagram.
2. Use the formula for area of a trapezium.
   Area = $\frac{1}{2}$ × (a + b) × h
   144 = $\frac{1}{2}$ × (18 + b) × 12
3. Solve the equation to find the value of b (the other parallel side).
   144 = 6 × (18 + b)
   24 = 18 + b   Divide both sides by 6.
   6 = b   Subtract 18 from both sides.
4. The other parallel side has length 6 cm.

▲ The word trapezium originates from the shape made by the ropes and bar of an old-fashioned flying trapeze.

Exercise 7.2A

1 Calculate the perimeter and area of each shape.
State the units of your answers.

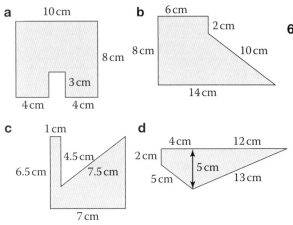

a 10 cm, 8 cm, 3 cm, 4 cm, 4 cm

b 6 cm, 2 cm, 8 cm, 10 cm, 14 cm

c 1 cm, 4.5 cm, 6.5 cm, 7.5 cm, 7 cm

d 4 cm, 12 cm, 2 cm, 5 cm, 5 cm, 13 cm

2 Pete is making a mobile out of shapes like this.
He cuts the shape out of a piece of card that is 30 cm × 20 cm.
What is the area of the card left over?

12 cm, 9 cm, 4 cm, 5 cm, 4 cm, 15 cm

3 Caroline has drawn a sandcastle.
What is the area of her castle and flag?
Start by dividing the shape into parts.

2 cm, 15 cm, 8 cm, 11 cm, 3 cm, 3 cm, 11 cm, 21 cm

4 Use x and y axes from 0 to 6 on centimetre square paper.

a Plot and join these points to make an arrow.
(0, 3), (3, 6), (5, 6), (3, 4), (6, 4)
(6, 2), (3, 2), (5, 0), (3, 0), (0, 3)

b Find the area of the arrow.

c Use a different method to check your answer to part **b**.

5 A trapezium has area 132 cm².
One of the parallel sides is 12 cm.
The parallel sides are 4 cm apart.
Find the length of the other parallel side.

6 Ryan has two congruent rectangles.
The sides of the rectangles are 2 cm and 6 cm.

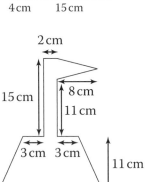

a Ryan arranges the shape to make this compound shape.
Find the perimeter of the shape.

b Ryan wants to arrange the shapes to make a compound shape with perimeter 28 cm.
Draw three ways that Ryan can arrange the rectangles.

7 a Draw three different parallelograms that have area 36 cm².
Label the base and perpendicular height on your drawings.

b Draw three different triangles that have area 24 cm².
Label the width and height on your drawings.

c Draw three different trapeziums that have area 20 cm².
Label the two parallel sides and the perpendicular height on your drawings.

8 Jeannie has these facts about an allotment.

● The allotment is a rectangle with area 144 m².

● Fencing for the allotment costs £10 per metre.

Jeannie wants to build a fence all around her allotment.
She has a budget of £400.
Does Jeannie have enough money to build the fence?
Explain how you decide.

9 Find the area of the ▢ = 1 cm² shaded triangle.

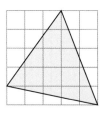

7.3 Transformations 1

- You describe a **reflection** using the **mirror line** or reflection line.

Choose a point on the object to find the corresponding point on the image.

Corresponding points are **equidistant** from the mirror line.

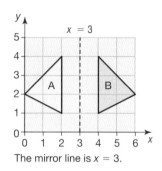

The mirror line is $x = 3$.

To describe a **rotation** you give

- the **centre of rotation** – the point about which it **turns**
- the angle or measure of turn
- the direction of turn – either **clockwise** or **anticlockwise**.

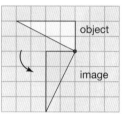

The shape rotates 90° anticlockwise about the dot.

- You can describe a **translation** by specifying the distance moved right or left, and then up or down.

You can write the translation in a column like this: $\binom{\text{right}}{\text{up}}$.

$\binom{5}{3}$ means 5 right and 3 up. $\binom{-2}{-4}$ means 2 left and 4 down.

Left and down are negative directions.

- In a reflection, rotation or translation the object (original shape) and the image (new shape) are **congruent**.

Corresponding angles and lengths are the same in the image and the object.

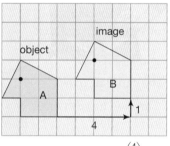

The shape is translated by $\binom{4}{1}$.

EXAMPLE

Describe the transformation that maps shape A on to

a   shape B

b   shape C

c   shape D.

Maps means changes.

a   In a reflection the mirror line bisects the line joining corresponding points on the object and image.

Shape B is a reflection of shape A in the line $x = 4$.

b   The vertex (1, 1) does not move during the rotation, so it must be the centre of rotation.

Shape C is a rotation of shape A through 180° about (1, 1).

c   Shape D is a translation of shape A by the vector $\binom{-6}{-1}$.

Geometry   Working in 2D

Exercise 7.3S

1 Copy the diagrams and draw the mirror lines.

a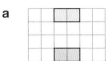

b

2 Copy the diagrams. Reflect the shapes in both mirror lines to create a pattern.

a **b**

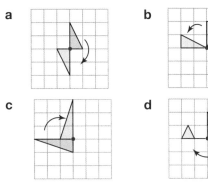

3 State the angle and direction of turn for each rotation (green shape to blue shape).

a **b**

c **d**

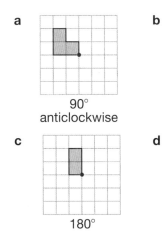

4 Copy these shapes onto square grid paper. Rotate the shapes through the given angle and direction about the dot (•).

a
90°
anticlockwise

b
90°
clockwise

c
180°

d
90°
clockwise

5 Which of these shapes are translations of the green shape?

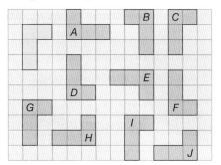

6 Match each translation with one of these vectors.

$$\begin{pmatrix} 5 \\ 3 \end{pmatrix} \begin{pmatrix} 2 \\ -4 \end{pmatrix} \begin{pmatrix} 5 \\ -3 \end{pmatrix} \begin{pmatrix} 3 \\ 1 \end{pmatrix} \begin{pmatrix} -4 \\ 2 \end{pmatrix} \begin{pmatrix} -3 \\ 1 \end{pmatrix}$$

a 3 right, 1 up   **b** 5 right, 3 up

c 3 left, 1 up   **d** 5 right, 3 down

e 2 right, 4 down   **f** 4 left, 2 up

7 Copy these shapes onto square grid paper. Translate the shapes according to each vector.

a

b

$\begin{pmatrix} 1 \text{ right} \\ 2 \text{ up} \end{pmatrix}$ $\begin{pmatrix} 1 \\ 0 \end{pmatrix}$

c

d

$\begin{pmatrix} 0 \\ -1 \end{pmatrix}$ $\begin{pmatrix} 0 \\ -1 \end{pmatrix}$

e

f

$\begin{pmatrix} 2 \\ 0 \end{pmatrix}$ $\begin{pmatrix} 1 \\ -1 \end{pmatrix}$

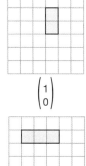
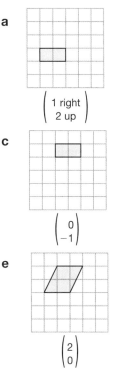
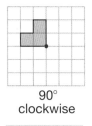
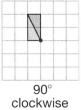
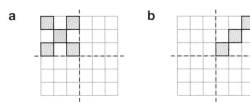
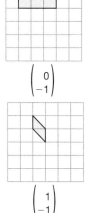

Q 1099, 1113, 1115, 1127   SEARCH

7.3 Transformations 1

- Reflections, rotations and translations are all transformations.
- To describe a reflection, you draw or give the equation of the mirror line.
- To describe a rotation, you give the centre and the angle of rotation.
- To describe a translation, you give the distance and direction or the vector.

> A reflection flips a shape over. A rotation turns a shape. A translation is a sliding movement.

HOW TO

① Check that the object and image are congruent.
Then decide which type of transformation is involved.

② Give a full description of the transformation.

EXAMPLE

A regular pentagon is divided into five isosceles triangles.
The centre of the pentagon is marked with a dot (•).
The green triangle is rotated about the dot onto the yellow triangle.

a State whether the green and yellow triangles are congruent or similar.

b Calculate the angle and direction of the rotation.

a ① Congruent – same size and same shape.

b ② The five angles at the dot total 360°.
One angle at the dot is 360° ÷ 5 = 72°.

③ Rotation is 72° clockwise about the dot.

EXAMPLE

In this diagram, triangle *A* undergoes three pairs of transformations.

Triangle *A* is reflected in the line $x = 0$ (the *y*-axis) to triangle *E*.
Then triangle *E* is reflected in the line $x = -5$ to triangle *F*.
What single transformation maps triangle *A* onto triangle *F*?

① The combination of reflections is a **translation**.

② A translation by the vector $\begin{pmatrix} -10 \\ 0 \end{pmatrix}$ maps *A* onto *F*.

> The combination of reflections is a translation.

Exercise 7.3A

1 Give the equation of the mirror line for each reflection.

a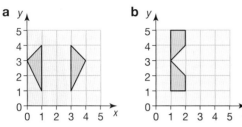

b

2 Describe fully the transformation that maps

a W to X **b** W to Y **c** W to Z.

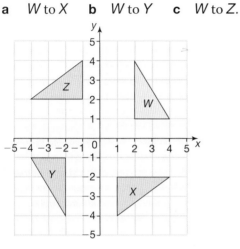

3 A regular octagon is divided into eight isosceles triangles. The centre of the octagon is marked with a dot (•). The green triangle is rotated about the dot onto the yellow triangle. Calculate the angle and direction of the rotation.

4 Copy this diagram.

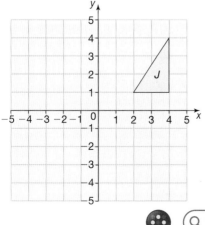

4 **a** Reflect triangle J in the x-axis. Label it K.

b Rotate triangle K 180° about centre (0, 0). Label it L.

c Describe fully the single transformation that takes triangle L to triangle J.

5 Copy this diagram.

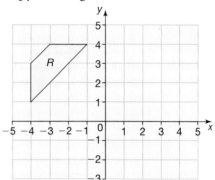

a Translate trapezium R by the vector $\begin{pmatrix} 5 \\ -3 \end{pmatrix}$. Label the image S.

b Translate trapezium S by the vector $\begin{pmatrix} -1 \\ 2 \end{pmatrix}$. Label the image T.

c Describe fully the single transformation that takes trapezium T to trapezium R.

*6 Sue says a rotation of 90° clockwise about (1, 4) maps A onto B. Is Sue correct? Explain your answer.

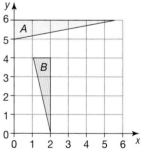

*7 Describe three *different* transformations that map A onto B.

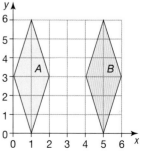

Q 1099, 1113, 1115, 1127 SEARCH

7.4 Transformations 2

In an **enlargement** the angles stay the same and the lengths increase in proportion.
The scale factor of an enlargement is the ratio of corresponding sides.

> A scale factor greater than 1 enlarges the shape.

● Scale factor = $\dfrac{\text{length of image}}{\text{length of original}}$

EXAMPLE

The green shape is an enlargement of the yellow shape.

Calculate the scale factor for each enlargement.

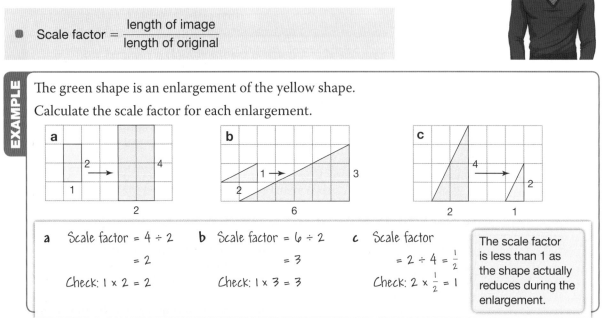

a Scale factor = 4 ÷ 2

= 2

Check: 1 × 2 = 2

b Scale factor = 6 ÷ 2

= 3

Check: 1 × 3 = 3

c Scale factor

= 2 ÷ 4 = $\frac{1}{2}$

Check: 2 × $\frac{1}{2}$ = 1

> The scale factor is less than 1 as the shape actually reduces during the enlargement.

The position of an enlargement is fixed by the **centre of enlargement**.

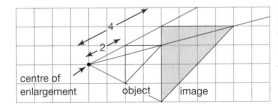

You multiply the distance from the centre to the object by the **scale factor**.
This gives the distance to the image along the same extended line.

The scale factor of the enlargement is 2.

> The **red** lines start from the centre and pass through corresponding **vertices** of the two shapes.

● To describe an enlargement, you give the scale factor and the centre of enlargement.

EXAMPLE

Draw the enlargement of the yellow shape, using scale factor 2 and *P* as the centre of enlargement.

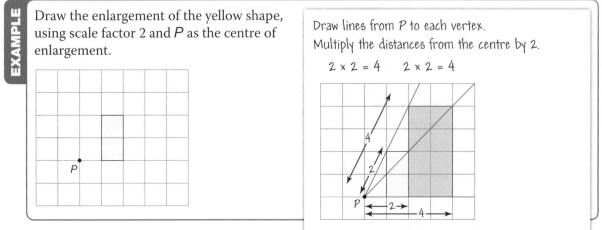

Draw lines from *P* to each vertex.
Multiply the distances from the centre by 2.

2 × 2 = 4   2 × 2 = 4

Exercise 7.4S

1 **a** Decide whether these rectangles are enlargements of the green rectangle. If so, calculate the scale factor.

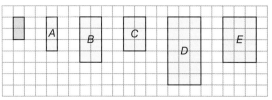

b List the rectangles that are similar to the green rectangle.

2 Copy each diagram on square grid paper. Enlarge each shape by the given scale factor using the given centre of enlargement.

a **b**

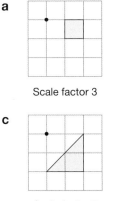

Scale factor 3   Scale factor 2

c **d**

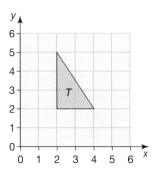

Scale factor 2   Scale factor 3

3 **a** Copy this diagram, but extend both axes to 16.

b Enlarge triangle *T* by scale factor 2, centre (0, 0). Label the image *U*.

c Enlarge triangle *T* by scale factor 3, centre (0, 0). Label the image *V*.

4 **a** Copy this diagram, but extend both axes from −7 to 7.

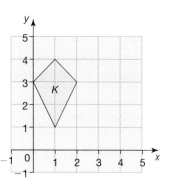

b Enlarge kite *K* by scale factor 2, centre (1, 3). Label the image *L*.

c Enlarge kite *K* by scale factor 4, centre (1, 3). Label the image *M*.

5 **a** Draw a grid with an *x*-axis from −3 to 5 and a *y*-axis from −2 to 10. Plot the points (1, 3) (1, 6) (3, 9) (4, 6). Join them to make quadrilateral *Q*.

b Enlarge quadrilateral *Q* by scale factor $\frac{1}{3}$, centre (4, 3). Label the image *R*.

c Enlarge quadrilateral *Q* by scale factor $\frac{1}{2}$, centre (−2, −2). Label the image *S*.

6 **a** Copy this diagram.

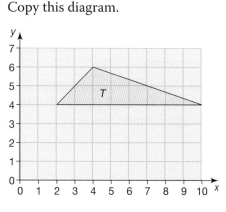

b Enlarge triangle *T* by scale factor $\frac{1}{2}$, centre (0, 0). Label the image *U*.

c Enlarge triangle *T* by scale factor $\frac{1}{2}$, centre (2, 2). Label the image *V*.

Q 1125   SEARCH

7.4 Transformations 2

RECAP

- In an enlargement, the object and image are similar.
- If the scale factor is less than one (but still postive), then the image is smaller than the object.
- To describe an **enlargement** you give the **scale factor** and the **centre of enlargement**.

HOW TO

① Draw construction lines to join corresponding vertices on the object and image.

② The construction lines meet at the centre of enlargement.

③ The scale factor of an enlargement is the ratio of corresponding sides.

$$\text{Scale factor} = \frac{\text{length of image}}{\text{length of original}}$$

To find the centre of enlargement, draw lines between corresponding points.

EXAMPLE

Find the centre of enlargement and calculate the scale factor of the enlargement from *A* to *B*.

① Draw the red lines to find the centre of enlargement.

② Centre of enlargement is $(-2, -1)$

③ Scale factor $= 3 \div 1 = 3$

EXAMPLE

Describe the enlargement that maps triangle *X* onto triangle *W*.

① Draw the red lines to find the centre of enlargement.

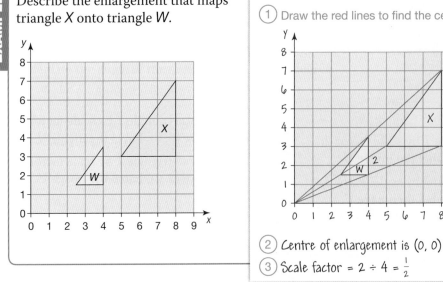

② Centre of enlargement is $(0, 0)$

③ Scale factor $= 2 \div 4 = \frac{1}{2}$

Exercise 7.4A

1 Copy each diagram on square grid paper. Shape *A* has been enlarged onto shape *B*. Find the centre of enlargement and calculate the scale factor for these enlargements.

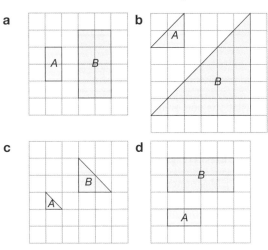

a   **b**

c   **d**

2 Describe fully the single transformation that maps

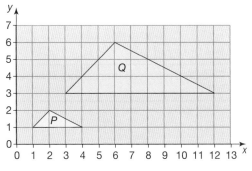

a triangle *P* onto triangle *Q*

b triangle *Q* onto triangle *P*.

3 Describe fully the single transformation that maps

a rectangle *R* onto rectangle *S*

b rectangle *S* onto rectangle *R*.

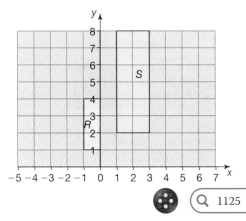

4

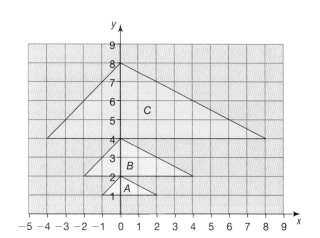

a Describe fully the single transformation that maps

   i triangle *B* onto triangle *A*

   ii triangle *B* onto triangle *C*

   iii triangle *A* onto triangle *C*.

b How many times bigger is the perimeter of triangle *C* than that of triangle *A*?

5 a Draw *x*- and *y*-axes from -7 to 5. Draw triangle *X* with vertices at $(-1, 1)$, $(0, 3)$ and $(3, 2)$.

b Enlarge triangle *X* by scale factor 2, centre $(0, 0)$. Label the image *Y*.

c Enlarge triangle *Y* by scale fractor $1\frac{1}{2}$, centre $(0, 0)$. Label the image *Z*.

d Describe the enlargement that will transform triangle *X* to triangle *Z*.

e Describe the enlargement that will transform triangle *Z* to triangle *X*.

f Comment on your answers to **d** and **e**.

6 The table gives the vertices of a kite and its image after a transformation.

Kite	(2, 0)	(3, 2)	(2, 3)	(1, 2)
Image	(6, 0)	(9, 6)	(6, 9)	(3, 6)

a What happens to the co-ordinates?

b Describe fully the transformation.

Summary

Checkout
You should now be able to...

	Test it Questions
✔ Use standard units of measure for length, measure line segments and angles.	1
✔ Use bearings.	1 – 2
✔ Interpret maps and scale drawings.	2
✔ Know and apply formulae to calculate the area of triangles, parallelograms and trapezia.	3
✔ Identify, describe and construct reflections, rotations, translations and enlargements.	4 – 6

Language | Meaning | Example

Language	Meaning	Example
Length	Length is a measure of distance.	Millimetres, centimetres, metres and kilometres are all measures of length. Length can be measured with a ruler.
Angle	The amount that one straight line is turned relative to another that it meets or crosses.	Angles are measured in degrees. One degree is $\frac{1}{360}$ th of a complete turn. Use a protractor to measure an angle.
Area	The amount of space occupied by a 2D shape.	Area = 12 units^2 Perimeter = 14 units
Perimeter	The total distance around the edges that outline a shape.	
Transformation	A geometric mapping that takes the points in an **object** to points in an **image**.	Rotation, reflection, translation, enlargement.
Translation	A transformation in which all the points in the object are moved the same distance and in the same direction.	
Reflection / Mirror line	A transformation that moves points to an equal distance on the opposite side of a mirror line.	
Rotation / Centre of rotation	A transformation that turns points through a fixed angle whilst keeping their distance from the centre of rotation fixed.	anticlockwise
Enlargement / Scale factor / Centre of Enlargement	A transformation that moves points a fixed multiple, the scale factor, of their distance from the centre of enlargement.	$\frac{1}{3}$
Invariant	Does not change under a transformation.	A mirror line under reflection.

Review

1 Measure the size of

   a angle *RTS*

   b side *RT*.

2 *A* is 2 km north of *B*. *C* is 3 km on a bearing of 065° from *B*. Make a scale drawing of *A*, *B* and *C* using the scale 2 cm to 1 km.

3 Work out the area of these shapes. Remember to state the units of your answers.

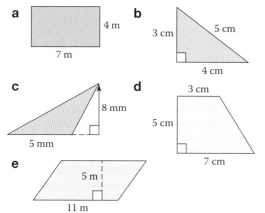

   a 4 m, 7 m

   b 3 cm, 5 cm, 4 cm

   c 8 mm, 5 mm

   d 3 cm, 5 cm, 7 cm

   e 5 m, 11 m

4 Copy this diagram.

   a Reflect triangle *A* in the *x*-axis and label the image *B*.

   b Rotate triangle *A* 180° about (0, 0) and label the image *C*.

   c Describe the transformation of the object *C* to the image *B*.

5 On a copy of this diagram

   a translate triangle *A* by the vector $\begin{pmatrix} -4 \\ -6 \end{pmatrix}$ and label the image *C*,

   b describe the transformation that maps *A* to *B*.

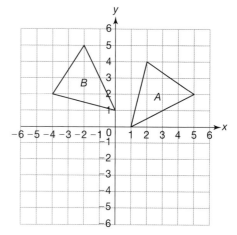

6 On a copy of this diagram

   a enlarge *A* by scale factor 3 from centre of enlargement (1, 9),

   b describe the transformation that maps *B* back onto *A*.

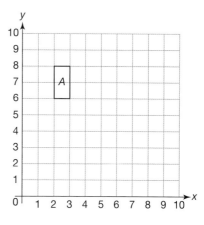

What next?

Score		
0 – 2		Your knowledge of this topic is still developing. To improve look at MyMaths: 1086, 1099, 1103, 1108, 1113, 1115, 1117, 1125, 1127, 1128, 1129, 1146
3 – 5		You are gaining a secure knowledge of this topic. To improve look at InvisiPens: 07Sa – q
6		You have mastered these skills. Well done you are ready to progress! To develop your exam technique look at InvisiPens: 07Aa – e

Assessment 7

1 a Karl says that to convert from cm to m you divide by 100.
   Marta says that you multiply by 100. Who is correct? [1]

   b Karl says that to convert from km to mm you divide by 1000 000.
   Marta says you divide by 10 000. Who is correct? [1]

2 a Draw a quadrilateral, *ABCD*, with sides *BC* = 6.5 cm, *AD* = *DC* = 7.5 cm and angles
   ∠*ADC* = 105° and ∠*BCD* = 60°. [2]

   b Measure   **i** *AB*   **ii** ∠*DAB*   **iii** ∠*ABC*. [3]

   c What type of quadrilateral is *ABCD*? [1]

   d Use a ruler to find the midpoint of each side. Join the midpoints to form another
   quadrilateral. What type of quadrilateral is this? [2]

3 Briony looks at these angles. She says that angle

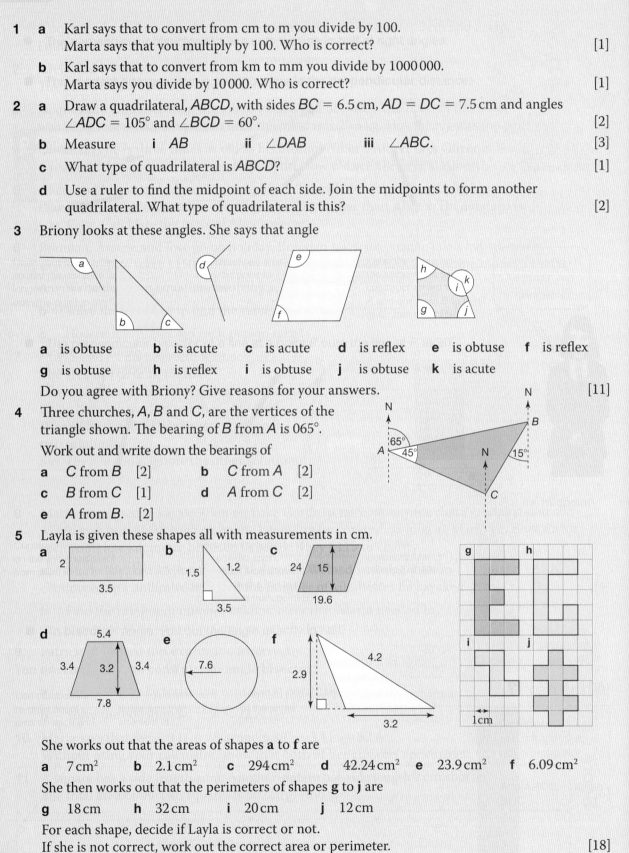

   a is obtuse   **b** is acute   **c** is acute   **d** is reflex   **e** is obtuse   **f** is reflex

   g is obtuse   **h** is reflex   **i** is obtuse   **j** is obtuse   **k** is acute

   Do you agree with Briony? Give reasons for your answers. [11]

4 Three churches, *A*, *B* and *C*, are the vertices of the
   triangle shown. The bearing of *B* from *A* is 065°.

   Work out and write down the bearings of

   a *C* from *B* [2]   **b** *C* from *A* [2]

   c *B* from *C* [1]   **d** *A* from *C* [2]

   e *A* from *B*. [2]

5 Layla is given these shapes all with measurements in cm.

   She works out that the areas of shapes **a** to **f** are

   a 7 cm²   **b** 2.1 cm²   **c** 294 cm²   **d** 42.24 cm²   **e** 23.9 cm²   **f** 6.09 cm²

   She then works out that the perimeters of shapes **g** to **j** are

   g 18 cm   **h** 32 cm   **i** 20 cm   **j** 12 cm

   For each shape, decide if Layla is correct or not.
   If she is not correct, work out the correct area or perimeter. [18]

6 Square patio slabs come in 3 sizes, 1 m × 1 m, 2 m × 2 m and 3 m × 3 m.
Farakh wants to build a square patio of side 7 m.

    a How many of the 1 m × 1 m slabs would Farakh need? [1]

    b Can Farakh build a square patio using just 2 m × 2 m or just 3 m × 3 m slabs? [3]
       Give reasons for your answer.

    c Can Farakh build a square patio using a mix of 2 m × 2 m and 3 m × 3 m slabs?
       Draw a diagram to explain your answer. [4]

7 Mark translates the red triangle. Match each translation to
the correct triangle on the diagram.

    a 5 right and 3 upwards

    b 4 left and 5 upwards

    c 3 left and 5 downwards

    d 5 right and −2 upwards

    e −2 to the right and −2 upwards

    f $\begin{pmatrix} 2 \\ 3 \end{pmatrix}$ **g** $\begin{pmatrix} -1 \\ 5 \end{pmatrix}$

    h $\begin{pmatrix} 7 \\ -8 \end{pmatrix}$ **i** $\begin{pmatrix} -1 \\ -7 \end{pmatrix}$ [9]

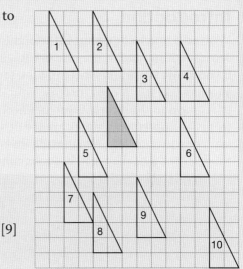

8 A rocket takes off from its launching pad and travels
25 miles horizontally and 45 miles vertically.

    a Show the path of the rocket on a squared grid. Take 1 square width/length
       as representing 10 miles. [2]

    b Write the vector which represents this translation. [1]

At this position the first stage is ejected and the next stage takes it a further
15 miles horizontally and 20 miles vertically.

    c Draw this new translation at the end of your first diagram. [2]

    d Write this new translation as a vector. [1]

    e Using your grid, find the **single** vector that represents the whole journey. [2]

    f How are the vectors in parts **b**, **d** and **e** related? [1]

9 a Peter transforms the triangle, *A*, into the triangle *B*.
       Describe this transformation fully.   [3]

    b Rotate triangle *A* 90° clockwise,
       about the origin. Label your triangle *C*.   [3]

    c Reflect triangle *C* in the line shown.
       Label your triangle, *D*.   [3]

    d Fully describe the transformation
       that maps *A* onto *D*.   [2]

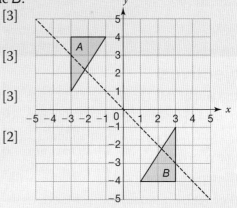

8 Probability

Introduction

When did you last look up at the night sky and see a shooting star? It is a rare event, although if you know when and where to look for meteor showers you will greatly increase your chances of seeing one. The world is full of uncertainty, from the unpredictable appearance of shooting stars or earthquakes, to manmade events like the result of a hockey match. Probability is the branch of mathematics that deals with the study of uncertainty and chance.

What's the point?

You apply probability whenever you weigh up everyday risks. For example, 'should I take an umbrella today?' A basic understanding of probability allows you to be more prepared for whatever life throws at you!

Objectives

By the end of this chapter you will have learned how to …

- Use experimental data to estimate probabilities and expected frequencies.
- Calculate theoretical probabilities and expected frequencies using the idea of equally likely events.
- Compare theoretical probabilities with experimental probabilities.
- Recognise mutually exclusive events and exhaustive events and know that the probabilities of mutually exclusive exhaustive events sum to 1.

Check in

1 Cancel these fractions to their simplest form.

a $\frac{12}{15}$ **b** $\frac{8}{10}$ **c** $\frac{5}{20}$ **d** $\frac{15}{25}$ **e** $\frac{10}{10}$

2 Order these decimals from smallest to largest.

a 0.25   0.2   0.3

b 0.7   0.8   0.75

c 0.85   0.8   1

3 Work out each of these fraction calculations.

a $\frac{1}{3} + \frac{2}{3}$ **b** $\frac{3}{10} + \frac{7}{10}$ **c** $1 - \frac{9}{10}$ **d** $1 - \frac{4}{5}$ **e** $1 - \frac{3}{4}$

4 Work out each of these decimal calculations.

a $1 - 0.2$ **b** $1 - 0.7$ **c** $1 - 0.9$

Chapter investigation

Two people can play an old game called 'Rock, Paper, Scissors'.

In the game you make a shape with your hand.

- Rock is a closed fist.
- Paper is an open palm with closed fingers.
- Scissors is two fingers held like scissor blades.

The players reveal their 'hand' simultaneously.

Rock beats scissors, paper beats rock, and scissors beats paper.

Is there a best strategy for playing this game?

- Probability measures how likely an **event** is to happen.
- All probabilities have a value from 0 (impossible) to 1 (certain).

You can use language such as **likely** and **unlikely** to describe probabilities without doing any calculations.

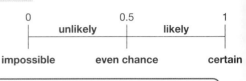

```
0            0.5            1
|—————————————|—————————————|
    unlikely       likely
impossible   even chance   certain
```

EXAMPLE

Describe the likelihood of the following events in words.

a You see a head if you toss a fair coin.

b The weather will be sunny for a barbecue in mid-July.

> The event in part **a** has a theoretical value which you can calculate.
>
> For the event in part **b** you can only estimate the probability.

a Even chance There are two (equally likely) outcomes.

b Likely You could make a better estimate nearer the date using a weather forecast.

The weather is sunny for 3 out of the 4 days just before a barbecue. Luke predicts that there is a 75% chance that the weather will be sunny for the barbecue.

He writes

$$P(S) = 75\%$$

P for probability   S for sunny   75% or 0.75 or $\frac{3}{4}$

> You can give probabilities as fractions, decimals or percentages.

Luke has produced an **estimated probability**, which is also called **relative frequency**.

This can be calculated from an experiment by performing **trials**.

- Estimated probability is called the **relative frequency**.
- Relative frequency $= \dfrac{\text{number of favourable trials}}{\text{total number of trials}}$

EXAMPLE

A drawing pin is thrown 10 times and lands point up in 7 cases.

a Estimate the probability that a drawing pin lands point up

The drawing pin is now thrown another 90 times and lands point up another 67 times.

b Calculate the relative frequency of the drawing pin landing point up using all the observations made so far.

> 0.74 is a better estimate because it is based on more information.

a $P(\text{point up}) = \dfrac{7}{10} = 0.7 = 70\%$

b $P(\text{point up}) = \dfrac{7 + 67}{10 + 90} = \dfrac{74}{100} = 0.74 = 74\%$ There were a total of 74 out of 100.

Exercise 8.1S

1 Use one of the words in the list to describe the probability of each event.

impossible   unlikely   even chance

likely   certain

Give a reason for your answers.

a Rafael and Sebastian have played 12 tennis matches against each other in the last year. Rafael won 6 out of 12 matches.

    i Sebastian wins the next match between the two players.

    ii Sebastian wins the next five matches between the two players.

b You score less than 7 when you throw a fair dice.

c The average temperature in Edinburgh in February is 4°C.
The average temperature in Sydney, Australia in February is 26°C.

    i The maximum temperature in Edinburgh was under 10°C on February 18th 1913.

    ii The maximum temperature in Sydney, Australia was at least 10°C on February 29th 1913.

d A set of cards is numbered 1 to 9. You pick a card at random and it is an even number.

2 Hari works at Quick-Fix garage.
Over the last month, 40 customers have come to the garage with a flat tyre caused by a puncture.
Hari records which tyre was punctured.

Tyre	Frequency
Front left	8
Front right	7
Back left	13
Back right	12

Estimate the probability that the next customer with a flat tyre has a puncture on their

a front left tyre   **b** front right tyre

c back left tyre   **d** back right tyre.

Give your answer as a

i fraction   **ii** decimal   **iii** percentage.

3 Rory wants to pick a red ball and can choose bag A or bag B.

Bag A	Colour	Red	Blue
	Frequency	1	2

Bag B	Colour	Red	Blue
	Frequency	2	8

a Which bag should Rory choose? Give reasons for your answer.

b Rory takes out two balls without replacing them in the bag. Rory wants to pick two reds. Which bag should he choose? Explain your answer.

4 There are 10 coloured balls in a bag. One ball is taken out and then replaced in the bag. The colours of the ball are shown in the table.

Colour	Red	Green	Blue
Frequency	9	14	27

a How many times was a ball taken out of the bag?

b Estimate the probability of taking out

    i a red ball   **ii** a green ball

    iii a blue ball.

c How many balls of each colour do you think are in the bag?

d How could you improve your estimate?

5 Toss a coin until you get a head, counting how many tosses you make, including the one on which the head appears.

a Do this 20 times and calculate the relative frequency from your observations for each of the values you have recorded.

b Comment on your result.

6 Throw an ordinary dice until you get a six, counting how many throws you make, including the one on which the six appears.

a Do this twenty times and calculate the relative frequency from your observations for each of the values you have recorded.

b Comment on your result.

Q 1209, 1210, 1211   SEARCH

8.1 Probability experiments

- You can estimate the probability of an event by conducting an experiment.
- Estimated probability is called the **relative frequency**.

$$\text{Relative frequency} = \frac{\text{number of favourable trials}}{\text{total number of trials}}$$

HOW TO

To estimate the probability of an event

① Think first in terms of words – likely / evens / unlikely.

② Calculate the relative frequency of the event.

③ Answer the question. Keep in mind that the estimated probability becomes more reliable as you increase the number of trials.

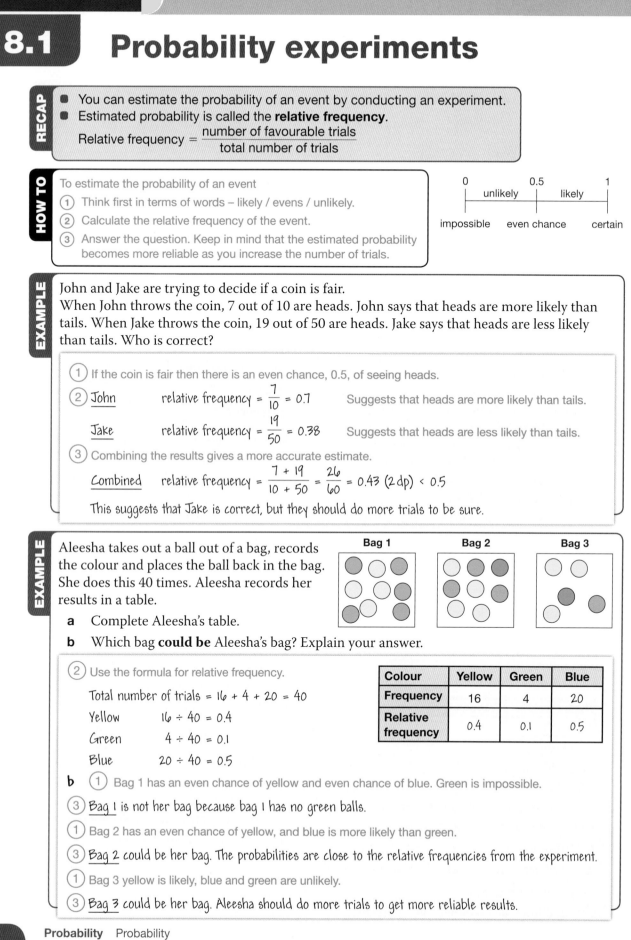

EXAMPLE

John and Jake are trying to decide if a coin is fair.
When John throws the coin, 7 out of 10 are heads. John says that heads are more likely than tails. When Jake throws the coin, 19 out of 50 are heads. Jake says that heads are less likely than tails. Who is correct?

① If the coin is fair then there is an even chance, 0.5, of seeing heads.

② John   relative frequency $= \dfrac{7}{10} = 0.7$ Suggests that heads are more likely than tails.

   Jake   relative frequency $= \dfrac{19}{50} = 0.38$ Suggests that heads are less likely than tails.

③ Combining the results gives a more accurate estimate.

   Combined   relative frequency $= \dfrac{7 + 19}{10 + 50} = \dfrac{26}{60} = 0.43 \ (2\,dp) < 0.5$

   This suggests that Jake is correct, but they should do more trials to be sure.

EXAMPLE

Aleesha takes out a ball out of a bag, records the colour and places the ball back in the bag. She does this 40 times. Aleesha records her results in a table.

Bag 1   **Bag 2**   **Bag 3**

a   Complete Aleesha's table.

b   Which bag **could be** Aleesha's bag? Explain your answer.

② Use the formula for relative frequency.

Total number of trials = 16 + 4 + 20 = 40

Yellow   16 ÷ 40 = 0.4

Green   4 ÷ 40 = 0.1

Blue   20 ÷ 40 = 0.5

Colour	Yellow	Green	Blue
Frequency	16	4	20
Relative frequency	0.4	0.1	0.5

b ① Bag 1 has an even chance of yellow and even chance of blue. Green is impossible.

③ Bag 1 is not her bag because bag 1 has no green balls.

① Bag 2 has an even chance of yellow, and blue is more likely than green.

③ Bag 2 could be her bag. The probabilities are close to the relative frequencies from the experiment.

① Bag 3 yellow is likely, blue and green are unlikely.

③ Bag 3 could be her bag. Aleesha should do more trials to get more reliable results.

Probability Probability

Exercise 8.1A

1 Elise spins this spinner 4 times. She says that if the spinner is fair then the spinner will land on each colour once. Is Elise correct?

2 A bag contains 100 coloured balls. Xavier, Yvonne and Zoe each select a ball from the bag, record if the ball is red and replace the ball. The table shows their results

	Number of trials	Number of red balls
Xavier	5	4
Yvonne	20	16
Zoe	100	95

a Xavier says that the probability of choosing a red ball is $\frac{4}{5}$. Criticise his statement.

b Zoe says that the bag must contain 95 red balls. Is she correct?

c **i** Explain why the most accurate estimate of the relative frequency of choosing a red ball is 0.92.

ii Does the bag contains 92 red balls?

3 Alik spins a spinner and records the result in a table. He adds each column when he lands on a new colour.

Colour	Red	White	Blue
Frequency	10	5	
Relative freq.		0.2	0.4

a Copy and complete Alik's table. What do you notice about the sum of the relative frequencies?

b How many of these spinners could be Alik's?

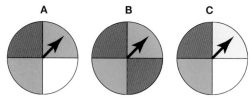

A   B   C

c Draw three possible spinners that could be Alik's spinner. You can divide your spinners into as many sectors as you like.

4 Marcie rolls a six-sided dice 60 times. Her results are shown in the tally chart.

Score	Frequency
1	⊪⊪⊪ ⊪⊪⊪ ⊪⊪⊪ II
2	⊪⊪⊪ ⊪⊪⊪ III
3	⊪⊪⊪ ⊪⊪⊪ ⊪⊪⊪ ⊪⊪⊪ II
4	⊪⊪⊪ III

a Which of these nets could be the net of Marcie's dice? Explain your answer.

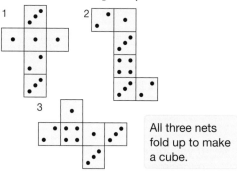

All three nets fold up to make a cube.

b Draw another net that **could** be Marcie's dice.

c Which net is most likely to be the net of Marcie's dice?

5 Alice and Aubrey are trying to decide if a spinner is fair.
The spinner is divided into three equal sections: black, white and green.
Alice spins the spinner 20 times. She lands on black 10 times and green 6 times.
Aubrey spins the spinner 30 times. She lands on green 11 times and white 12 times.
Is the spinner biased? Give your reasons.

*6 Keith thinks that fewer people are born on a Saturday or Sunday than on a weekday because any planned births will be scheduled during the week.

Keith collects information from 100 friends of different ages and finds that Saturday and Sunday seem to be as common as other days.
Suggest why the information he has collected does not mean that his idea about the days on which people are born is wrong.

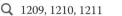

8.2 Expected outcomes

If you know the probability of an event, you can calculate how many times you **expect** an outcome to happen.

- Expected frequency = number of trials × probability of the event

EXAMPLE

Calculate the expected number of heads if you spin a fair coin

a 6 times   **b** 30 times

c Sam spins a coin 30 times. She get a head 21 times.
Sam decides that the coin is **biased** (unfair). Do you agree?

a $6 \times 0.5 = 3$ There is an even chance of spinning a head.

b $30 \times 0.5 = 15$

c The coin could be biased as 21 is larger than 15. However, Sam needs to spin the coin a lot more times before she can decide.

Each spin of the coin is a **trial**.

- If you increase the number of trials, the outcome of an experiment becomes closer to the **theoretical** (expected) outcome.

EXAMPLE

Estimate the number of fours you would expect see if you throw a fair dice 300 times.

There are 6 possible outcomes which are equally likely.

$$300 \times \frac{1}{6} = 50$$

'On average' there will be 50 of each of the numbers 1, 2, 3, 4, 5 and 6 in the 300 trials

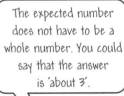

The expected number does not have to be a whole number. You could say that the answer is 'about 3'.

EXAMPLE

Calculate the expected number of twos if you throw a fair dice 20 times.

$$20 \times \frac{1}{6} = 50 = 3.33\ldots$$

You can give probabilities as decimals, fractions and percentages.

EXAMPLE

8% of men have red-green colour-blindness.
Women are rarely colour-blind.
A group of 65 men and 40 women are attending a meeting.
Estimate how many people in the group suffer from red-green colour-blindness.

$65 \times 0.08 = 5.2$ About 5 are likely to be colour-blind.

Only the men are likely to be colour-blind.

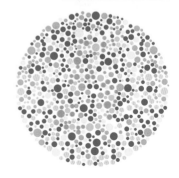

Exercise 8.2S

1 A spinner is made from a circle divided into 4 equal sectors, coloured red, blue, green and yellow. How many times would you expect to see the colour blue if the spinner is spun

    a   20 times   **b**   9 times?

2 The probability of a drawing pin landing point up when dropped is $\frac{3}{4}$.
Estimate how many times it will land point up if it is dropped

    a   20 times   **b**   7 times.

3 All the counters in a bag are either green or black. The probability of choosing a green counter is $\frac{1}{3}$.

A counter is taken from the bag and then replaced.
The experiment is carried out 20 times. Calculate the expected number of

    a   green counters

    b   black counters.

4 A spinner has 12 equal sections as shown in the diagram.

    a   Estimate how many times the arrow will land on a yellow section if it is spun 60 times.

    b   Estimate how many times the arrow will land on a blue section if it is spun 40 times.

5 Approximately 11% of people are left-handed. A group of 40 people are at a meeting.
What is the expected number of left-handed people attending the meeting?

6 When two fair coins are tossed the probability of getting two tails is $\frac{1}{4}$.

If 15 people each toss two coins, estimate the number of people who see two tails.

7 Three coins are thrown. The probability of the coins not being all heads or all tails is $\frac{3}{4}$.

If 25 people each throw three coins, estimate the number of people who do not see all heads or all tails.

8 A biased dice has the probabilities shown in the table. The dice is thrown 250 times and the outcomes recorded.
Calculate the expected numbers of each score that will be seen.

Score	1	2	3	4	5	6
Probability	0.1	0.2	0.25	0.15	0.1	0.2

9 A set of cards have the numbers 1 to 9 on them. They are shuffled and the top card turned over.

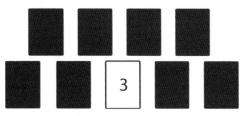

The number is recorded and the process repeated until 100 observations have been recorded.
Estimate how many times you will see

    a   an even number

    b   a square number

    c   a prime number

    d   a factor of 36.

10 Weather records show that approximately 30% of April days in a certain village have some rain. Estimate the number of days the village will have some rain in April next year.

11 A financial advisor claims that, on average, 70% of his recommended shares have increased in value. He has recommended 20 shares in the current year.
How many shares would you expect to increase in value over this year.

1211, 1264   SEARCH

8.2 Expected outcomes

- Expected frequency = number of trials × probability of the event
- This is an estimate of how many times 'on average' you will see the outcome if you repeat the experiment.

HOW TO

To estimate the number of times an event will happen

1. Find the total number of trials.
2. Find the probability the event will happen – this may be given to you, or you may need to work it out from the information provided.
3. Calculate the expected frequency.

What is actually seen can vary considerably when an experiment is repeated.

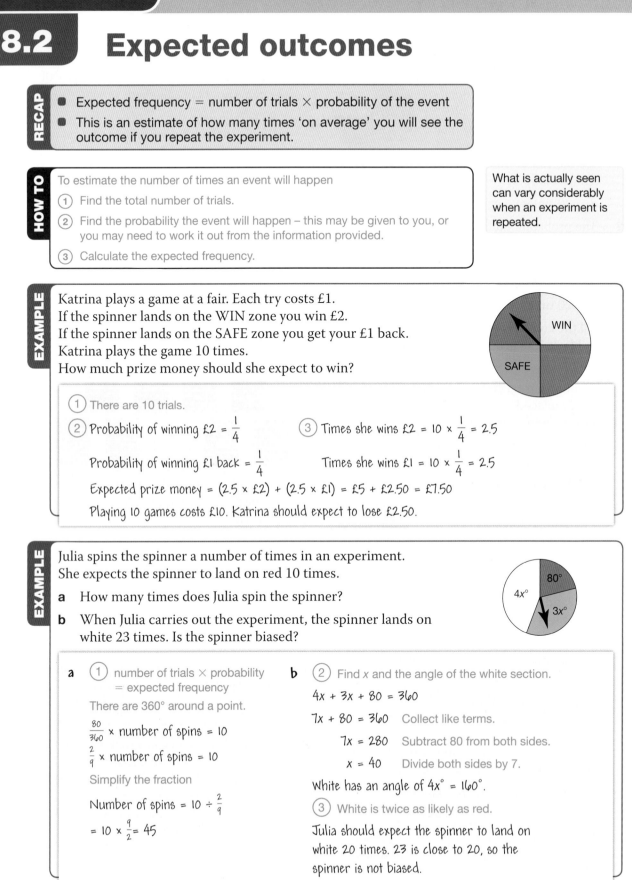

EXAMPLE

Katrina plays a game at a fair. Each try costs £1.
If the spinner lands on the WIN zone you win £2.
If the spinner lands on the SAFE zone you get your £1 back.
Katrina plays the game 10 times.
How much prize money should she expect to win?

1. There are 10 trials.

2. Probability of winning £2 = $\frac{1}{4}$

   3. Times she wins £2 = $10 \times \frac{1}{4}$ = 2.5

   Probability of winning £1 back = $\frac{1}{4}$

   Times she wins £1 = $10 \times \frac{1}{4}$ = 2.5

   Expected prize money = (2.5 × £2) + (2.5 × £1) = £5 + £2.50 = £7.50

   Playing 10 games costs £10. Katrina should expect to lose £2.50.

EXAMPLE

Julia spins the spinner a number of times in an experiment.
She expects the spinner to land on red 10 times.

a How many times does Julia spin the spinner?

b When Julia carries out the experiment, the spinner lands on white 23 times. Is the spinner biased?

a 1. number of trials × probability = expected frequency

There are 360° around a point.

$\frac{80}{360}$ × number of spins = 10

$\frac{2}{9}$ × number of spins = 10

Simplify the fraction

Number of spins = $10 \div \frac{2}{9}$

$= 10 \times \frac{9}{2}$ = 45

b 2. Find x and the angle of the white section.

$4x + 3x + 80 = 360$

$7x + 80 = 360$ Collect like terms.

$7x = 280$ Subtract 80 from both sides.

$x = 40$ Divide both sides by 7.

White has an angle of $4x° = 160°$.

3. White is twice as likely as red.

Julia should expect the spinner to land on white 20 times. 23 is close to 20, so the spinner is not biased.

Probability Probability

Exercise 8.2A

1 A biased coin is flipped 5 times.

The ratio of heads to tails is 2:3.

a The coin is flipped 30 times. Estimate the number of times you will see a head.

b The coin is flipped 40 times. Estimate the numbers of times you will see tails.

2 *ABCD* is a square and *E* and *F* are the midpoints of sides *AD* and *BC*.

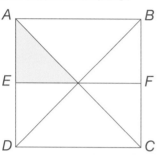

a If 40 points are selected at random in the square how many would you expect to see in the shaded area?

b If this shape was on the bottom of an open box and 40 small counters were dropped into the box, do you think you would see more, less, or about the same in the shaded area as in part **a**?

3 A spreadsheet generates a series of random digits. In a set of 150 random digits, estimate how many of them will be less than 4.

4 From looking at past results, Sanjiv knows that his favourite team have won 55% of their home league games and drawn 25%.

If the team plays 10 games, how many games would you expect the team to lose?

5 The table below shows the proportions of the hair colour of male and female students in Newcastle.

	Fair	Dark
Male	50%	50%
Female	60%	40%

A class in Newcastle has 18 boys and 15 girls. Estimate the number of dark haired students in the class.

6 The diagram shows a spinner with 5 coloured sides, but the spinner is not symmetrical.

Aisha thinks the probability of the spinner landing on a particular colour is proportional to the length of the side, and draws up a probability table.

Colour	Purple	Blue	Red	Brown	Green
Probability	$\frac{1}{4}$	$\frac{1}{6}$	$\frac{1}{4}$	$\frac{1}{12}$	$\frac{1}{4}$

If Aisha spins the spinner 60 times, estimate the number of times she will land on each colour (if her probabilities are right).

*7 A red bag has 5 white and 10 black balls in it, and a blue bag has 5 white and 5 black balls in it. All the balls are identical except for their colour.

a You choose a ball from the red bag at random and note its colour before returning it to the bag. If you do this 30 times, estimate how many times you will see a white ball.

b If you have used the blue bag instead, would you expect to see a white more often, less often or about the same?

*8 The coloured sections on this spinner are all equal.
Shawn is playing a game at a school fete which charges 50p a go. She wins a prize if her spinner lands on a blue sector.

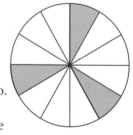

a If Shawn has 24 goes, how often would she be likely to win?

b If the prize is £1, how much profit is the fete expected to make from Shawn's 24 goes?

 1211, 1264 SEARCH

8.3 Theoretical probability

- A trial is an activity or experiement.
- An outcome is the result of a trial.
- An event is one or more outcomes of a trial.

> Rolling a dice is a trial.
> The outcomes of rolling a dice are 1, 2, 3, 4, 5, 6.
> Rolling an even number is an event.

The outcomes of some experiments are equally likely, for example, throwing a fair coin or rolling a fair dice.

- The **theoretical probability** of an event is

probability of event happening $= \dfrac{\text{number of outcomes in which event happens}}{\text{total number of all possible outcomes}}$

EXAMPLE

A fair dice is rolled and the number showing on the top is scored. Find the probability of these events.

a   3 is scored.

b   A factor of 12 is scored.

c   A prime number is scored.

> The probability of an event can be written as P(event).

There are 6 possible outcomes which are equally likely.

a $P(3) = \dfrac{1}{6}$ There is one 3.

b $P(\text{factor of } 12) = \dfrac{5}{6}$ Five outcomes are factors of 12: 1, 2, 3, 4 and 6.

c $P(3) = \dfrac{3}{6} = \dfrac{1}{2}$ Three outcomes are prime: 2, 3 and 5.

Theoretical probability is based on equally likely outcomes.

- You use experimental probability when the event is unfair or biased or when the theoretical outcome is unknown.

> The closer experimental probability is to theoretical probability the less likely it is that there is bias.

EXAMPLE

Sami threw a drawing pin 10 times and it landed point up 7 times. Cristiano threw a similar drawing pin 43 times and it landed point up 31 times.

a   Calculate the relative frequency of the pin landing point up for each person.

b   Which of these is the better estimate of the probability?

a   Sami $\dfrac{7}{10} = 0.7$ Cristiano $= \dfrac{31}{43} = 0.72$ (2 dp)

b   Cristiano used the largest sample so he has the better estimate.

- The experimental probability becomes a better estimate for the theoretical probability as you increase the number of trials.

Exercise 8.3S

1 A fair dice is rolled and the number showing on the top is scored.
Find these probabilities.

   a P(6)   **b** P(less than 3)

   c P(factor of 10)   **d** P(square number)

2 Are the outcomes in each case equally likely? If they are, give the probability of each outcome. If they are not, explain why.

   a The score showing when a fair die is thrown.

   b Whether a drawing pin lands point up or down when it is thrown in the air.

   c The day of the week a baby is born on.

   d The number of times I have to toss a fair coin until I see a head.

   e What the letter is, if a letter is chosen at random from a book.

3 All the counters in a bag are either green or black. At the start the probability a counter chosen at random from the bag is green is $\frac{1}{4}$. Counters are not replaced when they are taken out.

The first two counters taken out of the bag are green.

What is the least number of black counters that must still be in the bag?

4 A class of 24 students are each given a similar biased dice and asked to record how many times they saw a six when they threw their dice 20 times.

7	3	6	6	8	9	5	7
6	8	7	8	6	7	4	9
7	7	6	7	8	6	9	5

   a Give the highest and lowest relative frequencies found by any of the students.

   b Find the best possible estimate of the probability of a six showing on the biased dice from the results given.

5 For the following events give a value between 0 and 1 for the probability of the event described happening.

If you have to estimate the probability then write (est) after your answer.

   a You score a 4 when you throw a fair dice

   b A set of cards has the numbers 1-9 written on one of each of the cards. You pick a card at random and it is an even number.

   c You score less than 7 when you throw a fair dice.

   d In their 11th match of a league season the top team is playing at home to the bottom team. The match is a draw.

   e Your score is a factor of 60 when you throw a fair dice.

6 Michelle wonders how often there is a difference of more than 2 when you throw a pair of fair dice. She does an experiment, noting down how often it happens after each block of 10 throws.

5	4	5	3	2	4	6
3	3	3	5	4	2	4

   a Give the relative frequencies at the end of each block of 10 throws.

The first two are $\frac{5}{10} = 0.5$, and $\frac{9}{20} = 0.45$.

   b What is the best estimate Michelle has of the probability of getting a difference of more than 2?

7 Three friends were investigating how often a biased dice showed a six.
Sergio got 10 sixes from 40 throws.
Annalise got 8 sixes from 40 throws
Dominika got 21 sixes from 80 throws.
Calculate the best estimate of throwing a six.

8.3 Theoretical probability

- You use theoretical probability for fair activities, for example rolling a fair dice.
- For unfair or biased activities, or where you cannot predict the outcome, you use experimental probability (relative frequency).
- Relative frequency is the proportion of successful trials in an experiment.

Theoretical probability
$= \dfrac{\text{number of favourable outcomes}}{\text{total number of outcomes}}$

Relative frequency
$= \dfrac{\text{number of successful trials}}{\text{total number of trials}}$

HOW TO

To compare theoretical probability with relative frequency

1. Start by assuming that the activity is fair and calculate the theoretical probability of the event.
2. Use the results of the experiment to calculate the relative frequency.
3. If the relative frequency and theoretical probability are very different, then the activity could be biased.

The more trials you carry out, the more reliable the relative frequency will be.

EXAMPLE

Kaseem rolled a dice 50 times and in 14 of those he scored a 2.
Is the dice biased towards 2? Explain your answer.

① Theoretical probability of rolling a 2 $= \dfrac{1}{6} = 0.166\ldots$ If the dice is fair, then every outcome is equally likely.

② Relative frequency of rolling a 2 $= \dfrac{14}{50} = 0.28$ Converting to decimals makes it easier to compare.

③ The relative frequency is higher than the theoretical probability.
The dice appears to be biased towards 2.

EXAMPLE

A fair spinner is divided into blue, red and yellow sectors.
Valerie knows that the blue sector is one quarter of the spinner.
Valerie can't remember the size of angle x, but she knows that it is either 150° or 160°.
She spins the spinner 100 times and records her results in a table.

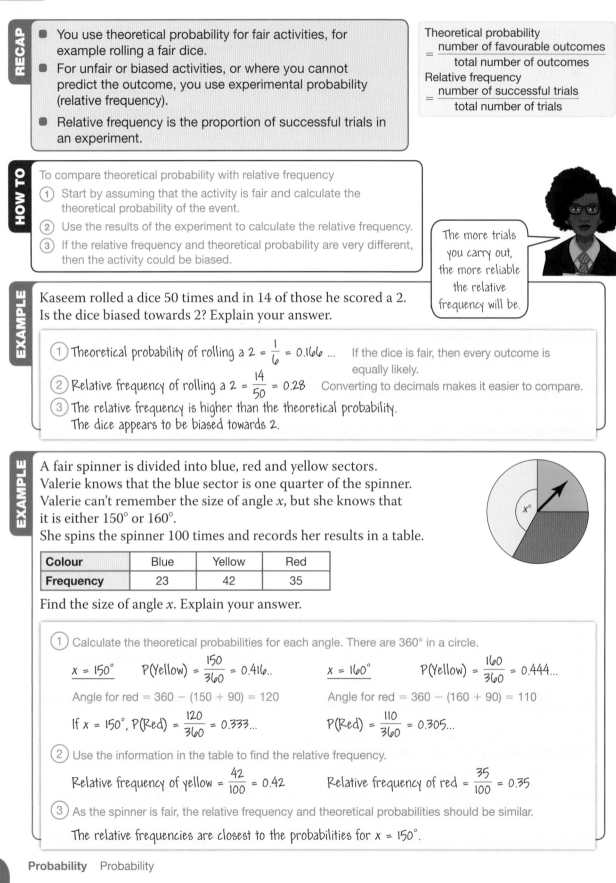

Colour	Blue	Yellow	Red
Frequency	23	42	35

Find the size of angle x. Explain your answer.

① Calculate the theoretical probabilities for each angle. There are 360° in a circle.

$x = 150°$ $P(\text{Yellow}) = \dfrac{150}{360} = 0.416\ldots$ $x = 160°$ $P(\text{Yellow}) = \dfrac{160}{360} = 0.444\ldots$

Angle for red $= 360 - (150 + 90) = 120$ Angle for red $= 360 - (160 + 90) = 110$

If $x = 150°$, $P(\text{Red}) = \dfrac{120}{360} = 0.333\ldots$ $P(\text{Red}) = \dfrac{110}{360} = 0.305\ldots$

② Use the information in the table to find the relative frequency.

Relative frequency of yellow $= \dfrac{42}{100} = 0.42$ Relative frequency of red $= \dfrac{35}{100} = 0.35$

③ As the spinner is fair, the relative frequency and theoretical probabilities should be similar.

The relative frequencies are closest to the probabilities for $x = 150°$.

Probability Probability

Exercise 8.3A

1 The table below shows the gender and hair colour of 30 students in a class.

	Fair	Dark
Male	7	9
Female	8	6

A student is chosen at random from the class. What is the probability that the student is

a a girl **b** a dark-haired boy.

2 Amber rolled a dice 100 times and in 71 of those rolls her number was a factor of 12.

Is the dice biased towards 5?

Give your reasons.

3 a Kaylee makes a three-digit number using the digits 1, 2 and 3. What is the probability that her number is greater than 230?

b Bridget makes a three-digit number using the digits 2, 5 and 9. What is the probability that her number is less than 500?

4 *ABCD* is a square and *E* and *F* are the midpoints of sides *AD* and *BC*.

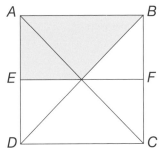

If a point is selected at random in the square what is the probability it is in the shaded area?

5 The square with sides 4 cm shown has four quarter circles shaded. A point is chosen at random in the square. Calculate the probability the point is in the shaded area.

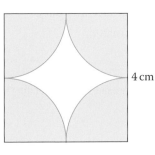

4 cm

6 Bag 1 contains 8 red balls and 12 green balls. Bag 2 contains 5 red balls and 7 green balls. The balls are identical apart from colour.

a You choose a ball from bag 1 at random. What is the probability you choose a green ball?

All the balls are now put into one bag and you choose a ball at random.

b What is the probability you choose a red ball?

7 The coloured sections on the spinner are all equal.

If the spinner lands on a green section, the score is doubled.

If the spinner lands on a blue section, the score is halved.

Shawn is playing a game with her friend Andrea who goes first, with the result shown.

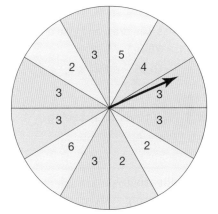

What is the probability that Shawn scores more than Andrea?

8 Iveta has a spinner divided equally into 19 sectors.

The sectors are white, black and red.

There are the same number of black and red sectors.

Iveta spins the spinner 80 times and records her results in a table.

Colour	White	Black	Red
Frequency	17	32	31

Explain why the spinner could be biased.

8.4 Mutually exclusive events

● Two events are **mutually exclusive** if they cannot both happen at the same time.

A set of events is exhaustive if they include all possible outcomes.

EXAMPLE

You throw an ordinary dice. Here are three possible events.

A: a prime number   B: an odd number   C: a square number

Which of these pairs of events are mutually exclusive?

a A and B   **b** A and C   **c** B and C

> A: 2, 3 and 5   B: 1, 3 and 5   C: 1 and 4
> **a** A and B both contain 3 and 5, so not mutually exclusive
> **b** A and C have nothing in common, so they are mutually exclusive
> **c** B and C both contain 1, so not.

● If a set of mutually exclusive events are also **exhaustive** then their probabilities sum to 1.

EXAMPLE

The probability that a hockey team wins a match is 0.65. The probability of a draw is 0.2. What is the probability that they lose the match?

> The outcomes win, draw and lose are mutually exclusive.
> 0.65 + 0.2 + P(lose) = 1
> P(lose) = 1 − (0.65 + 0.2) = 0.15

● Probability of an event happening = 1 − probability of event not happening
  P(A) = 1 − P(not A)

EXAMPLE

The probability that you are selected for a random body search by an electronic security monitoring system is 0.06. What is the probability you will not be selected?

> 1 − 0.06 = 0.94   The probabilities add to 1.

EXAMPLE

The probability that a child in a nursery class has a cat is 0.3.

  a Find the probability that the child does not have any pets.
  b Find the probability that the child has a hamster.

These events may not be mutually exclusive.

> **a** Not known, they can have other animals as pets.
> **b** Not known, a child may have either a cat or a hamster, both or neither!

Exercise 8.4S

1 For each of the following pairs of events say whether or not they are mutually exclusive.

   a P: a girl has red hair
       Q: she has brown eyes

   b F: a boy gets the same score on 2 dice
       G: his total score is odd

   c M: a girl gets the same score on 3 dice
       N: her total score is odd

   d X: a girl gets the same score on 3 dice
       Y: her total score is a multiple of 6.

2 A pack of cards contains 60 cards in four colours: black, red, green and blue.
There are 15 of each colour.
The black cards carry the numbers 1 to 15.
The red cards are multiples of 2.
The green cards are multiples of 3.
The blue cards are multiples of 6.
The top card is turned over.
For each pair of events, say whether or not they are mutually exclusive.

   a A: the card is black, and
       B: it is an even number

   b C: the card is red, and
       D: it is an odd number

   c E: the card is green, and
       F: it is a factor of 20

   d G: the card is blue, and
       H: it is a prime number

3 All the counters in a bag are either green or black. The probability a counter chosen at random from the bag is green is $\frac{1}{3}$.

   a Find the probability that a counter chosen at random is blue.

   b Find the probability that a counter chosen at random is black.

4 All the counters in a bag are either green, white or black.
The probability a counter chosen at random from the bag is green is $\frac{1}{3}$, and for white it is $\frac{1}{2}$.
What is the probability a counter chosen is

   a black   **b** not white.

5 A fair dice has a triangle, square, circle, rectangle, sphere and cylinder showing on the six sides. The dice is rolled.
Find the probability that

   a it shows a 3D shape

   b it does not show a 2D shape with straight edges.

6 Olympic, United and City are the names of 3 teams playing in a 5-a-side league. The probability Olympic will win the league is 0.2, which is the same as City.
United are twice as likely to win the league.
Find the probability that

   a United win the league

   b City do not win the league

   c the league is won by a team other than one of these three teams.

7 Seamons has two packs of 12 cards, each labelled 1 to 12. He removes all the multiples of 4 from one of the packs, combines the two packs and shuffles them.
He turns over the top card. Find the probability that the card is

   a a multiple of 6

   b an even number

   c not a prime number.

8 A biased dice has the probabilities shown in the table.

Score	1	2	3	4	5	6
Probability	0.1	0.2	x	0.3	0.1	0.1

Find the probability of getting

   a a three

   b an even number

   c not a prime number

*9 A bag contains at least green, blue and white balls. The probability a ball chosen at random is green is $\frac{1}{7}$. Blue is twice as likely to be chosen as green and white is twice as likely to be chosen as blue. Explain why you know there are no other colours in the bag.

1262, 1263   SEARCH

8.4 Mutually exclusive events

- Events are mutually exclusive if they cannot happen at the same time.
- If the events are also exhaustive, then the probabilities sum to 1.
- $P(A) = 1 - P(\text{not } A)$.

You could list the outcomes in a table.

HOW TO

① Read the question carefully and think about the possible outcomes. It often helps to list them.

② Use what you know about mutually exclusive and exhaustive events to calculate probabilities.

EXAMPLE

A red and a blue dice are thrown together and the highest score is recorded.
Find the probability that the score recorded is 5 or less.

① There are 36 possible outcomes, and these can be shown in a table.

	1	2	3	4	5	6
1	1	2	3	4	5	6
2	2	2	3	4	5	6
3	3	3	3	4	5	6
4	4	4	4	4	5	6
5	5	5	5	5	5	6
6	6	6	6	6	6	6

The complementary event is just another way of saying 1 minus the event.

② It is easier to find the **complementary** event

$$(5 \text{ or less}) = 1 - P(6)$$
$$= 1 - \frac{11}{36}$$
$$= \frac{25}{36}$$

EXAMPLE

There are a number of balls in a bag, identical except for their colour.

A ball is taken from the bag at random.

The probability of taking

- green is $\frac{1}{6}$
- blue is $\frac{1}{4}$
- black is $\frac{1}{2}$.

Are any other colours in the bag? Give your reasons.

② If the events are exhaustive then their probabilities add up to 1.

$$\frac{1}{6} + \frac{1}{4} + \frac{1}{2} = \frac{2}{12} + \frac{3}{12} + \frac{6}{12} = \frac{11}{12}$$

The probabilities add up to a number less than 1. There must be at least one other colour.

Exercise 8.4A

1 A red and a blue dice are thrown together and the difference between the scores is recorded.

   a Draw a table to show the possible outcomes.

   b What is the probability of a difference of
     i exactly one
     ii zero
     iii more than 1?

2 There are a number of balls in a bag, identical except for their colour.
A ball is taken from the bag at random. Serena says that the probability of taking a green is $\frac{1}{4}$, taking a blue is $\frac{1}{3}$ and taking a black is $\frac{1}{2}$.
Can Serena be correct? Give reasons for your answer.

3 There are a number of balls in a bag, identical except for their colour.
A ball is taken from the bag at random. Arinda says that the probability of taking a green is $\frac{1}{6}$, taking a blue is $\frac{1}{3}$, taking a white is $\frac{1}{8}$, and taking a black is $\frac{3}{8}$.

   a Can you tell if Arinda is correct? Give reasons for your answer

   b Can you tell if there are any other colours? Give reasons for your answer.

4 A bag of sweets contains 4 vanilla fudge pieces, 3 white chocolate caramels, 3 chocolate covered fudge pieces, and 2 caramel fudge pieces.
What is the probability that a sweet chosen at random contains

   a chocolate

   b fudge

   c caramel?

Why do these probabilities add to more than 1?

5 A set of cards with the numbers 1 to 10 is shuffled and a card chosen at random. Here are four possible events.
A a prime number   **B** a factor of 36
C an even number   **D** an odd number

   a List any pairs of mutually exclusive events.

   b List any pairs of exhaustive events.

   c Explain why A and C are not mutually exclusive.

6 The table below shows the gender and hair colour of the 28 students in a class.

	Fair	Dark	Red
Male	6	9	3
Female	7	6	0

A student is chosen at random from the class, and here are four possible events.
M the student is male
F the student is female
R the student has red hair
D the student has dark hair

Giving a reason, explain whether the following statements are true or false.

   a M and F are mutually exclusive.

   b M and F are exhaustive.

   c M and D are mutually exclusive.

   d F and R are mutually exclusive.

7 A red and a blue dice are thrown together and the blue score is subtracted from the red score.

   a Draw up a table to show the possible outcomes.

   b What is the probability the score is
     i -2 **ii** 5   **iii** less than 5?

8 The probabilities for a biased dice are shown in the table.

Score	1	2	3	4	5	6
Probability	0.1	0.2			0.2	0.1

Getting a 4 is three times as likely as the probability getting a 3.
Find the probability of getting a

   a 3   **b** 4   **c** prime number.

1262, 1263   SEARCH

Summary

Checkout

You should now be able to...

Test it

Questions

✔ Use experimental data to estimate probabilities and expected frequencies.	1 – 3
✔ Calculate theoretical probabilities and expected frequencies using the idea of equally likely events.	1 – 3
✔ Compare theoretical probabilities with experimental probabilities.	3
✔ Recognise mutually exclusive events and exhaustive events and know that the probabilities of mutually exclusive exhaustive events sum to 1.	4 – 7

Language	Meaning	Example
Trial	An activity or experiment.	Rolling a single dice.
Outcome	The result of a trial.	Rolling a single dice and getting a 2.
Event	One or more outcomes of a trial.	When you roll a single dice you can get a 1, 2, 3, 4, 5 or 6.
Impossible	It cannot happen. The probability of it happening is 0.	You roll a single dice and the number is even and odd.
Certain	It must happen. The probability of it happening is 1.	You roll a single dice and the number is less than 7.
Likely	It has a better chance of happening than not happening. The probability is more than 0.5 and less than 1.	You roll a dice and the number is greater than 1.
Unlikely	It has a worse chance of happening than not happening. The probability is more than 0 and less than 0.5.	You roll a dice and the number is greater than 4.
Even chance	It has exactly the same chance of happening as not happening. The probability of it happening is 0.5.	You flip a coin and get a head.
Relative frequency	The **experimental probability** of an outcome after several trials is $$\frac{\text{Number of times the outcome happened}}{\text{Number of times the activity was done}}$$	A drawing pin is thrown 100 times. It lands point down 72 times. The relative frequency of the pin landing point down is $\frac{72}{100} = 0.72$
Expected frequency	How many times you expect the outcome to happen. Number of trials × probability of the event	The expected frequency of getting a 5 when you roll a fair dice 40 times is $\frac{1}{6} \times 40 = 6.666... = 6.\dot{6}$ about 7 times.
Theoretical probability	The predicted value found by $$\frac{\text{Number of ways the outcome could happened}}{\text{Number of possible outcomes}}$$	There is 1 way of rolling a two on a fair dice. There are 6 different possible outcomes. The theoretical probability of rolling a two is $\frac{1}{6}$.
Bias/ Biased	All outcomes are not equally likely.	A dice has the numbers 1, 1, 2, 3, 4, 5 on its faces. It is more likely that you will roll a 1 than any of the other outcomes. The dice is biased.
Equally likely	All outcomes have the same probability of happening.	When you roll a fair dice all the outcomes {1, 2, 3, 4, 5, 6} are equally likely. They all have a probability of $\frac{1}{6}$.

Review

1 Jake counts the spots on a number of ladybirds he finds in a park.

Number of Spots	Frequency
2	8
7	20
10	12
other	10

a What is the relative frequency of ladybirds with 7 spots?

b Assuming there are 500 ladybirds in the park, estimate how many will have exactly 10 spots.

2 A spinner with four colours on it is repeatedly spun and the results recorded.

Outcome	Relative frequency
Blue	0.6
Red	0.16
Green	
Yellow	0.04

a What is the relative frequency of green?

b Use these results to estimate how many times the spinner would land on red if it were spun 150 times.

3 An unbiased dice is thrown 20 times and the number 5 occurred 2 times.

a What is the relative frequency of a 5?

b What is the theoretical probability of a 5?

The dice is now thrown an additional 100 times.

c What would you expect to happen to the relative frequency of a 5?

4 A bag contains 7 white and 3 black counters. A counter is selected at random.

What is the probability the counter is

a black

b red

c white or black?

5 A fair coin is flipped twice. What is the probability of getting 'heads' twice?

6 A cupboard contains lots of textbooks. The probability that a randomly selected book is a maths textbook is $\frac{3}{11}$. What is the probability that it is not a maths textbook?

7 Sam always takes either a cheese, tuna or chicken sandwich for lunch each day. The probability Sam takes a cheese sandwich is $\frac{1}{7}$. He is twice as likely to take a tuna sandwich as a cheese sandwich.

Calculate the probability of taking

a a tuna sandwich

b a chicken sandwich.

What next?

Score			
	0 – 3		Your knowledge of this topic is still developing. To improve look at MyMaths: 1209, 1210, 1211, 1262, 1263, 1264
	4 – 6		You are gaining a secure knowledge of this topic. To improve look at InvisiPens: 08Sa-e
	7		You have mastered these skills. Well done you are ready to progress! To develop your exam technique look at InvisiPens: 08Aa-e

Assessment 8

1   Amanda says that it is impossible to thrown a coin five times and see tails every time.
    Is she correct? Give a reason for your answer.   [1]

2   A fair coin is thrown and lands on heads twice. What is the probability of a head on
    the third throw? Explain your answer.   [2]

3   A dice is thrown 100 times and the following
    results are obtained.

Score	1	2	3	4	5	6
Frequency	14	17	22	18	15	14

    Calculate the relative frequency of scoring an odd number.   [2]

4   One in every eight children in a town has asthma. How many children would
    you expect to have asthma

    a   in the school of 1600 children   [1]

    b   in Green Lane where 14 children live   [1]

    c   in Mr Knowles' class of 34 children?   [1]

5   The sides of a biased spinner are labelled 1, 2, 3, 4 and 5.
    The probability that the spinner will land on
    each of the numbers 1, 3 and 4 is given in
    the table.

Number	1	2	3	4	5
Probability	x	0.14	0.26	0.22	x

    The probability that the spinner lands on either of the numbers 1 or 5 is the same.

    a   Valerie says that $x = 0.2$. Explain why Valerie wrong and find the value of x.   [3]

    Natalia spins the spinner 200 times.

    b   How many times should she expect the spinner to land on 3?   [1]

    c   How many times should Natalia expect the spinner to land on an odd number?   [2]

    d   How many times should Natalia expect the spinner to land on a prime number?   [2]

6   Adam and Ben buy identical packets of mixed vegetable seeds. They pick out the pea
    seeds from their packets. Their results are shown in the tables.

Adam's results

Number of packets	Number of seeds in each packet	Number of pea seeds in each packet
5	30	5, 5, 8, 3, 6

Ben's results

Number of packets	Number of seeds in each packet	Total number of pea seeds
10	30	51

    a   Write down an estimate of the probability of selecting a pea seed at random from a
        packet using

        i   Adam's results   ii   Ben's results.   [3]

    b   Whose results are likely to give the better estimate of the probability?
        Explain your answer.   [2]

7   Jessica buys 20 bottles of juice.
    There were four bottles each of apple, blueberry, grape, orange and pineapple.

    a   Jessica takes a bottle from the fridge at random. What is the probability that it
        is apple juice?   [1]

    b   Jessica's son drinks all of the orange juice. Jessica takes one bottle at random from the fridge.
        What is the probability that it is either grape or pineapple juice? Write your answer
        in its simplest form.   [2]

8 Harriet buys a box of cat food that contains 30 individual packets.
The probability that a tuna packet is chosen from the box is 0.3.
How many packets of tuna packets are in the box? [2]

9 The probability that the 8:45 train into Manchester Piccadilly station arrives early is 0.18.
The probability that the train arrives on time is 0.31.

    a A frustrated passenger complains that the train is late more often than it is on time or early. Is the passenger correct? [2]

    b Dave takes 225 journeys on this train during one year.
How many days can he expect to be on time? [2]

10 The table below shows the probabilities of selecting raffle tickets from a drum.
The tickets are coloured red, white or blue and numbered 1, 2, 3 or 4.
A Raffle Ticket is taken at random from the drum.

    a Calculate the probability that

		Number			
		1	**2**	**3**	**4**
Colour	**Red**	$\frac{2}{25}$	$\frac{1}{25}$	$\frac{3}{50}$	$\frac{1}{10}$
	White	$\frac{9}{50}$	$\frac{7}{50}$	$\frac{3}{25}$	$\frac{1}{50}$
	Blue	$\frac{3}{50}$	$\frac{4}{25}$	$\frac{1}{25}$	0

        i it is white and numbered 3 [1]

        ii it is numbered 1 [2]

        iii it is red [2]

        iv it is either blue or numbered 2. [3]

    b What raffle ticket is impossible to draw?
Explain your answer. [1]

11 At a busy road junction, the traffic lights operate as follows
red: 40 seconds   red and amber: 10 seconds   green: 40 seconds   amber: 10 seconds

Work out the probability of each event. Give your answer as a simplified fraction.

    a the lights show green [1] **b** there is an amber light showing [2]

    c there is not a red light showing [2] **d** either green or amber lights are showing [2]

12 a Are these pairs of events mutually exclusive?
Give reasons for your answers.

        i Red section and multiples of 3. [2]

        ii Blue section and prime numbers. [2]

        iii White section and multiples of 7. [2]

    b Grace adds 1 to every number on the spinner.
How does this change your answers to part **a**? [6]

13 During December the probability that it is snowing is 0.1. The probability that it is sunny is 0.3
Janice says that the probability that it is snowing or sunny in December is 0.4
Is Janice correct? You must give a reason for your answer. [2]

14 Rachel joins a regular pentagon, two squares and a triangle
to create a spinner.
Rachel assumes that the spinner is unbiased.

    a Find the probability of landing on the blue region of
the spinner. [4]

    b How does the assumption that the spinner is unbiased
affect your answer to part **a**? [1]

9 Measures and accuracy

Introduction

During the 1936 Olympic Games held in Berlin in Nazi Germany, the US athlete Jesse Owens broke the 100 m men's world record, with a time of 10.2 s, a record that remained unbroken for 20 years afterwards. He also took the gold medal in the long jump in the same games, and he is widely believed to have been the greatest athlete ever. It is said that Hitler was not impressed by Owens's victories, as he had hoped that the games would be a validation of Aryan supremacy.

The current 100 m world record is held by Usain Bolt of Jamaica, with a time of 9.58 s in 2009. Note the increased accuracy in the reporting of the time in 2009 compared with 1936.

What's the point?

Being able to rely on accurate decimal measurements is vital in athletics. In a 'photo-finish' with two competitors crossing the finishing line apparently together, 0.01 seconds can be the difference between gold and silver!

Objectives

By the end of this chapter you will have learned how to …

- Round numbers and measures to an appropriate degree of accuracy.
- Use approximation to make estimates.
- Check calculations using approximation and estimation.
- Use standard units of length, mass, volume, capacity, time and area.
- Use inequality notation to state error intervals and interpret limits of accuracy.

Check in

1 Work out these multiplications and divisions.

   a 40×10 **b** 14×1000 **c** 3.1×10 **d** 13.4×100

   e 6.3×1000 **f** $400 \div 10$ **g** $6000 \div 100$ **h** $430 \div 100$

   i $640 \div 1000$ **j** $3.1 \div 10$ **k** 0.034×100 **l** $0.78 \div 10$

2 What is the temperature on this thermometer?

$-10°C$ $0°C$ $10°C$

3 Gavin is running at 5 mph. He runs for 3 hours. How far does he travel?

Chapter investigation

Most of the quantities you can measure, such as length and mass, are based on the decimal number system; however time is not. During the French Revolution, it was proposed to replace the standard system of hours and minutes with French Revolutionary Time, in which time would be measured using powers of 10.

Investigate French Revolutionary Time and write a short report.

At what times would your school day start and finish using this system?

9.1 Estimation and approximation

● You can use **approximations** to make **estimates**.

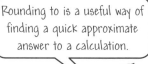

Rounding to is a useful way of finding a quick approximate answer to a calculation.

EXAMPLE

Find approximate answers to:

   a $12.3 - 8.9$ **b** $76.5 + 184.2$ **c** $20 - 14.53$

  a $12.3 - 8.9 \approx 12 - 9 = 3$ $12.3 - 8.9 = 3.4$

  b $76.5 + 184.2 \approx 80 + 200 = 280$ $76.5 + 184.2 = 260.7$

  c $20 - 14.53 \approx 20 - 15 = 5$ $20 - 14.53 = 5.47$

In a multi-stage calculation, remember the **order of operations**.

BIDMAS
Brackets
Indices (or powers)
Division or **M**ultiplication
Addition or **S**ubtraction

EXAMPLE

Estimate the answer to this calculation. $\dfrac{4.23 \times 5.89}{9.7}$

Round each of these numbers to the nearest whole number.

$$\frac{4.23 \times 5.89}{9.7} \approx \frac{4 \times 6}{10} = \frac{24}{10} = 2.4$$

● You can make estimates by rounding to one significant figure.

EXAMPLE

Estimate the answers to these calculations.

  a $\dfrac{8.93 \times 28.69}{0.48 \times 6.12}$ **b** $\dfrac{17.4 \times 4.89^2}{0.385}$

  a $\dfrac{8.93 \times 28.69}{0.48 \times 6.12} \approx \dfrac{9 \times 30}{0.5 \times 6}$ **b** $\dfrac{17.4 \times 4.89^2}{0.385} \approx \dfrac{20 \times 5^2}{0.4}$

           $= \dfrac{270}{3} = 90$ $= \dfrac{20 \times 25}{0.4} = \dfrac{500}{0.4} = \dfrac{5000}{4} = 1250$

You need to be careful when estimating powers.
For example, 1.3 is quite close to 1, but 1.3^7 is not close to 1^7.

EXAMPLE

Estimate the values of these calculations.

  a $\dfrac{3.27 \times 4.49}{1.78^2}$ **b** $\dfrac{\sqrt{2485}}{1.4^3}$ **c** $\dfrac{2.45^3}{2.5 - 2.4}$

  a 1.78^2 is 'a bit more than 3'   **b** $1.4^2 \approx 2$ so $1.4^3 \approx 2.8 \approx 3$ **c** $2.45 \approx 2.5$ and $2.5^2 \approx 6$,
        so it cancels with 3.27.   so $2.5^3 \approx 6 \times 2.5 \approx 15$

      $\dfrac{3.27 \times 4.49}{1.78^2} \approx 4.5$ $\dfrac{\sqrt{2485}}{1.4^3} \approx \dfrac{\sqrt{2500}}{3} \approx \dfrac{50}{3} \approx 17$ $\dfrac{2.45^3}{2.5 - 2.4} \approx \dfrac{15}{0.1} \approx 150$

Exercise 9.1S

1 Round each of these numbers to the nearest
  i 3 sf   ii 2 sf   iii 1 sf.

  a 8.3728   **b** 18.82

  c 35.84   **d** 278.72

  e 1.3949   **f** 3894.79

  g 0.008 372   **h** 2399.9

  i 8.9858   **j** 14.0306

  k 1403.06   **l** 140 306

2 Use rounding to estimate each of these calculations.

  a $43 + 189$ **b** $1563 + 28$

  c $1602 + 2654$ **d** $1451 + 6985$

3 Estimate answers to these calculations.

  a $4.88 + 3.07$ **b** $216 + 339$

  c $0.0049 + 0.003\,02$ **d** $43.89 - 28.83$

  e 3.77×0.85 **f** $44.66 \div 0.89$

4 Use rounding to estimate each of these calculations.

  a $27.6 + 21.7$ **b** $1623 - 897$

  c 32×2.1 **d** 2.9×23

  e 19.1×11.69 **f** 9×7.56

  g $14.25 \div 1.98$ **h** $16.3 \div 0.37$

5 By rounding all of the numbers to one significant figure, write a calculation that you could carry out mentally to estimate the answers to these calculations.

  a $355 \div 21$ **b** 39×43

  c $1053 \div 92$ **d** $4385 + 11\,655$

  e $108 + (2360 \div 52)$

6 Write a suitable estimate for each of these calculations. In each case clearly show how you estimated your answer.

  a 3.76×4.22 **b** 17.39×22.98

  c $54.31 \div 8.8$ **d** 4.98×6.12

  e $17.89 + 21.91$ **f** $34.8183 - 9.8$

7 a Irwin estimates the value of to

  $\dfrac{47.3 \times 18.9}{8.72}$ to be 100.

  Write three numbers Irwin could use to get his estimate.

7 b Sarah estimates the value of

  $\dfrac{21.4 \times 4.87}{49.8}$ to be 2.

  Write three numbers Sarah could use to get her estimate.

  c Alexandra estimates the value of

  $\dfrac{69.7 + 47.16}{4.13 \times 3.078}$ to be 10.

  Write four numbers Alexandra could use to get her estimate.

8 Write a suitable estimate for each of these calculations. In each case clearly show how you estimated your answer.

  a $\dfrac{4.59 \times 7.9}{19.86}$ **b** $\dfrac{5.799 \times 3.1}{8.86}$

  c $\dfrac{32.91 \times 4.8}{3.1}$ **d** $\dfrac{29.91 \times 38.3}{3.1 \times 3.9}$

  e $\dfrac{16.2 \times 0.48}{0.23 \times 31.88}$ **f** $\dfrac{45.15 \times 11.08}{0.46 \times 108.52}$

9 Write a suitable estimate for each of these calculations. In each case, clearly show how you estimated your answer.

  a $\{4.8^2 + (4.2 - 0.238)\}^2$

  b $\dfrac{63.8 \times 1.7^2}{1.78^2}$

  c $\{9.8^2 + (9.2 - 0.438)\}^2$

10 Use rounding to estimate each of these calculations.

  a $\dfrac{9.7^2 - 3.9^2}{2 \times 19}$ **b** $\dfrac{35.9 \times 1.9}{0.49}$

  c $\dfrac{5.8^2 - 1.5 \times 1.8^2}{7.59 + 6.89}$

11 Estimate these square roots mentally, to 1 decimal place.

  a $\sqrt{2}$ **b** $\sqrt{8}$ **c** $\sqrt{10}$

  d $\sqrt{15}$ **e** $\sqrt{20}$ **f** $\sqrt{26}$

  g $\sqrt{32}$ **h** $\sqrt{45}$ **i** $\sqrt{70}$

  Use a calculator to check your estimates.

12 Use rounding to estimate each of these calculations.

  a $\sqrt{2.03 \div 0.041}$ **b** $\sqrt{27.6 \div 0.57}$

  c $7.78 + \dfrac{4.79}{\sqrt{47.7 + 7.56}}$

 Q 1002, 1004, 1005, 1043   SEARCH

9.1 Estimation and approximation

HOW TO

1. Read the question carefully, and decide which calculation you need to carry out.
2. Estimate using approximations.
3. Round your answer to a suitable degree of accuracy.

EXAMPLE

Monty is planting a new rectangular lawn. His lawn is 4.8 m by 5.1 m. He needs between 45 and 60 grams of seed per square metre of lawn. Seed can be bought in 800 g packets.

a How many packets of seeds should he buy?   **b** Justify your answer to **a**.

a ① It is better to have too much than too little in this case, so round more up than down.

Area of lawn ≈ 5 × 5 ≈ 25 m²   ② Estimate by rounding to 1 sf.

③ Amount of seed ≈ 25 × 60 g ≈ 1500 g   Use the larger amount

He should buy 2 packets.

b Lowest amount = 4.8 × 5.1 × 45 = 1101 g so one packet is not enough.

Highest 25 × 60 = 1500 g, which is still less than 2 packets (1600 g).

EXAMPLE

Lewis hires a car to travel 118 miles. The hire charge for a small car is £39.98, plus 12 p per mile. A larger car costs £68.92, plus 15 p per mile. Estimate the difference in cost between the two cars.

① Small = 118 × 0.12 + 39.98   Large = 118 × 0.15 + 68.92

② Round all figures a suitable degree of accuracy.

≈ 120 × 0.12 + 40   ≈ 120 × 0.15 + 70

120 × 0.12 = 14.4   120 × 0.15 = 60 × 2 × 0.15

≈ 14 + 40   ≈ 60 × 0.3 + 70

≈ 54   ≈ 18 + 70

③ 88 − 54 = £34   ≈ 88

The large car is about £34 more.

> Your answers may not match other people's. It does not necessarily mean they are wrong.

EXAMPLE

A jellybean can be thought of as a cylinder with diameter 1.5 cm and length 2 cm.

Estimate the number of jellybeans that can fit in a 1 litre jar.

① Volume of a jellybean = $\pi r^2 l$ = $\pi \times 0.75^2 \times 2$

② $\pi \approx 3$, $0.75^2 = \left(\frac{3}{4}\right)^2 = \frac{9}{16} \approx 0.5$

Volume of a jellybean ≈ 3 × 0.5 × 2 ≈ 3 cm³

1 litre = 1000 cm³

③ Number of jellybeans ≈ 1000 ÷ 3 ≈ 300

Exercise 9.1A

1 **a** Write a calculation that you can do in your head to estimate the answers to these calculations.

    i $258 + 362$ **ii** $64 \div 27$

    iii 62.7×211.8 **iv** $96.7 - 64.8$

  b Explain carefully whether each of the calculations that you wrote in part **a** will produce an overestimate or an underestimate of the actual result.

  c Use a calculator to find the exact answers, and check your answers to part **b**.

2 Estimate which calculation is

   zero   more than 100   negative

  a $97.31 \times 2 - 26.3 \times 7.4$

  b $15.2^2 \times 7 \div 2$

  c $\dfrac{2.3^2 - 3.2 + 2.09}{0.7 \times 5 - 1.4 \times 3.8}$

3 Dave is making a garden patio.

The patio is 3.2 m by 4.8 m.

The patio slabs are 0.8 m by 0.5 m.

Estimate the number of patio slabs that Dave needs.

4 A grain of rice can be thought of as a cylinder with a 1 mm radius and 8 mm length.

A 1 kg bag of rice measures 18 cm × 12 cm × 6 cm.

  a Estimate the number of grains of rice in a 1 kg bag.

  b The weight of 1000 grains of rice is approximately 22 grams. Does your estimate in part **a** agree with this statement?

5 Caroline has 490 MB of space left on her camera's SD card. She is taking photographs which are 6.448 MB in size.

  a Approximately how many can she take?

  b She can send up to 30 MB as an attachment to an email. How many photos can she send?

6 Gosia is making jam to fill some 4 oz jars she bought at a car boot sale. She knows that 1 oz = 28.3495 ml.

  a Estimate how many of her jars she can **fill** if she makes 550 ml of jam.

  b About how much jam would she need to make to fill one more jar.

7 Paul's mobile phone gives a summary of his bank account. On 28 February he had about £2000 in his account.

06/03	ATM TOWN	−£100.00
06/03	DC DIRECT DEBIT	−£156.45
06/03	DDEBIT	−£99.30
06/03	CASH TRM MAR 06	−£80.00
04/03	CHQ IN	+£84.37
02/03	ABC & G	+£59.21
01/03	SALARY	+£1758.64
28/02	BALANCE	£2058.63

  a Approximately how much was paid in during the first week of March? (+)

  b Approximately how much was paid out? (−)

  c Approximately how much money did he have in the bank on 6 March?

*8 Greta is driving from London to Newcastle, a distance of 278 miles. She estimates she can drive about 60 miles every hour. Her diesel car does 43 miles on one gallon of diesel.

  a Estimate how long the journey takes.

  b She thinks 1 gallon is about 5 litres. Estimate how many litres she thinks she will use.

  c One litre of diesel costs £1.29. Approximately what will the journey from London to Newcastle cost Greta?

  d In fact 1 gallon = 4.54609 litres. Will she use more or less diesel than she estimated?

Q 1002, 1004, 1005, 1043   SEARCH

9.2 Calculator methods

You can use a scientific calculator to carry out more complex calculations that involve decimals.

EXAMPLE

Use your calculator to work out this calculation

$$3.46 + 2.9 \times 4.8$$

Give your answer to 1 decimal place.

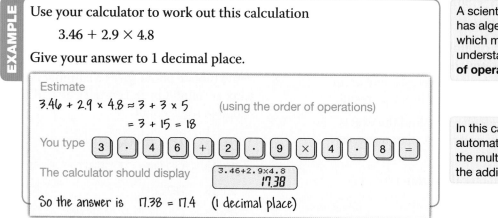

Estimate

$3.46 + 2.9 \times 4.8 \approx 3 + 3 \times 5$ (using the order of operations)

$= 3 + 15 = 18$

You type [3] [.] [4] [6] [+] [2] [.] [9] [×] [4] [.] [8] [=]

The calculator should display
```
3.46+2.9x4.8
17.38
```

So the answer is   17.38 = 17.4   (1 decimal place)

> A scientific calculator has algebraic logic, which means that it understands the **order of operations**.
>
> In this case the calculator automatically works out the multiplication before the addition.

A scientific calculator has **bracket** keys.

EXAMPLE

a   Use your calculator to work out $(3.9 + 2.2)^2 \times 2.17$.

Write all the figures on your calculator display.

a   Estimate

$(3.9 + 2.2)^2 \times 2.17 \approx (4 + 2)^2 \times 2$ (using the order of operations)

$= 6^2 \times 2$

$= 36 \times 2$

$= 72$

You type [(] [3] [.] [9] [+] [2] [.] [2] [)] [x^2] [×] [2] [.] [1] [7] [=]

The calculator should display
```
(3.9+2.2)2x2.17
80.7457
```
= 80.7457

1.4 You can use the bracket keys on a scientific calculator to do calculations where the **order of operations** is not obvious.

EXAMPLE

a   Use a calculator to work out the value of

$$\frac{21.42 \times (12.4 - 6.35)}{(63.4 + 18.9) \times 2.83}$$

Write all the figures on the calculator display.

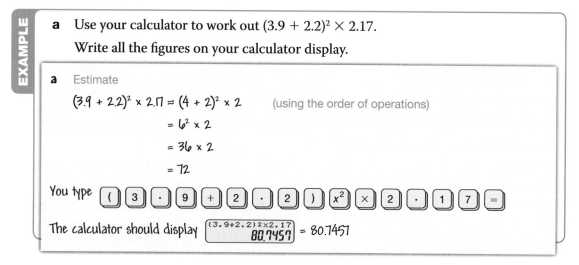

a   Rewrite the calculation as $(21.42 \times (12.4 - 6.35)) \div ((63.4 + 18.9) \times 2.83)$

Type this into the calculator:
```
(21.42x(12.4-6.35))÷((63.4+18.9)x2.83)
```
→
```
(21.42x(12.4-
0.556401856
```

So the answer is 0.556 401 856.

Estimate: $\dfrac{20 \times (12 - 6)}{(60 + 20) \times 3} = \dfrac{120}{240} = 0.5$

Exercise 9.2S

1 Use your calculator to work these out. In each case, first write an estimate for your answer. Write the answer from your calculator to 1 decimal place.

   a $3.4 + 6.2 \times 2.7$

   Estimate $\qquad 3.4 + 6.2 \times 2.7 \approx 3 + 6 \times 3$
   $$= 3 + 18$$
   $$= \underline{\qquad}$$

   Using the calculator $\quad 3.4 + 6.2 \times 2.7 = \underline{\qquad}$

   b $1.98 \times 11.7 - 4.6$

   c $7.8 + 19.3 \div 4.12$

   d $2.09 \times 2.87 + 3.25 \times 1.17$

   e $13.67 \div 1.75 + 3.24$

   f $1.2 + 3.7 \times 0.5$

2 Use your calculator to work out each of these calculations. Write all the figures on your calculator display.

   a $(2.3 + 5.6) \times 3^2$

   b $2.3^2 \times (12.3 - 6.7)$

   c $(2.8^2 - 2.04) \div 2.79$

   d $7.2 \times (4.3^2 + 7.4)$

   e $11.33 \div (6.2 + 8.3^2)$

   f $(2.5^2 + 1.37) \times 2.5$

3 Calculate each of these, giving your answer to 1 decimal place.

   a $\dfrac{5.4 + 3.8}{4.5 - 2.9}$ **b** $\dfrac{3.8 - 1.67}{4.3 - 2.68}$

   c $\dfrac{12.4 + 5.8}{14.5 - 3.9}$ **d** $\dfrac{13.08 - 2.67}{2.13 + 2.68}$

4 Use a calculator to evaluate these, giving each answer to 2 sf.

   a $3.2 \times (2.8 - 1.05)$

   b $2.8^2 \times (9.4 - 0.083)$

   c $16 \div (5.1^2 \times 7.2)$

   d $(3.8 + 8.9) \times (2.2^2 - 7.6)$

   e $1.8^3 + 4.7^3$

   f $52 \div (4.6 - 1.8^2)$

5 Use a calculator to evaluate these. Give each answer to 2 dp.

   a $\dfrac{4.59 \times 7.9}{19.86}$ **b** $\dfrac{5.799 \times 3.1}{8.86}$

   c $\dfrac{32.91 \times 4.8}{3.1}$ **d** $\dfrac{29.91 \times 38.3}{3.1 \times 3.9}$

   e $\dfrac{16.2 \times 0.48}{0.23 \times 31.88}$ **f** $\dfrac{45.15 \times 11.08}{0.46 \times 108.52}$

6 Use a calculator to evaluate these. Give each answer to 2 dp.

   a $\{4.8^2 + (4.2 - 0.238)\}^2$

   b $\dfrac{63.8 \times 1.7^2}{1.78^2}$

   c $\{9.8^2 + (9.2 - 0.438)\}^2$

7 Compare your answers to questions **5** and **6** with the estimates from Exercise 9.1S questions **8** and **9**.

8 Use your calculator to work out each of these. Write all the figures on your calculator display.

   a $\dfrac{462.3 \times 30.4}{(0.7 + 4.8)^2}$ **b** $\dfrac{13.58 \times (18.4 - 9.73)}{(37.2 + 24.6) \times 4.2}$

9 Use your calculator to work out each of these. Write all the figures on your calculator.

   a $\dfrac{165.4 \times 27.4}{(0.72 + 4.32)^2}$

   b $\dfrac{(32.6 + 43.1) \times 2.3^2}{173.7 \times (13.5 - 1.78)}$

   c $\dfrac{24.67 \times (35.3 - 8.29)}{(28.2 + 34.7) \times 3.3}$

   d $\dfrac{1.45^2 \times 3.64 + 2.9}{3.47 - 0.32}$

   e $\dfrac{12.93 \times (33.2 - 8.34)}{(61.3 + 34.5) \times 2.9}$

   f $\dfrac{24.7 - (3.2 + 1.09)^2}{2.78^2 + 12.9 \times 3}$

10 Use a calculator to evaluate these. Give each answer to 2 dp.

   a $\sqrt{2.03 \div 0.041}$ **b** $\sqrt{27.6 \div 0.57}$

   c $7.78 + \dfrac{4.79}{\sqrt{47.7 + 7.56}}$

   Compare your answers with the estimates from Exercise 9.1S question **12**.

Q 1043, 1932, 1933 SEARCH

9.2 Calculator methods

- Use your calculator to work out more complex calculations.
- Scientific calculators apply the order of operations.

> You can toggle between fractions and decimal answers on your calculator. Try this for yourself using your calculator.

HOW TO

1. Take time to understand the situation in the question. Estimate the answer before calculating.

2. Use your calculator to work out the answer. Check that the answer agrees with your estimate.

3. Interpret the calculator display in the context of the question.

EXAMPLE

Put brackets in this expression so that its value is 16.26.

$$1.4 + 3.9 \times 2.2 + 4.6$$

1. Estimate the answer by rounding to 1 sf.
   Put brackets in the expression $1 + 4 \times 2 + 5$ so that the answer is roughly 16.

   $(1 + 4) \times 2 + 5 = 5 \times 2 + 5 = 10 + 5 = 15$

2. Type $(1.4 + 3.9) \times 2.2 + 4.6$ into your calculator.

   $(1.4 + 3.9) \times 2.2 + 4.6$

3. The calculator should display

   ```
   (1.4+3.9)×2.2+4.6
                16.26
   ```

   $= 16.26$ ✓ the correct answer

EXAMPLE

The diagram shows a box in the shape of a cuboid.

Saleem builds boxes of different sizes. He charges £7.89 for each m³ of a box's volume. Work out Saleem's charge for building this box.

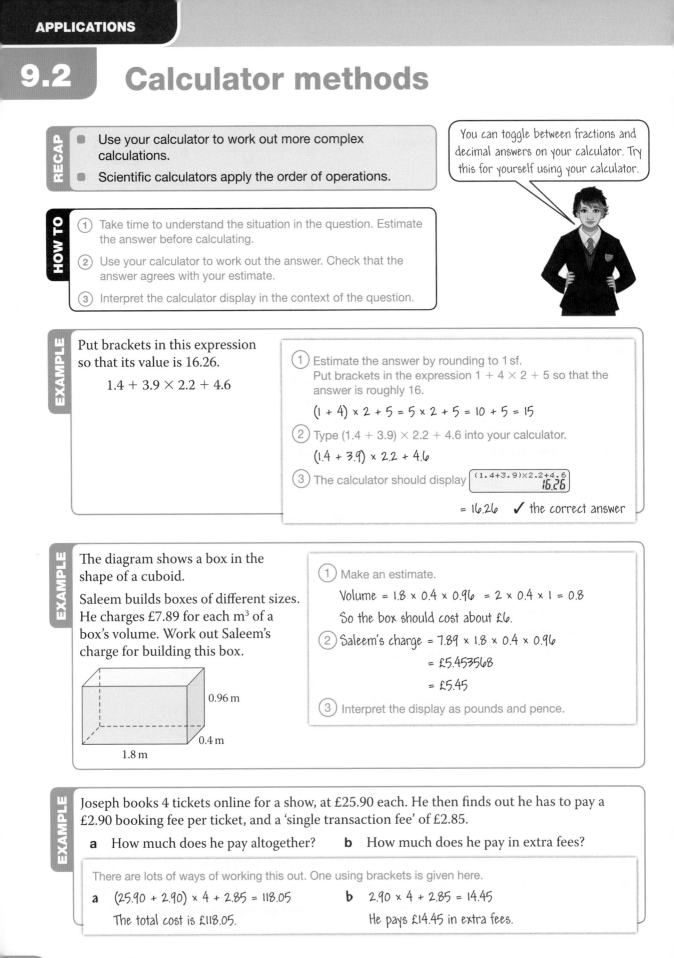

0.96 m

0.4 m

1.8 m

1. Make an estimate.

   Volume = $1.8 \times 0.4 \times 0.96 = 2 \times 0.4 \times 1 = 0.8$

   So the box should cost about £6.

2. Saleem's charge = $7.89 \times 1.8 \times 0.4 \times 0.96$

   $= £5.453568$

   $= £5.45$

3. Interpret the display as pounds and pence.

EXAMPLE

Joseph books 4 tickets online for a show, at £25.90 each. He then finds out he has to pay a £2.90 booking fee per ticket, and a 'single transaction fee' of £2.85.

a How much does he pay altogether?   **b** How much does he pay in extra fees?

There are lots of ways of working this out. One using brackets is given here.

a $(25.90 + 2.90) \times 4 + 2.85 = 118.05$ **b** $2.90 \times 4 + 2.85 = 14.45$

The total cost is £118.05.   He pays £14.45 in extra fees.

Exercise 9.2A

1 Put brackets into each of these expressions to make them correct.

a $2.4 \times 4.3 + 3.7 = 19.2$

b $6.8 \times 3.75 - 2.64 = 7.548$

c $3.7 + 2.9 \div 1.2 = 5.5$

d $2.3 + 3.4^2 \times 2.7 = 37.422$

e $5.3 + 3.9 \times 3.2 + 1.6 = 24.02$

f $3.2 + 6.4 \times 4.3 + 2.5 = 46.72$

2 Put brackets into each of these expressions to make them correct.

a $3.4 \times 2.3 + 1.6 = 13.26$

b $3.5 \times 2.3 - 1.04 = 4.41$

c $2.6 + 6.5 \div 1.3 = 7$

d $1.4^2 - 1.2 \times 2.3 = 1.748$

e $2.4^2 \div 1.8 \times 3.2 + 1.6 = 15.36$

f $3.2 + 5.3 \times 2.4 - 1.2 = 10.2$

3 Work out each of these using your calculator. In each case give your answer to an appropriate degree of accuracy.

a Véronique puts carpet in her bedroom. The bedroom is in the shape of a rectangle with a length of 4.23 m and a width of 3.6 m. The carpet costs £6.79 per m².

   i Calculate the floor area of the bedroom.

   ii Calculate the cost of the carpet which is required to cover the floor.

b Calculate $\frac{1}{3}$ of £200.

4 Elif has 110.5 hours of music downloaded on her computer.
She accidentally deletes 80.15 hours of her music!

How much music does Elif have remaining on her computer hard drive?

Give your answer

a in hours and minutes

b in minutes.

5 Barry sees a mobile phone offer.

Vericheep Fone OFFER

Monthly fee £12.99
FREE – 200 texts every month
FREE – 200 voice minutes every month

Extra text messages   3.2p each
Extra voice minutes   5.5p each

Barry decides to see if the offer is a good idea for him.

His current mobile phone offers him unlimited texts and voice minutes for £22.99 per month.

a In February, Barry used 189 texts and 348 voice minutes.
Calculate his bill using the new offer.

b In March, Barry used 273 texts and 219 voice minutes.
Calculate his bill using the new offer.

c Explain if the new offer is a good idea for Barry.

6 Aliona books 8 tickets online for a show, at £19.85 each. She then finds out he has to pay a £3.25 booking fee per ticket, and a 'single transaction fee' of £2.20.

a How much does she pay altogether?

b How much does she pay in extra fees?

*7 a The speed of sound depends, amongst other things, on the temperature of the air it travels through. If t stands for temperature in degrees Celsius, then the speed of sound, in feet per second, is

$$\sqrt{(273 + t)} \times 1087 \div 16.52$$

   i What is the speed at 0 °C?

   ii What is the speed at 18 °C?

b To convert feet per second to miles per hour, you multiply by 15, then divide by 22. Convert your answers to miles per hour.

9.3 Measures and accuracy

● You can measure **length**, **mass** and **capacity** using metric units.

A paper clip weighs 1 gram.

● Length is a measure of distance.

10 mm = 1 cm   100 cm = 1 m   1000 m = 1 km

● Mass is a measure of the amount of matter in an object.

1000 g = 1 kg   1000 kg = 1 tonne

● Capacity measures the amount of liquid in a 3D shape.

1000 ml = 1 litre   100 cl = 1 litre   1000 cm³ = 1 litre

Compound measures describe one quantity in relation to another.

● **Speed** = $\dfrac{\text{total distance travelled}}{\text{total time taken}}$ Units such as m/s; km/h

● **Density** = $\dfrac{\text{mass}}{\text{volume}}$ Units such as g/cm³; kg/m²

Use the triangle to work out which calculation to use.

EXAMPLE

Find the density of a piece of wood with cross-section area 42 cm², length 12 cm and mass 693 g.

Volume = 42 × 12 = 504 cm³
Density = mass ÷ volume
Density = 693 ÷ 504
        = 1.375 g/cm³

M = D × V
D = M ÷ V
V = M ÷ D

● Measurements are not exact. Their accuracy depends on the precision of the measuring instrument and the skill of the person making the measurement.

EXAMPLE

The mass of a meteorite is given as 235.6 g. Find the lower and upper bounds of the mass.

The mass is given correct to the nearest 0.1 g.
Lower bound is 235.55 g. Upper bound is 235.65 g.
If the mass of the meteorite is m then

235.55 ≤ m < 235.65

For continuous data, the upper bound is *not* a possible value of the data.

1.2

2.3 cm has an **implied accuracy** of 1 decimal place.

3.50 m has an implied accuracy of 2 decimal places.

If all measurements are to 1 dp, give your answers to 1 or 2 dp.

Exercise 9.3S

1 Choose one of these metric units to measure each of these items.

> millimetre gram   millilitre centimetre
> kilogram   centilitre metre   tonne
> litre   kilometre

   a your height

   b amount of tea in a mug

   c your weight

   d length of a suitcase

   e weight of a suitcase

   f distance from Paris to Madrid

   g quantity of drink in a can

   h amount of petrol in a car

   i weight of an elephant

   j weight of an apple.

Write the appropriate abbreviation next to your answers.

> Learn these equivalences.
>
> Length: 10 mm = 1 cm,
> 100 cm = 1 m, 1000 m = 1 km.
> Mass: 1000 g = kg,
> 1000 kg = 1 tonne.
> Capacity: 1000 ml = 1 litre
> 100 cL = 1 litre, 1000 cm³ = 1 litre

2 Convert these measurements to the units shown.

   a 20 mm = ___ cm **b** 400 cm = ___ m

   c 450 cm = ___ m **d** 4000 m = ___ km

   e 0.5 cm = ___ mm **f** 4.5 kg = ___ g

   g 6000 g = ___ kg **h** 6500 g = ___ kg

   i 2500 kg = ___ t **j** 3 litres = ___ ml

3 Cheryl takes 4 hours to walk 10 miles. Calculate her average speed, giving the units of your answer.

4 A cyclist travels at 15 mph for $2\frac{1}{2}$ hours. How many miles is the journey?

5 I can usually drive at an average speed of 60 mph on the motorway. How long will a 150-mile journey take?

6 A cube of side 2 cm weighs 40 grams.

   a Find the density of the material from which the cube is made, giving your answer in g/cm³.

   b A cube of side length 2.6 cm is made from the same material. Find the mass of this cube, in grams.

7 A type of emulsion paint has a density of 1.95 kg/litre. Find

   a the mass of 4.85 litres of the paint

   b the number of litres of the paint that would have a mass of 12 kg.

8 Each of these measurements was made correct to one decimal place. Write the upper and lower bounds for each measurement.

   a 5.8 m   **b** 16.5 litres

   c 0.9 kg   **d** 6.3 N

   e 10.1 s   **f** 104.7 cm

   g 16.0 km   **h** 9.3 m/s

9 Find the upper and lower bounds of these measurements, which were made to varying degrees of accuracy.

   a 6.7 m   **b** 7.74 litres

   c 0.813 kg   **d** 6 N

   e 0.001 s   **f** 2.54 cm

   g 1.162 km   **h** 15 m/s

10 Find the maximum and minimum possible total weight of

   a 12 boxes, each of which weighs 14 kg, to the nearest kilogram

   b 8 parcels, each weighing 3.5 kg.

11 Find the upper and lower bounds of these measurements, which are correct to the nearest 5 mm.

   a 35 mm   **b** 40 mm

   c 110 mm   **d** 45 cm

 Q 1006, 1067, 1121, 1246, 1968 SEARCH

9.3   Measures and accuracy

RECAP

- You can measure **length**, **mass** and **capacity** using metric units.
- A compound measure, such as speed or density, connects two different measurements.
- The upper bound is the smallest value that is greater than or equal to any possible value of the measurement.
- The lower bound is the greatest value that is smaller than or equal to any possible value of the measurement.

$$\text{Speed} = \frac{\text{distance}}{\text{time}}$$

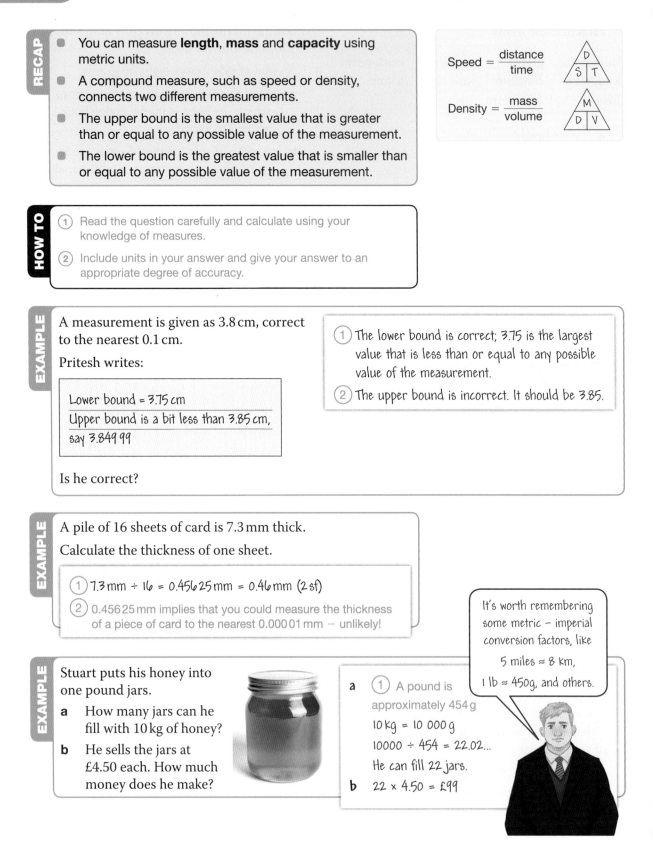

$$\text{Density} = \frac{\text{mass}}{\text{volume}}$$

HOW TO

1. Read the question carefully and calculate using your knowledge of measures.
2. Include units in your answer and give your answer to an appropriate degree of accuracy.

EXAMPLE

A measurement is given as 3.8 cm, correct to the nearest 0.1 cm.

Pritesh writes:

> Lower bound = 3.75 cm
> Upper bound is a bit less than 3.85 cm,
> say 3.849 99

Is he correct?

1. The lower bound is correct; 3.75 is the largest value that is less than or equal to any possible value of the measurement.

2. The upper bound is incorrect. It should be 3.85.

EXAMPLE

A pile of 16 sheets of card is 7.3 mm thick.

Calculate the thickness of one sheet.

1. 7.3 mm ÷ 16 = 0.456 25 mm = 0.46 mm (2 sf)

2. 0.456 25 mm implies that you could measure the thickness of a piece of card to the nearest 0.000 01 mm — unlikely!

EXAMPLE

Stuart puts his honey into one pound jars.

a How many jars can he fill with 10 kg of honey?

b He sells the jars at £4.50 each. How much money does he make?

It's worth remembering some metric – imperial conversion factors, like
5 miles ≈ 8 km,
1 lb ≈ 450 g, and others.

a 1. A pound is approximately 454 g
10 kg = 10 000 g
10000 ÷ 454 = 22.02...
He can fill 22 jars.

b 22 × 4.50 = £99

Exercise 9.3A

1 The table shows information about some journeys Shaun made in one week.
Copy and complete the table. Remember to show the units.

Distance	Time taken	Average speed
120 km	$1\frac{1}{2}$ hours	
250 miles	4 hours	
4 km		16 km/hour
	24 seconds	5 m/s
300 m	15 seconds	
0.4 m	160 seconds	
3 km		24 m/s
	20 minutes	60 km/h

2 Talal walks 5.4 km in an hour.
What is his speed in

    a m/h   **b** m/min   **c** m/s?

3 Work these out, giving your answers to a suitable degree of accuracy.

    a The total weight of two people weighing 68 kg and 73 kg.

    b The area of a rectangle with sides 2.2 m and 3.8 m.

    c The cost of 2.37 m of material at £5.75 per metre.

    d The time taken to travel 1 mile at a speed of 80 mph.

4 A completed jigsaw puzzle measures 24 cm by 21 cm (both to the nearest centimetre). Find the maximum and minimum possible area of the puzzle. Show your working.

5 A bag contains 98 g of sugar; then 34 g of sugar are taken out. Find the upper and lower bounds for the amount of sugar remaining in the bag, if the measurements are all correct to the nearest gram.

6 The maximum load a van can carry is 450 kg. The van is used to carry boxes that weigh 30 kg to the nearest 1 kg. Find the maximum number of boxes that the van can safely carry. Show your working.

7 A metal rod in the shape of a cuboid is measured to weigh 3925 kg.
The dimensions of the rod are measured as 25 cm × 25 cm × 8.0 m.

Marina is trying to decide what metal the rod is made from.
The densities of five different possible metals are shown in the table.

Metal	Density kg/m³
Titanium	4500
Tin	7280
Steel	7850
Brass	8525
Copper	8940

    a Which metal is the rod made from?

    b How much would the rod weigh if it was made from copper?

8 Yvonne gets 5 litre bottles of milk each week. Her neighbour Ella gets 6 pints. One pint is approximately 568 ml.

    a How many millilitres does Ella get?

    b How much more than Ella does Yvonne get?

    c Ella's milkman stops delivering milk in pint bottles. How many litre bottles must she get to have at least 6 pints?

9 Fruit and vegetables are often sold by the pound (lb) and the kilogram (kg, also a kilo).

How much do these purchases cost?

    a 2 kg of tomatoes at £2.18 per kilo.

    b Half a kilo of carrots at £1.32 per kilo.

    c 3 kg of new potatoes at 77 p per kilo.

    d 3 lb of apples at 80 p per pound.

    e Half a pound of mushrooms at £2.80 per pound.

Summary

Checkout
You should now be able to...

Test it
Questions

	Test it Questions
✔ Round numbers and measures to an appropriate degree of accuracy.	1
✔ Use approximation to make estimates.	2, 3
✔ Check calculations using approximation and estimation.	3
✔ Use standard units of length, mass, volume, capacity, time and area.	4 – 9
✔ Use inequality notation to state error intervals and interpret limits of accuracy.	10, 11

Language   Meaning   Example

Language	Meaning	Example
Approximation	A less accurate value of a number, usually obtained by rounding, which is easier to work with.	$\pi = 3.141592654...$ $= 3.1$ (1 dp) $= 3$ (1 sf)
Estimate	A calculation made using approximate values or a judgement. Use the symbol $\approx$ to show approximations have been used.	$\pi \times 1.9^2 \approx 3 \times 2^2 = 12$ Exact $= 11.34114...$
Significant figures (sf)	Describes the relative importance of digits in a number. A number can be rounded to a given number of significant figures.	$256.46 = 260$ (2 sf)
Mass	A measure of the amount of matter in an object. **Weight** measures the force of gravity on a mass.	The mass of a bag of sugar is 1 kg.
Capacity	A measure of the amount of fluid that a 3D shape will hold.	Volume $= 8000\,cm^3$ Capacity $= 8$ litres
Volume	A measure of the amount of 3D space occupied by an object.	
Speed	A measure of the distance travelled by an object in a certain time.	The speed limit on UK motorways is 70 mph.
Density	A measure of the amount of matter in a certain volume.	The density of iron is 7.87 g/cm³.
Accuracy	How close a measured or calculated quantity is to the exact value.	$1.239648 = 1.2$ is accurate to 1 dp $= 1.24$ is accurate to 2 dp
Implied accuracy	The accuracy of a value implied from the number of significant figures or decimal places given.	1.3 has implied accuracy of 1 dp. This means that the actual value could be between 1.25 and 1.35.

Review

1   Round each of these numbers to the nearest

    i   3 sf   ii   2 sf   iii   1 sf

    a   8752   b   14.981

    c   0.0682   d   0.5092

2   Estimate the answers to these calculations.

    a 18.5×11.9 b $\dfrac{105^2 + 265}{95}$

    c $\dfrac{3.4^2 + 5.9 \times 1.9^2}{15.33 - 8.91}$

    d $\sqrt{62.9} \div 4.21$

3   Use your calculator to work these out.
    In each case first write an estimate for
    your answer. Write all the figures on your
    calculator display.

    a $9.6 - 3.7 \div 2.49$ b $4.8 \times (3.2^2 + 6.7)$

    c $\dfrac{5.92 + 3.78}{12.51 - 3.29}$ d $\dfrac{326.2 \times 4.15}{(5.2 - 2.9)^2}$

4

| m² | cm | litres | cm³ | kg |

    Choose from the box a unit of

    a   length   b   volume

    c   capacity   d   mass

    e   area.

5   Convert these lengths to metres.

    a   350 cm   b   145 mm

    c   0.2 km   d   932 cm

6   Simon starts running at 09:50 and stops at
    11:42, how long did he run for?

7   Isaac is paid £14.50 per hour.

    a   How much will he be paid if he works
        300 minutes?

    b   If Isaac is paid £137.75, how many hours
        has he worked for?

8   A car travels at 20 m/s. How long does it
    take to travel

    a   100 m   b   0.5 km?

9   A 30 cm³ cylinder weighs 150 grams.

    a   Find the density of the cylinder giving
        your answer in g/cm³.

    b   A cube is made of the same material as
        the cylinder. It weighs 135 grams. Find

        i   the volume of the cube

        ii   its side length.

10 Helen cuts wood of length, x, which is
    90 cm measured correctly to the nearest
    centimetre. Write an inequality to describe
    the range of values of x.

11 Amaan's mass is measured and lies
    in the interval $59.5 \text{ kg} < m \leqslant 60.5 \text{ kg}$.
    How accurate was the measurement?

What next?

Score	0 – 4	Your knowledge of this topic is still developing.
		To improve look at MyMaths: 1002, 1004, 1005, 1006, 1043, 1067, 1121, 1246, 1736, 1737
	5 – 9	You are gaining a secure knowledge of this topic.
		To improve look at InvisiPens: 09Sa – f
	10 – 11	You have mastered these skills. Well done you are ready to progress!
		To develop your exam technique look at InvisiPens: 09Aa – d

Assessment 9

1 David makes the following incorrect statements. Rewrite the statements correctly.

    a   32.34 to 3 sf is 32   [1]   **b**   5678 to the nearest 100 is 5600   [1]

    c   0.203762 to 1 sf is 0   [1]   **d**   312 to the nearest 10 is 31   [1]

    e   294.82 to 2 sf is 294.82   [1]   **f**   256 133 to the nearest 1000 is 256   [1]

2 Anna, Bart and Christian make these estimates for the following calculations. For each calculation, decide who makes the best estimate.

		Anna	Bart	Christian	
a	5.8×6.5	30	36	40	[1]
b	20.2×5.6	100	110	120	[1]
c	7.9×8.8	56	66	72	[1]
d	$6.6 \div 1.3$	5	6	9	[1]
e	$7.9 \div 0.91$	7	8	9	[1]

3 The rate at which a person blinks, on average, is about 6 times per minute.
Mia says that a good estimate for the number of times a person blinks in one day is 9000.
Use your calculator to explain how she got that estimate.   [2]

4 Rebecca says '*the square root of 46 is bigger than one integer but smaller than another*'.

    a   What are these two integers?   [1]

    b   Write down the squares of these two numbers.   [1]

    c   Use your answer to estimate the square root of 46 to the nearest integer.   [1]

5 **a**   Patrick estimates the value of this calculation to be $1\frac{1}{3}$, find his mistakes and give a sensible estimate for the calculation.   [3]

$$\frac{5.8^2 - 1.5 \times 1.8^2}{7.59 + 6.89} \approx \frac{6^2 - 2 \times 2^2}{8 + 7} \approx \frac{36 - 15}{15} = 1\frac{1}{3}$$

    b   Seth estimates the value of this calculation to be 0.0775

$$\frac{28.9 + \sqrt{0.827}}{(23.4 - 17.6)^2} \approx 0.0775$$

        Use your calculator to find the difference between the exact value and Seth's estimate.   [2]

6 A chocolate fountain was eaten at a wedding reception.
Everyone ate some and none was left over!
The fountain weighed 3.56 kg and each person ate, on average, 88 g of chocolate.
Estimate the number of people at the reception.   [2]

7 Here is a sketch of Mr Digweed's garden.

    a   Find the area of the garden.   [4]

6.8 m   9.6 m   ←8.5 m→

    He has a load of soil delivered and spreads it evenly over the whole garden.

    b   The depth of the soil is 12.5 cm. What volume of soil was delivered?   [3]

    c   The new soil weighs 0.3 g/cm³.

        i   Calculate its weight in kg/m³.   [2]

        ii   Calculate the weight in tonnes, of soil delivered.   [2]

8 Naomi writes these expressions. Put brackets into each expression to make them correct.

    a $4.6 + 5.3 \times 2.6 = 18.38$ [1]   **b** $14.9 - 6.8 \div 2.5 = 12.18$ [1]

    c $3.4 \times 1.6 + 5.9 - 2.8 = 8.54$ [2]   **d** $2.6 + 7.56 \div 1.8 - 0.72 = 6$ [2]

    e $12.3 - 5.2 \times 1.6 + 3.4 \times 2 = 14.76$ [2]

9

tonne	kilogram	gram	kilometre	metre	centimetre	millimetre	litre	centilitre	millilitre

Estimate the following values using an appropriate unit from the list in each case.

    a   Capacity of a paddling pool.   [2]   **b**   Mass of a full suitcase.   [2]

    c   Mass of an eyeliner pencil.   [2]   **d**   Length of a rugby pitch.   [2]

    e   Weight of a plane.   [2]   **f**   Amount in a dose of cough mixture.   [2]

    g   Thickness of a DVD case.   [2]   **h**   Distance run in a marathon.   [2]

    i   Length of a hairbrush.   [2]   **j**   Amount of tea in one cup.   [2]

10 Here is a calculator display. $\boxed{25.3}$

    a   If the unit is £ what is the written answer?   [1]

    b   If the unit is metres write the answer in metres and centimetres.   [1]

    c   If the unit is kilograms write the answer in kilograms and grams.   [1]

    d   If the unit is centilitres write the answer in centilitres and millilitres.   [1]

    e   If the unit is hours write the answer in hours and minutes.   [2]

11 **a**   Bradley cycles 165 km in $2\frac{1}{2}$ hrs. What is his average speed?   [2]

    b   Lewis' car travels 9500 m in 3.8 min. What is its average speed in km/h?   [2]

    c   A train travels for $2\frac{3}{4}$ hrs at an average speed of 175 km/h. How far does it go?   [2]

    d   A plane travels for $7\frac{1}{3}$ hrs at an average speed of 465 mph.
        How far has it flown?   [2]

    e   A snail crawls travels 14 m at 47 m/h. How long does his journey take?   [2]

12 **a**   An cricket ball has a mass of 158 grams and volume of 195 cm³?
        Would it float in water? Explain your answer.   [3]

    b   A gold bracelet has a mass of 44 g and a density of 19.3 g/cm³. What is its volume?   [2]

    c   A concrete girder measuring 25 m by 15 m by 6 m has a density of 2.4 g/cm³.
        What is the mass of the bar in tonnes?   [3]

13 Write down the upper and lower limits of these measurements.

    a   The distance from the moon to the earth varies from 221 460 to 251 970 miles
        the nearest ten miles.   [1]

    b   Farakh's leg is 90 cm long to the nearest 10 cm.   [1]

    c   A box of chocolates weighs 455 g to the nearest 5 g.   [1]

    d   In July 2013 Mo Farah beat Steve Cram's 28 year old British 1500 m record with
        a time of 3 min 28.8 s to the nearest 0.1 s.   [1]

    e   The number of tea bags in a box is 240 to the nearest 4 bags.   [1]

    f   A lorry weights 28 tonnes to the nearest tonne.   [1]

    g   A walking stick has a length of 586 mm to the nearest mm.   [1]

14 A rectangular tray measures 45 cm by 24 cm by 0.3 cm correct to the nearest cm or 0.1 cm.
Its mass is 76 g correct to the nearest gram.

    a   Calculate the biggest and smallest possible values of its volume.   [2]

    b   Calculate the biggest and smallest possible values of its density.   [3]

10 Equations and inequalities

Introduction

Maths is not all about whether something is equal to something else. Sometimes it can be useful to know when a quantity needs to be less than or more than a particular value. An example could be a recipe for a soft drink, where the proportion of cane sugar needs to be within certain bounds to conform to standards and to customers' appetites. Some companies are very secretive about the formula for their particular branded soft drink. All you are allowed to know is that the proportions of ingredients lie within a certain range. Inequalities are statements that allow us to work mathematically with this kind of restriction.

What's the point?

In the food and drink industries it is important to be able to work within restrictions on particular ingredients, which are often imposed by health legislation. In the pharmaceutical industry, it can be critical as too much of one particular chemical can have harmful effects.

Objectives

By the end of this chapter you will have learned how to …

- Derive and solve simple linear equations.
- Solve quadratic equations algebraically by factorising.
- Derive and solve two linear simultaneous equations in two variables.
- Find approximate solutions to two linear simultaneous equations using a graph.
- Solve linear inequalities in one variable and represent the solution on a number line.

Check in

1 Copy and complete these equations.

    a $4 \times 3 = 12$ $12 \div \cdot = 4$ $12 \div \cdot = 3$

    b $7 \times 5 = 35$ $35 \div \cdot = 7$ $35 \div \cdot = 5$

2 Simplify these expressions.

    a $4x \div 4$ **b** $3m \div 3$ **c** $6n \div 2$ **d** $8p \div 4$

3 Work out the value of each expression when $y = 3$.

    a $2y$ **b** $4y - 1$ **c** y^2 **d** $4y \div 2$

4 Work out the value of these expressions when $x = 3$ and $y = 4$.

    a $x + y$ **b** $x - y$ **c** $x \times y$ **d** $\dfrac{3x}{y}$

Chapter investigation

This L shape is drawn on a 10×10 grid numbered from 1 to 100. It has 5 numbers inside it. We can call it L_{35} because the largest number inside it is 35.

What is the sum total of the numbers inside L_{35}?

Find a connection between the L-number and its total.

1	2	3	4	5	6	7	8	9	10
11	12	13	14	15	16	17	18	19	20
21	22	23	24	25	26	27	28	29	30
31	32	33	34	35	36	37	38	39	40
41	42	43	44	45	46	47	48	49	50
51	52	53	54	55	56	57	58	59	60
61	62	63	64	65	66	67	68	69	70
71	72	73	74	75	76	77	78	79	80
81	82	83	84	85	86	87	88	89	90
91	92	93	94	95	96	97	98	99	100

10.1 Solving linear equations 1

- An **equation** includes letter and number terms and an equals sign.

$x + 2 = 5$ is an equation.

- You can use the **balance method** to solve an equation.
  You do the same to each side to keep the equation balanced.

> The **inverse** of +5 is −5. Use the inverse operation to get *x* on its own.

Start with the equation
$x + 5 = 11$

Subtract 5 from each side
$x + 5 - 5 = 11 - 5$

The solution is $x = 6$.

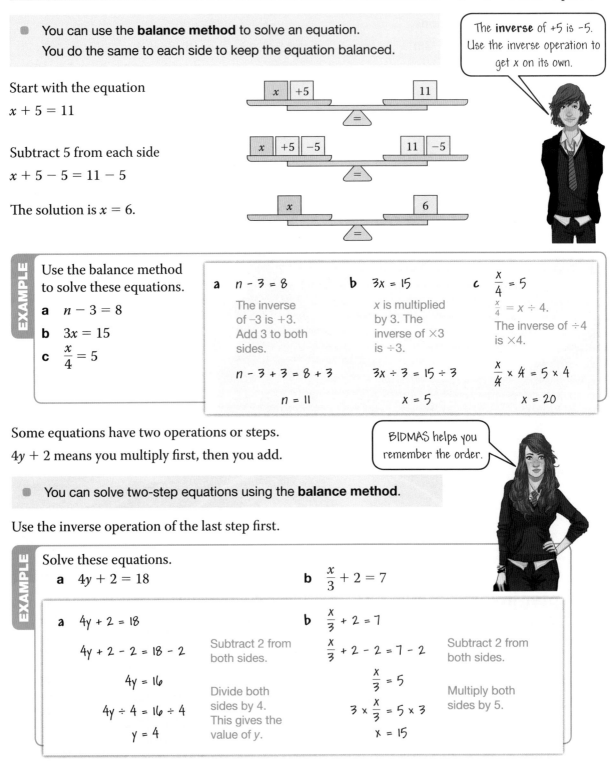

EXAMPLE

Use the balance method to solve these equations.

a $n - 3 = 8$
b $3x = 15$
c $\dfrac{x}{4} = 5$

a $n - 3 = 8$

The inverse of −3 is +3. Add 3 to both sides.

$n - 3 + 3 = 8 + 3$

$n = 11$

b $3x = 15$

x is multiplied by 3. The inverse of ×3 is ÷3.

$3x \div 3 = 15 \div 3$

$x = 5$

c $\dfrac{x}{4} = 5$

$\dfrac{x}{4} = x \div 4$.

The inverse of ÷4 is ×4.

$\dfrac{x}{4} \times 4 = 5 \times 4$

$x = 20$

Some equations have two operations or steps.

$4y + 2$ means you multiply first, then you add.

> BIDMAS helps you remember the order.

- You can solve two-step equations using the **balance method**.

Use the inverse operation of the last step first.

EXAMPLE

Solve these equations.

a $4y + 2 = 18$
b $\dfrac{x}{3} + 2 = 7$

a $4y + 2 = 18$

$4y + 2 - 2 = 18 - 2$

Subtract 2 from both sides.

$4y = 16$

$4y \div 4 = 16 \div 4$

Divide both sides by 4. This gives the value of *y*.

$y = 4$

b $\dfrac{x}{3} + 2 = 7$

$\dfrac{x}{3} + 2 - 2 = 7 - 2$

Subtract 2 from both sides.

$\dfrac{x}{3} = 5$

$3 \times \dfrac{x}{3} = 5 \times 3$

Multiply both sides by 5.

$x = 15$

- You can check a solution by **substituting** the value back into the equation.

Algebra Equations and inequalities

Exercise 10.1S

1 Use number facts to work out the missing numbers in these calculations.

a $6 + \Box = 10$ **b** $2 + \Box = 7$

c $9 = 4 + \Box$ **d** $14 = \Box + 6$

e $8 - \Box = 2$ **f** $12 - \Box = 6$

g $5 = \Box - 10$ **h** $7 = \Box - 5$

2 Solve these equations.

a $x + 5 = 8$ **b** $x + 3 = 14$

c $x + 8 = 13$ **d** $x + 3 = 18$

e $x + 9 = 0$ **f** $5 + x = 11$

g $x - 4 = 7$ **h** $x - 9 = 6$

3 Solve these equations.

a $5x = 15$ **b** $3x = 21$

c $6x = 18$ **d** $4x = 36$

e $28 = 4x$ **f** $7x = 28$

g $50 = 25x$ **h** $4x = 10$

4 Solve these equations.

a $\dfrac{s}{5} = 5$ **b** $\dfrac{t}{12} = 3$

c $\dfrac{u}{2} = 4$ **d** $\dfrac{v}{7} = 3$

e $\dfrac{v}{5} = 9$ **f** $\dfrac{w}{3} = 8$

g $10 = \dfrac{x}{5}$ **h** $3 = \dfrac{y}{9}$

5 Solve these equations.
Check your answers using substitution.

a $a + 7 = 11$ **b** $g - 7 = 8$

c $\dfrac{c}{4} = 9$ **d** $d - 5 = -2$

e $3e = 21$ **f** $9 + f = 5$

g $4g = 0$ **h** $\dfrac{h}{4} = 2.5$

6 Solve these equations.
Check your answers using substitution.

a $9 + r = 15$ **b** $11 - s = 14$

c $-15 = 3t$ **d** $\dfrac{u}{4} = -10$

e $v - 8 = -6$ **f** $6w = 8$

g $4 = \dfrac{x}{9}$ **h** $17 = 12 - y$

i $-4k = 10$ **j** $-10 = 4 - p$

7 Solve these equations.

a $2x + 3 = 13$ **b** $2x + 5 = 9$

c $2x + 8 = 20$ **d** $2x + 1 = 21$

8 Solve these equations.
Check your answers by substitution.

a $2m - 4 = 16$ **b** $3n - 5 = 10$

c $5 = 2p + 9$ **d** $5 + 3r = 26$

e $13 = 2s - 1$ **f** $4t + 1 = 1$

g $25 - 3x = 10$ **h** $19 - 2y = 7$

9 Solve these equations.

a $4m + 3 = 5$ **b** $4q - 2 = 2$

c $8 - 6p = 5$ **d** $4r + 2 = 12$

e $12 = 8q + 20$ **f** $6s - 6 = -3$

10 Solve these equations.

a $3a - 5 = 25$ **b** $2b + 9 = 27$

c $6e + 11 = 23$ **d** $\dfrac{f}{5} + 3 = 6$

e $\dfrac{g}{7} - 8 = 2$ **f** $\dfrac{i}{6} + 4 = 9$

g $\dfrac{j}{5} - 8 = 4$ **h** $3k - 20 = -5$

i $8m - 5 = 7$ **j** $3p - 8 = -14$

11 Solve these equations.
Check your answers using substitution.

a $3x + 9 = 24$ **b** $5 + 7x = 40$

c $4x - 8 = 24$ **d** $5x - 17 = 13$

e $2x + 7 = 18$ **f** $\dfrac{x}{10} + 6 = 11$

g $\dfrac{y}{3} - 2 = 4$ **h** $\dfrac{z}{4} - 4 = 6$

i $15 - 4y = 3$ **j** $22 - 6p = 10$

12 Solve these equations.
Check your answers using substitution.

a $5m - 6 = 29$ **b** $18 + 2p = 4$

c $10 - 3n = 4$ **d** $14 = 6q - 16$

e $7 = 29 + 2r$ **f** $4k - 36 = -12$

g $-11 = 6m - 41$ **h** $\dfrac{a}{2} + 2 = 6$

Q 1154, 1395, 1925   SEARCH

10.1 Solving linear equations 1

RECAP

- You can use the balance method to solve one-step and two-step equations.
- Do the same to each side to keep the equation balanced until you find the value of the unknown letter.

$5x + 12$ 32

take off 12   take off 12

$5x$ 20

divide by 5   divide by 5

x 4

HOW TO

Use linear equations to solve problems

① Write an equation for the problem in the question.

② Use the balance method to solve the equation.

③ Give the solution and answer the question.

④ Check the answer by substituting the solution into the equation.

EXAMPLE

I think of a number.

I divide the number by 6 and subtract 10.

The answer is −6.

What is my number?

> Use BIDMAS to help you write the equation. You divide first, then subtract.

① Write an equation for the problem in the question.

Let my number be x. $\frac{x}{6} - 10 = -6$

② Use the balance method to solve the equation.

$\frac{x}{6} - 10 + 10 = -6 + 10$ Add 10 to both sides.

$\frac{x}{6} = 4$

$\frac{x}{6} \times 6 = 4 \times 6$ Multiply both sides by 6.

$x = 24$

③ My number is 24.

④ Check the answer: $24 \div 6 - 10 = 4 - 10 = -6$ ✓

EXAMPLE

The perimeter of this rectangle is 24 cm.

a Write an equation for the perimeter of the rectangle.

b Hence find x, the length of the rectangle.

4 cm

x cm

① Form an equation for the perimeter of the rectangle.

The perimeter is the distance around the edge of the rectangle.

a $4 + x + 4 + x = 24$

    $2x + 8 = 24$

② Use the balance method to solve the equation.

b $2x + 8 - 8 = 24 - 8$ Subtract 8 from both sides.

    $2x = 16$

    $2x \div 2 = 16 \div 2$ Divide both sides by 2.

    $x = 8$

③ The length of the rectangle is 8 cm.

④ Check the answer: $4\,cm + 8\,cm + 4\,cm + 8\,cm = 24\,cm$ ✓

Exercise 10.1A

1 Write each 'think of a number' problem as an equation.

Solve the equation to find the number.

a I think of a number and subtract 12. The answer is 11.

b I think of a number and divide by 5. The answer is 8.

2 One of these equations has a different solution to the other two. Which is it?

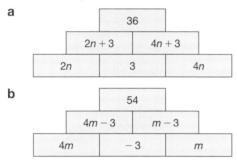

$\frac{m}{2} + 12 = 16$ $6p + 12 = 16$ $\frac{n}{4} + 11 = 13$

3 In each wall, add two bricks to find the number on the brick above.
Write and solve equations to find the unknown letter in each wall.

a

36	
$2n + 3$	$4n + 3$
$2n$	3	$4n$

b

54	
$4m - 3$	$m - 3$
$4m$	-3	m

4 Find the missing side length for each shape.

a

3 cm, x cm, 4 cm   Perimeter = 12 cm

b

y cm, 3 cm, y cm, 4 cm, 7 cm   Perimeter = 26 cm

c

8 cm, 8 cm, 8 cm, 8 cm, 8 cm, z cm, z cm   Perimeter = 48 cm

What do you notice about shape **c**?

5 Match each equation in box A with its solution in box B.

There are four 'spare' solutions. Make up an equation with each of these spare solutions.

Box A	Box B	
$2x + 6 = 9$	$x = 1$	$x = -1$
$3x + 1 = 7$	$x = 2$	$x = -2$
$3x + 15 = 6$	$x = 3$	$x = \frac{1}{2}$
$4x - 3 = -5$	$x = 4$	$x = \frac{3}{2}$
$4x + 11 = 3$		
$4x + 7 = 19$	$x = 5$	$x = -\frac{1}{2}$
$5x - 12 = 13$	$x = 6$	$x = -3$
$7x + 8 = 15$		

6 Sarah, Josh and Millie bring sandwiches to a picnic. Sarah brings y sandwiches. Josh brings twice as many sandwiches as Sarah. Millie brings 4 less than Josh.

There are 11 sandwiches at the picnic. Work out the number of sandwiches each person brings to the picnic.

7 Four people go out for a meal. The meal costs £x each. The drinks cost £15.

The total bill comes to £65.

Write and solve an equation to find x, the cost of one meal.

8 The perimeter of this shape is 30 mm. Find the length of each side.

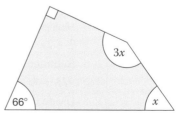

$5 - x$

$2x + 6$

9 This shape is a quadrilateral. Work out the value of x.

$3x$

$66°$

x

10.2 Solving linear equations 2

- You can **expand brackets** in an algebraic expression.
  - You multiply each term inside the bracket by the term outside.

$$3(x + 2) = 3 \times x + 3 \times 2 = 3x + 6$$

EXAMPLE

Solve the equations.

a $\quad 5(a + 3) = 10$

b $\quad 2(z - 2) = 1$

a $\qquad 5(a + 3) = 10$

Expand the brackets.

$5a + 15 = 10$

Subtract 15 from both sides.

$5a + 15 - 15 = 10 - 15$

Divide both sides by 5.

$5a = -5$

$a = -1$

b $\qquad 2(z - 2) = 1$

Expand the brackets.

$2z - 4 = 1$

Add 4 to both sides.

$2z = 5$

Divide both sides by 2.

$z = \dfrac{5}{2}$ or $2\dfrac{1}{2}$

- You can solve equations that have an unknown on both sides.

- Subtract the smallest unknown term from both sides.

- Solve using the balance method.

$4x + 2 = 3x + 5$ $\qquad$ The 3x term has the smallest number of x.

$4x - 3x + 2 = 3x - 3x + 5$ $\quad$ Subtract 3x from both sides.

$x + 2 = 5$

$x - 2 = 5 - 2$ $\qquad$ Subtract 2 from both sides.

$x = 3$

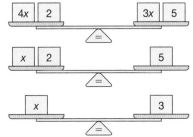

- You can solve equations involving **fractions** using the balance method.

- Arrange the equation so the fraction is on its own on one side of the equation.

- Multiply both sides by the denominator (bottom number).

- Solve using the balance method.

EXAMPLE

Solve the equations.

a $\quad \dfrac{5 - 2x}{3} = 7$

b $\quad 2 + \dfrac{3y}{5} = 8$

a $\qquad \dfrac{5 - 2x}{3} = 7$

Multiply both sides by 3.

$5 - 2x = 21$

Subtract 5 from both sides.

$-2x = 16$

Divide both sides by -2.

$x = -8$

b $\qquad 2 + \dfrac{3y}{5} = 8$

Subtract 2 from both sides.

$\dfrac{3y}{5} = 6$

Multiply both sides by 5.

$3y = 30$

Divide both sides by 3.

$y = 10$

Exercise 10.2S

1 Expand the brackets.

   a $2(a + 3)$ **b** $4(b - 5)$

   c $3(2c + 1)$ **d** $5(-d + 10)$

   e $8(2e - 1)$ **f** $10(2 - f)$

2 Expand and solve.

   a $2(a + 3) = 10$ **b** $4(b - 5) = 12$

   c $3(2c + 1) = 27$ **d** $5(-d + 10) = 25$

   e $8(2e - 1) = 24$ **f** $10(2 - f) = 20$

3 Expand and solve.

   a $3(x + 1) = 15$ **b** $4(s - 2) = 16$

   c $2(t - 3) = 0$ **d** $4(-v + 1) = 28$

4 Expand and solve.

   a $3(a + 4) = -6$ **b** $2(b - 5) = -12$

   c $-4(6 - c) = -16$ **d** $3(2d - 3) = -21$

5 Solve.

   a $4(e + 1) = 10$ **b** $3(f - 2) = 24$

   c $8(2 - g) = 10$ **d** $-4(h - 6) = 26$

6 Solve these equations.

   a $4m + 2 = 3m + 7$ **b** $6p - 5 = 5p - 2$

   c $3t + 2 = 2t + 5$ **d** $3n - 11 = 2n - 4$

   e $4q + 2 = 5q - 6$ **f** $5s - 2 = 4s + 6$

7 Solve these equations.

   a $2s + 5 = 3s + 8$ **b** $4t - 2 = 5t + 2$

   c $6u + 10 = 5u + 8$ **d** $4v - 6 = -15 - 5v$

8 Find the value of the unknown in each of these equations.

   a $2a + 14 = 6a - 6$

   b $4b - 2 = 6b + 6$

   c $3c - 4 = c + 1$

   d $5d + 15 = -6 - 2d$

9 Solve these equations.

   a $4x + 3 = 18 + 2x$

   b $4x + 10 = 2x + 4$

   c $8x + 15 = 12x + 14$

   d $6x - 4 = 10x + 2$

10 Solve these equations.

   a $2(r + 6) = 5r$ **b** $6(s - 3) = 12s$

   c $4(2t + 8) = 24t$ **d** $5(v - 1) = 6v$

11 Solve these equations.

   a $2(a + 5) = 7a - 5$

   b $3(b - 2) = 5b - 2$

   c $2(c + 6) = 5c - 3$

   d $3d + 8 = 2(d + 2)$

12 Solve these equations.
Check your answers using substitution.

   a $3(2x - 4) = 7x - 18$

   b $2(3y + 2) = 5y - 2$

   c $4(2z + 1) = 6z + 15$

   d $-4(6m + 1) = -17m - 18$

13 Solve these equations.
Check your answers using substitution.

   a $2(e + 3) = 4e - 1$

   b $4f + 3 = 2(f + 2)$

   c $4(2g + 1) = 6g + 1$

   d $3(2h + 3) = 5h + 8$

14 Solve these equations.

   a $\dfrac{x}{3} = 3$ **b** $\dfrac{m}{4} = -2$

   c $\dfrac{-n}{3} = 6$ **d** $\dfrac{m}{5} = 4$

15 Find the value of the unknown in each of these equations.

   a $\dfrac{s}{3} + 5 = 8$ **b** $4 - \dfrac{t}{2} = 1$

   c $\dfrac{u}{5} + 7 = 5$ **d** $16 = \dfrac{v}{4} + 13$

16 Solve these equations.
Check your answers using substitution.

   a $\dfrac{2x}{3} + 5 = 9$ **b** $\dfrac{3y}{2} - 5 = 4$

   c $3 - \dfrac{2z}{5} = -3$ **d** $\dfrac{3q}{2} + 5 = -7$

17 Solve these equations.

   a $\dfrac{x + 5}{3} = 2$ **b** $\dfrac{x - 3}{4} = 2$

   c $\dfrac{x + 9}{2} = -4$ **d** $\dfrac{10 - x}{4} = 1$

18 Solve these equations.
Check your answers using substitution.

   a $\dfrac{2x + 1}{5} = 5$ **b** $\dfrac{3x - 2}{4} = 4$

   c $\dfrac{11 - 2x}{3} = -1$ **d** $\dfrac{31 - 3x}{4} = 4$

Q 1182, 1928   SEARCH

10.2 Solving linear equations 2

RECAP

- Do the same to each side to keep the equation balanced until you find the value of the unknown letter.
- To expand a bracket, you multiply each term inside the bracket by the term outside.

HOW TO

1. Read the question and form an equation.
2. Simplify the equation by
   - expanding brackets
   - subtracting the smaller unknown term from both sides
   - multiplying both sides by the denominator of the fraction.
3. Use the balance method to solve the equation. Give your answer in the context of the question.
4. Check the answer by substituting the solution into the equation.

EXAMPLE

Lucy thinks of a number.

She adds 8 and then divides by 3. Her answer is the number she thought of.

Find Lucy's number.

1. Form an equation.

   If the number is n then $\dfrac{n + 8}{3} = n$

2. Multiply both sides by 3. $n + 8 = 3n$

   Subtract n from both sides. $8 = 2n$

   Divide both sides by 2. $4 = n$

3. Lucy's number is 4.

EXAMPLE

This square pattern is made from rectangular tiles.

Each tile has length $x + 3$ and width x.

The pattern is 10 tiles wide and 4 tiles long.

Find the dimensions of the square.

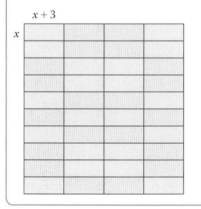

1. The pattern is a square so length = width.

   Length = $4(x + 3) = 4x + 12$

   Width = $10x$

   $4x + 12 = 10x$

2. Use the balance method to solve the equation.

   $12 = 6x$ Subtract $4x$ from both sides.

3. $x = 2$ Divide both sides by 6.

   The dimensions of the square are 20 units long by 20 units wide.

Exercise 10.2A

1 Charlie thinks of a number.

He multiples the number by 2 and subtracts 6. His answer is the number he thought of.

Find Charlie's number.

2 Karena thinks of a number.

She divides the number by 5 and adds 7. Her answer is 10.

Find Karena's number.

3 Adam thinks of a number.

He subtracts 5 from the number and then multiplies by 3. His answer is 15.

Find Adam's number.

4 The diagram shows a square with sides $2y + 5$ cm.

The perimeter of the square is 28 cm.

Find the value of y.

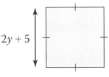

5 The area of the rectangle is 8 cm².

Find the value of z.

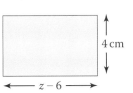

6 a Choose one expression from each set of cards.

b Write them as an equation: Expression from set 1 = expression from set 2

c Solve your equation to find the value of x.

d Repeat for different pairs of expressions.

Set 1

| $2(x+3)$ | $4(2x-1)$ | $3(4x+1)$ |

Set 2

| $3x-2$ | $4x+1$ | $8x-3$ |

7 The triangle and the square have equal perimeter. Find the value of x.

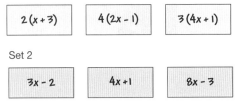

8 Rectangle B and square A have equal area.

Find the value of x.

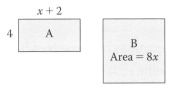

9 A blouse has m buttons. A shirt has $m + 2$ buttons.

Three shirts have the same number of buttons in total as four blouses.

How many buttons are there on a shirt?

10 a I think of a number, multiply by 2 and add 7. I get the same answer when I multiply the number by 4 and subtract 13. Find my number.

b I think of a number, multiply by 5 and subtract 8. I get the same answer when I double the number and add 10. Find my number.

c I think of a number, multiply by 3 and add 4. I get the same answer when I multiply by 5 and add 12. Find my number.

11 The perimeter of this equilateral triangle is $2x + 6$.

The length of one side of the triangle is 8 cm.

Find the value of x.

12 The perimeter of this square is $4 + x$.

The length of one side is 10 cm.

Find the value of x.

*13 The area of this trapezium is 65 cm².

a Find the value of x.

b Explain why the perimeter of the trapezium is 36 cm.

Hint: What do you notice about the length of the parallel sides?

10.3 Quadratic equations

- A **quadratic** equation contains an x^2 term as the highest power, for example
$$x^2 + 5x + 6 = 0.$$

Many quadratic equations can be solved by **factorising**.

For example, solve the equation $x^2 + 5x + 6 = 0$.

$x^2 + 5x + 6 = 0$ The two factors will each start with x.

$(x + \Box)(x + \Box) = 0$ Now find two numbers with a **sum** of 5 (for 5x) and a **product** of 6.

$(x + 3)(x + 2) = 0$ Check: $(x + 3)(x + 2) = x^2 + 3x + 2x + 6$
$$= x^2 + 5x + 6.$$

x^2 term, so equation is a quadratic.

The two factors have a product of zero. This means that at least one of them is equal to zero.

Either $x + 3 = 0$ or $x + 2 = 0$

If $x + 3 = 0$, $x = -3$ and if $x + 2 = 0$, $x = -2$ Check: $(-3)^2 + 5(-3) + 6 = 9 - 15 + 6 = 0$

The **solutions** are $x = -3$ and $x = -2$.   and $(-2)^2 + 5(-2) + 6 = 4 - 10 + 6 = 0$

- The key to solving quadratic equations is that the product of the two factors is zero, so one or other (or both) of the factors is zero. This leads to the two solutions.

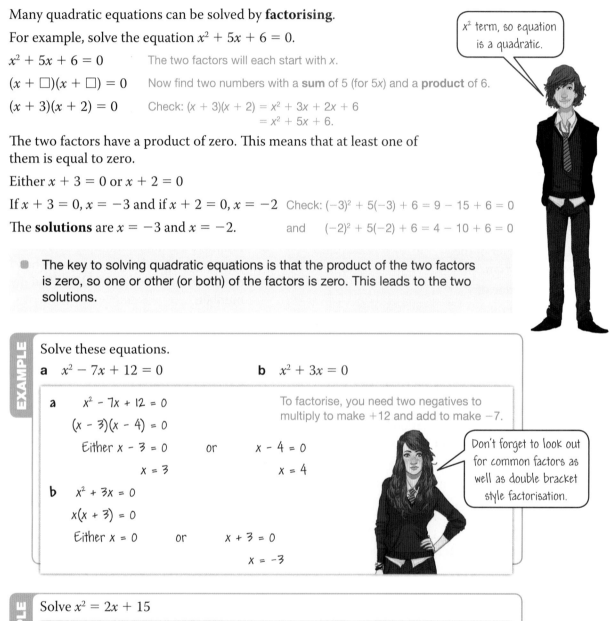

EXAMPLE

Solve these equations.

a $x^2 - 7x + 12 = 0$ **b** $x^2 + 3x = 0$

a $x^2 - 7x + 12 = 0$ To factorise, you need two negatives to multiply to make $+12$ and add to make -7.

$(x - 3)(x - 4) = 0$

Either $x - 3 = 0$ or $x - 4 = 0$

$x = 3$ $x = 4$

b $x^2 + 3x = 0$

$x(x + 3) = 0$

Either $x = 0$ or $x + 3 = 0$

$x = -3$

Don't forget to look out for common factors as well as double bracket style factorisation.

EXAMPLE

Solve $x^2 = 2x + 15$

First make the left-hand side of the equation equal to zero.

$x^2 - 2x - 15 = 0$ Find two numbers with sum -2 and product -15.

$(x + 3)(x - 5) = 0$

Either $x + 3 = 0$ or $x - 5 = 0$

$x = -3$ $x = 5$

Algebra   Equations and inequalities

Exercise 10.3S

1 **a** Find two numbers with a sum of 8 and a product of 12.

    b Find two numbers with a sum of 11 and a product of 24.

    c Find two numbers with a sum of 13 and a product of 36.

    d Find two numbers with a sum of 16 and a product of 55.

    e Find two numbers with a sum of -7 and a product of 12.

    f Find two numbers with a sum of -10 and a product of 24.

    g Find two numbers with a sum of 3 and a product of -28.

    h Find two numbers with a sum of -2 and a product of -15.

2 Factorise these quadratic expressions.

    a $x^2 + 8x + 12$ **b** $x^2 + 11x + 24$

    c $x^2 + 13x + 36$ **d** $x^2 + 16x + 55$

    e $x^2 - 7x + 12$ **f** $x^2 - 10x + 24$

    g $x^2 + 3x - 28$ **h** $x^2 - 2x - 15$

3 Solve these quadratic equations by factorising them into two brackets.

    a $x^2 + 7x + 12 = 0$ **b** $x^2 + 8x + 12 = 0$

    c $x^2 + 10x + 25 = 0$ **d** $x^2 + 5x - 14 = 0$

    e $x^2 - 4x - 5 = 0$ **f** $x^2 - 5x + 6 = 0$

    g $x^2 - 10x + 21 = 0$ **h** $x^2 = 3x + 40$

4 Solve these quadratic equations.

    a $0 = x^2 + 18x + 17$ **b** $0 = x^2 + 15x + 26$

    c $x^2 + 2x - 15 = 0$ **d** $x^2 + 6x - 16 = 0$

    e $0 = x^2 + 3x - 18$ **f** $0 = x^2 + 9x - 22$

5 Solve these quadratic equations.

    a $x^2 - x - 6 = 0$

    b $x^2 - x - 12 = 0$

    c $x^2 - 2x - 15 = 0$

    d $0 = x^2 - 6x - 16$

    e $0 = x^2 - 13x - 30$

    f $0 = x^2 - 3x - 28$

6 Solve these quadratic equations by factorising them into a single bracket.

    a $x^2 - 8x = 0$ **b** $x^2 + 4x = 0$

    c $x^2 - 6x = 0$ **d** $y^2 + 5y = 0$

    e $x^2 = 9x$ **f** $x^2 - 12x = 0$

    g $2x^2 + 8x = 0$ **h** $6x - x^2 = 0$

7 First factorise these quadratic equations, by using the common factor, then solve them.

    a $x^2 - 3x = 0$ **b** $x^2 + 8x = 0$

    c $2x^2 - 9x = 0$ **d** $3x^2 - 9x = 0$

    e $x^2 = 5x$ **f** $x^2 = 7x$

    g $12x = x^2$ **h** $4x = 2x^2$

    i $6x - x^2 = 0$ **j** $9y - 3y^2 = 0$

    k $0 = 7w - w^2$

8 Solve these quadratic equations by factorising.

    a $x^2 + 7x + 12 = 0$ **b** $x^2 + 8x + 12 = 0$

    c $x^2 + 10x + 25 = 0$ **d** $x^2 + 2x - 15 = 0$

    e $x^2 + 5x - 14 = 0$ **f** $x^2 - 4x - 5 = 0$

    g $x^2 - 5x + 6 = 0$ **h** $x^2 - 12x + 36 = 0$

    i $x^2 = 8x - 12$ **j** $0 = 5x - 6 - x^2$

    *k $x(x + 10) = -21$

9 Solve these quadratic equations.

    a $x^2 + 9x + 20 = 0$ **b** $x^2 + 13x + 12 = 0$

    c $x^2 + 8x - 20 = 0$ **d** $x^2 + 6x - 27 = 0$

    e $0 = x^2 - 10x + 24$

    f $0 = x^2 - 12x + 35$

    *g $2x^2 + 12x + 10 = 0$

    *h $3x^2 - 18x + 27 = 0$

> To solve parts **g** and **h**, start by looking for a number that divides each term.

*10 Solve these quadratic equations.

    a $x^2 - x = 20$ **b** $x^2 + 11x = 12$

    c $8 = x^2 + 2x$ **d** $x^2 + 4x = 21$

    e $x^2 + 45 = 14x$ **f** $12x = x^2 + 32x$

11 Solve these quadratic equations by factorising using the difference of two squares.

    a $x^2 - 16 = 0$ **b** $x^2 - 64 = 0$

    c $y^2 - 25 = 0$ **d** $9x^2 - 4 = 0$

    e $4y^2 - 1 = 0$ **f** $x^2 = 169$

    g $4x^2 = 25$ **h** $36 = 9y^2$

6.4

Q 1169, 1950   SEARCH

10.3 Quadratic equations

- **Quadratic** equations contain a squared term as the highest power, for example $x^2 + 7x - 9 = 0$.
- Many quadratic equations can be solved by rearranging so that one side equals zero and then **factorising**.

> Don't forget to look out for common factors as well as double bracket style factorisation.

HOW TO

① Use the information in the question to form a quadratic equation.

② Rearrange the quadratic so that it equals zero.

③ Solve the quadratic by factorising.

④ Check that your answers make sense. You may need to reject one solution, depending on the context.

EXAMPLE

Two numbers have a product of 105 and a difference of 8.

If the larger number is x,

a show that $x^2 - 8x - 105 = 0$

b solve this equation to find the two numbers.

a ① The two numbers are x and $x - 8$.
So $x(x - 8) = 105$

② $x^2 - 8x - 105 = 0$ (as required)

b $x^2 - 8x - 105 = 0$

③ Factorise the quadratic. Find two numbers that add to -8 and multiply to give -105. The two numbers are 7 and -15.

$(x + 7)(x - 15) = 0$

Either $x + 7 = 0$ or $x - 15 = 0$

So $x = -7$ and $x - 8 = -15$ or $x = 15$ and $x - 8 = 7$.

④ Two answers for x lead to two answers for $x - 8$.

The two numbers are -7 and -15 or 7 and 15.

EXAMPLE

If the area of the trapezium is 144 cm², show that $x^2 + 10x = 144$ and find the value of x.

① Area of trapezium $= \frac{(a + b)}{2}h$

$A = \frac{1}{2}(x + 10) \times 2x = x(x + 10)$

$144 = x(x + 10)$

$x^2 + 10x = 144$ (as required)

② $x^2 + 10x - 144 = 0$

③ Factorise the quadratic.
Find two numbers that add to 10 and multiply to give -144.
The two numbers are -8 and 18.

$(x - 8)(x + 18) = 0$

Either $x - 8 = 0$ so $x = 8$

Or $x + 18 = 0$ so $x = -18$

④ The trapezium cannot have sides with negative lengths.
The only solution is $x = 8$ cm.

Exercise 10.3A

1 a Vicky is trying to solve
$(x + 4)(x - 2) = 7$. Here is her attempt.

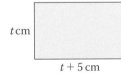

$(x + 4)(x - 2) = 7$
Either $x + 4 = 7$ or $x - 2 = 7$
            $x = 3$ or $x = 9$

What is wrong with her method?

b Can you solve this equation correctly
to show that the two solutions should
really be $x = 3$ and $x = -5$?

2 Simplify and then solve this equation.

$$x^2 + 4x - 5 = 2x(x - 1)$$

3 a Two numbers which differ by 4 have a
product of 60. Find the numbers.

b Two numbers which differ by 8 have a
product of -12. Find the numbers.

4 a Solve the equation $x^2 + 3x - 40 = 0$

b Two numbers have a product of 80. The
larger number is 6 more than twice the
smaller number.
Find two possibilities for the pair of
numbers.

5 a Solve the equation $x^2 + 5x - 50 = 0$

A rectangle has sides t cm and $t + 5$ cm. The
area of the rectangle is $50\,\text{cm}^2$.

t cm

$t + 5$ cm

b Show that t satisfies the equation
$t^2 + 5t - 50 = 0$

c Explain why there is only one possible
value for t.

6 A triangle has angles $x°$, $6x°$ and
$(x^2 + 10)°$. Find the value of x.

7 Callum has three numbers.

$40 - x^2$ x^2 $x^2 - x$

The mean of the three numbers is 20.
Find two possibilities for Callum's numbers.

8 A quadrilateral has interior angles $100°$,
$3x°$, $(x^2)°$ and $80°$ as shown on the diagram.

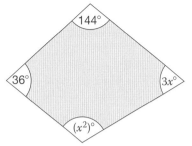

144°

36° $3x°$

$(x^2)°$

Find the value of x and say what type of
quadrilateral it is.

9 The area of the trapezium is $36\,\text{cm}^2$.
Find the value of x.

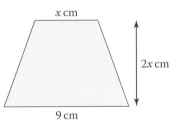

x cm

$2x$ cm

9 cm

10 The sides of a right-angled triangle are
x cm, $x + 1$ cm and 5 cm.

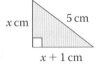

x cm   5 cm

$x + 1$ cm

a Show that x satisfies the equation
$x^2 + x - 12 = 0$

b Find the perimeter of the right-angled
triangle.

11 If this square and rectangle have equal areas,
find the side length of the square.

$8x - 8$

$3x - 2$

x

12 a Simplify and solve this equation
$(x - 4)(x - 8) = 2x - 17$

b How many times do the lines
$y = x^2 - 12x + 32$ and $y = 2x - 17$
intersect? Find the point(s) of intersection.

10.4 Simultaneous equations

● Two equations that have the same **solution** are called **simultaneous** equations.

EXAMPLE

Solve the simultaneous equations $3x + 2y = 12$ and $3x + 8y = 30$.

$3x + 2y = 12$ (1)   Label the equations (1) and (2).

$3x + 8y = 30$ (2)   The x terms are identical so you can **eliminate** them.

$(3x + 8y) - (3x + 2y) = 30 - 12$ Subtract equation (1) from equation (2).

$\qquad\qquad 8y - 2y = 18$

$\qquad\qquad\qquad 6y = 18$

$\qquad\qquad\qquad\ y = 3$

$\qquad 3x + 6 = 12$ Substitute 3 for y in equation (1).

$\qquad\quad 3x = 6$

$\qquad\quad\ x = 2$ Check: $3 \times 2 + 2 \times 3 = 12$ and
$\qquad\qquad\qquad\qquad\qquad\qquad\quad 3 \times 2 + 8 \times 3 = 30$

Write the equations one under the other so that you can compare them.

Subtract because the x terms have the same sign (SSS).

If the terms have a different sign, then you add.

EXAMPLE

Solve $4x - 3y = 5$ (1)
$\qquad\quad 8x + 3y = 1$ (2)

$(4x - 3y) + (8x + 3y) = 5 + 1$ Add the equations to eliminate the y-terms.

$\qquad\qquad 12x = 6$

$\qquad\qquad\ x = \dfrac{1}{2}$

$8 \times \dfrac{1}{2} + 3y = 1$ Substitute $\dfrac{1}{2}$ for x in equation (2).

$\qquad 4 + 3y = 1$

$\qquad\quad 3y = -3$ so $y = -1$ Check in equation (1):

$\qquad\qquad\qquad\qquad\qquad 4 \times \dfrac{1}{2} - 3 \times (-1) = 5$

You can solve **simultaneous** equations on a graph.

● The solution to simultaneous equations is where their graphs intersect.

EXAMPLE

Solve
$3x - y = 2$ and
$2x + y = 8$

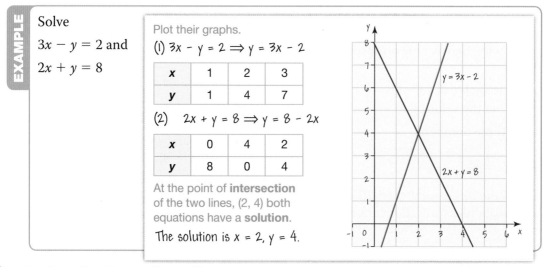

Plot their graphs.

(1) $3x - y = 2 \Rightarrow y = 3x - 2$

x	1	2	3
y	1	4	7

(2) $2x + y = 8 \Rightarrow y = 8 - 2x$

x	0	4	2
y	8	0	4

At the point of **intersection** of the two lines, (2, 4) both equations have a **solution**.

The solution is $x = 2$, $y = 4$.

Algebra   Equations and inequalities

Exercise 10.4S

1 Solve these pairs of simultaneous equations by subtracting one equation from the other.

a $3x + y = 15$
$x + y = 7$

b $6x + 2y = 6$
$4x + 2y = 2$

c $x + 5y = 19$
$x + 7y = 27$

d $5x + 2y = 16$
$x + 2y = 4$

e $m + 3n = 11$
$m + 2n = 9$

f $4x + 3y = -5$
$7x + 3y = -11$

2 Solve these pairs of simultaneous equations by adding one equation to the other.

a $3x + 2y = 19$
$8x - 2y = 58$

b $5x + 2y = 16$
$3x - 2y = 8$

c $7a - 3b = 24$
$2a + 3b = 3$

d $2x + 3y = 19$
$-2x + y = 1$

e $4x - 7y = 15$
$2x + 7y = 4\frac{1}{2}$

f $6p - 2q = -2$
$6p + 2q = 26$

3 Solve these pairs of simultaneous equations by either adding or subtracting in order to eliminate one variable.

a $x + y = 3$
$3x - y = 17$

b $5x - 2y = 4$
$3x + 2y = 12$

c $5a + b = -7$
$5a - 2b = -16$

d $3v + w = 14$
$3v - w = 10$

e $20p - 4q = 32$
$7p + 4q = 22$

f $3x - 2y = 11$
$3x + 4y = 23$

4 Solve these simultaneous equations.

$a = 2b + 7$
$a + b - 1 = 10$

$6w = 38 - 2v$
$5w = 6 + 2v$

5 Solve these simultaneous equations.

a $4x + 4y = 16$
$x + 4y = 13$

b $3x + 2y = 19$
$x + 2y = 9$

c $5p + 3q = 31$
$5p + q = 17$

d $4a + 3b = 17$
$4a + 5b = 15$

6 Solve these simultaneous equations.

a $4x - 4y = 20$
$x - 4y = 2$

b $x - 2y = 11$
$3x - 2y = 25$

c $5p - 3q = 27$
$5p - q = 29$

d $a - 2b = 5$
$4a - 5b = 23$

7 Use the graph to solve these pairs of simultaneous equations.

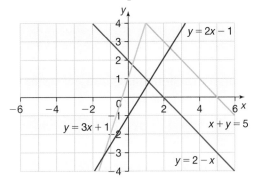

a **i** $y = 2x - 1$ and $x + y = 5$

ii $y = 3x + 1$ and $x + y = 5$

iii $y = 2 - x$ and $y = 2x - 1$

iv $y = 3x + 1$ and $y = 2 - x$

b Using the graph, explain why the simultaneous equations $x + y = 5$ and $y = 2 - x$ have no solution.

8 Plot graphs to solve these simultaneous equations.

a $y = 2x + 1$
$x + y = 10$

b $y = 3x - 2$
$x + y = 2$

9 Use the graphs to find an approximate solution for these simultaneous equations.

$x + 3y = 4$
$y = 2x + 2$

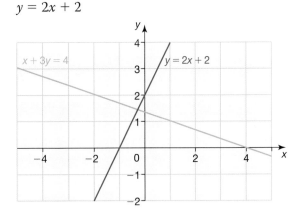

10 By drawing graphs, find **approximate** solutions for these simultaneous equations.

a $x + 3y = 4$
$y = 2x - 2$

b $x + 2y = 6$
$y = 3x + 1$

c $y = 3x - 1$
$y = x - 2$

d $y = 3x - 7$
$x = y + 2$

🔍 1175, 1176, 1319   SEARCH

10.4 Simultaneous equations

- You can solve simultaneous equations by eliminating one of the two variables.
- You can also solve simultaneous equations by plotting a graph. The solution is where the two lines intersect.

Sometimes the **coefficients** of the x-terms (or the y-terms) in **simultaneous** equations are not the same.

You have to adjust the equations before you can **eliminate** the x-terms (or the y-terms).

If I multiply equation (2) by 5, I'll have $-5y$.

$$3x + 5y = 4 \quad (1)$$
$$2x - y = 7 \quad (2)$$
$$3x + 5y = 4$$
$$10x - 5y = 35$$

Now you can add the equations to eliminate the y-terms.

HOW TO

1. Use the information in the question to form a pair of simultaneous equations.
2. Solve the simultaneous equations using elimination or by drawing a graph.
3. Give your answers and check that they make sense.

EXAMPLE

In a sweet shop, I spend £3.20 on three cans of soft drink and four bars of chocolate. The next day, I buy a can of soft drink and two bars of chocolate for £1.30. How much does each item cost?

1. Need to find the price of <u>both</u> items. Give all the variables letters.

    Let cost of cans of drink be c and cost of bars of chocolate be b.

    Write the information as equations using algebra.

    $3c + 4b = 320$ (1) $3c + 4b = 320$ (1)
    $c + 2b = 130$ (2) $\xrightarrow{\times 2}$ $2c + 4b = 260$ (3)

    If I multiply equation (2) by 2, I'll have $4b$.

2. You can eliminate the b-terms by subtracting (3) − (1):

    $c = 60$

    Substitute $b = 60$ into equation (2) to find b.

    $$c + 2b = 130$$
    $$60 + 2b = 130$$
    $$2b = 70$$
    $$b = 35$$

    You could multiply equation (2) by 3 and eliminate the c terms.

3. Check your answer. $3c + 4b = 320$ correct
    $3 \times 60 + 4 \times 35 = 320$

    Answer the question. I can of drink costs 60p and I chocolate bar costs 35p.

Exercise 10.4A

1 Which pairs of equations have the solution $x = 2$ and $y = 7$?

a
$$x + y = 9$$
$$x - y = -5$$

b
$$2x + y = 11$$
$$3x - y = -1$$

c
$$2x + 2y = 15$$
$$4x - y = 1$$

d
$$x + 2y = 16$$
$$x - 2y = 8$$

2 Solve these simultaneous equations.

a $2x + y = 8$
  $5x + 3y = 12$

b $3x + 2y = 19$
  $4x - y = 29$

c $8a - 3b = 30$
  $3a + b = 7$

d $2v + 3w = 12$
  $5v + 4w = 23$

e $9p + 5q = 15$
  $3p - 2q = -6$

f $3x - 2y = 11$
  $2x - y = 8$

3 Make as many pairs of simultaneous equations as you can using these four cards. Solve your pairs.

$$2x + y = 12$$

$$y - x = 15$$

$$3x - 4y = 7$$

$$2y + 3x = 19$$

4 Solve these problems by using simultaneous equations.

a How much does a lemon cost?

3 lemons
4 oranges
£1.27

4 oranges
5 lemons
£1.61

b The perimeter of this triangle is 30 cm. How long is the base?

5 For each question, set up a pair of simultaneous equations and solve them to find the required information.

a Two numbers have a sum of 23 and a difference of 5. What numbers are they?

b Two numbers have a difference of 6. Twice the larger plus the smaller number also equals 6. What numbers are they?

6 Use simultaneous equations to find the value of each symbol in the puzzle.

7 Tickets for a theatre production cost £3.50 per child and £5.25 per adult. 94 tickets were sold for a total of £365.75. How many children attended the production?

8 Use a graphical method to solve these problems.

a Twice one number plus three times another is 4. Their difference is 2. What are the numbers?

b The sum of the ages of James and Isla is 4. The difference between twice Isla's age and treble James' age is 3. How old are they?

9 a By plotting graphs if necessary, explain why the simultaneous equations $y = 2x - 1$ and $y = 2x + 4$ have no solution.

b Is it possible to have a pair of simultaneous equations with more than one solution?

10 Ellen tries to solve the simultaneous equations

$$3y = 7x - 6 \qquad 6y = 13x + 24$$

using a graphing method.

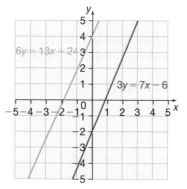

Ellen says the simultaneous equations do not have a solution.

Explain why Ellen is wrong, and find the solution to the simultaneous equations.

 🔍 1175, 1176, 1319   SEARCH

10.5 Inequalities

- In an **inequality**, the left-hand and right-hand sides are not necessarily equal.
- An inequality usually has a range of values.

You can show inequalities on a number line.

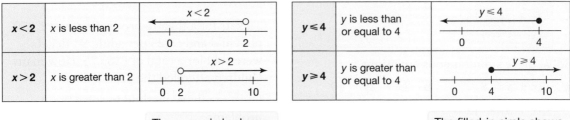

The open circle shows that 2 is not included.

The filled-in circle shows that 4 is included.

Inequalities can have more than one term, for example $3x + 4 < 19$.

You can solve an inequality to find a set of values for x.

Just use the balance method, as for solving an equation, by treating the inequality sign like an equals sign.

EXAMPLE

a Solve the inequality
$3x + 4 < 19$

b Show the solution set on a number line.

a Using the balance method:

$$3x + 4 < 19$$
$$3x + 4 - 4 < 19 - 4$$
$$3x < 15$$
$$x < 5$$

b

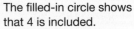

The solution set is $x < 5$.

- You can solve two-sided inequalities to find the solution set.
  - You can give the **integer** values in the solution set.

An integer is a whole number such as −1, 4, 10.

EXAMPLE

$-4 < 2n \leqslant 6$ where n is an integer.

a Show all the possible values of n on a number line.
b Write all the possible values of n.

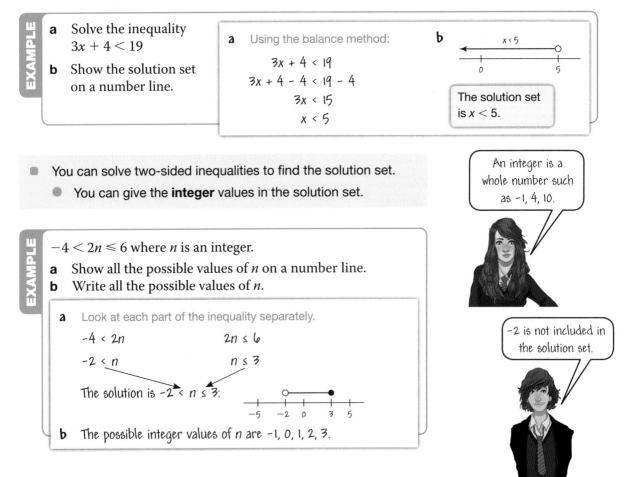

a Look at each part of the inequality separately.

$-4 < 2n$ $2n \leqslant 6$

$-2 < n$ $n \leqslant 3$

The solution is $-2 < n \leqslant 3$:

b The possible integer values of n are −1, 0, 1, 2, 3.

−2 is not included in the solution set.

Exercise 10.5S

1 Show these inequalities on a number line.

   a $x < 1$ **b** $x \geqslant 1$

   c $x \geqslant 5$ **d** $x < -2$

   e $x < 1.5$ **f** $x > -4$

   g $x \leqslant 3$ **h** $x \leqslant -1.5$

2 Match each inequality to a number line.

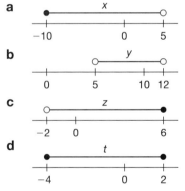

   i $x < -3$

   ii $x \leqslant 3$

   iii $x - 1 \leqslant -3$

   iv $x \geqslant -3$

   v $x - 1 > -3$

   vi $x + 2 < -2$

3 Write the inequalities shown by these number lines.

   a (x, -10, 0, 5)

   b (y, 0, 5, 10, 12)

   c (z, -2, 0, 6)

   d (t, -4, 0, 2)

4 Show these inequalities on number lines like this.

   (−5 −4 −3 −2 −1 0 1 2 3 4 5)

   a $-3 \leqslant x < 2$ **b** $4 \geqslant n > -1$

   c $1 < y \leqslant 3$ **d** $-4 \leqslant m < 0$

5 n is an integer.

   $-4 \leqslant n < 5$

   a Show all the possible values of n on a number line.

   b Write all the possible values of n.

6 **a** If $x > 5$, what can you say about

      i $2x$ **ii** $4x$?

   b If $y \leqslant 6$, write an inequality for

      i $3y$ **ii** $5y$.

   c If $x \geqslant -4$, write an inequality for $5x$.

   d If $m < -3$, write an inequality for $6m$.

7 Solve these inequalities and show the solution sets on number lines.

   a $2x \leqslant 4$ **b** $2x < 10$

   c $3x > -6$ **d** $4x \geqslant -16$

8 Copy and complete these.

   a If $3x + 2 > 11$ then $3x > \square$ and $x > \square$

   b If $7x - 4 > 31$ then $7x > \square$ and $x > \square$

   c If $2x + 9 \leqslant 11$ then $2x \leqslant \square$ and $x \leqslant \square$

   d If $5x - 3 \geqslant 12$ then $5x \geqslant \square$ and $x \geqslant \square$

9 Solve each of these inequalities.

   Show each solution on a number line.

   a $x + 7 \leqslant 12$ **b** $x - 2 \geqslant 4$

   c $3x + 5 \geqslant 11$ **d** $2x - 5 < 3$

   e $5x + 1 \geqslant -4$ **f** $6x - 2 \leqslant 16$

   g $3x + 2 > 11$ **h** $2x - 9 \leqslant -5$

10 Solve these inequalities.

   a $5x - 4 \geqslant 11$ **b** $4x + 7 > 15$

   c $3x + 5 \geqslant 5$ **d** $4x + 7 > 3$

   e $5x + 6 \geqslant -4$ **f** $3x - 8 \leqslant 4$

   g $4x + 20 \geqslant 10$ **h** $2x + 23 < 8$

11 $2 \leqslant 2x < 10$

   x is an integer.

   Write all the possible values of x.

12 In these inequalities, y is an integer.

   i Show each solution on a number line.

   ii Write all the possible values of y.

   a $-6 < 2y \leqslant 4$

   b $-3 \leqslant 3y < 15$

   c $3 < 2y \leqslant 10$

   d Write down the possible values of y that satisfy all three inequalities.

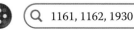 1161, 1162, 1930   SEARCH

10.5 Inequalities

RECAP

- An **inequality** is a mathematical statement including one of these symbols:

  $<$ $>$ $\leqslant$ $\geqslant$

  less than   more than   less than or equal to   more than or equal to

- You can solve an inequality by rearranging and using inverse operations, in a similar way to solving an equation.

$$
\begin{array}{l}
3x + 2 > 5x - 1 \\
2 > 2x - 1 \\
3 > 2x \\
1.5 > x
\end{array}
$$

Compare with

$$
\begin{array}{l}
3x + 2 = 5x - 1 \\
2 = 2x - 1 \\
3 = 2x \\
1.5 = x
\end{array}
$$

Use an 'empty' circle for $<$ and $>$

Use a 'filled' circle for $\leqslant$ and $\geqslant$.

- The solution to an inequality can be a range of values, which you can show on a number line:

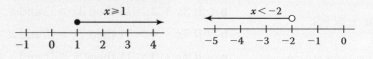

$x \geqslant 1$

$x < -2$

- If you multiply or divide an inequality by a positive number, the inequality remains true.

- If you multiply or divide an inequality by a negative number you need to reverse the inequality sign to keep it true.

$4 < 6$ and $8 < 12$ and $2 < 3$

$4 < 6$ but $-2 > -3$
$5 > 2$ but $-15 < -6$

HOW TO

(1) Use the information in the question to form an inequality. The inequality could be one-sided or two-sided.

(2) Use the balance method to solve the inequality. Remember that multiplying or dividing by a negative number changes the direction of the inequality sign.

(3) Give the range of values for the answer.

EXAMPLE

The area of the rectangle is less than 24 cm². Explain why x must satisfy $4 < x < 12$.

12 − x cm

3 cm

Think about what happens if x is greater than 12. If x = 14 then the length is −2cm, which is impossible.

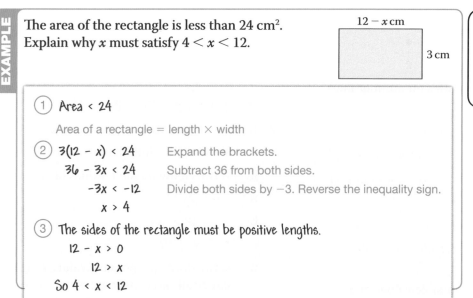

(1) Area < 24

    Area of a rectangle = length × width

(2) 3(12 − x) < 24   Expand the brackets.

    36 − 3x < 24   Subtract 36 from both sides.

      −3x < −12   Divide both sides by −3. Reverse the inequality sign.

        x > 4

(3) The sides of the rectangle must be positive lengths.

    12 − x > 0

      12 > x

  So 4 < x < 12

Algebra   Equations and inequalities

Exercise 10.5A

1 If x can take the possible integer values $-1, 0, 1, 2, 3$, which of these could be true?

a $x > -2$ **b** $-1 \leqslant x \leqslant 3$

c $-2 < x < 4$ **d** $-1 \leqslant x < 3$

e $-2 < x \leqslant 3$ **f** $-1 < x < 3$

2 Max, Natalie, Owen and Pritesh try to solve the inequality $10 - 4x < 26$.

Max	**Natalie**
$10 - 4x < 26$	$10 - 4x < 26$
$-4x < 16$	$10 < 4x + 26$
$x > 4$	$-16 < 4x$
	$-4 < x$

Owen	**Pritesh**
$10 - 4x < 26$	$10 - 4x < 26$
$-4x < 16$	$-4x < 16$
$x < -4$	$x > -4$

Who is correct?

Explain what went wrong in each incorrect answer.

3 Solve these inequalities.

a $-3x > 9$ **b** $-5x \geqslant 20$

c $-x + 1 < 9$ **d** $-2x - 10 \leqslant 0$

e $-4x + 3 > -13$ **f** $5 - 6x \leqslant 23$

4 a The angle $(2x + 30)°$ is obtuse.

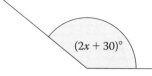

$(2x + 30)°$

Find the range of values that x can take.

b The angle $(5y - 45)°$ is reflex.

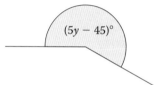

$(5y - 45)°$

Find the range of values that y can take.

5 The area of the square is less than the area of the rectangle.

Find the range of possible values that x can take and represent your solution on a number line.

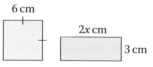

6 cm

$2x$ cm

3 cm

6 a The area of this rectangle exceeds its perimeter. Write an inequality and solve it to find the range of values of x.

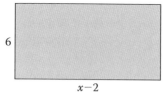

6

$x - 2$

b Given that x is an integer, find the smallest possible value that x can take.

7 Solve the inequality $-2 < 3x - 1 \leqslant 5$ and represent the solution on a number line.

8 Find the range of values of x that satisfy *both* inequalities.

a $3x + 6 < 18$ and $-2x < 2$

b $10 > 5 - x$ and $3(x - 9) < 27$

9 Explain why it is not possible to find a value of y such that $3y \leqslant 18$ and $2y + 3 > 15$.

10 The triangle is isosceles.

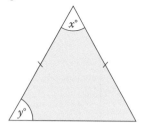

$x°$

$y°$

Angle x is less than $30°$.

a Explain why $180 - 2y < 30$.

b Complete this statement for the range of possible values that y can take.
$y > \square$

c Explain why y must be less than $90°$.

🔍 1161, 1162, 1930   SEARCH

Summary

Checkout

You should now be able to...

	Test it Questions
✓ Derive and solve simple linear equations.	1 – 3
✓ Solve quadratic equations algebraically by factorising.	4 – 5
✓ Derive and solve two linear simultaneous equations in two variables.	6 – 7
✓ Find approximate solutions to two linear simultaneous equations using a graph.	8
✓ Solve linear inequalities in one variable and represent the solution on a number line.	9 – 10

Language	Meaning	Example
Balance method	A method for solving an equation by performing the same operation on each side.	$x - 6 = 7$ $x - 6 + 6 = 7 + 6$ add 6 to both sides $x = 13$
Quadratic	A quadratic expression contains a square term such as x^2 as the highest power.	$6x^2 - 7x - 3$
Factorising	Writing an expression as two or more different expressions multiplied together.	$3x^2 + 6x \quad = 3(x^2 + 2x)$ $= 3x(x + 2)$ $x^2 - 3x - 10 = (x + 2)(x - 5)$
Solve **Solution**	Find a value for the unknown variable that will make the equation true.	$4x - 3 = 9$ is true when $x = 3$. $x = 3$ is the solution to the equation.
Simultaneous equations	Two or more equations that are true at the same time for the same values of the variables.	$y = 3x - 2$ and $2x + y = 8$ are both true when $x = 2$ and $y = 4$.
Inequality	A comparison of two quantities that are not equal.	$5 < 9$ 5 is less than 9 $x \geq 6$ the value of x is greater than or equal to 6

Review

1 Solve these equations.

    a $a - 13 = 35$ **b** $9b = 54$

    c $5c + 8 = 43$ **d** $2d - 8 = 25$

    e $\frac{e}{5} = 20$ **f** $7f + 19 = 5$

2 Solve these equations with unknowns on both sides.

    a $2x + 5 = x + 11$

    b $5x - 7 = 9x - 15$

    c $20 - 3x = 34 - x$

3 Vicky is x years old. Her mum is 24 years older and the sum of their ages is 42.

    a   Write an equation in x for the sum of their ages.

    b   Solve your equation to find the ages of Vicky and her mum.

4 Solve these quadratic equations by factorising.

    a $x^2 + 8x + 15 = 0$

    b $x^2 - 6x + 5 = 0$

    c $x^2 + x - 6 = 0$

    d $x^2 - 64 = 0$

    e $x^2 - 12x = 0$

5 Solve this quadratic equation.

    $x^2 - 8x = 20$

6 Solve these pairs of simultaneous equations.

    a $2x + 3y = 21$ **b** $5a - b = 7$

          $2x - y = 1$ $10a + b = 8$

    c $x + 4y = 21$ **d** $5v + 3w = 1$

          $2x - y = 15$ $4v - 5w = 23$

7 The combined mass of two boxes of chocolates and one bottle of water is 3.75 kg. The mass of three boxes of chocolates and two bottles of water is 6 kg.

    a   Write simultaneous equations to describe this situation.

    b   Find the mass of a box of chocolates and the mass of a bottle of water.

8 Plot a graph to solve these simultaneous equations.

    $x + y = 4$

    $y = 2x + 1$

9 Show these inequalities on number lines.

    a $x > 4$ **b** $x \leqslant 6$ **c** $1 < x \leqslant 6$

10 Solve these inequalities. Show each solution on a number line.

    a $3x + 7 > 22$ **b** $5x - 2 \leqslant 13$

    c $2x + 9 \geqslant 12 - x$

         $8 - 2x < 10$

What next?

Score			
	0 – 4		Your knowledge of this topic is still developing.
			To improve look at MyMaths: 1154, 1161, 1162, 1169, 1175, 1176, 1181, 1182, 1319, 1395, 1925, 1928, 1930
	5 – 8		You are gaining a secure knowledge of this topic.
			To improve look at InvisiPens: 10Sa – m
	9 – 10		You have mastered these skills. Well done you are ready to progress!
			To develop your exam technique look at InvisiPens: 10Aa – g

Assessment 10

1 A square of side 7 cm has the same area as a rectangle with sides 5 cm and $2s$ cm.
Find and solve an equation in s. [3]

2 Avel's dad is 4 times as old as Avel.
The sum of their ages is 55. How old is Avel? [2]

3 Jacques and Gilles went up the hill to fetch a pail of water. The mass of the water
was 22 kg more than the pail. In total the mass was 27.5 kg. How heavy was the pail? [2]

4 Oliver says he is three times as old as his brother Albert. Albert says Oliver is
6 years older than he is. They are both right. How old are Oliver and Albert? [4]

5 Alice bought a cake and cut it into 3 pieces. Tweedledum's piece was 36 g heavier
than Tweedledee's. Tweedledee's piece was 22 g lighter than Alice's. The total mass
was 454 g. Calculate the mass of Alice's piece. [4]

6 Brendan, Arsene and José go on holiday. Brendan takes €x, Arsene takes half as much
as José and José takes €150 more than Brendan. They took €2000 spending money in total.

   a Write down an equation involving x. [1]

   b Solve the equation to find the value of x. [3]

   c How much money did each person take? [2]

7 *PQRS* and *PYZS* are rectangles.
The area of *YQRZ* is 45 cm².
Grace works out that the value of a is 5 cm.
Show how she worked out the value of a.
When Grace solved her equation, she got
two solutions.
Say why she could ignore the other solution. [8]

8 Mario attempted to kayak 30 km on Lake Garda to raise money for charity. Sophia said she
would give him €5 for each complete kilometre he covered, as long as Mario gave her €2 for
each complete kilometre he failed to kayak. They €115 to charity.

   a Mario kayaked k kilometres in the event. Write down an equation
involving k and solve it to find the number of kilometres he kayaked. [4]

   b Find the minimum number of kilometres that Mario needed to
complete to be sure of making a contribution to charity. [3]

9 Patrick factorised this expression incorrectly.
$w^2 - w - 72 = (w - 9)(w - 8)$

   a Show how to factorise the expression correctly. [2]

   b Solve the equation $w^2 - w - 72 = 0$ [1]

10 Maria says that $89^2 - 11^2 = 7800$ and $6.89^2 - 3.11^2 = 37.8$
Without using your calculator, show that both of her answers are correct. [6]

11 A square has a side length s cm. Another square has side length 2 cm shorter.
The total area of the two squares is 202 cm². Find s. [6]

12 Two consecutive odd numbers have a product of 63.
There are two 'sets' of answers. What are they? [4]

13 The formula $S = \dfrac{n(n + 1)}{2}$ represents the sum of the numbers $1 + 2 + 3 + \cdots\cdots + n$.
The total of the numbers 1 to n is 5050. What is n? [4]

14 Eoin hits a cricket ball straight upwards. The formula $h = 20t - 5t^2$ represents
its height, h, above the ground, t seconds after he throws it.
Find the two times when the height of the ball is 15 m above the ground.
Say why there are two possible answers. [5]

15 The formula $2d = n(n - 3)$ represents the number of diagonals, d, in a polygon
with n sides. A dodecagon has 54 diagonals. How many sides does it have? [4]

16 Batman and Robin are 'b' and 'r' years old. Their ages sum to 46 and Robin is
10 years younger than Batman. How old are they? [4]

17 On a coach trip, an adult ticket is £a and a child ticket £c.
Tickets for 2 adults and 2 children are £22. Tickets for 2 adults and 5 children are £35.50.
How much is   **a**   an adult's ticket   [3]   **b**   a child's ticket? [1]

18 A magic goose lays brown eggs and once a year a gold egg! 5 brown eggs and 1 gold
egg have a mass of 150 g, while 8 brown and 1 gold egg have a mass of 210 g.
Find the weight of a gold egg. [4]

19 Liz and Graham had dinner. Graham's meal was £2.50 more expensive than Liz's.
The final bill was £40.50. How much did each meal cost? [4]

20 Five packets of 'doggibix' and three packets of 'cattibix' cost £5.49.
Three packets of 'doggibix' and one packet of 'cattibix' cost £2.79.
Find the cost of one packet of   **a**   Doggibix   [4]   **b**   Cattibix. [1]

21 A DVD costs £D and a CD £c. 3 DVDs and 2 CDs cost £45.83 in total.
1 DVDs and 3 CDs cost £26.92. How much does each cost? [5]

22 Farmer Scott can buy 2 sheep and 6 cows, or 10 sheep and 2 cows, for £3500.
What is the price of   **a**   a sheep   [4]   **b**   a cow? [1]

23 Titus Lines is going fishing and needs bait. He can buy 5 maggots and 6 worms for
38p or 6 maggots and 12 worms for 60p. How much are maggots and worms? [5]

24 Shaun has incorrectly represented the inequality $-1 < x \leqslant 3$ on a number line as shown.

Draw the expression correctly on a number line. [3]

25 Craig incorrectly says that all of the integer solutions to $14 \geqslant 4c \geqslant -3$ are $c = 4, 3, 2$
and 1. List all of the integer solutions. [3]

26 A triangle has angles x, y and z, $x > 35$, $y > 61$. Write an inequality for z. [2]

27 **a**   If $a > 1$ write down the inequality for $\dfrac{1}{a}$. [1]

    b   Is it always true that $b^2 > b$? Give reasons for your answer. [2]

Abigail, Mike, Juliet and Raheem, have completed their business plan, and decide they want to locate their restaurant in an area near the railway station in Newton-Maxwell. To choose the ideal location, they need to set boundaries based on proximity to the high street and competitor restaurants. They also start planning other key considerations: designing promotional material, what tables they need to buy and how many, and contacting potential suppliers.

Task 1 – Location

The friends make a list of conditions that the location must meet (see below). Raheem draws a scale map on 5 mm square paper.

a What is the scale of the map?

b Draw an accurate copy of the map and shade the areas that meet all three conditions.

They decide on a location that is on a bearing of 315° from T and 014° from A.

c On your copy of the map, label their desired location with R.

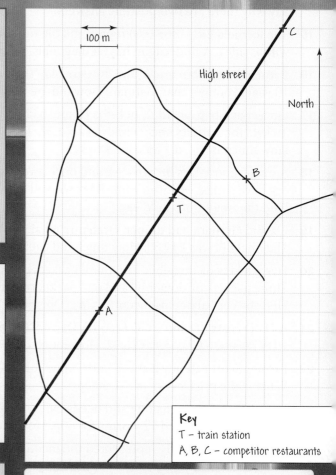

```
Conditions for location
```

- No more than 500 m from the railway station
- At least 200 m from each competitor restaurant
- No more than 150 m from the high street

Task 2 – The restaurant logo

The diagram shows the dimensions used for a logo to appear on the restaurant's business cards.

a Find the area of the logo.

For use on A5 flyers, the logo is enlarged by a scale factor of 2.5

b Find the area of the enlarged logo.

The logo is the only part of the business card and the flyer that uses coloured ink.

c A colour ink cartridge lasts long enough to print 4000 business cards. How many A5 flyers would you expect to print using one colour ink cartridge?

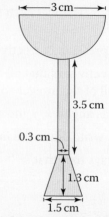

▲ Juliet's logo for use on their promotional material. It consists of a semicircle, a rectangle and a trapezium, as shown in the second figure.

FIRE

Fire regulations

- Total number of staff and non-staff must not exceed 32.
- Ratio of non-staff to staff must not exceed 8:1.

Table	Shape of top	Dimensions
Style A	Circular	Diameter = 1.4 m
Style B	Regular octagon	Side length = 58 cm
		Length between opposite sides = 1.4 m

▲ Possible styles of table for restaurant.

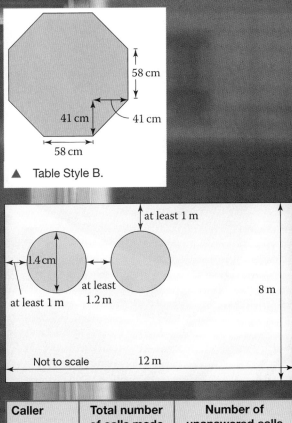

▲ Table Style B.

Not to scale

Caller	Total number of calls made	Number of unanswered calls
Abigail	30	10
Mike	22	8
Juliet	40	18
Raheem	18	8

▲ Number of calls made in the first week.

Task 3 – Safety regulations

The restaurant must adhere to fire safety regulations.

Find the maximum number of customers allowed in the restaurant at any time.

Task 4 – Choosing tables

They narrow down their choice of table based on the maximum number of customers allowed at one time.

a Show that the regular octagon in the diagram measures 1.4 m across.

b Without doing any calculations, which of the table tops (style A or style B) has the larger area?

c Calculate the area of each table top.

Style A is cheaper, so they decide to buy that one.

d Is this a reasonable decision? Give reasons for your answer.

Task 5 – Number of tables

The dining space is rectangular and measures 12 m by 8 m. Abigail draws a sketch to indicate the gap required from each wall, and between each table.

How many tables of Style A can fit in the dining space?

Task 6 – Calling potential suppliers

Each member of the team makes telephone calls to potential suppliers. They logged how many calls they made in the first week, and how many went unanswered.

a Which person had the highest proportion of unanswered calls?

b Estimate how many unanswered calls you would expect out of the next 50 calls made by the team.

c What assumptions have you made in answering part **b**?

11 Circles and constructions

Introduction

The invention of the wheel was most definitely a landmark event in human technological development, giving people the ability to travel at speed. However it was the use of gears and cogs on a massive scale during the Industrial Revolution that really accelerated advancement, not just in technology but also in social and economic development.

What's the point?

Without mankind's understanding of circles, and how their properties can be exploited in marvellous ways, we would still be living in largely agricultural communities in a pre-industrial state, with no computers, mobile phones, cars, ...

Objectives

By the end of this chapter you will have learned how to ...

- Identify and apply circle definitions, properties and formulae.
- Construct triangles.
- Use the standard ruler and compass constructions.
- Solve loci problems.

Check in

1   Using a protractor, measure these angles.

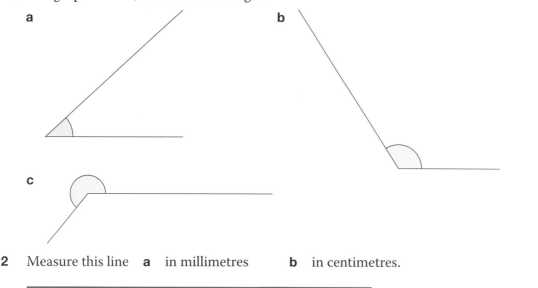

a

b

c

2   Measure this line   **a**   in millimetres   **b**   in centimetres.

3   Calculate the area of this rectangle.
State the units of your answer.

4 cm

←——— 7 cm ———→

Chapter investigation

Sketch a circle, and draw a straight line through it. How many pieces have you divided the circle into?

Draw a second straight line thought the circle. What is the maximum number of pieces you can divide the circle into?

Continue drawing straight lines through the circle (the lines can cut at any angle, and the circle can be as big as you want it to be). Is there a relationship between the number of cuts and the maximum number of pieces? Investigate.

11.1 Circles 1

These are the names of the different parts of a circle.

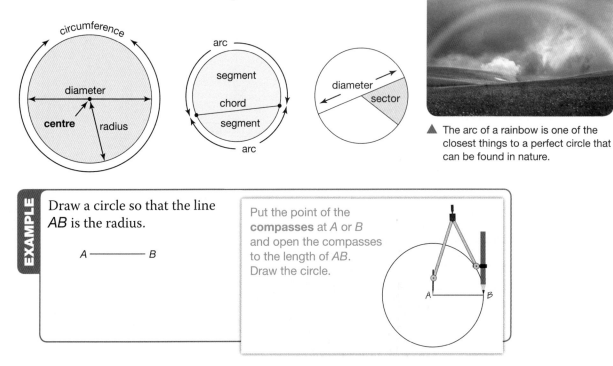

▲ The arc of a rainbow is one of the closest things to a perfect circle that can be found in nature.

EXAMPLE

Draw a circle so that the line *AB* is the radius.

A ——————— B

Put the point of the **compasses** at *A* or *B* and open the compasses to the length of *AB*. Draw the circle.

There is a formula to work out the circumference of a circle.

π is a Greek letter, pronounced 'pi', and its approximate value is 3.14.

● $C = \pi \times \text{diameter} = \pi d = 2\pi r$

$d = 2 \times r$

There is a formula to work out the area of a circle.

● Area of circle $= \pi \times \text{radius} \times \text{radius}$
    $= \pi \times r \times r$ or πr^2

r^2 means $r \times r$

EXAMPLE

Find **i** the circumference **ii** the area of each circle.

a $r = 5\,\text{cm}$

b $d = 32\,\text{mm}$

Round your answers to a sensible degree of accuracy. 3 sf is usually good practice.

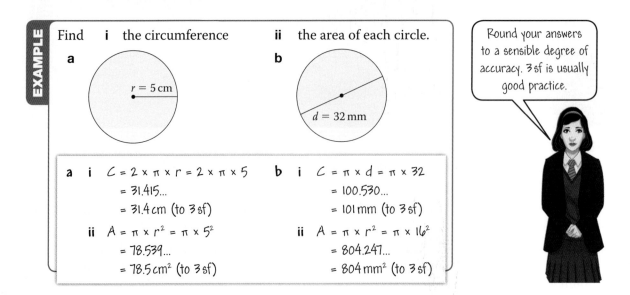

a i $C = 2 \times \pi \times r = 2 \times \pi \times 5$
     $= 31.415...$
     $= 31.4\,\text{cm}$ (to 3 sf)

ii $A = \pi \times r^2 = \pi \times 5^2$
     $= 78.539...$
     $= 78.5\,\text{cm}^2$ (to 3 sf)

b i $C = \pi \times d = \pi \times 32$
     $= 100.530...$
     $= 101\,\text{mm}$ (to 3 sf)

ii $A = \pi \times r^2 = \pi \times 16^2$
     $= 804.247...$
     $= 804\,\text{mm}^2$ (to 3 sf)

Geometry   Circles and constructions

Exercise 11.1S

1 Measure

   a the diameter of the circle

   b the radius of the circle.

2 **a** Draw a circle with a radius of 4 cm.

   b Draw a chord of length 5 cm inside the circle.

3 Draw a circle with a diameter of 10 cm.

4 **a** Draw a 4 cm line *AB*.

A ——————————— B
           4 cm

   b Draw a circle so that *AB* is the diameter.

   c Find the radius of the circle.

5 Calculate the circumferences of these circles. State the units of your answers.

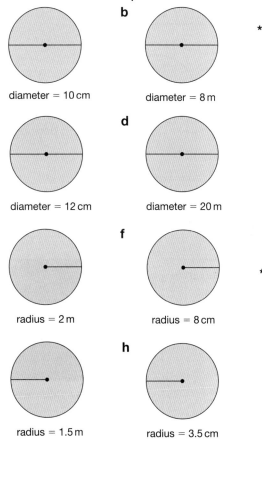

  a diameter = 10 cm   **b** diameter = 8 m

  c diameter = 12 cm   **d** diameter = 20 m

  e radius = 2 m   **f** radius = 8 cm

  g radius = 1.5 m   **h** radius = 3.5 cm

6 Find the circumferences of these circles.

   a radius = 12 mm

   b radius = 23 cm

   c diameter = 105 mm

   d diameter = 1.2 cm

   e radius = 3.6 cm

   f diameter = 125 cm

7 Find the areas of the circles in question **5**.

8 Find the areas of the circles in question **6**.

*9 Calculate the diameter of a circle, if its circumference is

   a 18.84 cm

   b 15.7 m

   c 28.26 cm

   d 47.1 m

   e 314 cm

*10 Find the area of each semicircle.

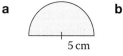

  a 5 cm   **b**

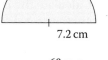

    7.2 cm

11 A circular hole is cut in a square card. Find the area of card that is left.

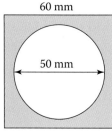

60 mm

50 mm

*12 This pendant is made from an isosceles trapezium and a semi-circle. The circular hole has diameter 4 mm.

Find the area of the pendant.

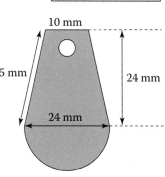

10 mm

25 mm

24 mm

24 mm

11.1 Circles 1

- Use the formula $C = \pi d$ or $C = 2\pi r$ to find the circumference of a circle.
- Use $A = \pi r^2$ to find the area.

- Area of a semicircle $= \frac{1}{2} \times$ area of whole circle

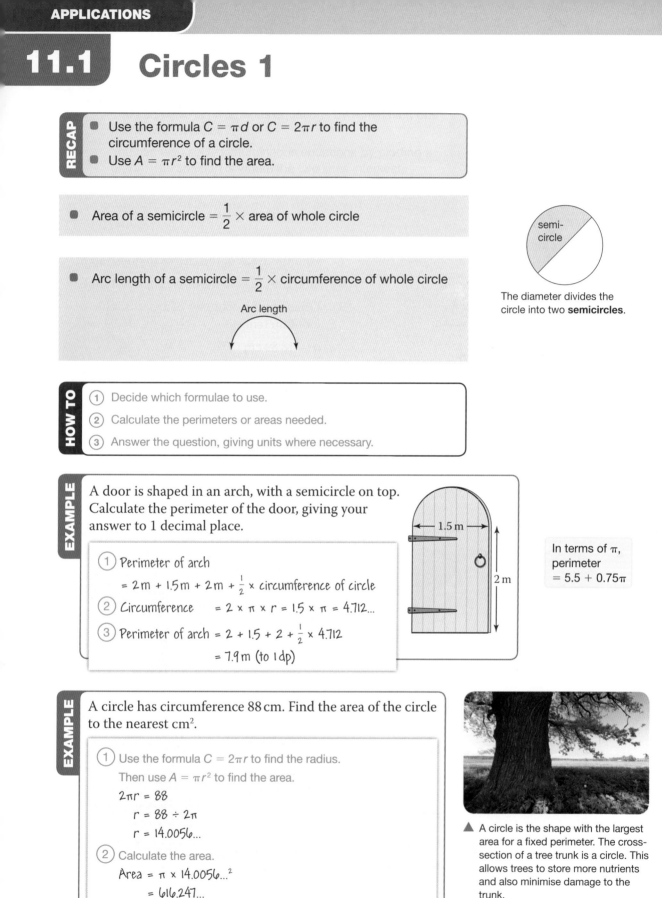

semi-circle

The diameter divides the circle into two **semicircles**.

- Arc length of a semicircle $= \frac{1}{2} \times$ circumference of whole circle

Arc length

HOW TO

① Decide which formulae to use.

② Calculate the perimeters or areas needed.

③ Answer the question, giving units where necessary.

EXAMPLE

A door is shaped in an arch, with a semicircle on top. Calculate the perimeter of the door, giving your answer to 1 decimal place.

① Perimeter of arch

$= 2\,m + 1.5\,m + 2\,m + \frac{1}{2} \times$ circumference of circle

② Circumference $= 2 \times \pi \times r = 1.5 \times \pi = 4.712...$

③ Perimeter of arch $= 2 + 1.5 + 2 + \frac{1}{2} \times 4.712$

$= 7.9\,m$ (to 1 dp)

1.5 m

2 m

In terms of π, perimeter $= 5.5 + 0.75\pi$

EXAMPLE

A circle has circumference 88 cm. Find the area of the circle to the nearest cm².

① Use the formula $C = 2\pi r$ to find the radius. Then use $A = \pi r^2$ to find the area.

$2\pi r = 88$

$r = 88 \div 2\pi$

$r = 14.0056...$

② Calculate the area.

Area $= \pi \times 14.0056...^2$

$= 616.247...$

③ $= 616\,cm^2$ (to the nearest cm²)

▲ A circle is the shape with the largest area for a fixed perimeter. The cross-section of a tree trunk is a circle. This allows trees to store more nutrients and also minimise damage to the trunk.

Geometry Circles and constructions

Exercise 11.1A

1 Shamin is cutting out circles for an art project.
She has squares of card that are 4.2 cm wide.

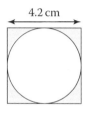
4.2 cm

a What is the area of the biggest circle she can cut out?

b What area of card is left?

2 A circular helipad (for landing a helicopter) has a radius of 14 m.
The cost of building the helipad is £85 per square metre.
Find the cost of building the helipad to the nearest £100.

3 A flowerbed in the park is semicircular. It has a radius of 2 m.

a Work out the area of the flowerbed.

Percy the park keeper wants to plant flowers that each need an area of 0.3 m².

b How many of these flowers can Percy plant in the flowerbed?
What space does he have left?

4 Viaduct arches have straight sides 50 m high.
The arch at the top is a semicircle with diameter 8 m.

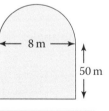
8 m
50 m

A spider crawls from ground level on one side, around the arch, and back down the other side.
Work out how far it crawls.

> Give your answer in terms of π and to 1 dp.

5 A standard running track is 400 m in length.
Design a running track which includes

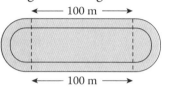
100 m
100 m

- two straight sections of 100 m each
- two semicircular sections at opposite ends.

Give your answer as a scale drawing, with dimensions clearly marked.

6 Two circles have the same centre.
One has a radius of 3.5 cm and the other has a radius of 2.5 cm.
Find the area of the shaded region.

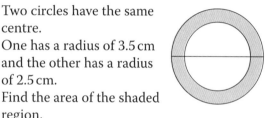

7 A round hole has a circumference of 44 cm.
Work out the radius of the hole, to 1 decimal place.

*8 A circle has area 200 cm².
Find the radius of the circle.
Give your answer to the nearest cm.

r

9 The diameter of the wheels on Maisy's bike is 60 cm. Maisy says the wheels go round over 500 times for every kilometre she travels.
Is Maisy correct? Show your workings.

*10 Show that the area of a circle with diameter 70 cm is more than three-quarters of the area of a square with sides 70 cm.

*11 A rectangular sheet of metal is 1.2 metres long and 80 centimetres wide.
A badge-making machine cuts circles of metal from this sheet.

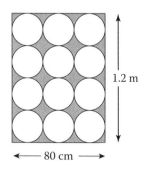
1.2 m
80 cm

a How many circles of diameter 4 cm can be cut from the sheet?

b Sally says that more than 20% of the metal is wasted.
Is Sally correct?
You must show your working.

11.2 Circles 2

- Area of a semicircle = $\frac{1}{2}$ × area of whole circle.

- Perimeter of a semicircle = $\frac{1}{2}$ × circumference of whole circle + diameter.

semi-circle

The diameter divides the circle into two **semicircles**.

EXAMPLE

Calculate the perimeter of this semicircle.

← 6 cm →

Circumference of whole circle = 2 × π × r

= 37.699 ... cm

Perimeter of semicircle = $\frac{1}{2}$ × circumference of whole circle + diameter

= ($\frac{1}{2}$ × 37.699...) + (2 × 6)

= 18.849 ... + 12

= 30.8 cm (to 1 dp)

An **arc** is a fraction of the circumference of a circle.

circumference

r

arc

A **sector** is a fraction of a circle, shaped like a slice of pie.

diameter

sector

To calculate the area and arc length of a sector, you use the angle at the centre of the sector.

- Arc length = $\frac{\theta}{360°}$ × circumference of whole circle

= $\frac{\theta}{360°}$ × $2\pi r$

- Sector area = $\frac{\theta}{360°}$ × area of whole circle

= $\frac{\theta}{360°}$ × πr^2

θ is the Greek letter 'theta'.

θ

EXAMPLE

Find the arc length and area of this sector.

O

106° 10 cm

A B

Substitute the values in the formula.

Arc length of sector = $\frac{106°}{360°}$ × 2 × π × 10

= 18.50049... using a calculator

= 18.5 cm (to 3 sf)

Area of sector = $\frac{106°}{360°}$ × π × 10^2

= 92.502... using a calculator

= 92.5 cm² (to 3 sf)

Geometry Circles and constructions

Exercise 11.2S

1 Find the area of each semicircle.

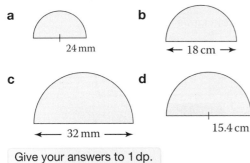

a   24 mm

b   ← 18 cm →

c   ← 32 mm →

d   15.4 cm

> Give your answers to 1 dp.

2 Find the area of each semicircle.

a   radius = 12 cm

b   radius = 2.3 cm

c   diameter = 12.9 m

d   diameter = 22.3 cm

e   radius = 9.5 mm

f   diameter = 3.39 cm

3 Work out the perimeter of each semicircle in question **1**.

4 Work out the perimeter of each semicircle in question **2**.

5 This **quarter-circle** has a radius of 8 cm.
Calculate

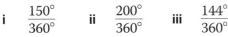

a   the length of arc *AB*

b   the area of sector *OAB*.

6 a Simplify

   i $\dfrac{150°}{360°}$ **ii** $\dfrac{200°}{360°}$ **iii** $\dfrac{144°}{360°}$

b   Simplify, leaving each in terms of π

   i $\dfrac{60°}{360°} \times 2\pi \times 24$

   ii $\dfrac{120°}{360°} \times 2\pi \times 15$

   iii $\dfrac{80°}{360°} \times \pi \times 6^2$

   iv $\dfrac{135°}{360°} \times \pi \times 12^2$

7 Find the area of each sector.

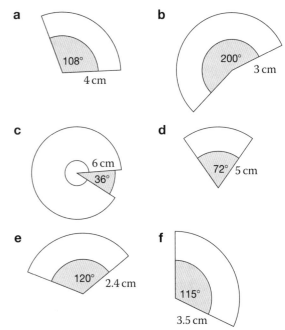

a   108°   4 cm

b   200°   3 cm

c   6 cm   36°

d   72° 5 cm

e   120° 2.4 cm

f   115°   3.5 cm

> In this exercise give your answers to 1 decimal place where appropriate.

8 Find the arc length of each sector in question **7**.

9 Copy and complete the table for four sectors.

	angle	radius	arc length	area of sector
a	40°	18 mm		
b	135°	1.6 m		
c	252°	3.5 cm		
d	312°	0.46 km		

10 Find the perimeter of each sector.

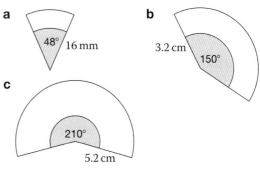

a   48° 16 mm

b   3.2 cm   150°

c   210°   5.2 cm

 1118, 1952   SEARCH

11.2 Circles 2

- Perimeter of a semicircle $= \frac{1}{2} \times$ circumference of whole circle $+$ diameter.
- Arc length $= \frac{\theta}{360°} \times$ circumference of whole circle $= \frac{\theta}{360°} \times 2\pi r$
- Sector area $= \frac{\theta}{360°} \times$ area of whole circle $= \frac{\theta}{360°} \times \pi r^2$

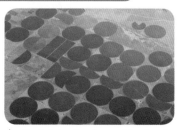

HOW TO

(1) Draw a diagram if it helps. Remember that a sector is a fraction of a whole circle.

(2) Decide which area, arc length or perimeter formula to use.

(3) Work out the answer to the question, giving any units.

▲ Circle irrigation is used in dry climates to water crops by rotating a sprinkler around a pivot. This forms crop circles and also conserves water.

EXAMPLE

The diagram shows the shape of a window.

ABC is an equilateral triangle.

AC is an arc of a circle, centre *B*.

BC is an arc of a circle, centre *A*.

Amy says the perimeter of the window is less than 4 metres. Is Amy correct?

(1) Each angle in an equilateral triangle is 60°.
The diagram shows sector *ABC*.

(2) Use arc length $= \frac{\theta}{360°} \times 2\pi r$

Arc BC $= \frac{60°}{360°} \times 2\pi \times 120$

$= 125.663...$ cm

Arc AC is also $125.663...$ cm

Perimeter $= 125.663... + 125.663... + 120$

$= 371.327...$ cm

$= 371$ cm (nearest cm)

$4\,m = 400\,cm$

| 1 m = 100 cm |

(3) The perimeter is less than 400 cm so Amy is correct.

EXAMPLE

A circle has a radius of 28 cm. A sector of this circle has an area of 200 cm².

Calculate the angle of the sector to the nearest degree.

(1) The area of the sector is proportional to the size of the angle.

(2) $\frac{\text{Sector area}}{\text{Area of whole circle}} = \frac{\text{angle}}{360°}$ Rearrange the formula to make the angle the subject.

Angle $= \frac{\text{Sector area}}{\text{Area of whole circle}} \times 360°$ Area of whole circle $= \pi \times 28^2 = 2463.0086...$

$= \frac{200}{2463.0086} \times 360°$

$= 29.232...$

(3) Angle $= 29°$ (nearest degree)

Exercise 11.2A

1 A square card has sides of length 24 cm.
A quarter-circle of radius 24 cm is cut from the square.

Find the area that is left.

24 cm

2 Find the area of this incomplete annulus.

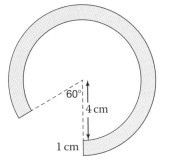

60°

4 cm

1 cm

> An annulus is the region between two concentric circles, shown here as the purple ring.

3 Jan wants to lay a patio in a corner of her garden. The patio is shaped like the sector of a circle.

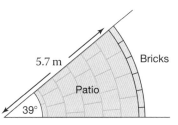

5.7 m

Bricks

Patio

39°

a What is the area of the patio?

b She wants to lay a brick path around the curved edge of the patio.
Each brick is 30 cm long.

Calculate the number of bricks she needs.

4 The diagram shows the landing sector for a javelin. The distance between the arcs is 10 m.

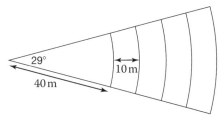

29°

10 m

40 m

The groundsman says that the total length of the lines is over 300 metres.
Is the groundsman correct? Show your working.

*5 A sector of radius 3 cm is cut from one corner of an equilateral triangle.

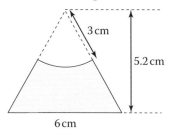

3 cm

5.2 cm

6 cm

Find the area of the remaining shape.

6 The radius of a circle is 30 cm.
An arc of this circle is 75 cm long.
Calculate the angle θ.

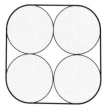

75 cm

30 cm

θ

7 The area of a sector is 8 m².
The radius of the circle is 2 m.
Calculate the angle of the sector.

*8 A circle has diameter 20 cm.
The arc length of a sector of this circle is 10 cm.
Calculate the area of the sector.

9 A continuous belt is needed to go around four identical oil drums as shown. The diameter of each drum is 24 inches. Work out the length of the belt. Give your answer in terms of π.

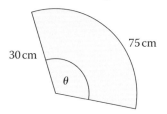

*10 Two quarter-circles are used to draw the leaf on this square tile.
Show that

20 cm

a the perimeter of the leaf is 20π cm

b the area of the leaf is 200(π − 2) cm².

11.3 Constructions

You will always **construct** identical triangles if you know...

two sides and the angle between them (SAS)

or two angles and a side (ASA)

or right angle, the hypotenuse and a side (RHS)

or three sides (SSS).

You will need a ruler and a protractor for SAS, ASA and RHS triangles.

You will need a ruler and compasses for SSS triangles.

EXAMPLE

a Construct the triangle *ABC* so that angle *C* = 90°, *AB* = 6 cm and *BC* = 3 cm.

b Construct the triangle *PQR* with lengths *PR* = 6 cm, *QR* = 8 cm and *PQ* = 10 cm.

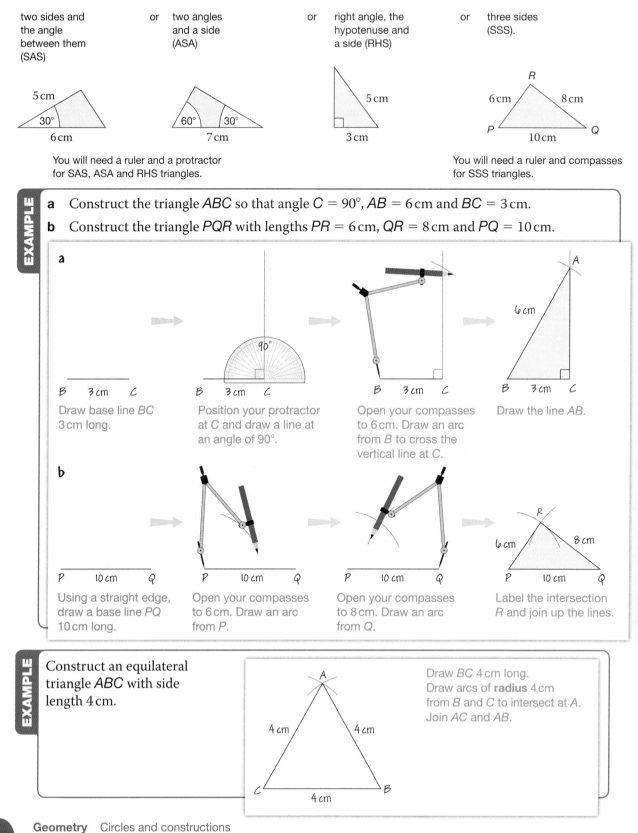

a

Draw base line *BC* 3 cm long.

Position your protractor at *C* and draw a line at an angle of 90°.

Open your compasses to 6 cm. Draw an arc from *B* to cross the vertical line at *C*.

Draw the line *AB*.

b

Using a straight edge, draw a base line *PQ* 10 cm long.

Open your compasses to 6 cm. Draw an arc from *P*.

Open your compasses to 8 cm. Draw an arc from *Q*.

Label the intersection *R* and join up the lines.

EXAMPLE

Construct an equilateral triangle *ABC* with side length 4 cm.

Draw *BC* 4 cm long.
Draw arcs of **radius** 4 cm from *B* and *C* to intersect at *A*.
Join *AC* and *AB*.

Geometry Circles and constructions

Exercise 11.3S1

The triangles in this exercise have been sketched by hand.
Draw them accurately using the measurements given.

1 Make accurate drawings of these triangles (SAS). Measure the unknown length in each triangle.

a

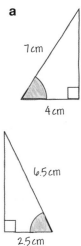

b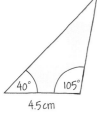

c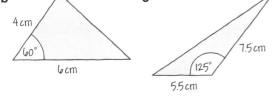

> Leave your **construction lines** on your drawing to show your method.

2 Make accurate drawings of these triangles (ASA). Measure the two unknown lengths in each triangle. State the units of your answers.

a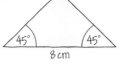

b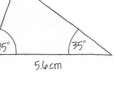

c

3 Make accurate drawings of these triangles (SSS). Measure the marked angle in each triangle.

a

b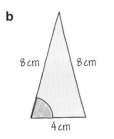

c

4 Make accurate drawings of these triangles (RHS). Measure the unknown length and the marked angle in each triangle.

a

7 cm

4 cm

b 8.5 cm

6 cm

c 6.5 cm

2.5 cm

5 Use a ruler and compasses or a protractor to construct these triangles.

a	Sides 8 cm, 4 cm, 7 cm	(SSS)
b	3 cm, 30°, 4 cm	(SAS)
c	Sides 10 cm, 7.5 cm, 6 cm	(SSS)
d	8 cm, 2 cm, 90°	(RHS)
e	Sides 6 cm, 9 cm, 5 cm	(SSS)
f	45°, 4 cm, 45°	(ASA)

> It helps to draw a rough sketch first.

6 Construct isosceles triangles with sides

    a 7 cm, 7 cm, 5 cm

    b 5 cm, 5 cm, 7 cm

7 **a** Construct an equilateral triangle *ABC* with sides 5 cm.

    b Construct a second equilateral triangle with base *AB* and sides 5 cm to get a rhombus.

    c Follow the steps in **a** and **b** to construct a rhombus with sides 3.5 cm.

8 **a** Construct a triangle with sides 5 cm, 12 cm, 13 cm.

    b What type of triangle is this?

9 **a** Construct a triangle *ABC* with sides *AB* = 3 cm, *BC* = 4 cm, *CA* = 5 cm.

    b Construct triangle *ADC* with side *CA* from the triangle in part **a**, and side *AD* = 4 cm and side *DC* = 3 cm.

    c What special type of quadrilateral is this?

 1089, 1090   SEARCH

11.3 Constructions

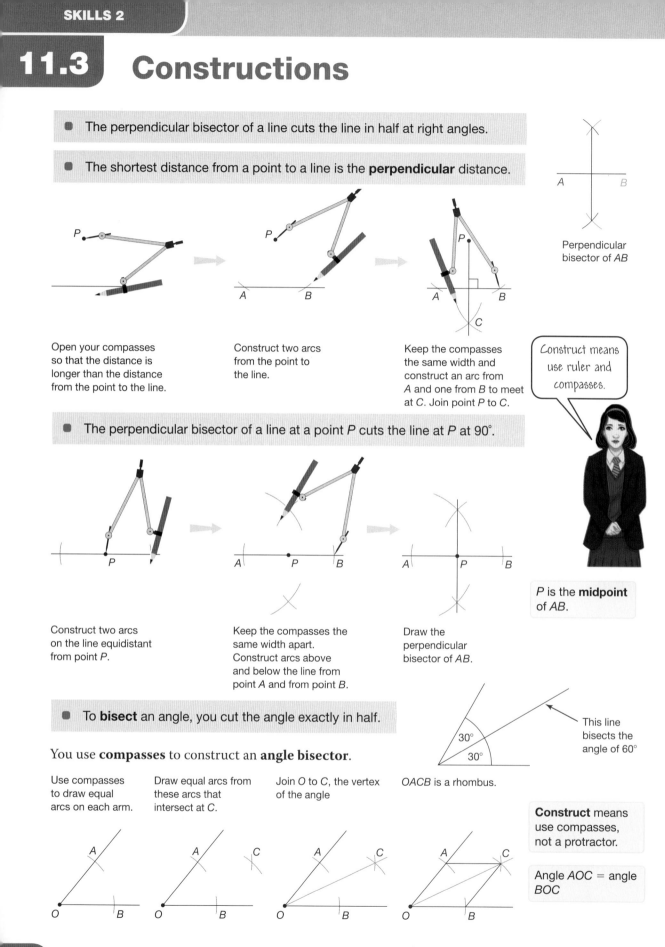

● The perpendicular bisector of a line cuts the line in half at right angles.

● The shortest distance from a point to a line is the **perpendicular** distance.

A | B

Perpendicular bisector of AB

Open your compasses so that the distance is longer than the distance from the point to the line.

Construct two arcs from the point to the line.

Keep the compasses the same width and construct an arc from A and one from B to meet at C. Join point P to C.

Construct means use ruler and compasses.

● The perpendicular bisector of a line at a point P cuts the line at P at 90°.

P is the **midpoint** of AB.

Construct two arcs on the line equidistant from point P.

Keep the compasses the same width apart. Construct arcs above and below the line from point A and from point B.

Draw the perpendicular bisector of AB.

● To **bisect** an angle, you cut the angle exactly in half.

You use **compasses** to construct an **angle bisector**.

This line bisects the angle of 60°

30°
30°

Use compasses to draw equal arcs on each arm.

Draw equal arcs from these arcs that intersect at C.

Join O to C, the vertex of the angle

$OACB$ is a rhombus.

Construct means use compasses, not a protractor.

Angle AOC = angle BOC

Geometry   Circles and constructions

Exercise 11.3S2

For all constructions use a pencil and do not rub out your construction lines.

1 a Draw a line *AB*, so that *AB* = 8 cm.

b Using compasses, construct the perpendicular bisector of *AB*.

c Label the midpoint of *AB* as *M*.

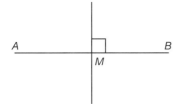

d Measure the length *AM*.

2 a Draw a line of length 64 mm.

b Construct the perpendicular bisector of the line.

c Check by measuring that the perpendicular bisector passes through the midpoint of the line.

3 Trace these lines and the points marked *X*. For each, use ruler and compasses to construct a perpendicular from the point *X* to the line.
Check your constructions using a protractor.

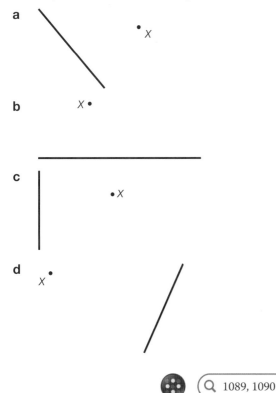

4 Trace these lines and mark the point *X*. For each, use ruler and compasses to construct the perpendicular from the point *X* on the line. Show all construction lines. Check your constructions using a protractor.

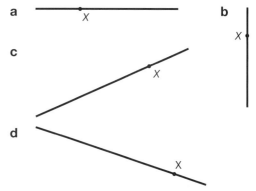

5 Trace these angles.
Construct the angle bisector of each angle, using ruler and compasses.

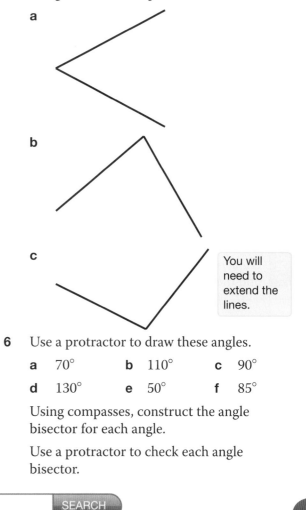

You will need to extend the lines.

6 Use a protractor to draw these angles.

| a | 70° | b | 110° | c | 90° |
| d | 130° | e | 50° | f | 85° |

Using compasses, construct the angle bisector for each angle.

Use a protractor to check each angle bisector.

11.3 Constructions

- A ruler and a protractor can be used to accurately construct
  - SAS triangles
  - ASA triangles
  - RHS triangles
- A pencil, pair of compasses and a ruler can be used to construct
  - SSS triangles
  - the perpendicular bisector of a line segment
  - a perpendicular to a line from or at a given point
  - an angle bisector
  - an angle of 60°
- The perpendicular is the shortest distance from a point to a line.

HOW TO

(1) Draw a diagram or copy a given diagram.

(2) Decide what you need to construct.

(3) Use a pencil, pair of compasses and ruler to carry out the construction.

(4) Answer any questions asked.

The longest side of a **right-angled** triangle is called the **hypotenuse**.

EXAMPLE

Using compasses, construct an angle of 60°.

(2) Start by constructing an equilateral triangle.

(3) Draw a line.

Draw an arc from A crossing the line at B.

Draw an arc from B crossing the first arc at C.

Draw a line from A to C.

EXAMPLE

The scale diagram shows a pipeline PQ and a new house H.

The builder needs to lay a pipe to connect H to PQ.

Using pencil, compasses and ruler only, find the length of the shortest possible pipe.

House, H

pipeline

Q

P

Scale
1 cm represents 50 m

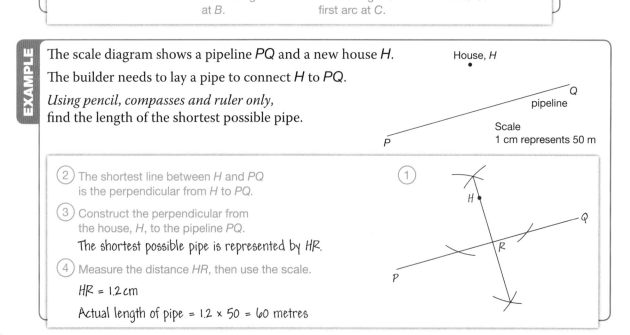

(2) The shortest line between H and PQ is the perpendicular from H to PQ.

(3) Construct the perpendicular from the house, H, to the pipeline PQ.

The shortest possible pipe is represented by HR.

(4) Measure the distance HR, then use the scale.

HR = 1.2 cm

Actual length of pipe = 1.2 × 50 = 60 metres

Geometry Circles and constructions

Exercise 11.3A

1 Construct these rhombuses using compasses and ruler.

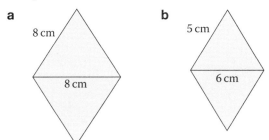

a 8 cm 8 cm

b 5 cm 6 cm

> Use a *pencil, ruler and compasses only* for each construction.

2 a Explain why you cannot construct a triangle with sides 9 cm, 4 cm, 3 cm.

b Explain what happens when you try and construct a triangle with sides 9 cm, 4 cm, 5 cm.

> Try to construct the triangles to see what happens.

3 Without drawing the construction, write whether these sets of three sides will make a triangle.

a Sides 5 cm, 5 cm, 9 cm

b Sides 2 cm, 2 cm, 2 cm

c Sides 29 cm, 26 cm, 4 cm

d Sides 22 cm, 12 cm, 10 cm

e Sides 20 cm, 7 cm, 9 cm

f Sides 14 cm, 8 cm, 6 cm

g Sides 15 cm, 60 mm, 100 mm

h Sides 120 mm, 8 cm, 9 cm

4 a Construct a triangle ABC such that $BC = 12$ cm, $AC = 7$ cm and $\angle B = 30°$.

> This triangle is called SSA because you have two sides and the non-included angle.

b Now try to draw a second, **different** triangle with the same measurements. (Move the position of A.)

c Are SSA triangles unique?

5 Construct a regular hexagon of side 4 cm.

> Start by drawing a suitable circle.

6 The diagram shows the position of a new factory, F, and the nearest main road, MN.

a Copy the diagram.

b Find the shortest possible length for a road from F to MN.

N
road
factory, F
M Scale 1 cm represents 100 m

7 Jamie says that the construction of an angle bisector works because of congruent triangles.
Is this true? Give your reasons.

8 Sue says that, when you construct a perpendicular bisector, the arcs drawn from one end of the line do not need to have the same radius as those drawn from the other end.
Is Sue correct? Give your reasons.

9 For each part use x and y axes from −6 to 6.

a i Join the points A (2, 6) and B (−4, −2).

ii Construct the perpendicular bisector of AB.

iii Write down the coordinates of M, the mid-point of AB.

b i Draw the line $y = x$.

ii Construct the perpendicular from the point (3, −5) to the line $y = x$.

iii Write down the coordinates of the point where the perpendicular meets $y = x$.

> **Did you know...**
>
>
>
> In constructions, Euclid used a straight edge with no markings, rather than a ruler. He could not bisect a line by just measuring so he had to construct the midpoint.

 🔍 1089, 1090   SEARCH

11.4　Loci

There are two different ways to think of a **locus**.

> **Loci** is the plural of locus.

- A set of points that follow one or more rules.
- The path followed by a moving point.

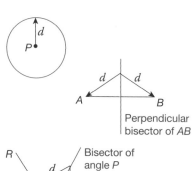

- The locus of points that are equidistant from a point, *P*, is a circle with centre *P*.

- The locus of points that are equidistant from two given points, *A* and *B*, is the **perpendicular bisector** of *AB*.

Perpendicular bisector of *AB*

- The locus of points that are equidistant from line segments *PQ* and *PR* is the **angle bisector** of angle *P*.

Bisector of angle *P*

A locus may be one or more points, a line or a region.

Sometimes the rules for a locus give overlapping regions.

EXAMPLE

In rectangle *ABCD* sketch the locus of points that are less than 2 cm from *A* and nearer to *B* than to *C*.

Sketch a circle, centre *A*, radius 2 cm. Use a dotted line to show that it is not included in the locus. Shade *inside* the circle to show that these are the points that are *less than* 2 cm from *A*.

Draw a line to show the points that are equidistant from *B* and *C* (the perpendicular bisector of *BC*). Then shade the region of points nearer to *B*.

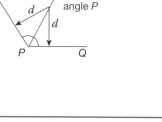

The locus is the overlapping region

> When the locus is a region, draw the boundary first.

EXAMPLE

A point *P* moves so that its distance from *AB* is always equal to its distance from *CD*.

Using a *pencil, ruler and compasses only*, construct the locus of *P*.

Construct the bisector of each angle.

The locus is in two parts.

> Label the answer clearly.

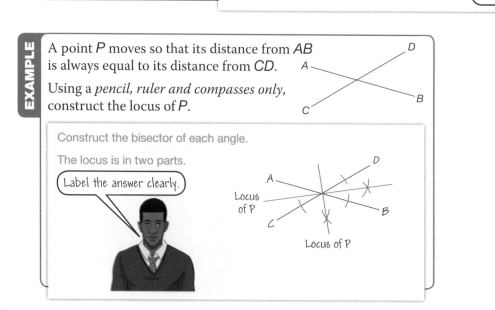

Exercise 11.4S

Unless told otherwise, use a pencil, ruler and compasses only for each construction.

1 Sketch each of the following loci.

   a The locus of points that are 2 cm from a fixed point P.

   b The locus of points that are less than 3 cm from a fixed point Q.

   c The locus of points that are more than 4 cm from a fixed point R.

2 Draw sketches to show the locus of a point that moves so that

   a its distance from line PQ is equal to its distance from a parallel line RS

   $P \xrightarrow{\hspace{3cm}} Q$

   $R \xrightarrow{\hspace{3cm}} S$

   b its distance from line segment XY is always equal to a fixed length d

   $\overline{\begin{matrix} X \qquad\qquad Y \end{matrix}}$

   c it is outside square $ABCD$ and a distance d from its sides.

3 **a** Draw a line, AB, that is 7.8 cm long.

   b Construct the locus of points that are equidistant from A and B.

4 **a** Use a protractor to draw angle $PQR = 76°$.

   b Construct the locus of points that are equidistant from PQ and QR.

5 **a** Draw a line, AB, of length 6.4 cm.

   b Draw the locus of points that are 3 cm from AB.

6 Using a line XY, 7 cm long.

   a Draw the locus of points that are 4 cm from X.

   b Draw the locus of points that are 5 cm from Y.

   c Shade the area containing the points that are less than 4 cm from X and less than 5 cm from Y.

7 **a** Draw a rectangle $ABCD$ with the dimensions shown.

   b A point P moves so that it is always outside the rectangle and 2 cm from it. Draw the locus of P.

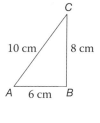

8 Draw an accurate diagram to show the locus of points that are more than 3.5 cm, but less than 4.5 cm from a fixed point P.

9 **a** Draw a square with sides of length 6 cm. Label the vertices A, B, C and D.

   b On your diagram show the locus of points inside $ABCD$ that are less than 3 cm from A and more than 4 cm from BC.

10 **a** Draw a right-angled triangle ABC with the dimensions shown.

   b On your diagram show the locus of points inside ABC that are more than 7 cm from C and nearer to A than to B.

11 **a** Construct an equilateral triangle, ABC, with sides of length 10 cm.

   b Draw the locus of points that are equidistant from AB and BC.

   c Draw the locus of points that are equidistant from AB and AC.

   d Mark the point that is equidistant from AB, BC and AC. Label this point X.

*12 Draw sketches to show

   a the locus of a point P on the circumference of a coin as it rolls along the ruler

   b the locus of a vertex Q of the square card as it is rotated against the ruler.

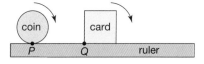

11.4 Loci

- The locus of a point which is a constant distance from another point is a circle.

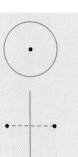

- The locus of a point that is **equidistant** from two other fixed points is the **perpendicular bisector** of the line joining the fixed points.

- The locus of a point at a constant distance from a fixed line is a parallel line.

- The locus of a point equidistant from two intersecting lines is the angle bisector of the lines.

HOW TO

① Decide what you need to construct.

② Use a pencil, pair of compasses and ruler to carry out the construction.

③ Answer any questions asked.

EXAMPLE

Some treasure is positioned 2 m from C, but **equidistant** from A and B.
How many possible locations of the treasure are there?

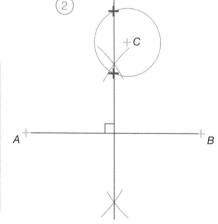

① The treasure is a fixed distance from C.
Draw a circle, centre C, radius 2 m.
Join A and B.

The treasure is equidistant from A and B.
Construct the perpendicular bisector of AB.
Mark where the line crosses the circle.

③ There are two possible locations of the treasure.

EXAMPLE

P and Q are two points 2.5 cm apart on a line.

P Q
·   ·

Shade in the region that satisfies all these conditions:

- Right of the perpendicular to the line PQ at point P.

- Closer to P than to Q.

- More than 1 cm from P.

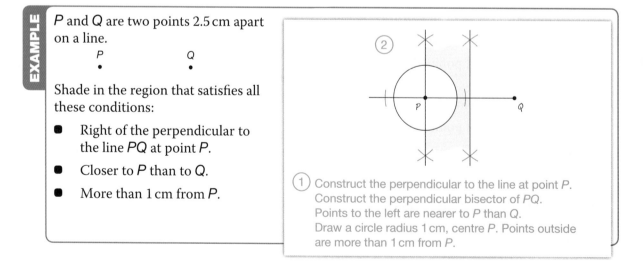

① Construct the perpendicular to the line at point P.
Construct the perpendicular bisector of PQ.
Points to the left are nearer to P than Q.
Draw a circle radius 1 cm, centre P. Points outside are more than 1 cm from P.

Geometry   Circles and constructions

Exercise 11.4A

1 Two straight roads, *AB* and *CD*, intersect at *X*.
Harry rides his mountain bike from *X* on a course that is equidistant from *XD* and *XB*.
Draw an accurate diagram to show Harry's route.

2 Two radio beacons, *A* and *B*, are 80 km apart. The bearing of *B* from *A* is 240°

  a Using a scale of 1 cm to represent 10 km, draw a diagram to show *A* and *B*.

  b An aircraft flies so that it is equidistant from *A* and *B*. Draw the aircraft's route.

  c Beacon *A* can transmit signals up to 50 km.
Show clearly where on its route the aircraft can receive signals from *A*.
How far is this?

3 The sketch shows Ed's lawn.

  a Draw a scale diagram of Ed's lawn. Use 1 cm to represent 2 m.

  b Ed wants to put a clothes drier in his lawn. The drier must be at least 3 m from the house and at least 4 m from a tree at *C*.
Show on your diagram where the drier could be placed.

4 A lifeboat *L* is 10 km from another lifeboat *K* on a bearing of 045°. They both receive a distress call from a ship. The ship is within 7 km of *K* and within 5 km of *L*.

  a Draw a scale drawing to show the positions of *K* and *L*.

  b On your diagram shade the area in which the ship could be.

5 Meera's lawnmower cable can reach 12 m from the power point at *P*.

  a Use a scale diagram to show which parts of the lawn Meera can cut.

  b What is the minimum length of cable Meera needs to reach all of the lawn?

6 An island has 2 mobile phone masts, *A* and *B*. Mobile phone mast *A* has a range of 30 km. Mobile phone mast *B* has a range of 40 km.

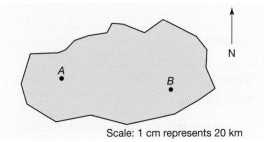

Scale: 1 cm represents 20 km

  a On a copy of the map, show the area covered by the mobile phone masts.

  b Mobile phone mast *B* is to be replaced. Suggest ways in which the coverage of the island could be improved by the new mobile phone mast.

*7 Pat wants to tether her goat on this grass using a 4 metre long chain.

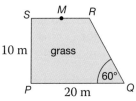

  a Draw a scale drawing to show the grass the goat can reach if it is tethered to

    i *P*   **ii** *Q*

    iii *R*   **iv** *S*

    v *M*, the mid-point of *RS*.

  b **i** Which point allows the goat to reach the greatest area of grass?

    ii Show that your answer to part **bi** is correct by calculating the areas reached from each point.

 1147   SEARCH

Summary

Checkout

You should now be able to...

✔	Identify and apply circle definitions, properties and formulae.	1 – 5
✔	Construct triangles.	6
✔	Use the standard ruler and compass constructions.	7 – 9
✔	Solve loci problems.	10

Language — Meaning — Example

Language	Meaning	Example
Circle	The 2D shape formed by all the points that are the same distance away from the centre point.	circumference, diameter, centre, radius
Diameter	A straight line joining two points on a circle that pass through its centre.	
Radius **Radii (plural)**	A line drawn from the centre of the circle to the circumference.	
Circumference	The perimeter of a circle.	
Arc	Any part of the circumference.	arc, sector, θ, chord, segment, tangent
Chord	A straight line joining two points on the circumference of a circle.	
Tangent	A line which touches the circle at one point only.	
Segment	The shape enclosed by an arc and a chord.	
Sector	The shape enclosed by two radii and an arc.	
Construct	Draw something accurately using compasses and a ruler.	Equilateral triangle
Construction lines	Lines drawn during a construction that are not part of the final object.	
Bisect	Cut into two parts of the same shape and size.	30°, angle bisector, 30°, A, B, M
Angle bisector	A line which bisects an angle.	
Perpendicular bisector	A line which bisects another line at right angles.	
Locus **Loci (plural)**	A set of points which satisfy a set of rules. The path followed by a moving point.	The set of points less than 1 cm from P is the interior of a circle. •P

Review

1 What term is used to describe

    a a line segment that joins the centre to a point on the circumference of a circle

    b a line segment that joins two points on the circumference of a circle?

2 Calculate the area of these circles.

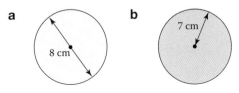

    a 8 cm   **b** 7 cm

3 Calculate the circumference of the circles in question **2**.

4 Calculate

    a the area

    b the perimeter of this semi–circle.

    12 m

5 Calculate

    a the arc length

    b the area of the sector.

    60°
    4 cm

6 Use a ruler and compasses or a protractor to construct these triangles.

    a Sides 5 cm, 4 cm, 7 cm   (SSS)

    b 5 cm, 70°, 6 cm   (SAS)

    c 6 cm, 3 cm, 90°   (RHS)

    d 35°, 7 cm, 25°   (ASA)

7 **a** Draw a line exactly 7 cm long.

    b Use a ruler and pair of compasses to accurately construct the perpendicular bisector of this line.

8 Trace this line and the point marked x. Use a ruler and a pair of compasses to accurately construct a perpendicular from the point x to the line

    • x

9 **a** Use a protractor to draw an angle of 70°.

    b Use a ruler and a pair of compasses to accurately construct an angle bisector of this angle.

10 A dog is attached to a lamp post by a rope which is 2 m long and can rotate freely around the post.

    Use a scale of 3 cm : 1 m to draw to scale the loci of the furthest points the dog can reach.

What next?

Score		
0 – 3		Your knowledge of this topic is still developing. To improve look at MyMaths: 1083, 1088, 1089, 1090, 1118, 1147
4 – 8		You are gaining a secure knowledge of this topic. To improve look at InvisiPens: 11Sa – g
9 – 10		You have mastered these skills. Well done you are ready to progress! To develop your exam technique look at InvisiPens: 11Aa – e

Assessment 11

1 **a** Draw a circle with radius 5 cm. [1]

    b Draw a circle with diameter 90 mm. [1]

    c Draw a chord of length 7 cm across each circle. [2]

    d Kay says she can draw a chord 11 cm long across each circle.
       Is she right? Give reasons for your answer. [1]

2 **a** Paul calculates the radii of three different circles from their circumferences.
       Find the missing radius and match the correct radius to the correct circumference.
       Show your working.

       Radii   **i** x cm   **ii** 0.320 cm   **iii** 50.0 cm

       Circumferences **A** 61.58 cm   **B** 314.16 cm   **C** 2.01 cm [3]

    b He then calculates the radii of three different circles from their areas. Find the missing
       radius and match the correct radius to the correct area. Show your working.

       Radii   **i** 5.08 m   **ii** 7.50 m   **iii** y m

       Areas   **A** 176.7 m^2 **B** 81.1 m^2 **C** 0.407 m^2 [3]

3 Bengt works out the shaded areas in these shapes to 3 sf.

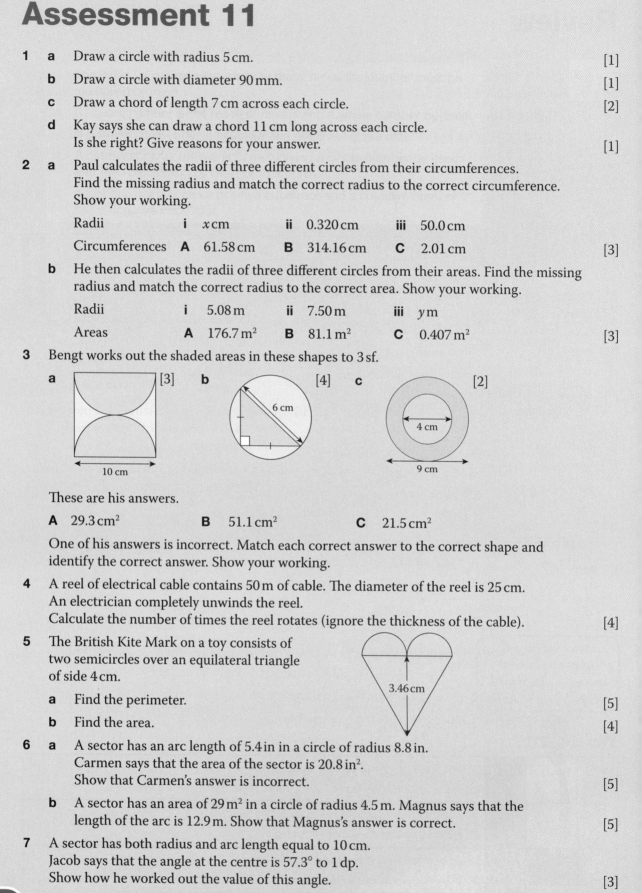

    a [3]   **b** [4]   **c** [2]

    These are his answers.

    A 29.3 cm^2 **B** 51.1 cm^2 **C** 21.5 cm^2

    One of his answers is incorrect. Match each correct answer to the correct shape and
    identify the correct answer. Show your working.

4 A reel of electrical cable contains 50 m of cable. The diameter of the reel is 25 cm.
    An electrician completely unwinds the reel.
    Calculate the number of times the reel rotates (ignore the thickness of the cable). [4]

5 The British Kite Mark on a toy consists of
    two semicircles over an equilateral triangle
    of side 4 cm.

    a Find the perimeter. [5]

    b Find the area. [4]

6 **a** A sector has an arc length of 5.4 in in a circle of radius 8.8 in.
       Carmen says that the area of the sector is 20.8 in^2.
       Show that Carmen's answer is incorrect. [5]

    b A sector has an area of 29 m^2 in a circle of radius 4.5 m. Magnus says that the
       length of the arc is 12.9 m. Show that Magnus's answer is correct. [5]

7 A sector has both radius and arc length equal to 10 cm.
    Jacob says that the angle at the centre is 57.3° to 1 dp.
    Show how he worked out the value of this angle. [3]

8 A football club has this logo.

   a Use a ruler and compasses only to accurately draw this logo. [4]

   b Measure and record the sizes of ∠*WXY* and ∠*WZY*. [2]

   c What sort of angle is ∠*WXY*? [1]

   d Measure and record the length of *ZX*. [1]

9 **a** Construct an isosceles triangle with sides 10 cm, 10 cm and 7 cm. [3]

   b Using compasses only bisect the smallest angle. [3]

   c Measure the angle between the perpendicular bisector and the opposite side. [1]

   d Measure the perpendicular height of the triangle. [1]

10 Hugo's garden is a rectangle *ABCD* where *AB* = 20 m and *BC* = 8 m. He has a playhouse 4 m by 3 m, in the bottom left hand corner *D* with the 4 m side of the play house against the 8 m side of the garden.

   a Using a scale of 1 cm to 1 m draw the outline of Hugo's garden. [2]

   b Using compasses only construct the playhouse. [3]

Hugo has a swing 3.7 m from side *AD* and 3 m from side *AB*.

   c Show the position of the swing. [2]

Hugo's puppy is tethered at the top right hand corner *B* on a lead. The lead will just touch the closest corner of the playhouse to corner *B*.

   d Draw an arc on your diagram showing the maximum distance the puppy can reach. [2]

   e Measure and write down the length of the puppy's lead. [1]

   f Can the puppy reach the swing? Give reasons for your answer. [2]

11 Three 'UFO' hunters are out searching. Alan is 10 miles due north of Brian and Clive is due east of Brian. ∠*BCA* = 60°.

   a Construct triangle *ABC* using only a ruler and compass. [8]

Use a scale of 1 cm to represent 1 mile.

Alan and Clive see a 'UFO' 7 miles from each of them moving along their perpendicular bisector. The 'UFO' is south and east of Alan and north and east of Clive. 2.5 seconds later Brian sees it lying on the angle bisector of ∠*BCA*.

   b Use only a ruler and compasses to mark the positions of the UFO on your diagram. [7]

   c Measure and record how far the 'UFO' has travelled in this 2.5 second internal. [1]

   d Calculate an estimate of the speed of the UFO. [2]

255

12 Ratio and proportion

Introduction

Colour theory is based on the idea that all possible colours can be created from three 'primary colours'. Video screens use the additive colours red, green and blue. Artists often use red, yellow and blue as the basis for mixing paints. Whilst the printing industry uses the subtractive colours cyan, magenta and yellow. The ratio in which these primary colours are mixed determines the colour of the result.

What's the point?

An understanding of ratio is essential for artists in mixing colours on a palette to achieve a desired result. Furthermore, and understanding of proportion is essential for artists in ensuring that the elements of their artwork are in the 'right proportions'.

Objectives

By the end of this chapter you will have learned how to …

- Use fractions and percentages to describe a proportion.
- Divide a quantity in a given ratio and reduce a ratio to its simplest form.
- Use scale factors, scale diagrams and maps.
- Solve problems involving percentage change.

Check in

1 Convert these fractions to decimals.

   a $\frac{3}{4}$ **b** $\frac{2}{5}$ **c** $1\frac{1}{4}$

2 Convert these decimals to fractions in their simplest form.

   a 0.6   **b** 0.25   **c** 0.2

3 The table shows the favourite types of sandwiches in the school canteen.
What proportion of the class surveyed chose ham sandwiches?
Write your answer in its simplest form.

Type	Frequency
Cheese	12
Salad	8
Ham	10
Total	30

4 10 litres of white paint cost £12.
Work out the cost of 20 litres of paint.

Chapter investigation

The ratio of height to head circumference for the average person is said to be around 3 : 1.
Investigate.

12.1 Proportion

- A **proportion** is a part of the whole. It is usually written using a fraction or a percentage.

You can use fractions to express one number as a proportion of another number.

EXAMPLE

a What proportion of this shape is shaded?

b What proportion of 360° is 90°?

c What proportion of 20 is 32?

a The whole is divided into 16 equal parts.
11 parts are shaded.
The proportion shaded is $\frac{11}{16}$

b 360° is the whole.
$\frac{90}{360} = \frac{1}{4}$ 90° is $\frac{1}{4}$ of 360°.

c 20 is the whole.
$\frac{32}{20} = \frac{8}{5}$ 32 is $\frac{8}{5}$ of 20.

- A percentage can be expressed as a fraction or a decimal.

EXAMPLE

Change these percentages into fractions and decimals.

a 24%   **b** 3%

a $24\% = \frac{24}{100} = 0.24$ **b** $3\% = \frac{3}{100} = 0.03$

You can express one quantity as a percentage of another in three steps.

1 Write the first quantity as a fraction of the second.

2 Convert the fraction to a decimal by division.

3 Convert the decimal to a percentage by multiplying by 100.

> Percentage is per-cent which means parts per hundred.

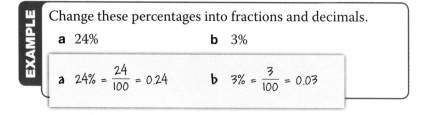

EXAMPLE

Skye took two tests. In German she scored 35 out of 50, and in French she scored 60 out of 80. Write these scores as percentages.

German

$35 \text{ out of } 50 = \frac{35}{50}$
$= 35 \div 50$
$= 0.7$
$= 70\%$

French

$60 \text{ out of } 80 = \frac{60}{80}$
$= 60 \div 80$
$= 0.75$
$= 75\%$

- You can compare proportions by converting them to percentages or decimals.

Exercise 12.1S

1 Write the proportion of each of these shapes that is shaded. Write each of your answers as a fraction in its simplest form.

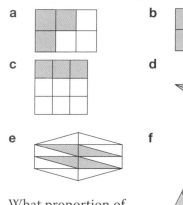

a

b

c

d

e

f

2 What proportion of

a 15 is 5
b 20 is 10
c 10 cm is 3 cm
d 12 kg is 3 kg
e £30 is £12
f £8 is £16
g 200 is 25
h 6.4 is 0.4
i £1.44 is 12p
j 1.2 m is 30 cm?

3 Write these percentages as fractions of 100.

a 26%   **b** 71%   **c** 2%   **d** 102%

4 Write these percentages as fractions and decimals.

a 38%   **b** 46%   **c** 82%   **d** 7%

5 Write these fractions as percentages.

a $\frac{80}{100}$ **b** $\frac{7}{10}$ **c** $\frac{3}{4}$
d $\frac{12}{50}$ **e** $\frac{30}{20}$ **f** $\frac{26}{25}$

6 Write the proportion of each of these shapes that is shaded. Write each of your answers as

i a fraction in its simplest form
ii a percentage (to 1 dp as appropriate).

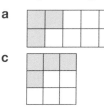

a

b

c

d

7 What proportion of each word is made up of vowels? Give your answers as

i fractions in their lowest terms
ii percentages.

7 **a** JAGUAR   **b** PANTHER
c TIGER   **d** LEOPARD
e PAPERBACK   **f** PERPETUATE
g STAGGERS   **h** MEIOSIS

> Vowels are the letters a, e, i, o and u.

8 Find these proportions, giving your answers as

i fractions in their lowest terms
ii percentages.

a 7 out of every 20
b 8 parts in a hundred
c 6 out of 20
d 75 in every 1000
e 18 parts out of 80
f 9 parts in every 60
g 20 out of every 3
h 20 out of 8
i 412 parts in a hundred
j 6400 in every thousand

9 Stefan is looking at the vegetables he has planted in his garden. The broad beans are 50 cm tall. The sweetcorn is 100 cm tall. What proportion of the height of the sweetcorn are the broad beans?

10 Bobby and Wendy are comparing their weights. Bobby is 60 kg. Wendy is 48 kg. What proportion of the weight of Bobby is Wendy?

11 Rudi and Colin are comparing their wages. Rudi earns £250 per week. Colin earns £400 per week.
What proportion of Colin's wage does Rudi earn?

12 Four candidates stood in an election.
A received 19 000 votes
B received 16 400 votes
C received 14 800 votes
D received 13 200 votes

Write each of these results as a percentage of the total number of votes.
Give your answers to 1 dp.

Q 1015, 1029, 1037, 1961   SEARCH

12.1 Proportion

- A proportion is part of the whole.
- You can use percentages, fractions and decimals to describe proportions.
- To change a fraction to a decimal, divide the top of the fraction by the bottom.
- To change a decimal to a percentage, multiply by 100.

HOW TO

1. Decide which value is the 'whole'.
2. Write the proportion as a fraction.
3. Simplify, or convert to decimals or percentages for comparison.

EXAMPLE

a Complete these equivalent fractions.

$$\frac{5}{10} = \frac{3}{\boxed{?}}$$

$$\frac{5}{15} = \frac{3}{\boxed{?}}$$

$$\frac{5}{20} = \frac{3}{\boxed{?}}$$

b Julianne has a bag with black and red counters.
There are n counters in total.
There are 5 black counters in the bag.
Julianne takes 6 counters from the bag.
3 of the remaining counters are black.
The **proportion** of black counters in the bag has not changed.

Explain why $\frac{5}{n} = \frac{3}{n-6}$

c Use your answer to part **a** to find the value of n.

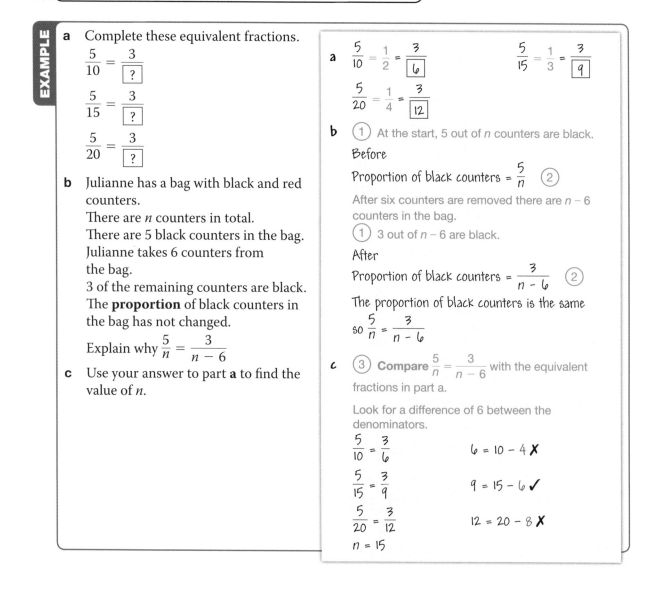

a $\frac{5}{10} = \frac{1}{2} = \frac{3}{\boxed{6}}$ $\frac{5}{15} = \frac{1}{3} = \frac{3}{\boxed{9}}$

$\frac{5}{20} = \frac{1}{4} = \frac{3}{\boxed{12}}$

b ① At the start, 5 out of n counters are black.

Before
Proportion of black counters = $\frac{5}{n}$ ②

After six counters are removed there are $n - 6$ counters in the bag.

① 3 out of $n - 6$ are black.

After
Proportion of black counters = $\frac{3}{n-6}$ ②

The proportion of black counters is the same

so $\frac{5}{n} = \frac{3}{n-6}$

c ③ **Compare** $\frac{5}{n} = \frac{3}{n-6}$ with the equivalent fractions in part a.

Look for a difference of 6 between the denominators.

$\frac{5}{10} = \frac{3}{6}$ $6 = 10 - 4$ ✗

$\frac{5}{15} = \frac{3}{9}$ $9 = 15 - 6$ ✓

$\frac{5}{20} = \frac{3}{12}$ $12 = 20 - 8$ ✗

$n = 15$

Exercise 12.1A

1 Put these in order of size starting with the smallest first.

a 72% $\dfrac{3}{4}$ 0.74

b 0.29   31% $\dfrac{3}{10}$

c 0.8   83% $\dfrac{7}{8}$ 0.78

d $\dfrac{2}{5}$ 0.39   41% $\dfrac{3}{7}$

e $\dfrac{8}{10}$ $\dfrac{3}{4}$ 83%   0.84

f 28% $\dfrac{3}{11}$ 0.3 $\dfrac{7}{24}$

2 Sunita took three tests. In Maths she scored 48 out of 60, in English she scored 39 out of 50 and in Science she scored 55 out of 70.

   a In which subject did she do

      i the best

      ii the worst?

   b Suggest a way in which you could give Sunita a single overall score for her three tests.

3 20% of the cars in a car park are red.
40 cars are not red.
How many cars are in the car park?

4 Peter is driving 220 miles from Bristol to York.
Peter drives at an average speed of 45 miles per hour for the first hour.
He drives at an average speed of 60 miles per hour for the next two hours.
Peter says that
"After three hours I will have completed three-quarters of the journey."
Do you agree with Peter?
Explain your answer.

5 Jada compares the amount of sugar in three different types of fruit juice.
  ● Tropical 22.4 g of sugar in 180 ml
  ● Pineapple 28.8 g of sugar in 250 ml
  ● Blackcurrant 23.4 g of sugar in 200 ml
Jada says that tropical juice has the highest proportion of sugar. Is she correct? Show your workings.

6 a Complete these equivalent fractions.

$$\dfrac{16}{\boxed{?}} = \dfrac{18}{45} \qquad \dfrac{18}{45} = \dfrac{20}{\boxed{?}}$$

   b A company has 45 employees at the start of 2015.
18 employees are female.
By 2016, two more female employees have joined the company.
The **proportion** of female employees has not changed.
Use your answer to part **a** to find the number of total employees in 2016.

7 Jenni compares the proportion of shaded area of two shapes.

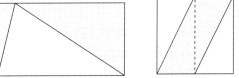

Shape A   Shape B

She says that shape A has the largest proportion of shaded area.
Do you agree with Jenni?
Explain your answer.

8 What proportion of the shape is shaded?

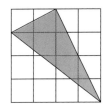

*9 A dairy farmer takes 200 cheeses to sell at a market. In the first hour she sells 20% of the cheeses. In the second hour she sells 15% of those that are left.

   a How many cheeses has she sold in total?

   b What percentage of the original number of cheeses did she sell in the first 2 hours?

12.2 Ratio

You can compare the size of two quantities using a **ratio**.
2 : 3 means '2 parts of one quantity, compared with 3 parts of another'.

> Check that the measurements have the same units.

- You can simplify a ratio by dividing both parts by the same number. When a ratio cannot be simplified any further it is in its **simplest form**.

EXAMPLE

Henry the snake is only 30 cm long. George the snake is 90 cm long. Express this as a ratio.

Compare the two lengths.

= Henry's length : George's length

= 30 cm : 90 cm

= 30 : 90

Express the ratio 30 : 90 in its simplest form.

$$30 : 90$$
$$\div 10 \qquad \div 10$$
$$= 3 : 9$$
$$\div 3 \qquad \div 3$$
$$= 1 : 3$$

> The ratio 1 : 3 means that 90 is three times bigger than 30.

You can divide a quantity in a given ratio.

EXAMPLE

Sean and Patrick share £355 in the ratio 3 : 7.
How much money do they each receive?

Find the total number of parts of the whole amount.

Total number of parts = 3 + 7 = 10 parts

Find the value of one part.

Each part = £355 ÷ 10 = £35.50

Sean receives 3 parts = 3 × £35.50 = £106.50

Patrick receives 7 parts = 7 × £35.50 = £248.50

Check your answer by adding up all the parts. They should add up to the amount being shared!

£106.50 + £248.50 = £355

> A simplified ratio does not contain any units in the final answer.

- You can express a **ratio** in the form 1 : n using division. This is often called a **scale**.

Maps and plans are drawn to scale. You can solve problems involving scales by multiplying or dividing by the scale.

In this scale diagram, 1 cm on the drawing represents 100 cm on the real house.

Corresponding lengths are multiplied by the same **scale factor**. You can write the scale factor as a ratio.

You can write this scale factor as 1 cm represents 100 cm or 1 : 100.

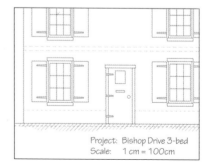

Project: Bishop Drive 3-bed
Scale: 1 cm = 100cm

Exercise 12.2S

1 Write each of these ratios in its simplest form.

   a   2:6   **b**   15:5

   c   6:18   **d**   4:28

   e   5:50   **f**   30:6

   g   24:8   **h**   2:30

2 Write the number of blue squares to the number of red squares as a ratio in its simplest form for each of these shapes.

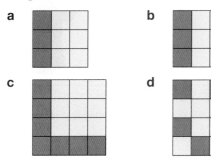

3 Simplify each of these ratios.

   a   4:2   **b**   3:9

   c   21:14   **d**   18:16

   e   45:60   **f**   48:36

   g   80:72   **h**   96:120

4 Solve each of these problems.

   a   Divide £90 in the ratio 3:7.

   b   Divide 369 kg in the ratio 7:2.

   c   Divide 103.2 tonnes in the ratio 5:3.

   d   Divide 35.1 litres in the ratio 5:4.

   e   Divide £36 in the ratio 1:2:3.

5 Solve each of these problems.
Give your answers to 2 decimal places where appropriate.

   a   Divide £75 in the ratio 8:7.

   b   Divide £1000 in the ratio 7:13.

   c   Divide 364 days in the ratio 5:2.

   d   Divide 500 g in the ratio 2:5.

   e   Divide 600 m in the ratio 5:9.

6 Write each of these ratios in its simplest form.

   a   40 cm:1 m   **b**   55 mm:8 cm

   c   3 km:1200 m   **d**   4 m:240 cm

   e   700 mm:42 cm   **f**   12 mins:450 secs

7 Divide the amounts in the ratios indicated.

   a   £500   2:5:3   Hint for **c**:
   b   360°   4:1:3   Convert to metres.
   c   2 km   9:4:7   1 km = 1000 m

8 Express each of these ratios as a ratio in the form $1:n$ (a scale).

   a   2:6   **b**   3:12

   c   10:20   **d**   8:40

   e   3:6   **f**   5:15

   g   4:20   **h**   12:36

   i   30:60   **j**   9:45

   k   45:90   **l**   20:120

9 A map has a scale of 1:50 000 or 1 cm represents 50 000 cm. Calculate in metres the actual distance represented on the map by these measurements.

   a   2 cm   **b**   8 cm

   c   10 cm   **d**   0.5 cm

   e   14.5 cm   **f**   0.01 cm

10 A map has a scale of 1:5000.

   a   What is the distance in real life of a measurement of 6.5 cm on the map?

   b   What is the distance on the map of a measurement of 30 m in real life?

11 A map has a scale of 1:500.

   a   What is the distance in real life of a measurement of 10 cm on the map?

   b   What is the distance on the map of a measurement of 20 m in real life?

12 A map has a scale of 1:5000.

   a   What is the distance in real life of a measurement of 4 cm on the map?

   b   What is the distance on the map of a measurement of 600 m in real life?

13 A map has a scale of 1 : 2000.

   a   What is the distance in real life of a measurement of 5.8 cm on the map?

   b   What is the distance on the map of a measurement of 3.6 km in real life?

Q   1036, 1038, 1039, 1103   SEARCH

12.2 Ratio

- A ratio allows you to compare two amounts.
- 5:2 means 5 parts of one thing compared with 2 parts of another.
- You can simplify a ratio by diving both parts by the same number.
- You can divide a quantity in a given ratio.
- Real-life lengths are reduced in proportion on a scale diagram or map. The scale of the diagram gives the relationship between the drawing and real life.

> You can write a map scale as a ratio 1:n.

HOW TO

1. Set out the problem to show the ratio or proportion.
2. Multiply or divide to keep the ratios the same.
3. Write the answer.

EXAMPLE

A paint mix uses red and white paint in a ratio of 1:12.
How much white paint will be needed to mix with 1.8 litres of red paint?

1. Set out the problem as equivalent ratios.

red : white

× 0.3 (3:8) × 0.3   2.4 ÷ 8 = 0.3
        ?:2.4

2. 3 × 0.3 = 0.9
3. 0.9 ml of red paint is needed.

EXAMPLE

Kara and Mia share money in the ratio 3:7.
Mia receives £168 more than Kara.
How much money did they share?

1. Mia gets 4 more parts than Kara.

   4 parts = £168
2. 1 part = £42
   Total number of parts = 3 + 7 = 10
   10 parts = £420
3. They shared £420.

EXAMPLE

In this scale drawing 1 cm represents 3 m.
Calculate the area of the front.
State your units.

3 cm

7 cm

1. Change 3 m to cm.   3 m = 300 cm
   Write the scale.   ×300

       1 cm   300 cm

               ÷300

2. Multiply by the scale.   3 cm × 300 = 900 cm = 9 m
                              7 cm × 300 = 2100 cm = 21 m
3. Calculate the answer.   Area of front = 9 m × 21 m = 189 m²

Ratio and proportion   Ratio and proportion

Exercise 12.2A

1 The ratio of the length of a car to the length of a van is $2:3$.
The car has a length of 240 cm.

    a Express the length of the car as a percentage of the length of the van.

    b Calculate the length of the van.

2 The ratio of the weight of Dave to Morgan is $6:5$. Morgan has a weight of 85 kg.

    a Express the weight of Dave as a percentage of the weight of Morgan.

    b Calculate the weight of Dave.

3 A metal alloy is made from copper and aluminium.
The ratio of the weight of copper to the weight of aluminium is $5:3$.

    a What weight of the metal alloy contains 45 grams of copper?

    b Work out the weight of copper and the weight of aluminium in 184 grams of the metal alloy.

4 Siobhan and Ralph shared £700 in the ratio $2:3$.
Siobhan gave a quarter of her share to Karen.
Ralph gave a fifth of his share to Karen.
What fraction of the £700 did Karen receive?

5 Using this scale drawing of the Eiffel Tower, calculate

    a the height

    b the width of the base.

Scale: 1cm represents 60 m

6 On the plan of a house, a door measures 3 cm by 8 cm.
The plan scale is 1 cm represents 25 cm.
Calculate the dimensions of the real door.

7 This is a scale drawing of a volleyball court. The drawing has a scale of $1:200$. Calculate

    a the actual distances marked x, y and z

    b the area of the court.

State the units of your answers.

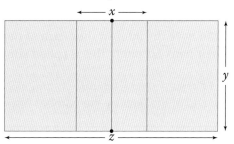

8 The ratio of red counters to blue counters in a bag is $3:7$. What is the probability of choosing a blue counter?

9 Nick and Jess share money in the ratio $2:5$. Nick gets £120 less than Jess.

    a How much money did they share?

    b How much money did they both receive?

10 Marnie, Hannah and Jessa share money between them.
Hannah gets twice as much as Marnie.
Jessa gets one-third of Hannah's amount.

    a Write the ratio of Marnie: Hannah: Jessa in the form $m:h:j$ where m, h and j are whole numbers and the ratio is in its simplest form.

    b Hannah receives \$120. How much more money did Marnie receive than Jessa?

11 The ratio of the internal angle to the external angle in a regular polygon is $3:1$.
How many sides does the polygon have?

12 The angles P and Q in this trapezium are in the ratio $11:25$.
Find the size of angles P and Q.

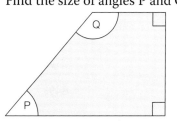

12.3 Percentage change

You often need to calculate a percentage of a quantity.

- A quick method, especially when using a calculator, is to multiply by the appropriate decimal number.

To find 38% of a quantity, multiply by 0.38.

Banks and building societies pay **interest** on money in an account. The interest is always written as a **percentage**.

People can have the interest they earn at the end of each year paid out of their bank account. This is called **simple interest**.

- To calculate simple interest you multiply the interest earned at the end of the year by the number of years.

EXAMPLE

Calculate the simple interest on £3950 for 4 years at an interest rate of 5%.

Calculate the interest for one year.

Don't forget to estimate.

5% of £3950

$\approx$ 5% of £4000

= £200

Interest each year = 5% of £3950

$= \frac{5}{100} \times 3950$

$= 0.05 \times 3950 = £197.50$

Total amount of simple interest after 4 years $= 4 \times £197.50$

$= £790$

- To calculate a **percentage increase**, work out the increase and add it to the original amount.
- To calculate a **percentage decrease**, work out the decrease and subtract it from the amount.

Percentages are used in real life to show how much an amount has increased or decreased.

You can calculate a percentage increase or decrease in a single calculation.

EXAMPLE

a In a sale all prices are reduced by 16%. A pair of trousers normally costs £82. What is the sale price of the pair of trousers?

b Last year, Leanne's Council Tax bill was £968. This year the local council have raised the bill by 16%. How much is Leanne's new bill?

a Sale price = (100 − 16)% of the original price

= 84% of £82

$= \frac{84}{100} \times 82$

= 0.84 × 82

= 68.88

= £68.88

−16%   +16%

84%   100%   116%

b New bill = (100 + 16)% of the original bill

= 116% of £968

$= \frac{116}{100} \times £968$

= 1.16 × £968

= £1122.88

Ratio and proportion Ratio and proportion

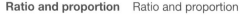

Exercise 12.3S

1   Write a decimal number equivalent to each percentage.

   a   50%   **b**   60%   **c**   25%

   d   8.5%   **e**   0.15% **f**   0.01%

2   Calculate these amounts without using a calculator.

   a   10% of £400   **b**   10% of 2600 cm

   c   5% of 64 kg   **d**   25% of 80 m

   e   50% of 380p   **f**   5% of £700

   g   25% of 12 kg   **h**   20% of £31

3   Calculate these percentages, giving your answer to 2 decimal places where appropriate.

   a   45% of 723 kg   **b**   25% of $480

   c   23% of 45 kg   **d**   21% of 28 kg

   e   17.5% of £124 **f**   34% of 230 m

4   Calculate these percentages, giving your answer to two decimal places where appropriate.

   a   7% of £3200   **b**   12% of £3210

   c   27% of €5400   **d**   3.5% of £2200

   e   0.3% of €4450 **f**   3.7% of £12 590

5   Which is larger: 10% of £350 or 15% of £200?

6   Which is larger: 5% of £40 or 8% of £28?

7   Calculate the simple interest paid on £4580

   a   at an interest rate of 4% for 3 years

   b   at an interest rate of 11% for 5 years

   c   at an interest rate of 4.6% for 4 years

   d   at an interest rate of 8.5% for 3 years.

8   Calculate the simple interest paid on

   a   an amount of £3950 at an interest rate of 10% for 2 years

   b   an amount of £6525 at an interest rate of 8.5% for 2 years

   c   an amount of £325 at an interest rate of 2.4% for 7 years

   d   an amount of £239.70 at an interest rate of 4.25% for 13 years.

9   Write the decimal number you must multiply by to find these percentage increases.

   a   20%   **b**   30%   **c**   45%

10   Write the decimal number you must multiply by to find these percentage decreases.

   a   40%   **b**   60%   **c**   35%

11   Calculate these percentage changes.

   a   Increase £450 by 10%

   b   Decrease 840 kg by 20%

   c   Increase £720 by 5%

   d   Decrease 560 km by 30%

   e   Increase £560 by 17.5%

   f   Decrease 320 m by 20%

12   Calculate these amounts.

   a   Increase £250 by 10%

   b   Decrease £2830 by 20%

   c   Increase £17 200 by 5%

   d   Decrease £3600 by 30%

   e   Increase £3.60 by 17.5%

   f   Decrease £2500 by 20%

13   Calculate each of these using a mental or written method.

   a   Increase £350 by 10%

   b   Decrease 74 kg by 5%

   c   Increase £524 by 5%

   d   Decrease 756 km by 35%

   e   Increase 960 kg by 17.5%

   f   Decrease £288 by 25%

14   Calculate these percentage changes. Give your answers to 2 decimal places as appropriate.

   a   Increase £340 by 17%

   b   Decrease 905 kg by 42%

   c   Increase £1680 by 4.7%

   d   Decrease 605 km by 0.9%

   e   Increase $2990 by 14.5%

   f   Decrease 2210 m by 2.5%

Q   1060, 1237, 1302, 1934   SEARCH

12.3 Percentage change

RECAP

- You can find a percentage of an amount by multiplying the quantity by an equivalent decimal or fraction.
- You can calculate a percentage increase or decrease by calculating the percentage increase or decrease and then adding it to, or subtracting it from, the original amount.

HOW TO

To write one number as a percentage of another

1. Write the first number as a fraction of the other.
2. Convert the fraction to a percentage.

EXAMPLE

Find

a 13 as a percentage of 20   **b** 8 as a percentage of 23

a ① $\frac{13}{20} = \frac{13 \times 5}{20 \times 5} = \frac{65}{100} = 65\%$ ② Convert to a fraction with a denominator of 100.

b ① $\frac{8}{23} = (8 \div 23) \times 100\% = 34.8\%$ ② Use a calculator to convert to a percentage.

> If a price was reduced from £20 to £13, them the discount would be $(100 - 65)\% = 35\%$.

- In a **reverse percentage** problem, you are given an amount after a percentage change, and you have to find the original amount.

HOW TO

To calculate the original amount after a percentage change

1. Use the final amount to work out 1% of the original amount.
2. Find the original amount by multiplying by 100.

EXAMPLE

Find the original price of a denim jacket reduced by 15% to £32.30

① 100% − 15% = 85%
85% of original price = £32.30
1% of original price = £32.30 ÷ 85
② 100% of original price
= £32.30 ÷ 85 × 100 = £38

> You could do this in one step.
> 32.30 ÷ 85 × 100 = 32.30 ÷ 0.85

EXAMPLE

Following a 5% price increase, a car radio costs £168
How much did it cost before the increase?

① 100% + 5% = 105%
105% of original price = £168
1% of original price = £168 ÷ 105 = £1.60
② 100% of original price = £1.60 × 100 = £160

Exercise 12.3A

1 Use a written method to find

   a   12 as a percentage of 20

   b   36 as a percentage of 75

   c   24 as a percentage of 40

2 Use a calculator to find

   a   19 as a percentage of 37

   b   42 as a percentage of 147

   c   8 as a percentage of 209

3 **a**   A t-shirt is reduced from £25 to £20. Find the percentage discount.

   b   A coat is reduced from £80 to £64. Find the percentage discount.

4 **a**   Jenna's savings increase from £400 to £450. Calculate the interest rate.

   b   Amanda's antique vase increases in value from £600 to £720. Calculate the percentage increase in value.

5 A car costs £9750 when new. 5 years later it is sold for £4500. What is the average percentage loss each year?

6 Carys is trying to find the original price of an item that is on sale. Its sale price is £56 and it was reduced by 20%.

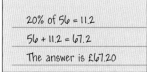

```
20% of 56 = 11.2
56 + 11.2 = 67.2
The answer is £67.20
```

Carys' working is shown here.

What is wrong with her working?

7 A book costs £4 after a 20% price reduction. How much did it cost before the reduction? Show your working.

8 These numbers are the results when some amounts were increased by 10%. For each one, find the original number.

   a   55   **b**   44

   c   88   **d**   121

9 Find the original cost of the following items.

   a   A vase that costs £7.20 after a 20% price increase.

   b   A table that costs £64 after a 20% decrease in price.

10 **a**   Francesca earns £350 per week. She is awarded a pay rise of 3.75%. Frank earns £320 per week. He is awarded a pay rise of 4%. Who gets the bigger pay increase? Show all your working.

   b   Bertha's pension was increased by 5.15% to £82.05. What was her pension before this increase?

11 Calculate the original cost of these items, before the percentage changes shown. Show your method. You may use a calculator.

   a   A hat that costs £46.50 after a 7% price cut.

   b   A skirt that costs £32.80 after a price rise of 6%.

12 A car manufacturer increases the price of a Sunseeker sports car by 6%. The new price is £8957. Calculate the price before the increase.

13 During 2005 the population of Camtown increased by 5%. At the end of the year the population was 14 280.

What was the population at the beginning of the year?

*14 To decrease an amount by 8%, multiply it by 0.92.
For a further decrease of 8%, multiply by 0.92 again, and so on.
Use this idea to calculate

   a   the final price of an item with an original price of £380, which is given two successive price cuts of 8%.

   b   the final price of an item with an original price of £2400, which is given three successive price cuts of 10%.

Q 1060, 1237, 1302, 1934   SEARCH

Summary

Checkout
You should now be able to...

	Test it Questions
✔ Use fractions and percentages to describe a proportion.	1 – 2
✔ Use ratio notation and simplify ratios.	3 – 6
✔ Use scale factors, scale diagrams and maps.	7 – 8
✔ Solve problems involving percentage change.	9 – 12

Language   Meaning   Example

Language	Meaning	Example
Proportion	A proportion is a part of the whole. Two quantities are in proportion if one is always the same multiple of the other.	If there are 6 eggs in a carton Total number of eggs = 6 × number of cartons
Ratio	A ratio compares the size of one quantity with the size of another.	Ratio of blue yellow squares = 2:5
Simplify (ratio)	Divide both parts by common factors.	6:9 = 2:3 (divide both parts by 3)
Scale	The ratio of the length of an object in a scale drawing to the length of the real object.	The Ordnance survey produce maps with scales such as: 1:100 000, 1:50 000 and 1:25 000.
Scale drawing	An accurate drawing of an object to a given scale.	
Percentage	A type of fraction in which the value given is the number of parts in every hundred.	$\frac{33}{100} = 33\%$
Simple interest	Interest that is calculated on the original amount only and not on any extra interest that has built up.	£100 saved in a bank account at 4% simple interest will earn 4% of 100 = £4 each year.
Percentage increase	An increase by a percentage of the original amount.	4.20 increased by 25% = 125% of 4.20 = 1.25 × 4.20 = 5.25
Percentage decrease	A decrease by a percentage of the original amount.	4.20 decreased by 25% = 75% of 4.20 = 0.75 × 4.20 = 3.15
Reverse percentage problem	Calculating the original amount from the final amount and the percentage added.	Quantity after a 15% decrease = 544 85% of original = 544 1% of original = 544 ÷ 85 = 6.4 Original quantity = 100 × 6.4 = 640

Review

1 **a** What fraction of the stars are blue?

b What percentage of the stars are yellow?

2 Mia scores $\frac{21}{35}$ on her Biology exam and 72% on her Physics exam. In which exam does she score the higher mark?

3 Write these ratios in their simplest form.

a 18 : 21 **b** 27 : 45 **c** 1 cm : 4 m

4 Share

a £35 in the ratio 1 : 4

b £90 in the ratio 4 : 5.

5 Emily has three times as many sweets as Ethan.

a Write a ratio of the number of sweets they have in the form Ethan : Emily.

b What fraction of all the sweets does Ethan have?

6 The ratio of butter to flour in a recipe for pastry is 2 : 5.

a How much butter and how much flour will be needed to make 840 g of pastry?

b How much flour will be needed to mix with 300 g of butter?

c What fraction of the mixture is flour?

7 **a** The rectangle is enlarged by scale factor 7. What is the length of the enlarged rectangle?

b The width of the enlarged rectangle is 21 cm What is the width of the original rectangle?

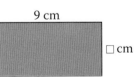

9 cm

□ cm

8 A map is drawn using the scale 1 : 200 000.

a How far in real life is a length of 5 cm on the map?

b A road is 1.5 km long, how long will it be on the map?

9 **a** Increase 50 by 14%.

b Decrease 150 by 29%.

10 A bank pays 1.5% per annum interest on a savings account. £4000 is paid in at the beginning of the year and no further deposits or withdrawals are made. How much is in the account after 1 year?

11 Tom spends £600 on a credit card. At the end of the month he has to pay £15 interest.

a How much interest has he paid as a percentage of the amount he borrowed?

Tom does not pay the £600 back but continues to pay £15 interest each month.

b What percentage of the original £600 has he paid in interest by the end of the first year?

12 **a** The price of a litre of petrol rose 5% to £1.26, what was it before the increase?

b A shirt is reduced in price by 25% and now costs £33, what did it cost originally?

What next?

<table>
<tr><td rowspan="6">Score</td><td>0 – 5</td><td></td><td>Your knowledge of this topic is still developing.
To improve look at MyMaths: 1015, 1029, 1036, 1037, 1038, 1039, 1060, 1103, 1237, 1302, 1934</td></tr>
<tr><td>6 – 10</td><td></td><td>You are gaining a secure knowledge of this topic.
To improve look at InvisiPens: 12Sa – j</td></tr>
<tr><td>11 – 12</td><td></td><td>You have mastered these skills. Well done you are ready to progress!
To develop your exam technique look at InvisiPens: 12Aa – c</td></tr>
</table>

Assessment 12

1 Four bottles contain four different amounts of water.
Write the missing proportions as simply as possible.

A	B	C	D
200 ml	150 ml	75 ml	25 ml

  a D holds ☐ of the amount C holds. [1]   **b** holds ☐ of the amount B holds. [1]

  c D holds ☐ of the amount A holds. [1]   **d** C holds ☐ of the amount B holds. [1]

  e C holds ☐ of the amount A holds. [1]   **f** B holds ☐ of the amount A holds. [1]

2 Nomsah received $\frac{13}{25}$ of the prize money from a completion.

  a Write proportion this as a percentage and a decimal. [1]

  b The total prize was 225 000 rand. How much did Nomsah win? [2]

3 Jeremy bought a car for £17 250. A year later its value had decreased by $\frac{6}{25}$.
How much was the car worth a year later? [3]

4 A coin is made from copper and brass in the ratio 24:10.
Write this ratio

  a in its simplest form [1]   **b** in the form $n:1$. [1]

5 Year 9 had to choose between geography and history in their option
choices for GCSE. The ratio of students choosing geography to history was 5:4.
There are 99 students in year 9. How many chose Geography? [4]

6 **a** A tin contains 63 biscuits, some made with chocolate and the rest 'plain', in the
ratio 3:4. Calculate the numbers of chocolate and plain biscuits in the tin. [4]

  b Another similar tin contains plain to milk biscuits in the ratio 5:6. There are
15 plain biscuits in the tin. How many biscuits are in the tin altogether? [4]

7 Debs makes cheese straws, using flour, butter and cheese in the ratio 24:6:9.

  a Write this ratio in its simplest form. [1]

Debs uses 160 g of flour in her mixture.

  b Work out the number of grams of butter and cheese Debs uses. [5]

8 **a** The cast for the school play has 24 boys and 20 girls.
Write this as a ratio in its simplest form. [1]

  b One boy joins the cast and 5 girls leave. Write the new ratio in its the form $n:1$. [3]

9 The ratio of girls to boys in a cycling squad is 4:7. There are 88 people in the squad.

  a How many of each gender are in the squad? [4]

The number of girls goes up by 12 and the number of boys increases by 37.5%.

  b Find the number of girls and boys in the squad after the change and show
that the ratio of girls to boys stays the same. [4]

10 On a street map of London the scale is written as 1:20 000.

  a How many metres in London is represented by a 1 cm on the map? [1]

  b On the map, Buckingham Palace is 5.85 cm from the statue of Eros in Piccadilly Circus.
How far apart are they in reality? Give your answer in km. [2]

  c A child is taken 4.15 km from Paddington Station to Great Ormond Street
Children's Hospital. How far apart are they on the map? [2]

11 A road atlas has a scale of 1 inch to 5 miles.

   a Gatwick Airport to Heathrow is 8.1 inches on the map. How far apart are they? [2]

   b Liverpool FC and Newcastle United FC are 128 miles as the crow flies.
      How far apart are they on the map? [2]

   c The straight-line distance from John O Groats to Lands End on the map is 120.4 inches.
      The actual distance by road is 838 miles. Calculate the difference in mileage
      between the actual distance and the straight-line distance. [3]

12 a Jeremy invests £2.50 at an interest rate of 4% for 5 years.
      Find the simple interest he receives. [2]

   b Jane invests £5200 an interest rate of 3.5% for 4 years.
      Find the simple interest she receives. [2]

13 a Ben says that to increase a value by 29.1% you multiply by 29.1
      Gavin says you multiply by 1.291 Who is correct? [1]

   b Gemma says that to decrease a value by 10% you multiply by 0.9
      Sarah says you multiply by 0.1 Who is correct? [1]

   c Steven says to increase 90 kg by 3% you multiply by 0.3 and add to 90 kg.
      Explain his error and show how to increase 90 kg by 3% correctly. [2]

   d Shayda says to decrease 550 ml by 45.4% you multiply by 0.454.
      Explain her error and show how to decrease 550 ml by 45.4% correctly. [2]

In questions **14–22** give your answers to 2 dp wherever appropriate.

14 A station has 45 steps up to the platform bridge. The first section has 20 steps.
    What percentage of the total number of steps does the first section take? [2]

15 Kate buys material for her wedding dress. The roll had 25 m on it originally and after
    the shop assistant had cut it there was 17.5 m of the material left in the roll.
    What percentage of the roll has Kate bought? [3]

16 Cowell bought a painting as an investment for £2 500 000. Some years later
    it was revalued as being worth 8.5% more than originally. What is its new value? [2]

17 In 2000 the average price of a new house was £86 100.
    In 2010 this average was 164 300.
    By what percentage had the 2010 value risen above the 2000 value? [3]

18 'FALSEPRINT', the film processor, sells prints various in sizes.
    They claim that their 7" by 5" print is more than 45% bigger than their 6" by 4" size.
    Are they correct? [4]

19 A river contains 250 crocodiles. After two years of hunting, the number of crocodiles
    decreases by 8% in the first year and 5.5% in the second.
    How many crocodiles, to the nearest crocodile, are there in the river after

   a 1 year [2]   b 2 years? [2]

20 Dani bought her a motocross bike for £2500.
    She sold the bike at a loss of 12%. What price did Dani receive for the bike? [2]

21 After a pay rise of 3.5% Osborne's monthly salary is £2566.80
    What was his monthly salary before his increase? [2]

22 Titus Lines, the angler, buys a new rod for £251.34 which the dealer had sold him
    making himself a profit of 6.5%. What price had the dealer originally paid for the rod? [2]

Revision 2

1 Andrea draws a quadrilateral with corners at the coordinates (1,2),(3,4), (4,7) and (3,10). Find the area of this quadrilateral. [6]

2 The diagram shows a company logo.
$BC = CD = 8$ cm,
$AC = 15$ cm and
$AE = 9$ cm.
Calculate the area of the logo. [4]

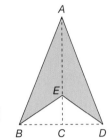

3 Town B is 5 km from Town A on a bearing of 135°. Town C is 12 km from Town A on a bearing of 045°.

    a Draw a map of the three towns using the scale 1 cm : 1 km. [2]

    b Write down the value of ∠BAC. [1]

    c Calculate the area of the triangle *ABC* enclosed by the three towns. [2]

4 A field in the shape of a trapezium has two parallel sides, 150 m and 176 m. The perpendicular distance between them is 243 m. Calculate the area of the field

    a in m² [2]

    b hectares. (1 hectare = 10 000 m²) [1]

5 Tabitha draws triangle *ABC* on a coordinate grid as shown. She then transforms it on to triangle *PQR* shown.

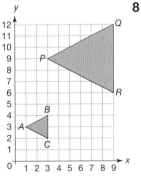

    a Describe the transformation that maps *ABC* to *PQR*. [3]

Tabitha then maps *PQR* to position $P'(2,7)$, $Q'(8,10)$, $R'(8,4)$.

    b Draw triangle $P'Q'R'$ on a copy of the grid and describe this transformation fully. [3]

    c Reflect the triangle *ABC* in the line $x = 5$. [2]

6 The accurate measure of the length of a year is 365.2596 days. An astronomer records four approximations for this value.

    A 365.37 days   **B** 360 days
    C 365.270 days   **D** 366.3 days

    a Which of these approximations is written to
        i 2 sf [1]
        ii the nearest 10 [1]
        iii the nearest 0.1? [1]

    b Which value is closest to the accurate figure and how close is it? [2]

7 Pepys' bookshelf can hold 38 books at an average thickness of 2.9 cm.

    a Estimate the length of the bookcase. [1]

The average weight of a book is 1100 g.

    b Estimate the total mass of the books on Pepys's bookshelf. [1]

    c Use your calculator to find the exact answers to parts **a** and **b**. [2]

    d Find the percentage error for each estimate and measurement. [3]

    e Which estimate of yours was the closest to the actual value? [1]

8 A cuboid measures 5 m × 3 m × 4 m. All measurements are correct to the nearest metre.

    a Calculate the upper and lower bounds for the
        i surface area [6]
        ii volume. [3]

    b Change 3.6 g/cm³ to kg/m³. [4]

    c Calculate the upper and lower bounds of the mass of the block if the density of the material is 3.6 g/cm³ [3]

9 A length of wire is 58 cm long. It can be bent into a rectangle x cm long. The area of the rectangle is 100 cm². Find the dimensions of the rectangle. [6]

10 A triangle has angles p, q and r, where $p > 21$ and $q > 59$. Write an inequality for r. [2]

11 A coffee machine can take only 10p and 50p coins. When emptied it had 41 coins totalling £10.10. How many of each value coin did the machine have? [5]

12 A badminton club has r racquets. Rajiv says there are more than 50 and Sarah says there are no more than 55. When they counted the racquets they found the number of racquets was a prime number. Find all the possible solutions. [2]

13 Rory attempted 6 rounds of golf (108 holes) to raise money for charity. His dad agreed to pay him £6 for each complete hole he won. Rory agreed to pay his dad £2.50 for each hole he lost. Rory raised £546 for his youth section as a result.

  a Find the number of holes that Rory won. [5]

  b Find the smallest number of holes that Rory needed to win to be sure of making a contribution to charity. [3]

14 A circle of radius 30 in has a central square hole of side length 5 in.

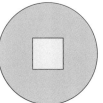

  a Find the circumference of the millstone. [2]

  b Calculate the area of the millstone. [4]

15 A drinks manufacturer cuts its bottle tops from a rectangular sheet, 35 cm by 30 cm. The manufacturer can cut 42 tops from each sheet.

  a Work out the diameter of each bottle top. [3]

  b Find the area of each bottle top. [2]

  c Calculate the total area of the 42 bottle tops. [1]

  d Find the percentage area of the sheet used for the bottle tops. [2]

16 A rectangular garden, $ABCD$ is 12 m by 9 m, where CD = 9 m is the side of the house. A window at the centre of CD has a modem in it which gives a WiFi signal range of 7.5 m.

  a Use a ruler and compasses to construct a diagram of the garden. [5]

  b Shade in the area of the garden within range of the WiFi. [2]

Ima has a deckchair 7.2 m from A and 5.2 m from B.

  c Is she within WiFi range? Give reasons for your answer. [3]

17 A tennis club has a ratio men : women as 7 : 6. There are 104 people in the club.

  a How many of each sex are there in the club? [3]

The number of men and women rises by $\frac{1}{4}$.

  b Does the ratio of men to women change? Give reasons for your answer. [2]

  c What are the new numbers of men and women in the squad? [1]

The same number of men leave as women join. The ratio is now 1 : 1.

  d How many people of each gender are in the club now? [2]

18 On a map 1 inch represents 3 miles on the ground.

  a From Bristol to Bath, as the crow flies, is 3.7 inches. How far are they apart on the ground? [1]

  b From York to Leeds, as the crow flies, is 22 miles. How far apart are they on the map? [1]

19 a A box of sweets has a total mass of 600 g. 16% of this mass is packaging and wrappings. What is the actual mass of the sweets? [2]

  b A daily newspaper claims an average daily sale of 1 456 739. What percentage increase would take its daily average sales to 1 500 000? [3]

  c The value of a new car is £18 000. It depreciates by 16% yearly. How much is the car worth at the end of 3 years? [3]

13 Factors, powers and roots

Introduction

Cryptography is the study of codes, with the aim of communicating in a secure way without messages being deciphered. Modern cryptography is often based on prime numbers. In particular finding very large numbers that can be written as the product of two not-quite-as-large prime factors. Finding what these two prime numbers are is a challenge, even with modern computers; but you need to find them in order to crack the code.

What's the point?

In the modern day, illegal computer-based syndicates use increasingly sophisticated techniques to access sensitive digitally-held information, including bank accounts. Prime number encryption increases the security of stored data, making it harder for the hackers.

Objectives

By the end of this chapter you will have learned how to …

- Use mathematical language to describe factors, multiples and primes.
- Use Venn diagrams or factor trees to systematically list the prime factors of a number.
- Use prime factor decomposition to calculate the HCF and LCM of two or more numbers.
- Write the HCF and LCM using product notation.
- Calculate positive integer powers and their roots.
- Recognise powers of 2, 3, 4 and 5.

Check in

1   Calculate

    a 7×7 **b** $2 \times 2 \times 2$ **c** $10 \times 10 \times 10$ **d** $32 \div 100$

2   Find the missing repeated number in this expression.

    ___ $\times$ ___ $= 25$

3   Write the first 6 prime numbers.

4   Write all the factors of 24.

Chapter investigation

77 is the product of two prime numbers: 7 and 11.

That is, $7 \times 11 = 77$

Is 702 the product of two prime numbers? If not, why not?

What about 703?

13.1 Factors and multiples

● The factors of a number are those numbers that divide into it exactly, leaving no remainder.

> Product means multiply together.

You can often write a number as the **product** of two factors.

The factor pairs of 24 are 1×24, 2×12, 3×8 and 4×6.

You can write the factors in a list: 1, 2, 3, 4, 6, 8, 12, 24.

You can use simple divisibility tests to find the factors of a number.

Factor	Test
2	the number ends in a 0, 2, 4, 6 or 8
3	the sum of the digits is divisible by 3
4	the last two digits of the number are divisible by 4
5	the number ends in 0 or 5

Factor	Test
6	the number is divisible by 2 *and* by 3
7	there is no simple check for divisibility by 7
8	half of the number is divisible by 4
9	the sum of the digits is divisible by 9
10	the number ends in 0

● A **prime number** is a number with only two factors. These are 1 and the number itself.

● The first ten prime numbers are 2, 3, 5, 7, 11, 13, 17, 19, 23, 29.

> You can think of multiples as being the numbers in the 'times tables'.

● The **multiples** of a number are those numbers that divide by it exactly, leaving no remainder.

The multiples of 18 are 18, 36, 54, 72, ...

● The **highest common factor** (HCF) of two numbers is the largest number that is a factor of both numbers.

● The **lowest common multiple** (LCM) of two numbers is the smallest number that they both divide into.

> Be systematic and listing the multiples and factors makes it easy to spot the HCF and LCM.

EXAMPLE

Find the HCF and LCM of 10 and 15.

The factors of 10 are: 1 2 5 10
The factors of 15 are: 1 3 5 15

1 and 5 are **common factors** of 10 and 15.
5 is the highest common factor of 10 and 15.

The first six multiples of 10 are: 10 20 30 40 50 60
The first six multiples of 15 are: 15 30 45 60 75 90

30 and 60 are **common multiples** of 10 and 15.
30 is the lowest common multiple of 10 and 15.

Exercise 13.1S

1 Write all the factor pairs for each number.

a 8   b 16   c 23   d 34

e 10   f 26   g 42   h 48

i 39   j 44   k 38   l 77

2 Write the first three multiples of each number.

a 7   b 9   c 12   d 15

e 17   f 25   g 30   h 32

i 45   j 50   k 18   l 60

3 Use the divisibility tests to answer each of these questions.
In each case explain your answer.

a Is 5 a factor of 135?

b Is 5 a factor of 210?

c Is 2 a factor of 321?

d Is 3 a factor of 231?

e Is 9 a factor of 91?

f Is 5 a factor of 95?

g Is 10 a factor of 710?

h Is 9 a factor of 321?

i Is 3 a factor of 451?

j Is 6 a factor of 98?

4 Write all the factor pairs of each of these numbers.

a 24   b 45   c 66   d 100

e 120   f 132   g 160   h 180

i 360   j 324   k 224   l 264

m 312   n 325   o 432   p 418

5 Write the first three multiples of each of these numbers.

a 17   b 29   c 42   d 25

e 47   f 35   g 90   h 120

i 95   j 208   k 144   l 111

6 Find the common factors of

a 10 and 20   b 12 and 15

c 20 and 25   d 8 and 20

e 21 and 28   f 30 and 40

g 12 and 28   h 9 and 36

i 24 and 30

7 Find the first two common multiples of

a 6 and 10   b 9 and 12

c 4 and 6   d 10 and 15

e 14 and 21   f 20 and 30

8 Find the highest common factor of

a 6 and 4   b 25 and 40

c 18 and 30   d 24 and 56

e 30 and 75   f 36 and 54

g 50 and 125   h 24, 36 and 72

i 30, 75 and 105   j 48, 112 and 144

9 Find the lowest common multiple of

a 6 and 4   b 5 and 8   c 12 and 18

d 15 and 25   e 14 and 21   f 30 and 75

10 a Copy the 1–100 number square.

b Follow these instructions.

● 1 is not a prime number so you can cross it out.

● 2 is the lowest prime number. Cross out all the multiples of 2 except for the number 2.

● 3 is the next number not crossed out. It is the next prime number. Cross out all the multiples of 3 except for the number 3.

● 5 is the next number not crossed out. It is the next prime number. Cross out all the multiples of 5 except for the number 5.

c Carry on until you have only prime numbers left.

d Make a list of all the prime numbers from 1 to 100.

1	2	3	4	5	6	7	8	9	10
11	12	13	14	15	16	17	18	19	20
21	22	23	24	25	26	27	28	29	30
31	32	33	34	35	36	37	38	39	40
41	42	43	44	45	46	47	48	49	50
51	52	53	54	55	56	57	58	59	60
61	62	63	64	65	66	67	68	69	70
71	72	73	74	75	76	77	78	79	80
81	82	83	84	85	86	87	88	89	90
91	92	93	94	95	96	97	98	99	100

1032, 1034, 1044   SEARCH

13.1 Factors and multiples

- The **highest common factor** (HCF) of two numbers is the largest number that is a factor of both numbers.
- The **lowest common multiple** (LCM) of two numbers is the smallest number that they both divide into.
- You can use divisibility tests to find the factors of a number.
- A prime number has two factors, 1 and itself.

HOW TO

① Read the question carefully; decide how to use your knowledge of factors and multiples.

② Be systematic – use divisibility tests and listing to help you to find multiples and factors.

③ Answer the question; make sure that you explain your answers fully.

EXAMPLE

n is an even positive number. $3n + 1$ is a multiple of 5. Find a possible value for n.

① Even numbers are multiples of 2.

② Using a table helps to organise the answer.

Find $3n + 1$ for each multiple of 2. Decide if the answer is a multiple of 5.

n	$3n + 1$	Multiple of 5?
2	$3 \times 2 + 1 = 6 + 1 = 7$	No
4	$3 \times 4 + 1 = 12 + 1 = 13$	No
6	$3 \times 6 + 1 = 18 + 1 = 19$	No
8	$3 \times 8 + 1 = 24 + 1 = 25$	Yes, $5 \times 5 = 25$

8 is a possible value for n.

EXAMPLE

A climber can see two mountains.

There is a beacon on top of each mountain.

One beacon flashes every 15 seconds and the other flashes every 25 seconds.

If the beacons flash together, how long will it be before they next flash together?

① The question is about multiples of numbers.

② Write a list of multiples for 15 and 25. Find the lowest common multiple.

Beacon 1: 0   15   30   45   50   ⑦⑤   90   ...

Beacon 2: 0   25   50   ⑦⑤   100   ...

③ The beacons will flash together after 75 seconds.

EXAMPLE

A large rectangular play area measures 36 metres by 42 metres.

The council wants to lay the play area with square concrete tiles.

Tiles are available with lengths of whole numbers of metres.

If the tiles are to be as large as possible, find the size of the tiles.

① The question is about factors of numbers.

② Write a list of factors for 36 and 42. Find the highest common factor.

Factors of 36: 1   2   3   4   ⑥   9   12   18   36

Factors of 42: 1   2   3   ⑥   7   14   21   42

③ The tiles are 6 metres by 6 metres.

Number   Factors, powers and roots

Exercise 13.1A

1 **a** Write a multiple of 20 that is bigger than 200.

   b Write a multiple of 15 that is between 100 and 140.

   c Write a multiple of 6 that is bigger than 70 but less than 100.

2 Jo's age this year is a multiple of 6.
Next year it will be a multiple of 5.
How old might Joe be?

3 n is a whole number.
$2n + 3$ is a prime number.
Find two possible values of n.

4 n is an odd number.
$4n - 5$ is a multiple of 3.
Find two possible values of n.

5 **a** Amos says "all odd numbers are prime numbers". Give two examples that show he is wrong.

   b Aya says "all prime numbers are odd numbers". Give an example to show she is wrong.

   c Arik says "take one off any multiple of 6 and you always get a prime number".

   i Give three examples where this is true.

   ii Give one example where it is false.

6 The warning light on a TV mast flashes every eight seconds.

   A nearby TV mast flashes every 14 seconds.

   If they flash together at a particular moment, how long will it be before they flash together again?

7 Jamie is tiling his bathroom wall.
The wall measures 100 cm by 160 cm.
He wants to tile the wall with square tiles.
Tiles are available in whole numbers of centimetres.
If the tiles are as large as possible, what size tile should Jamie buy?

8 Marisa is making flower arrangements for the tables in a restaurant.
Marisa has 36 roses and 60 daisies.
The flower arrangements must be the same.
She must make as many flower arrangements as possible.

8 She must use all of the flowers.
How many flower arrangements can Marisa make?

9 Simon is having a barbecue.
He buys burgers and burger buns.
The burgers come in packets of 10.
The burger buns come in packets of 24.
Simon needs to buy the same number of burgers as burger buns.
He wants to spend as little money as possible.
How many packets of burgers and burger buns does Simon need to buy?

10 Mai and Luca are playing a game.
They choose a number from 1 to 99 and find the sum of all of its factors.
Whoever has the highest total wins the game.

   a Mai chooses 32 and Luca chooses 55. Who wins the game?

   b Mai chooses 80. Suggest a number that Luca could choose to win the game.

11 **a** Copy and complete this table.

Numbers	Product	HCF	LCM
6 and 4	6 × 4 = 24	2	12
8 and 10	8 × 10 = ___		
12 and 18			
6 and 9			
15 and 20			
15 and 25			

   b Write anything you notice about the numbers in your table.

   c Write a quick way to find the LCM if you know the HCF.

*12 A perfect number is one where if you add all the factors which are less than the number itself, the total is that number. For example 6 is a perfect number, because its factors are 1, 2, 3, 6 and 1 + 2 + 3 = 6

   a Show that 28 is a perfect number.

   b Show that 120 is not a perfect number.

   c Show that 496 is a perfect number.

13.2 Prime factor decomposition

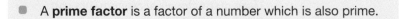

> ● A **prime factor** is a factor of a number which is also prime.

> Remember, 1 is not a prime number.

Factors of 28 are {1, 2, 4, 7, 14, 28}.

Prime factors of 28 are {2, 7}.

> ● Every whole number can be written as the product of its **prime factors**.

There are two common methods to find the prime factors.

Division by prime numbers

Divide the number by the smallest **prime number**.
Repeat dividing by larger prime numbers until you reach a prime number.

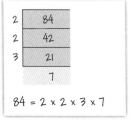

$84 = 2 \times 2 \times 3 \times 7$

Factor trees

Split the number into a **factor** pair.
Continue splitting until you reach a prime factor.

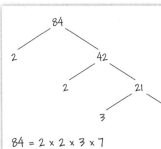

$84 = 2 \times 2 \times 3 \times 7$

> You can split the number into any factor pair, the answer will be the same in the end.

You can find the HCF and LCM of two numbers by writing their **prime factors** in a Venn diagram.

> ● The HCF is the product of the numbers in the intersection.
> ● The LCM is the product of all the numbers in the diagram.

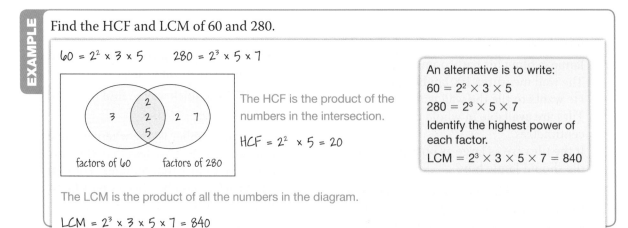

EXAMPLE

Find the HCF and LCM of 60 and 280.

$60 = 2^2 \times 3 \times 5$ $280 = 2^3 \times 5 \times 7$

The HCF is the product of the numbers in the intersection.

$HCF = 2^2 \times 5 = 20$

An alternative is to write:
$60 = 2^2 \times 3 \times 5$
$280 = 2^3 \times 5 \times 7$
Identify the highest power of each factor.
$LCM = 2^3 \times 3 \times 5 \times 7 = 840$

The LCM is the product of all the numbers in the diagram.

$LCM = 2^3 \times 3 \times 5 \times 7 = 840$

Exercise 13.2S

1 Write all the factor pairs of each of these numbers.

 a 18   **b** 14

 c 30   **d** 48

2 Find the highest common factor of

 a 6 and 4   **b** 12 and 18

 c 14 and 16 **d** 28 and 35

 e 30 and 54 **f** 56 and 64

 g 60 and 72 **h** 27 and 45

3 Find the lowest common multiple of

 a 6 and 4   **b** 6 and 8

 c 12 and 18 **d** 15 and 25

 e 21 and 28 **f** 26 and 39

 g 40 and 50 **h** 25 and 15

4 Copy and complete these calculations to show the different ways that 24 can be written as a product of its factors.

 a $24 = \square \times 2$ **b** $24 = 3 \times \square$

 c $24 = 2 \times 3 \times \square$ **d** $24 = 4 \times \square$

5 Each of these numbers has just two prime factors, which are not repeated. Write each number as the product of its prime factors.

 a 77   **b** 51

 c 65   **d** 91

 e 119   **f** 221

6 Copy and complete the factor tree to find the prime factor decomposition of 18.

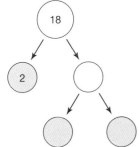

7 Work out the value of each of these expressions.

 a 2^3 **b** 3×5^2

 c $2^3 \times 5$ **d** $3^2 \times 7$

 e $2^2 \times 3^2 \times 5$ **f** $3^2 \times 7^2$

8 Express these numbers as products of their prime factors.

 a 18 **b** 24 **c** 40

 d 39 **e** 48 **f** 82

 g 100 **h** 144 **i** 180

 j 315 **k** 444 **l** 1350

9 For each number, find its prime factors and write it as the product of powers of its prime factors.

 a 36 **b** 120 **c** 34

 d 25 **e** 48 **f** 90

 g 27 **h** 60 **i** 72

10 Write the prime factor decomposition for each of these numbers.

 a 1052   **b** 2560

 c 630   **d** 825

 e 715   **f** 1001

 g 219   **h** 289

 i 2840   **j** 2695

 k 1729   **l** 3366

 m 9724   **n** 11 830

 o 2852   **p** 10 179

11 Find the highest common factor (HCF) of each pair of numbers, by drawing a Venn diagram or otherwise.

 a 35 and 20 **b** 48 and 16

 c 21 and 24 **d** 25 and 80

 e 28 and 42 **f** 45 and 60

12 Find the lowest common multiple (LCM) of each pair of numbers, by drawing a Venn diagram or otherwise.

 a 24 and 16 **b** 32 and 100

 c 22 and 33 **d** 104 and 32

 e 56 and 35 **f** 105 and 144

13 Find the LCM and HCF of each pair of numbers.

 a 180 and 420 **b** 77 and 735

 c 240 and 336 **d** 1024 and 18

 e 762 and 826 **f** 1024 and 1296

Q 1032, 1044   SEARCH

13.2 Prime factor decomposition

RECAP

- A number can be written as the product of its prime factors. This is called prime factor decomposition.
- You can use a factor tree to find the prime factors of a number.
- The **highest common factor** (HCF) of two numbers is the largest number that is a factor of them both.
- The **lowest common multiple** (LCM) of two numbers is the smallest number that they both divide into.
- You can find the HCF and LCM of two numbers by writing their **prime factors** in a Venn diagram.
- The HCF is the product of the numbers in the intersection.
- The LCM is the product of all the numbers in the diagram.

$$990 = 2 \times 3^2 \times 5 \times 11$$

HOW TO

① Read the question carefully; decide how to use your knowledge of factors and multiples.

② Be systematic – use listing, factor trees and/or Venn diagrams to help you to find multiples and factors.

③ Answer the question; make sure that you explain your answers.

EXAMPLE

Remi has four numbers given as prime factorisations.

$2^2 \times 3 \qquad 2 \times 3^2 \times 5 \qquad 2^2 \times 7 \times 11$

$5^2 \times 13$

Which of the numbers are

a a multiple of 6

b a multiple of 10

c a multiple of 8?

Give reasons for your answers.

① Can you make 6, 10 or 8 from the factors in each decomposition?

a ② Multiples of 6 have 2 and 3 as prime factors.
③ $2^2 \times 3$ and $2 \times 3^2 \times 5$ are multiples of 6.

b ② Multiples of 10 have 2 and 5 as prime factors.
③ $2 \times 3^2 \times 5$ is a multiple of 10.

c ② Multiples of 8 have 2^3 as a prime factor.
③ There are no multiples of 8.

EXAMPLE

The highest common factor of two numbers is 6.

The lowest common multiple is 540.

One of the numbers is 54.

Find the other number.

① Use a Venn diagram and prime factors.

② Find the prime factor decompositions of 6, 54 and 540.

$6 = 2 \times 3$

Write 2 and 3 in the intersection.

$54 = 2 \times 3 \times 3 \times 3$

Add 3 and 3 to the diagram.

The product of the numbers in the Venn diagram is 540.

$540 = 2 \times 3 \times 3 \times 3 \times 2 \times 5$

Add 2 and 5 to the diagram.

Factors   Other
of 54   number

③ Find the other number.

The other number is $= 2^2 \times 3 \times 5 = 60$

Exercise 13.2A

1 In each of these questions, Jack has been asked to write each of the numbers as the product of its prime factors.

   i Mark his work and identify any errors he has made.

   ii Correct any of Jack's mistakes.

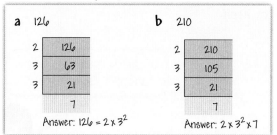

a 126

2	126
3	63
3	21
	7

Answer: $126 = 2 \times 3^2$

b 210

2	210
3	105
3	21
	7

Answer: $2 \times 3^2 \times 7$

2 Allison has four numbers given as prime factorisations.

$$2^2 \times 7 \qquad 2 \times 3^2 \qquad 2^3 \times 5 \qquad 2 \times 3 \times 5^2$$

Which of the above are

   a a multiple of 6

   b a multiple of 10

   c a multiple of 50

   d a multiple of 8 ?

3 The number 18 can be written as $2 \times 3 \times 3$. You can say that 18 has three prime factors.

   a Find three numbers with exactly three prime factors.

   b Find five numbers with exactly four prime factors.

   c Find four numbers between 100 and 300 with exactly five prime factors.

   d Find a two-digit number with exactly six prime factors.

4 A wall measures 234 cm by 432 cm. What is the largest size of square tile that can be used to cover the wall, without needing to cut any of the tiles?

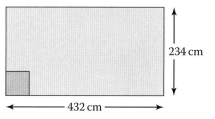

234 cm

432 cm

5 A cuboid is made from 210 small cubes. The prime factor decomposition of 210 is $2 \times 3 \times 5 \times 7$.
One way of combining the prime factors is $(2 \times 3) \times 5 \times 7 = 6 \times 5 \times 7$. The dimensions of the cuboid could be $6 \times 5 \times 7$.

   a Find all the ways in which the prime factors of 210 can be combined to make three factors.

   b Using your answer to part **a**, list the dimensions of all the cuboids that could be made with 210 cubes.

6 A metal cuboid has a volume of 1815 cm³. Each side of the cuboid is a whole number of centimetres, and each edge is longer than 1 cm.

> Volume of a cuboid = length × width × height.

   a Find the prime factor decomposition of 1815.

   b Use your answer to part **a** to find all the possible dimensions of the cuboid.

7 Two hands move around a dial. The faster hand moves around in 24 seconds, and the slower hand in 30 seconds. If the two hands start together at the top of the dial, how many seconds does it take before they are next together at the top?

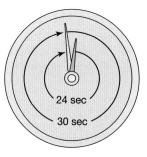

24 sec

30 sec

8 Usha says that if you double two numbers, you double their lowest common multiple.

   a Show that this is true for 6 and 9.

   b Give another example where it is true.

9 Two numbers have HCF = 15 and LCM = 90.
One of the numbers is 30.
What is the other number?

10 $108 = 2^2 \times 3^3$
How many factors does 108 have?

Q 1032, 1044   SEARCH

13.3 Powers and roots

- A **square number** is the result of multiplying a whole number by itself.

$1^2 = 1 \times 1 = 1 \qquad 2^2 = 2 \times 2 = 4 \qquad 3^2 = 3 \times 3 = 9$

- A **square root** is a number that when multiplied by itself gives a result equal to the given number.

The square roots of 225 are 15 and -15 because $15 \times 15 = 225$ and $-15 \times -15 = 225$.
You write $\sqrt{225} = 15$.

You use $\sqrt{}$ for square roots and $\sqrt[3]{}$ for cube roots.

- A **cube number** is the result of multiplying a whole number by itself and then multiplying by that number again.

$1^3 = 1 \times 1 \times 1 = 1 \qquad 2^3 = 2 \times 2 \times 2 = 8 \qquad 3^3 = 3 \times 3 \times 3 = 27$

- A **cube root** is a number that when multiplied by itself and then multiplied by itself again gives a result equal to the given number.

$\sqrt[3]{4913} = 17$ because $17^3 = 17 \times 17 \times 17 = 4913$

- A positive number has a positive cube root and a negative number has a negative cube root.

$\sqrt[3]{-125} = -5$ because $(-5)^3 = -5 \times -5 \times -5 = -125$

Take care when finding powers of negative numbers on your calculator. Remember to put brackets around the negative number.

- The **index** (or **power**) tells you how many times the number must be multiplied by itself.

$7^4 = 7 \times 7 \times 7 \times 7 = 2401$

$6^3 = 6 \times 6 \times 6 = 216$

Work out how to do this on your calculator.

13^5
371293

EXAMPLE

Use your calculator to find the missing powers.

a $\quad 18^\square = 5832$ b $\quad 1.7^\square = 8.3521$ c $\quad (-1.2)^\square = -1.728$

For each question test the different powers.

a $\quad 18^2 = 324$
Try again.
$18^3 = 5832$
Correct.

b $\quad 1.7^2 = 2.89$
Try again.
$1.7^3 = 4.913$
$1.7^4 = 8.3521$
Correct.

c $\quad$ Square numbers are always positive.
$(-1.2)^3 = -1.728$
Correct.

Did you know...

French mathematician and philosopher Rene Descartes was the first person to use index form to write powers of numbers.

Number Factors, powers and roots

Exercise 13.3S

1 Find these numbers.

   a the 4th square number

   b the 8th square number

   c the 20th square number

   d the 5th cube number

   e the 7th cube number

   f the 10th cube number

2 a Find a square number that is the sum of two cube numbers.

   b Find a cube number that is the sum of two square numbers.

3 Use your calculator to work these out.

   a 6^2 **b** 11^2 **c** 14^2 **d** 23^2

   e 31^2 **f** 47^2 **g** 4^3 **h** 6^3

   i 8^3 **j** 13^3 **k** 18^3 **l** 21^3

4 Use your calculator to work these out.

   a $3^2 + 2^2$ **b** $5^2 - 3^2$

   c $6^2 - 2^3$ **d** $4^3 + 4^2$

   e $10^2 - 8^2$ **f** $13^2 + 4^3$

   g $14^2 - 5^3$ **h** $6^3 - 13^2$

   i $12^2 + 13^2 + 14^2$ **j** $6^3 + 7^3 + 8^3$

5 Use the x^2 and x^3 function keys on your calculator to work out each of these. Give your answer to 2 decimal places as appropriate.

   a 2.5^2 **b** 49^2 **c** 3.2^3

   d 4.8^2 **e** 7.3^2 **f** 4.9^3

   g 1.2^2 **h** 0.5^2 **i** 9.9^2

   j 9.9^3 **k** $(5\,\text{cm})^2$ **l** $(4\,\text{m})^3$

6 Find these numbers without using a calculator.

   a $\sqrt{25}$ **b** $\sqrt{9}$ **c** $\sqrt{16}$

   d $\sqrt{1}$ **e** $\sqrt{4}$ **f** $\sqrt{64}$

7 Calculate these using a calculator, giving your answer to 2 dp as appropriate.

   a $\sqrt{40}$ **b** $\sqrt{61}$ **c** $\sqrt{180}$

   d $\sqrt{249}$ **e** $\sqrt{676}$ **f** $\sqrt{1000}$

8 Find these numbers without using a calculator.

   a $\sqrt[3]{8}$ **b** $\sqrt[3]{125}$ **c** $\sqrt[3]{-1}$

   d $\sqrt[3]{64}$ **e** $\sqrt[3]{1000}$ **f** $\sqrt[3]{-27}$

9 Calculate these using the $\sqrt[3]{\ }$ key on your calculator. Give your answers to 2 dp where appropriate.

   a $\sqrt[3]{40}$ **b** $\sqrt[3]{512}$ **c** $\sqrt[3]{3375}$

   d $\sqrt[3]{100}$ **e** $\sqrt[3]{24\,389}$ **f** $\sqrt[3]{-29791}$

10 Copy the question and find the missing powers.

   a $3^\square = 9$ **b** $5^\square = 25$

   c $4^\square = 64$ **d** $2^\square = 8$

   e $3^\square = 27$ **f** $10^\square = 1000$

11 Find the value of these numbers without using a calculator.

   a 3^4 **b** 1^5 **c** 2^7

   d 3^6 **e** 10^6 **f** $(-4)^4$

12 Use the y^x function key on your calculator to work these out.

   a 12^3 **b** 6^6 **c** 21^3

   d 16^5 **e** 13^3

13 Use your calculator to work out these powers. In each case copy the question and fill in the missing numbers.

   a $24^\square = 576$ **b** $1.5^\square = 3.375$

   c $(-2.25)^\square = 5.065$ **d** $(-0.5)^\square = -0.125$

   e $5^\square = 3125$ **f** $(-3)^\square = 729$

14 Use a calculator method to work these out, giving your answers to 2 decimal places where appropriate.

   a 15.5^3 **b** $(-2.7)^6$ **c** 2.1^{10}

   d 21.6^4 **e** 13.5^3 **f** $(-16.8)^5$

15 Use a mental, written or calculator method to work out each of these.

   a $2^4 + 3^2$ **b** $10^3 \div 5^2$

   c $8^6 - 13^3$ **d** $10^6 \div 5^3$

   e $\left(\dfrac{1}{4}\right)^4 + \left(\dfrac{1}{8}\right)^2$ **f** $\sqrt{\dfrac{4}{9}} + \sqrt[3]{\dfrac{1}{8}}$

   Q 1053, 1924   SEARCH

13.3 Powers and roots

- The index (or power) tells you how many times the number must be multiplied by itself.
- A square number is the result of multiplying a whole number by itself.
- A positive number has a negative and positive square root.
- A cube number is the result of multiplying a whole number by itself and then multiplying by that number again.
- A positive number has a positive cube root and a negative number has a negative cube root.

$6^2 = 6 \times 6 = 36$
$(-6)^2 = -6 \times -6 = 36$
$\sqrt{36} = 6$

$6^3 = 6 \times 6 \times 6 = 216$
$\sqrt[3]{216} = 6$
$(-6)^3 = -6 \times -6 \times -6$
$\quad = -216$
$\sqrt[3]{-216} = -6$

① Read the question carefully. You may need to use your knowledge from other areas of maths.

② Apply your knowledge of squares and cubes.

③ Answer the question and include any units.

Elsa has square tiles with area 81 cm². She uses six tiles to make this shape. Find the perimeter of the shape.

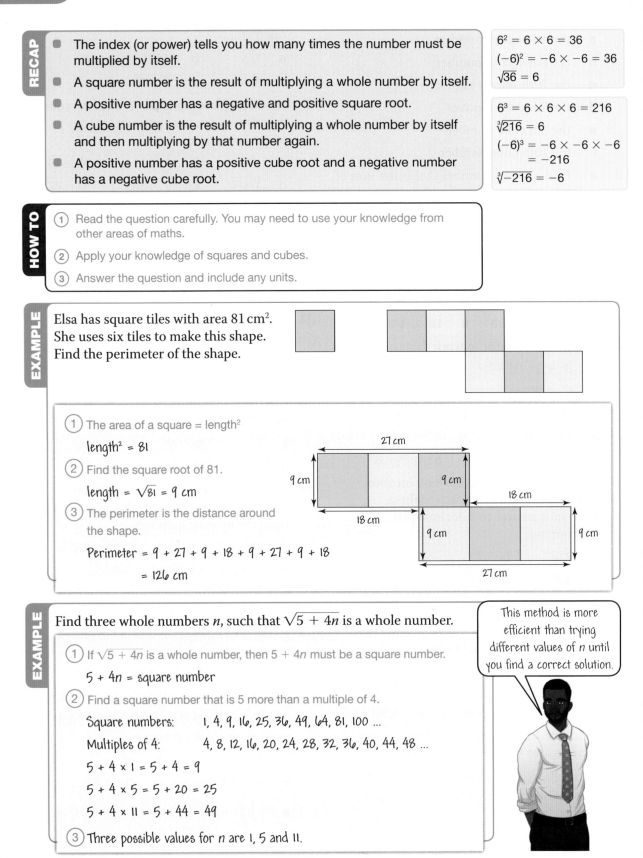

① The area of a square = length²

length² = 81

② Find the square root of 81.

length = √81 = 9 cm

③ The perimeter is the distance around the shape.

Perimeter = 9 + 27 + 9 + 18 + 9 + 27 + 9 + 18

= 126 cm

Find three whole numbers n, such that $\sqrt{5 + 4n}$ is a whole number.

This method is more efficient than trying different values of n until you find a correct solution.

① If $\sqrt{5 + 4n}$ is a whole number, then $5 + 4n$ must be a square number.

5 + 4n = square number

② Find a square number that is 5 more than a multiple of 4.

Square numbers: 1, 4, 9, 16, 25, 36, 49, 64, 81, 100 ...

Multiples of 4:   4, 8, 12, 16, 20, 24, 28, 32, 36, 40, 44, 48 ...

5 + 4 × 1 = 5 + 4 = 9

5 + 4 × 5 = 5 + 20 = 25

5 + 4 × 11 = 5 + 44 = 49

③ Three possible values for n are 1, 5 and 11.

Number   Factors, powers and roots

Exercise 13.3A

1 Some numbers can be represented as the sum of two square numbers.
For example $1^2 + 2^2 = 1 + 4 = 5$.
Try to find all the numbers less than 50 that can be represented as the sum of two square numbers.

2 Harry has mixed up his answers to these questions.

a Without using a calculator, match each of these questions to the correct answer.

Questions		Answers	
1	$\sqrt{169}$	A	9
2	$\sqrt[3]{343}$	B	8
3	$\sqrt{121}$	C	7
4	$\sqrt{81}$	D	4
5	$\sqrt{64}$	E	11
6	$\sqrt[3]{1000}$	F	13
7	$\sqrt{196}$	G	10
8	$\sqrt[3]{64}$	H	14

b Check your answers using your calculator.

See how many of the questions and answers you matched correctly.

3 Lewis has square tiles with area 144 cm². He uses nine tiles to make a larger square. Find the perimeter of the larger square.

4 Jack has a set of cubes with volume 216 cm³. Jack wants to build a tower by stacking the cubes on top of each other.
He wants the tower to be at least 1 metre tall. Find the minimum number of cubes that Jack needs to build the tower.

5 Do not use a calculator for these questions.

$\sqrt{95}$ lies between 9 and 10.

9 and 10 are consecutive numbers – they are integers that are in counting order.

a Between which two consecutive numbers does $\sqrt{150}$ lie?

b Between which two consecutive numbers does $\sqrt{300}$ lie?

c Between which two consecutive numbers does $\sqrt{80}$ lie?

6 $\sqrt{7} = 2.645751$
Calculate $(2.645751)^2$.
Explain why the answer is not 7.

7 Two consecutive numbers are multiplied together. The answer is 3192.
What are the two numbers?

8 The cube of a particular integer lies between 3000 and 4000.
Without using a calculator, find the integer. Show your workings.

9 In a school the teachers never reveal their ages but always give clues.
Mr Earle gave three clues.

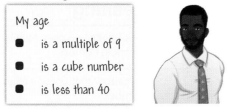

My age
● is a multiple of 9
● is a cube number
● is less than 40

a What is Mr Earle's age?

b Can you work out Mr Earle's age from two clues? Explain your answers.

10 Find a pair of prime numbers a, b such that $\sqrt{a^2-b}$ is a whole number.

11 Find a pair of whole numbers p, q such that $\sqrt{p^3-3q}$ is a whole number.

12 a James says that when you square a number the answer always gets bigger. Give two examples to show that this is not always true.

b Lauren says that when you find the square root of a number the answer always gets smaller.
Give two examples to show that this is not always true.

c Describe in your own words what happens to the size of a number when you find the square or the square root of it.

13 Do not use a calculator.

a Find the prime factor decomposition of 1444.

b Use your answer to find the square root of 1444.

Q 1053, 1924   SEARCH

Summary

Checkout

You should now be able to...

		Test it Questions
✓	Use mathematical language to describe factors, multiples and primes.	1 – 4
✓	Use Venn diagrams or factor trees to systematically list the prime factors of a number.	5 – 7
✓	Use prime factor decomposition to calculate the HCF and LCM of two or more numbers.	8, 9
✓	Write the HCF and LCM using product notation.	10
✓	Calculate positive integer powers and their roots.	11, 12
✓	Recognise powers of 2, 3, 4 and 5.	13

Language	Meaning	Example
Multiple	The original number multiplied by an integer (a whole number).	$2 \times 6 = 12$ $3 \times 6 = 18$ 12 and 18 are both multiples of 6.
Factor	A number that divides exactly into another number.	$15 = 3 \times 5$ 3 and 5 are both factors of 15.
Prime number	A number that has only two factors, itself and one.	$13 = 1 \times 13$
Prime factor	A factor that is a prime number.	13 is a prime factor of 52.
Prime factor decomposition	Writing a number as a product of its prime factors.	$52 = 2^2 \times 13$
Common factor	A factor that is shared by two or more numbers.	$30 = 2 \times 3 \times 5$ $70 = 2 \times 5 \times 7$ 2 and 5 are common factors of 30 and 70
Highest common factor (HCF)	The largest factor that is shared by two or more numbers.	Factors of 30 Factors of 70
Lowest common multiple (LCM)	The smallest multiple that is shared by two or more numbers.	$HCF = 2 \times 5 = 10$ $LCM = 3 \times 2 \times 5 \times 7 = 210$
Square number	To square a number multiply it by itself. Square numbers have integer square roots.	$1^2 = 1$ $2^2 = 4$ $3^2 = 9$ $4^2 = 16$
Square root	A number that when multiplied by itself is equal to the number underneath the square root symbol.	$4 \times 4 = 16$ $\sqrt{16} = 4$
Cube number	To cube a number multiply it by itself and then by itself again. Cube numbers have cube roots that are integers.	$1^3 = 1$ $2^3 = 8$ $3^3 = 27$ $4^3 = 64$
Cube root	A number that when multiplied by itself and then by itself again is equal to the number underneath the cube root symbol.	$4 \times 4 \times 4 = 64$ $\sqrt[3]{16} = 4$

Review

1   List all the factors of these numbers.

   a   12   **b**   19   **c**   48

2   List the first 5 multiples of these numbers.

   a   5   **b**   13   **c**   18

3   State whether or not each of these is a prime and explain your answers.

   a   25   **b**   29   **c**   57

4 | 1 | 2 | 3 | 9 | 18 | 54 | 72 |

   List the numbers in the box that are

   a   prime   **b**   factors of 36

   c   multiples of 6.

5   Copy and complete this factor tree to show the prime factors of 72.

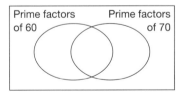

6   Find the prime factor decomposition of these numbers.

   a   18   **b**   28

7   **a**   Copy and complete this Venn diagram to show the prime factors of 60 and 70.

Prime factors of 60		Prime factors of 70

7   **b**   Use the diagram to work out the

      i   highest common factor

      ii   lowest common multiple

      of 60 and 70.

8   Find the highest common factor of these pairs of numbers.

   a   8 and 12   **b**   27 and 54

   c   7 and 9   **d**   11 and 13

9   Find the lowest common multiples of these pairs of numbers.

   a   3 and 4   **b**   6 and 8

   c   11 and 9   **d**   11 and 13

10 **a**   Write these numbers as products of their prime factors, using product notation.

      i   54   **ii**   200

   b   Find the LCM and HCF of 54 and 200. Write your answers using product notation.

11 Calculate the value of these roots.

   a $\sqrt{16}$ **b** $\sqrt{81}$

   c $\sqrt[3]{8}$ **d** $\sqrt[3]{1000}$

12 Calculate the value of the following expressions.

   a 3^2 **b** 7^2

   c 5^3 **d** 2^5

13 What is the value of x in these equations?

   a $4^x = 64$ **b** $5^x = 125$

What next?

Score	0 – 5		Your knowledge of this topic is still developing.
			To improve look at MyMaths: 1032, 1034, 1044, 1053, 1924
	6 – 10		You are gaining a secure knowledge of this topic.
			To improve look at InvisiPens: 13Sa – f
	11 – 13		You have mastered these skills. Well done you are ready to progress!
			To develop your exam technique look at InvisiPens: 13Aa – h

Assessment 13

1   Leo says that all of the pairs of factors of 42 are 2×21 and 3×14.
    Is he correct? If not, write down all of the pairs of factors of 42.   [2]

2   **a**   Sara says that 3454 is not divisible by 6. Why is her statement true?   [2]

    b   Write a single digit on the end of 10 to make a number that is divisible by

        i 2   **ii** 3   **iii** 4   **iv** 5   **v** 6   **vi** 7   **vii** 8   **viii** 9   **ix** 10.   [9]

3   **a**   Is 4 a factor of 110? Say how you know.   [2]

    b   I am thinking of a number that is greater than 4 and less than 20.
            My number is a factor of 110.
            Write down all the possible integer answers.   [3]

4   Gareth says that the HCF of 54 and 99 is 3.
    Is he correct? If not, show how to work out the HCF of these two numbers.   [2]

5   Hannah says that the LCM of 36 and 60 is 180.
    Show how she worked this out.   [3]

6   At a party it was discovered that John, Paul, George and Ringo had birthdays
    on the 6th, 15th, 21st and 30th of the month. Mick joined the group and
    it was discovered that his birthday was a factor of everyone else's.
    What day of the month was Mick born on?   [2]

7   Trains from Birmingham go to Glasgow, Liverpool and Leeds. Trains to Glasgow
    leave once an hour, to Liverpool every 45 minutes and to Leeds every 25 minutes.
    At 6 am three trains leave for all these destinations.
    When will trains to all these destinations next leave at the same time?   [5]

8   A 'perfect' number is one where all its factors, except for the number itself, add up to that
    number. The factors of 6 are 1, 2, 3 and 6 and $1 + 2 + 3 = 6$ so 6 is the first perfect number.
    The next perfect number is 28. Show that 28 is a 'perfect' number.
    Show your working.   [3]

9   Henry knows an alternative method to find the HCF of two numbers.

    1. Write down the numbers side by side.

    2. Cross out the largest and write underneath it the 'difference' between the two.

    3. Repeat step 2 until the two numbers left are the same.

    4. The remaining number is the HCF of the two original numbers.

    Try Henry's method with the following numbers

        a   16 and 28   [3]   **b**   30 and 66   [3]   **c**   252 and 588.   [3]

10 Isa and Josh both write 19 800 as a product of its prime factors using index notation.

    Isa writes $19\,800 = 2^3 \times 3^2 \times 5^2 \times 11$ Josh writes $19\,800 = 2 \times 3^3 \times 5 \times 11$
    Who is correct? Show your workings.   [3]

11 Amanda says that

        a 4^5 is greater than 5^4 **b** 2^{10} is greater than 10^2 **c** 2^4 is greater than 4^2.

    Decide for each pair of values if she is correct or not. Give reasons for your answers. [6]

12 The number grid below has 4 columns. The *prime numbers* in the grid are *shaded*.

Column A	Column B	Column C	Column D
1	2	3	4
5	6	7	8
9	10	11	12
13	14	15	16
17	18	19	20
21	22	23	24
25	26	27	28
29	30	31	32
33	34	35	36
37	38	39	40

    a 33 is not shaded. Explain why 33 is **not** a prime number. [2]

    b Hasan says that 100 will eventually be in Column A.
       Is he right? Give reasons for your answer. [1]

    c There is only one prime number in column B.
       Will there ever be another? Explain how you know. [2]

    d What will be the next two prime numbers in Column C? [2]

    e There are no prime numbers in column D.
       Will there ever be a prime number in column D. Give reasons for your answer. [1]

13 Tope says that all square numbers over 1 can be written as the sum of two prime numbers.

    a Show this is true for the number 25. [2]

    b Show there are two ways of doing this for the number 16. [2]

    c Show there are four ways of doing this for the number 36. [4]

14 a Kerry says that to work out $3^3 \times 3^2$ you multiply the indices. Glen says you
       add the indices. Who is correct? Use the correct rule to work out $3^3 \times 3^2$. [2]

    b Giorgia says that to work out $14^8 \div 14^2$ you subtract the indices. Jonas says you
       divide the indices. Who is correct? Use the correct rule to work out $14^8 \div 14^2$. [2]

15 a Sam says numbers have two square roots. George says some numbers
       have no square roots. Who is right? Give reasons for your answer. [2]

    b Amelia joins in the conversation and says that all numbers have 2 cube roots.
       Is she right? Give reasons for your answer. [2]

16 Ms Connell took her class to the monkey house at the zoo, but a naughty
child opened the cage door and let the monkeys out with the children.
Ms Connell counted about 70 heads and tails but there were twice as many
tails as heads. She also knew that the number of children was a prime number.
How many children were in Ms Connell's class? [4]

17 Arthen thinks that all the prime numbers can be found by substituting $p = 0, 1, 2$, etc. into the
formula $P = 41 - p + p^2$.

    a Check Arthen's formula for $p = 0, 3$, and 6. [4]

    b Write down one value which shows that Arthen is not correct. Give your reasons. [2]

14 Graphs 1

Introduction

Kinematics is the topic within maths that deals with motion. By writing equations and drawing graphs that describe the relationship between distance, speed and acceleration, it is possible to calculate things such as the speed of a vehicle at a particular time during its journey.

What's the point?

Having the mathematical tools to describe how objects move means that we can better understand our world, which is in continual motion.

Objectives

By the end of this chapter you will have learned how to …

- Work with coordinates in all four quadrants.
- Plot straight-line graphs including diagonal, vertical and horizontal lines.
- Identify gradients and intercepts of straight lines graphically and algebraically.
- Use the form $y = mx + c$ to identify parallel lines.
- Use one point and the gradient of the line to find its equation.
- Use two points to find the equation of a line.
- Interpret the gradient of a straight line graph as a rate of change.
- Plot and interpret graphs involving distance, speed and acceleration.

Check in

1 Work out the value of each expression when $x = 2$.

    a $x + 6$ **b** $3x - 1$ **c** $2x + 4$ **d** $\frac{x}{4} + 3$

2 Substitute $x = -1$ into these expressions and evaluate them.

    a $2x + 1$ **b** $3x - 2$ **c** $4x + 5$ **d** $2x - 5$

3 **a** Draw a coordinate grid with x and y-axes from -5 to 5, using squared paper.

    b Plot these coordinates on your grid.

      i $(0, 4)$ **ii** $(2, -5)$ **iii** $(-4, 1)$ **iv** $(-2, -1)$

Chapter investigation

The points $(2, 4)$ and $(5, 13)$ both lie on the same straight line.

Find the equation of this line.

Does the point $(21, 18)$ lie on this same line? How can you tell without drawing the line?

14.1 Drawing straight-line graphs

- The graphs of linear equations such as $y = 2x + 3$ are straight lines.

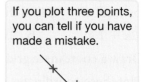

If you plot three points, you can tell if you have made a mistake.

One point must be wrong.

To plot a graph of a function:

- Draw up a table of values
- Calculate the value of y for each value of x
- Draw a suitable grid
- Plot the (x, y) pairs and join them with a straight line.

a Draw the graph of $y = 2x + 3$.

b Use the graph to find **i** the value of x when $y = 7$ **ii** the value of y when $x = 3\frac{1}{2}$.

Construct a table of values.

Choose four or five values, including negative values and zero.

a

x	−2	−1	0	1	2
y	−1	1	3	5	7

$(-2, -1)$ $(-1, 1)$ $(0, 3)$ $(1, 5)$ $(2, 7)$

Then plot the points and draw the line.

Make sure that your grid includes the smallest and largest y-values.

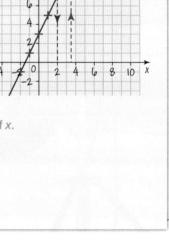

b i Find $y = 7$ on the y-axis. Draw a horizontal line to the graph line. Draw a vertical line to the x-axis. Read off the value of x.

$x = 2$

ii Find $x = 3\frac{1}{2}$ on the x-axis. Draw a vertical line to the graph line. Draw a horizontal line to the y-axis. Read off the value of y.

$y = 10$

A straight line can be **diagonal**, **vertical** or **horizontal**.

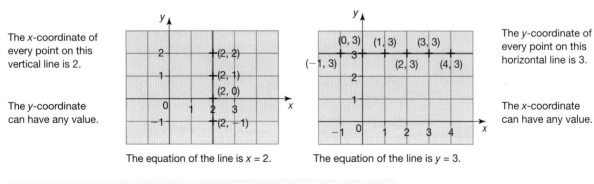

The x-coordinate of every point on this vertical line is 2.

The y-coordinate can have any value.

The equation of the line is $x = 2$.

The y-coordinate of every point on this horizontal line is 3.

The x-coordinate can have any value.

The equation of the line is $y = 3$.

- Horizontal lines have equations of the form $y = c$.
- Vertical lines have equations of the form $x = c$.

c stands for a number.

Exercise 14.1S

1 Write the coordinates of the points in each diagram.

a

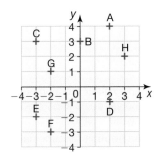

b

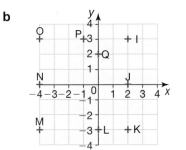

2 a Plot these sets of points on a copy of this grid.

   i $(3, 2)$ $(-1, 2)$ $(-1, -1)$

   ii $(1, 2)$ $(-1, 0)$ $(3, 0)$

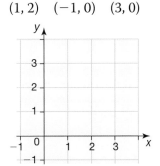

  b Join each set of points in order.

  c What is the name of each shape?

3 Write the equations of these horizontal and vertical graphs.

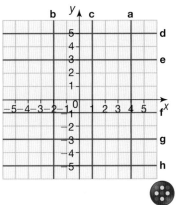

4 Copy the grid from question **3** without the graphs. Draw these graphs on the same grid.

  a $y = 4$ **b** $x = 5$ **c** $y = -2$

  d $x = -2$ **e** $x = -4$

5 Use your graphs from question **4** to find the coordinates of the points where these pairs of graphs cross.

  a $x = -2$ and $y = 4$

  b $x = 5$ and $y = -2$

  c $x = -4$ and $y = 4$

6 Write the coordinates of the points where these pairs of lines cross.

  a $x = 2$ and $y = -3$ **b** $x = 1$ and $y = 6$

  c $x = 3$ and $y = -1$

7 a Copy and complete the table of values for $y = 2x + 5$.

x	−3	−1	0	1	3
y	−1			7	

  b Draw the graph of $y = 2x + 5$ on a copy of the grid.

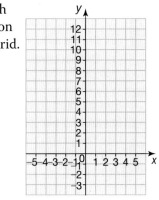

8 Draw the graphs of these functions.

  a $y = 3x - 2$ **b** $y = -2x + 4$

  c $y = \frac{1}{2}x + 3$ **d** $y = 5 - x$

9 Draw a graph of $y = -2x + 3$. Use your graph to find

  a the value of y when $x = 0.5$

  b the value of x when $y = 0$.

10 Draw a graph of $y = 4x + 1$. Use your graph to find

  a the value of y when $x = -0.5$

  b the value of x when $y = 7$.

🔍 1093, 1394, 1395, 1396   SEARCH

14.1 Drawing straight-line graphs

- The graphs of linear equations are straight lines. A straight line can be
  - Diagonal, for example $y = 2x + 3$ and $y = -3x - 6$
  - Horizontal, for example $y = -3$ and $y = 8.2$
  - Vertical, for example $x = -3.5$ and $x = 17$

The equation $y = 3x - 4$ gives y **explicitly** in terms of x.
The equation $2x + 3y = 6$ gives y **implicitly** in terms of x.

An **explicit** function has the variables separated by the = sign.

An **implicit** function can have the variables on the same side of the = sign.

HOW TO

To draw a graph of an implicit function

- Draw up a table of values   or
  for $x = 0$ and $y = 0$
- Rearrange the equation to make y the subject.

For $2x + 3y = 6$:

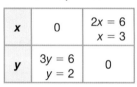

x	0	2x = 6 x = 3
y	3y = 6 y = 2	0

For $2x + 3y = 6$:

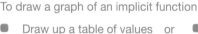

$3x = 6 - 2y$

$y = \dfrac{6 - 2x}{3}$

x	−2	−1	0	1	2
y	$\dfrac{10}{3}$	$\dfrac{8}{3}$	2	$\dfrac{4}{3}$	$\dfrac{2}{3}$

HOW TO

To use a graph to solve a real-life problem

1. Draw up a table of values. Calculate the values of y for at least three x-values. You may decide to write a formula for the situation in the question first.
2. Draw a suitable grid, label the axis and plot the graph.
3. Use your graph to give the answer in the context of the question.

EXAMPLE

Plot a graph to represent the total cost of hiring a party venue, if the owner charges £100 hire fee and £5 per guest.
Use the graph to estimate the number of guests if the total bill is £285.

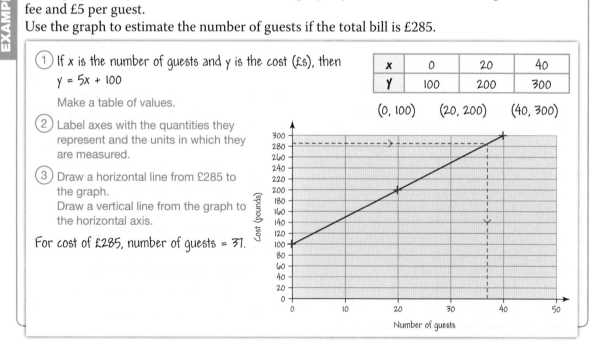

1. If x is the number of guests and y is the cost (£s), then
   y = 5x + 100
   Make a table of values.

x	0	20	40
Y	100	200	300

(0, 100)   (20, 200)   (40, 300)

2. Label axes with the quantities they represent and the units in which they are measured.

3. Draw a horizontal line from £285 to the graph.
   Draw a vertical line from the graph to the horizontal axis.

For cost of £285, number of guests = 37.

Exercise 14.1A

1 Plot and label these points on a square grid.

A (3, 3) B (−2, 3) C (−4, −1)

A, B and C are three corners of a parallelogram. Write the coordinates of the fourth corner, D.

2 **a** Which point with integer coordinates fits these clues?
Above $y = -1$, below $y = 3$,
below $y = 2x + 1$, above $y = 2 - x$
and left of $x = 2$.

   b Write your own clues to describe the point (3, 4).

3 **a** The point (2, 5) lies on the graph $y = 2x + 1$. Does (3, 8) lie on this graph? Explain your answer.

   b The point (2, 5) lies on the graph $y = 2x + 1$. Name another point that lies on this line.

4 **a** Copy and complete this table of values for pounds to New Zealand dollars.

Pounds	0	1		10
NZ dollars		2.4		

   b Write three pairs of coordinates for the conversion graph.

   c Draw the graph by copying the axes and plotting the points.

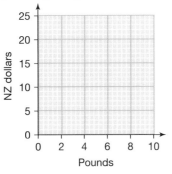

   d Use your graph to work out which is more:

   i £6 or 12 NZ dollars?

   ii 18 NZ dollars or £7?

5 Rearrange these equations to make y the subject. Which is the odd one out?

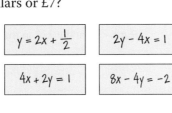

$y = 2x + \frac{1}{2}$ $2y - 4x = 1$

$4x + 2y = 1$ $8x - 4y = -2$

6 Draw a graph of each of these functions given in implicit form.

   a $x + y = 5$ **b** $y - x = 3$
   c $2x + y = 6$ **d** $5x + y = 10$

7 I have a mobile phone. Each month, I pay £5 line rental. For every hour I then spend on the phone, I pay £7.

   a Copy and complete this table of values to show my total phone bill for different lengths of time spent on the phone.

x (hours on phone)	1	2	3	4	5
y (total bill £)	12				

   b Plot a graph to show hours against total bill.

   c Use your graph to find the approximate cost if I spend 3 hours and 15 minutes on the phone one month.

   d What is the equation of the graph?

8 Would you prefer to get £3 per week pocket money, with 20p for every chore done (such as washing up) or £5 per week pocket money with 15p for every chore done? Use line graphs to report on your favoured option and to find how many chores you would need to do to receive the same amount under both options.

9 A campsite charges £15 per night per tent, plus an extra £3 per person.

   a Construct a table of charges and, hence, a graph to show the cost of the campsite depending upon how many people stay in the tent. (The largest tent available is one that sleeps 15 people.)

   b Use your graph to calculate how many people stayed in the tent if the total cost was £36.

   c Explain why the total cost could never be £50.

   d Suggest an equation for your graph, stating clearly the meaning of any letters you use.

   e Use your equation to work out the cost of pitching a new Supertent that sleeps 27 people.

Q 1093, 1394, 1395, 1396 SEARCH

14.2 Equation of a straight line

Gradient measures steepness.

The graph $y = 2x + 1$ has gradient 2.
For every 1 unit across, the graph goes up 2 units.

The graph $y = 2x + 1$ has y-intercept 1.
It crosses the y-axis at $(0, 1)$.

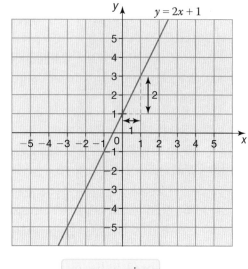

- Diagonal lines have an equation of the form
  $$y = mx + c \quad \text{where } m \text{ is the gradient and}$$
  $$c \text{ is the } y\text{-intercept.}$$

If the equation is not in the form $y = ...$, rearrange it first,
for example

$$3x + 2y = 12 \implies 2y = -3x + 12 \implies y = -\frac{3}{2}x + 6$$

Now you can see that the gradient is $-\dfrac{3}{2}$ and the intercept is 6.

$$\text{Gradient} = \frac{\text{rise}}{\text{run}}$$

EXAMPLE

Find the gradient and
intercept of the lines

a $x + y = 4$

b $2x - 5y = 10$

a $x + y = 4$
$y = 4 - x$

$y = -x + 4$

gradient = -1, intercept = 4

b $2x - 5y = 10$
$2x - 10 = 5y$ Divide by 5.

$\dfrac{2}{5}x - 2 = y$

gradient = $\dfrac{2}{5}$, intercept = -2

- **Parallel** lines have the same gradient.

EXAMPLE

Find the equation of a line parallel to $y = 4x + 5$.

The line $y = 4x + 5$ has gradient 4.

A line that is parallel to it will have the same gradient.

$y = 4x + 1$ is parallel to the line $y = 4x + 5$.

There are many
other possible
equations. Any
equation
$y = 4x + \text{something}$
is parallel.

- If you know the gradient and a point on the line you can find
  the equation of the line.

EXAMPLE

What is the equation of a line
with gradient 8 that passes
through the point $(2, 7)$?

Gradient = 8, so equation is $y = 8x + c$.
The line goes through $(2, 7)$ so,
$7 = 8 \times 2 + c$ Put $x = 2$ and $y = 7$ in the equation $y = 8x + c$.
$7 = 16 + c$
$c = -9$
The equation of the line is $y = 8x - 9$.

Algebra Graphs 1

Exercise 14.2S

1 Name these lines without plotting the points.

    a Line 1: $(-5, 5)\ (-2, 5)\ (5, 5)\ (9, 5)$

    b Line 2: $(-8, -3)\ (-8, 3)\ (-8, 6)\ (-8, 10)$

2 Write the gradient and y-intercept for each of these equations.

    a $y = 3x - 1$ **b** $y = 2x + 5$

    c $y = 4x - 3$ **d** $y = \frac{1}{2}x + 2$

    e $y = 5x + 1$ **f** $y = -3x + 7$

3 Match the equations to the graphs.

    a $y = 3x + 1$ **b** $y = 3$

    c $y = 2x + 4$ **d** $x = -2$

    e $y = x + 3$

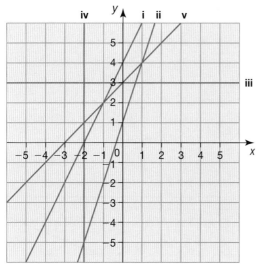

4 For each graph

    i write its gradient and intercept

    ii write the equation of the line.

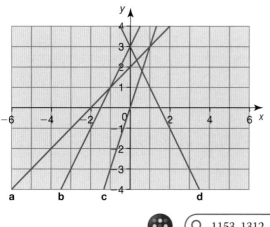

5 Copy and complete the table.

Gradient	Intercept	Equation
3	5	
5	-2	
-2	7	
$\frac{1}{2}$	9	
$-\frac{1}{4}$	-3	
0	4	
1	0	

6 Rearrange these equations in the form $y = mx + c$.

    a $x + y = 5$ **b** $y - x = 3$

    c $x + y = -2$ **d** $x - y = 3$

    e $2x + y = 6$ **f** $5x + y = 9$

    g $3x + y = -2$ **h** $y - 2x = 5$

    i $2y + x = 4$ **j** $2y - x = 8$

    k $2x + 4y = 16$ **l** $6x + 2y = 8$

7 List the equations from question **6** whose graphs are parallel to

    a $y = x$ **b** $y = -x$

    c $y = -3x$ **d** $y = -\frac{1}{2}x$

8 Write the equation of a straight line that is parallel to $y = 3x - 1$.

9 Write the equation of a straight line that is parallel to $y + 4x = 2$.

10 Find the equations of the six lines described in the table.

a	Gradient of 7 and intercepts y-axis at (0, 5)	**b**	Gradient of $\frac{1}{2}$ and passes through (0, 3)
c	Parallel to a line with gradient 4 and passing through (3, 8)	**d**	Gradient of 3 and passing through (4, 7)
e	Gradient of -2 and cutting through (4, -3)	**f**	Parallel to $y = \frac{1}{4}x - 1$ and passing through (0, -2)

Q 1153, 1312, 1314, 1957   SEARCH

14.2 Equation of a straight line

- You can write the equation of any straight line in the form $y = mx + c$, where
  - m is the gradient of the line
  - c is the y-intercept.

HOW TO

You can use the equation of a straight line to solve problems.

1. Find the gradient of the line. Use the graph in the question or draw a sketch.

2. Find the y-intercept either using the graph or by substituting a known point into the equation $y = mx + c$.

3. Give your answer in the context of the question.

EXAMPLE

Find the equation of the line joining $(1, 2)$ and $(4, 3)$.

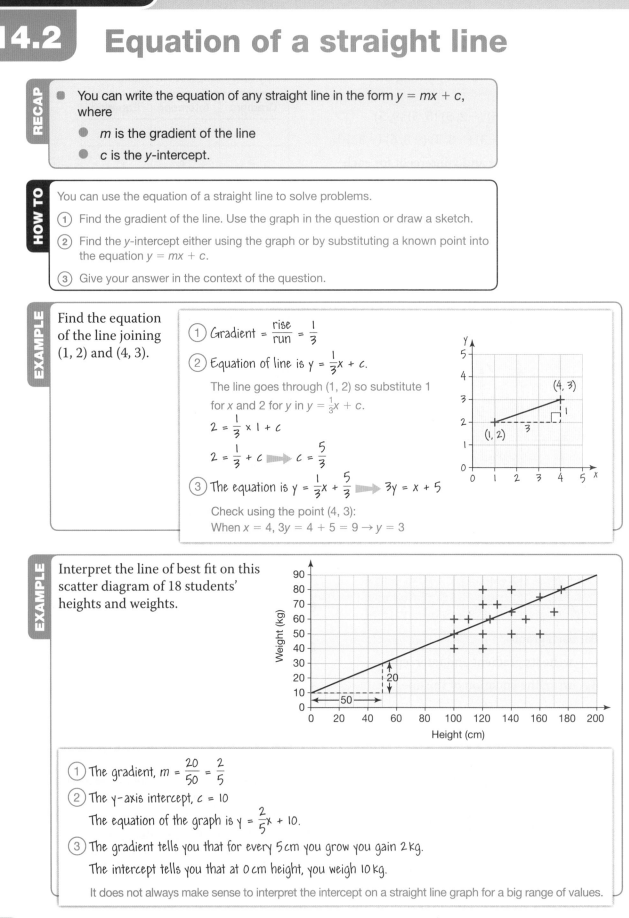

1. Gradient $= \dfrac{\text{rise}}{\text{run}} = \dfrac{1}{3}$

2. Equation of line is $y = \dfrac{1}{3}x + c$.

   The line goes through $(1, 2)$ so substitute 1 for x and 2 for y in $y = \dfrac{1}{3}x + c$.

   $2 = \dfrac{1}{3} \times 1 + c$

   $2 = \dfrac{1}{3} + c \implies c = \dfrac{5}{3}$

3. The equation is $y = \dfrac{1}{3}x + \dfrac{5}{3} \implies 3y = x + 5$

   Check using the point $(4, 3)$:
   When $x = 4$, $3y = 4 + 5 = 9 \rightarrow y = 3$

EXAMPLE

Interpret the line of best fit on this scatter diagram of 18 students' heights and weights.

1. The gradient, $m = \dfrac{20}{50} = \dfrac{2}{5}$

2. The y-axis intercept, $c = 10$

   The equation of the graph is $y = \dfrac{2}{5}x + 10$.

3. The gradient tells you that for every 5 cm you grow you gain 2 kg.

   The intercept tells you that at 0 cm height, you weigh 10 kg.

   It does not always make sense to interpret the intercept on a straight line graph for a big range of values.

Algebra Graphs 1

Exercise 14.2A

1 Find the equations of the three lines described in the table.

a	b	c
Passing through (0, 1) and (1, 5)	Passing through (0, 2) and (5, 7)	Passing through the midpoint of (1, 7) and (3, 13) with a gradient of 8

2 **a** What is the gradient of the line joining (0, 5) to (12, 41)?

   b What is the equation of the line joining (0, 5) to (12, 41)?

   c Repeat **a** and **b** for the points (3, 10) and (5, 6).

3 Here are the equations of several lines.

| **A** $y = 3x - 2$ | **B** $y = 4 + 3x$ | **C** $y = x + 3$ |

| **D** $y = 5$ | **E** $2y - 6x = -3$ | **F** $y = 3 - x$ |

   a Which three lines are parallel to one another?

   b Which two lines cut the y-axis at the same point?

   c Which line has a zero gradient?

   d Which line passes through (2, 4)?

   e Which pair of lines are reflections of one another in the y-axis?

4 Find the equations of these lines.

   a A line parallel to $y = -4x + 3$ and passing through $(-1, 2)$.

   b A line parallel to $2y - 3x = 4$ and passing through (6, 7).

5 True or false?

   a $y = 2x - 1$ and $y = 2 + 2x$ are parallel.

   b $y = 3x - 4$ and $y = 6 - x$ both pass through (2, 4).

   c $x = 6$ and $y = 2x - 8$ never meet.

   d $y = x + 1$ and $y = x^2$ meet once.

6 For each graph, find its equation in the form $y = mx + c$ and interpret the meaning of m and c, deciding if it is sensible to interpret c.

a

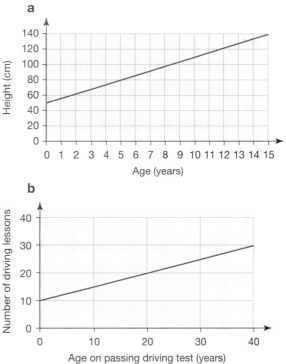

b

7 Mrs Harman gives her class a geography test. The results (%) for Paper 1 and Paper 2 for 10 students are shown.
Mrs Harman plots a scatter diagram and draws the line of best fit.
The points (50, 40) and (60, 50) lie on the line of best fit.

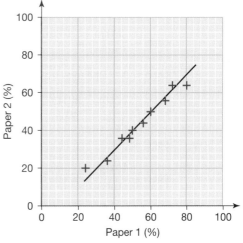

Interpret the line of best fit on the scatter diagram.

 Q 1153, 1312, 1314, 1957   SEARCH

14.3 Kinematic graphs

A distance-time graph shows information about a journey.

Time is always displayed on the horizontal axis and distance on the vertical axis.

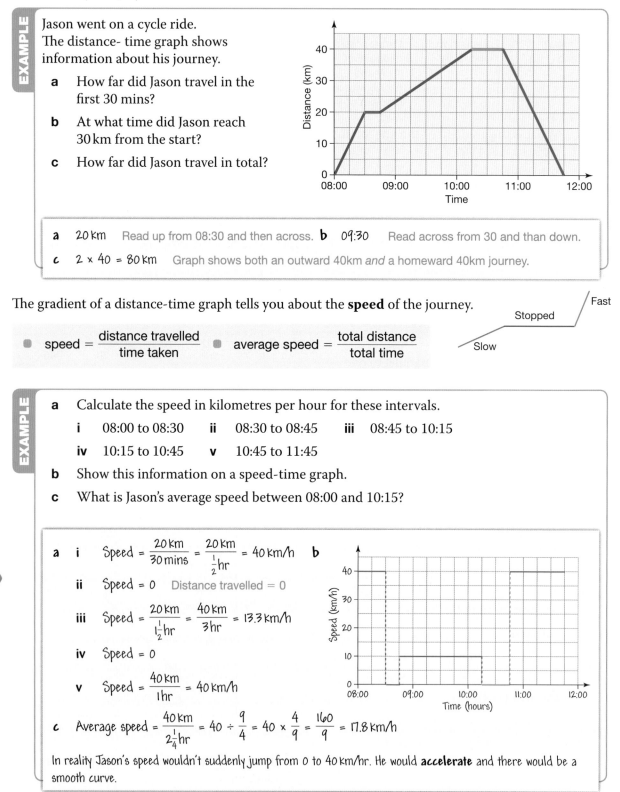

EXAMPLE

Jason went on a cycle ride.
The distance- time graph shows information about his journey.

a How far did Jason travel in the first 30 mins?

b At what time did Jason reach 30 km from the start?

c How far did Jason travel in total?

---

a 20 km   Read up from 08:30 and then across.   **b** 09:30   Read across from 30 and than down.

c 2 × 40 = 80 km   Graph shows both an outward 40km *and* a homeward 40km journey.

The gradient of a distance-time graph tells you about the **speed** of the journey.

- speed = $\dfrac{\text{distance travelled}}{\text{time taken}}$
- average speed = $\dfrac{\text{total distance}}{\text{total time}}$

Fast / Stopped / Slow

EXAMPLE

a Calculate the speed in kilometres per hour for these intervals.

   i   08:00 to 08:30   **ii**   08:30 to 08:45   **iii**   08:45 to 10:15

   iv   10:15 to 10:45   **v**   10:45 to 11:45

b Show this information on a speed-time graph.

c What is Jason's average speed between 08:00 and 10:15?

---

a **i**   Speed = $\dfrac{20\,km}{30\,mins} = \dfrac{20\,km}{\frac{1}{2}\,hr} = 40\,km/h$ **b**

   ii   Speed = 0   Distance travelled = 0

   iii   Speed = $\dfrac{20\,km}{1\frac{1}{2}\,hr} = \dfrac{40\,km}{3\,hr} = 13.3\,km/h$

   iv   Speed = 0

   v   Speed = $\dfrac{40\,km}{1\,hr} = 40\,km/h$

c Average speed = $\dfrac{40\,km}{2\frac{1}{4}\,hr} = 40 \div \dfrac{9}{4} = 40 \times \dfrac{4}{9} = \dfrac{160}{9} = 17.8\,km/h$

In reality Jason's speed wouldn't suddenly jump from 0 to 40 km/hr. He would **accelerate** and there would be a smooth curve.

Exercise 14.3S

1 The distance-time graph shows information about Lisa's coach journey.

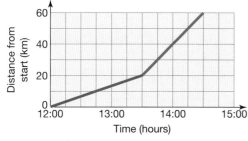

a How far does she travel between
   i 12:00 and 13:30
   ii 13:30 and 14:30?

b How long does it take to travel
   i 10 km from the start
   ii 50 km from the start?

2 Tamera is riding a bike. Information about her journey is shown in the graph.

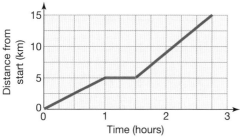

a Tamera's journey starts at 8 a.m. At what time does her journey finish?

b What happens after Tamera has cycled for one hour?

3 Mark sets off from home in his car at 2 pm. He stops to get petrol, and then continues on his journey. Mark returns home later in the afternoon. The distance-time graph shows more information about this journey.

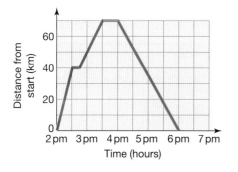

3 **a** How far was the petrol station from Mark's home?

b For how long did Mark wait at the petrol station?

c After stopping for petrol, how long did Mark take to drive to his destination?

d At what time did Mark set off on his journey home?

e What was Mark's speed during the first section of his journey?

f Between what times was Mark travelling at his greatest speed? How do you know?

g How far did Mark travel
   i between 4 pm and 5 pm
   ii between 3 pm and 5 pm
   iii in total?

4 Work out the overall average speed for each of these journeys.

a
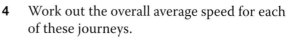

b

*c

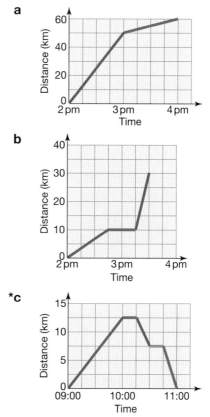

5 Draw speed-time graphs for the three journeys shown in question **4**.

6 Bertha flies an aeroplane directly from London to Lisbon at 200 km/h for 3 hours. Draw a distance-time graph to show this.

14.3 Kinematic graphs

- Information about a journey can be shown on a distance-time or speed-time graph.
- The gradient of a line on a distance-time graph is the speed.

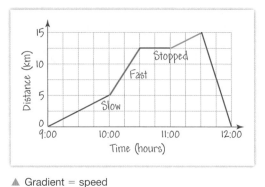

▲ Gradient = speed

HOW TO

22.1

1. Decide what to do and where to find the information needed.
2. Use the formula $\text{speed} = \dfrac{\text{distance}}{\text{time}}$
3. Give your answer and include any units.

EXAMPLE

a   Linda goes on a bicycle ride. Draw a distance-time graph for her journey.

b   What was Linda's average speed on her return journey?

c   If Linda stopped for a 30 min rest on her return journey, would your answer to part **b** change?

- Linda started her journey at 11:00.
- She travelled 15 km in the first 30 mins.
- She travelled at a constant speed of 20 km/hr for the next 2 hours.
- She then stopped for 1 hour.
- She returned home at 17:00.

a   First point is (11:00, 0 km)   ①

Second point is (11:30, 15 km)   ②

$20 \text{ km/hr} = \dfrac{\text{distance}}{2 \text{ hr}}$

Axes need to include 11:00 to 17:00 and 0 to 55 km.

distance = 2 × 20 = 40 km

Third point is (13:30, 55 km)
Fourth point is (14:30, 55 km)
Last point is (17:00, 0 km)

③

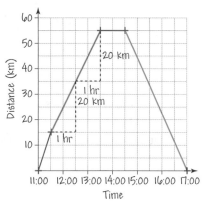

b   Average speed $= \dfrac{55 \text{ km}}{2\frac{1}{2} \text{ hr}} = 55 \div 2.5 = 22 \text{ km/hr}$ ②

c   No, the distance travelled and time are the same.

The gradient of a speed-time graph is the **acceleration**. It tells you how the speed is changing.

- $\text{Acceleration} = \dfrac{\text{Change in speed}}{\text{Time taken}}$

EXAMPLE

The graph shows the speed-time graph for a train leaving a station.

a   What is the acceleration in m/s²?

b   If the train accelerates, at the same rate, for one minute what will be its final speed?

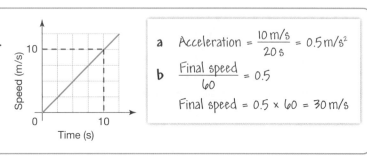

a $\text{Acceleration} = \dfrac{10 \text{ m/s}}{20 \text{ s}} = 0.5 \text{ m/s}^2$

b $\dfrac{\text{Final speed}}{60} = 0.5$

Final speed = 0.5 × 60 = 30 m/s

Exercise 14.3A

1 James walks 15 km in 3 hours. He rests for 30 minutes and then walks home at a constant speed of 6 km/h. Draw a distance-time graph to show his journey.

2 Jodie visited her grandmother. Use the information given to draw a distance-time graph for her journey.

- Jodie sets out at 14:00.
- She travels 400 m in the first 5 mins before realising she has forgotten a present.
- She goes straight back home, travelling at the same speed.
- Jodie quickly picks up the present and takes 25 mins to travel the 2 km to her gran's house.
- She spends 30 mins at her grans.
- She goes straight back home at a speed of 0.833 m/s.

3 A train leaves a station at 11 a.m. and travels at a constant speed of 80 km/h. A second train leaves the same station 30 minutes later. It travels at 140 km/h.

   a Construct a distance-time graph to show this.

   b At what time are the two trains the same distance from the starting station?

4 The graph shows information about a race between Asif and Sian.

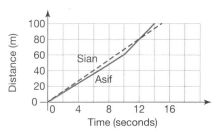

   a Who was quickest at the start of the race?

   b What happened 12 seconds into the race?

   c Who won the race?

5 Explain why each of these distance-time graphs are impossible.

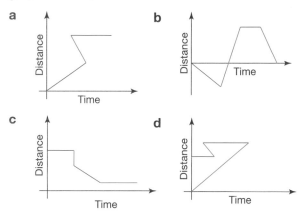

6 Match each distance-time graph.

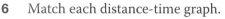

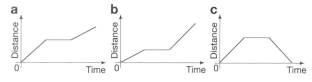

to its matching speed-time graph.

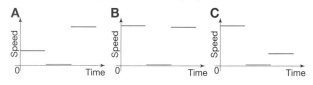

7 A rocket accelerates in two stages as shown in the speed-time graph.

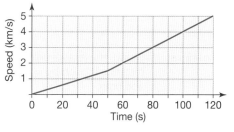

   a Calculate the acceleration, in km/s², for

   i stage 1   **ii** stage 2.

   b Calculate the average acceleration for the whole journey.

Summary

Checkout

You should now be able to...

Test it

Questions

✔ Work with coordinates in all four quadrants.	**1**
✔ Plot straight-line graphs including diagonal, vertical and horizontal lines.	**2 – 4**
✔ Identify gradients and intercepts of straight lines graphically and algebraically.	**5, 6**
✔ Use the form $y = mx + c$ to identify parallel lines.	**7**
✔ Use one point and the gradient of the line to finds its equation.	**8**
✔ Use two points to find the equation of a line.	**9**
✔ Interpret the gradient of a straight line graph as a rate of change.	**10**
✔ Plot and interpret graphs involving distance, speed and acceleration.	**10**

Language   Meaning   Example

Language	Meaning	Example
Coordinate grid	An **origin** and a set of **x** and **y** **axes** that allow you to specify a point.	$(1, -3)$ is a point one unit right and three units down from the origin $(0, 0)$.
Gradient	A measure of the slope of a line on a graph found by dividing the change in y by the change in x.	Gradient, $m = \frac{2}{1} = 2$
y-intercept	The point at which a straight line graph crosses the y-axis	
$y = mx + c$	The standard way to write the equation of a straight line. m = gradient, c = y-intercept	y-intercept, $c = -2$ $y = 2x - 2$
Distance–time graph	Shows the relationship between distance moved by an object and time.	Speed $= \frac{12}{6} = 2$ m/s
Speed	How for an object travels in a unit of time. It is given by the gradient of a line on a distance–time graph.	
Acceleration	How much an object's speed changes in a unit of time.	0 to 10 m/s in 2s is an acceleration of $10 \div 2 = 5$ m/s².

Review

1 Write down the coordinates of points A, B, C, D and E.

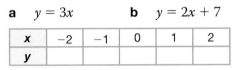

2 Copy and complete this table of values for each equation.

    a $y = 3x$ **b** $y = 2x + 7$

x	−2	−1	0	1	2
y					

3 Draw appropriate axes and plot the lines for each of the equations in question **2**.

4 Copy the grid from question **1** without the points. Draw these graphs on the same grid.

    a $y = 3$ **b** $x = -2$

5 **a** What is the gradient of the line shown?

    b What is the equation of the line?

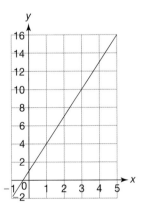

6 Complete a table of values and plot the graphs of these equations for values of x between -1 and 4. For each line state the gradient and y-intercept.

    a $y = 3x - 1$ **b** $y = 20 - 5x$

7 Write down the equation of any line that is parallel to the line with equation $y = 3x + 4$.

8 What is the equation of the line that has gradient 7 and crosses through the y-axis at the point $(0, -4)$?

9 What is the equation of the line that passes through the points $(0, 2)$ and $(3, 14)$?

10 The distance–time graph shows a bus journey from the depot to the station and back again.

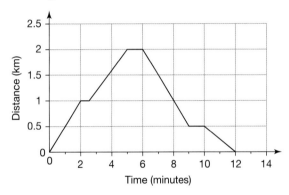

    a What is the total distance travelled by the bus?

    b What does the gradient of the curve represent in this context?

    c What is the average speed of the bus over the whole 12 minute journey? Give your answer in km/h.

What next?

Score			
	0 – 4		Your knowledge of this topic is still developing. To improve look at MyMaths: 1093, 1153, 1312, 1314, 1322, 1323, 1394, 1395, 1396
	5 – 8		You are gaining a secure knowledge of this topic. To improve look at InvisiPens: 14Sa – k
	9 – 10		You have mastered these skills. Well done you are ready to progress! To develop your exam technique look at InvisiPens: 14Aa – f

Assessment 14

1   **a**   Draw a grid with x-axis from -2 to 3 and y-axis from -2 to 3.

Plot these sets of points on a copy of this grid.

$(-1, 2)$ $(2, 2)$ $(2, 1)$ $(1, 1)$

$(1, -1)$ $(0, -1)$ $(0, 1)$ $(-1, 1)$ [4]

  b   Join the set of points in order.
  What letter have you drawn?   [1]

2   **a**   Complete this table for the four graphs.   [8]

x	-1	0	1	2	3	4
A $y = 2x + 1$						
B $y = 7 - x$						
C $y = \frac{1}{2}x + 1$						
D $2x + y = 8$						

  b   Draw a grid with x-axis from -1 to 4. Draw all four graphs on the
  same axes. Label each graph.   [8]

  c   Write down the coordinates of the points of intersection of these graphs.

  i   A and B   **ii**   A and C   **iii**   B and C   **iv**   B and D   [4]

3   Draw a grid with the x- and y-axes from 0 to 6.
    Draw these graphs on the same grid.

  a $x = 5$ **b** $x = 2$ **c** $y = 1$ [4]

  d   Plot the points $(0, 6)$ and $(6, 0)$ on the grid. Join the points with a straight line.
  What is the equation of the line?   [2]

  e   Find the area of the triangle enclosed by the four lines?   [2]

4   Carli says that $(-2, -1)$, $(0, 2)$ and $(3, 9)$ are all points on the line $y = 2x + 3$.
    Is Carli correct? Explain your answer.   [3]

5   **a**   Does the point $(-4, 5)$ lie on the line $3y - 4x = 30$?   [1]

  b   Does the point $(7, 6)$ lie on the line $y = 1 - x$ or $y = x - 1$? [1]

6   Work out the gradient and y-intercept for each of the following straight lines.

  a $y = 2x + 7$ [2]   **b** $y = 9 + 4x$ [2]   **c** $y = 6x - 11$ [2]

  d $y = 12 - 4x$ [2]   **e** $4x - 7y = 14$ [3]   **f** $15x + 14y = 35$ [3]

7   The gradient and y-intercept of four lines are
    shown in the table.
    Find the equations of each line in the
    form $y = mx + c$.   [4]

	gradient	y-intercept
a	1	1
b	-1	1
c	2	6
d	-4	13

8 a The sketch shows Demelza's journey on Monday morning.

Match each stage in Demelza's journey, A to B, B to C, C to D, D to E and E to F, with the correct description.

   i Demelza drives from her home to work.

   ii Demelza leaves her car in the carpark at her work.

   iii Demelza returns home to pick up her daughter's maths homework.

   iv Demelza stops to answer a phone call.

   v Demelza starts her journey from home to work. [4]

b On which section of her journey is she travelling fastest? Explain your answer. [2]

9 Jess goes for a walk.

- The park is 600 m away and the walk takes 10 minutes.
- Jess stays in the park for an hour.
- Jess walks to a shop which is a further 100 m away. The walk to the shop takes 5 minutes.
- Jess spends 10 minutes in the shop.
- Jess walks home. The walk home takes 15 minutes.
- Jess walks at a constant speed for each stage of her journey.

Draw a distance–time graph of Jess' journey. [10]

10 Calculate the gradients of the straight lines which pass through the following pairs of points.

   a $(1, 5)$ and $(5, 9)$ [2]   **b** $(6, 7)$ and $(8, -9)$ [2]

   c $(-3, 5)$ and $(-4, 6)$ [2]   **d** $(-11, 0)$ and $(5, -8)$ [2]

11 Find the equation of each line in the form $y = mx + c$.

   a Gradient $= -5$, passing through $(0, -61)$ [1]

   b Gradient $= 3$, passing through $(1, 2\frac{1}{2})$ [2]

   c Gradient $= \frac{1}{3}$, passing through $(3, -1)$ [2]

   d Gradient $= \frac{1}{4}$, passing through $(0, -\frac{3}{4})$ [1]

12 Find the equation of each line in the form $y = mx + c$.

   a Line parallel to $y - 7x - 9 = 0$ that intercepts the y-axis at $(0, 5)$. [2]

   b Line parallel to $3x + y - 77 = 0$ that intercepts the y-axis at $(0, -7.25)$. [2]

   c Line parallel to $2x + 6y = 15$ that intercepts the y-axis at $(0, 2\frac{2}{3})$. [2]

13 a State the gradient of each line (A–D) on the grid. [5]

   b Match each equation to one of the lines A, B, C and D drawn on the graph. [3]

   i $5x + 3y = 20$ **ii** $2y - 3x = 9$

   iii $2x - y = 0$ **iv** $2x + 3y = 0$

15 Working in 3D

Introduction

In the Cave of Crystal Giants in Mexico, vast crystals of selenite grow to lengths of up to 11 metres, weighing up to 55 tons – the largest crystals to have yet been discovered. The cave was only discovered because there was a mine nearby – it is highly likely that there are even more awesome geological structures lying somewhere undiscovered beneath our feet!

What's the point?

We live in a three-dimensional world, and 3D geometry allows us to describe the wonderful things that we can see in the natural world, as well as helping us to devise increasingly sophisticated structures in the manmade world.

Objectives

By the end of this chapter you will have learned how to ...

- Identify the numbers of faces, edges and vertices of 3D shapes.
- Construct and interpret plans and elevations of 3D shapes.
- Use standard units for volume, cm³, m³.
- Calculate the volume of cuboids, cylinders and other prisms.
- Apply the formulae for volume and surface area of spheres, pyramids, cones and composite solids.

Check in

1   Calculate the area of each shape.

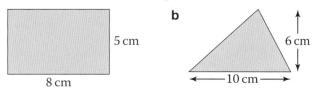

a

5 cm

8 cm

b

6 cm

10 cm

2   Give the mathematical name for these distances in a circle.

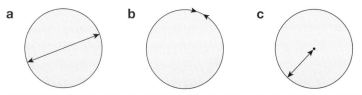

a   b   c

Chapter investigation

A cylindrical drinks container is to have a capacity of 1 litre.

Give a possible set of dimensions for the container.

Find another possible set of dimensions – which of your two sets has the lower surface area?

The manufacturer wants to minimise the amount of packaging used. Investigate different dimensions, to find the lowest surface area. Could the manufacturer sell this to consumers? Give reasons for your answer.

15.1 3D shapes

A **face** is a flat surface of a solid.
An **edge** is the line where two faces meet.
A **vertex** is a point at which three or more edges meet.

A **prism** has a constant cross-section.
You name a prism by the shape of its cross-section.

A **pyramid** has faces that meet at a common point.
You name a pyramid by the shape of its base.

vertex   face   edge

cylinder   triangular prism   square-based prism

square-based pyramid   triangular-based pyramid   cone

● A **net** is a 2D shape that can be folded to form a 3D shape.

EXAMPLE

The nets of two solids are shown.
Name the solid that can be made from each net.

a

b

a regular tetrahedron
b triangular prism

A regular tetrahedron has four faces – all equilateral triangles.

You can look at this car from different directions.

from above, ...   from the front ...   and from the side.

● A **plan** of a solid is the view from directly overhead (bird's eye view).
● An **elevation** is the view from the front or the side of the solid.

Plan   Front elevation   Side elevation

EXAMPLE

The plan and front elevation of a prism are shown.

a Draw a 3D sketch of the prism.

b Draw the side elevation on square grid paper.

Plan   Front elevation

Notice the extra bold line in the plan, when the level of the cubes alters.

a

b

Side elevation

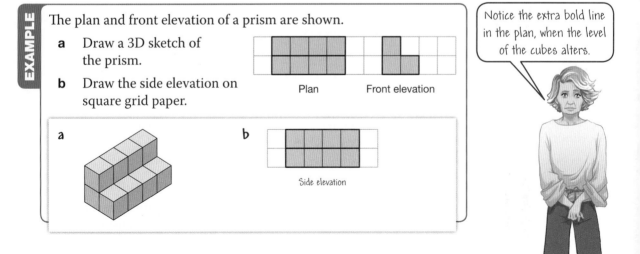

Exercise 15.1S

1 Give the mathematical name of each of these solids.

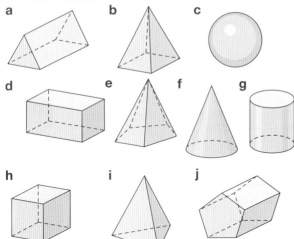

a   b   c

d   e   f   g

h   i   j

2 Draw

   a   a prism with a square cross-section

   b   a pyramid with a hexagonal base

   c   a tetrahedron.

3 This solid consists of eight triangles. It is called an octahedron.
Write

   a   the number of faces

   b   the number of edges

   c   the number of vertices of this solid.

4 These nets make cuboids. Copy the nets onto centimetre squared paper and cut them out to make the cuboids.

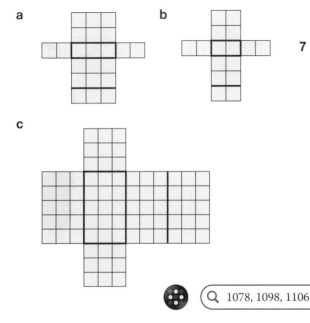

a   b

c

5 On square grid paper, draw the net for each cuboid.

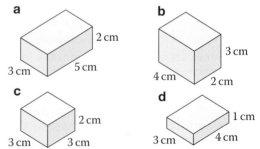

a   2 cm   3 cm   5 cm

b   3 cm   4 cm   2 cm

c   2 cm   3 cm   3 cm

d   1 cm   3 cm   4 cm

6 On square grid paper, draw the plan (P), the front elevation (F) and the side elevation (S) for each solid.

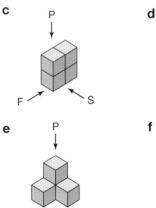

a   b

c   d

e   f

7 The plan, front elevation and side elevation are given for these solids made from cubes. Draw a 3D sketch of each solid and state the number of cubes needed to make it.

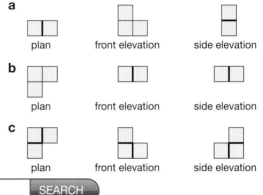

a   plan   front elevation   side elevation

b   plan   front elevation   side elevation

c   plan   front elevation   side elevation

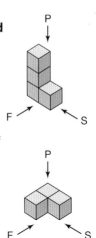

Q 1078, 1098, 1106   SEARCH

15.1 3D shapes

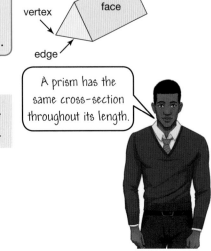

vertex · face · edge

A prism has the same cross-section throughout its length.

The area of the net is called the surface area.

- The **surface area** of a 3D shape is the total area of its **faces**.
- The **volume** of a 3D shape is the amount of space it takes up.

HOW TO

1. Use your knowledge of 3D solids to draw a net, the plan and elevations or a sketch of the solid.

2. Work out the answer, remember to include units and give an explanation if the question asks for it.

EXAMPLE

Find the surface area of the cuboid.

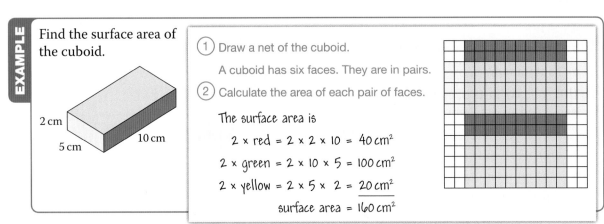

2 cm, 5 cm, 10 cm

1. Draw a net of the cuboid.

   A cuboid has six faces. They are in pairs.

2. Calculate the area of each pair of faces.

   The surface area is

   $2 \times \text{red} = 2 \times 2 \times 10 = 40 \text{ cm}^2$

   $2 \times \text{green} = 2 \times 10 \times 5 = 100 \text{ cm}^2$

   $2 \times \text{yellow} = 2 \times 5 \times 2 = 20 \text{ cm}^2$

   surface area $= 160 \text{ cm}^2$

EXAMPLE

Here are the plan and front elevation of a prism.
The front elevation shows the cross-section of the prism.

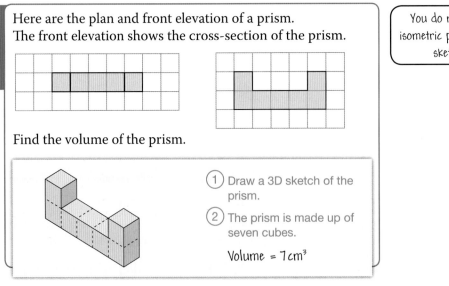

Find the volume of the prism.

You do not need isometric paper for a sketch.

1. Draw a 3D sketch of the prism.

2. The prism is made up of seven cubes.

   Volume $= 7 \text{ cm}^3$

Exercise 15.1A

1 State whether each arrangement of squares is a net of a cube.

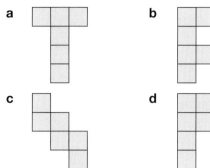

a   b

c   d

2 Sketch **six** different nets of a cube.

3 Sketch the plan (P), the front elevation (F) and the side elevation (S) for this solid.

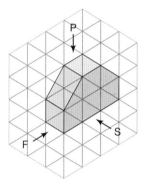

> This solid is drawn on an **isometric grid**.

4 **a** Calculate the volume, for each solid. Each cube represents $1\,\text{cm}^3$.

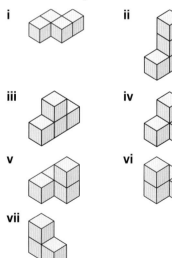

i   ii

iii   iv

v   vi

vii

b All these solids fit together to make a cube. Find the volume of the cube.

c What are the dimensions of the cube?

5 The diagrams show the plan and the front elevation of different solids.

   i Draw a sketch of each solid.

   ii Find the volume of each solid.

a

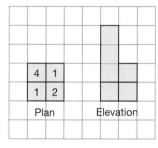

4	1
1	2

Plan   Elevation

b

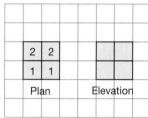

2	2
1	1

Plan   Elevation

> The numbers on the plan tell you the number of cubes in each column.

6 Draw the net of a cube with volume $8\,\text{cm}^2$.

*7 Tina wants to make this box for soap. It is in the shape of a prism.

The cross section is a trapezium.

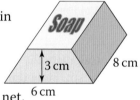

4 cm

Soap

3 cm   8 cm

6 cm

   a Draw an accurate net.

   b Tina wants to decorate all the edges with lace. Is $\frac{1}{2}$ metre of lace enough?

   c Tina also wants to glue a flower to each vertex. How many flowers does she need?

*8 Ben wants to use this card to make an open-topped box in the shape of a cube.

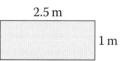

2.5 m

1 m

He wants the box to be as big as possible.

   a Draw a sketch of a net he could use.

   b How long will the sides of the box be?

  Q 1078, 1098, 1106   SEARCH

15.2 Volume of a prism

The **volume** of a 3D shape is the amount of space it takes up.

Volume is measured in cubic units: cubic centimetres (cm^3), cubic metres (m^3).

> Liquids are usually measured in litres and millilitres.
> $1 m^3$ = 1000 litres
> 1 litre = 1000 cm^3
> 1 ml = 1 cm^3

● Volume of a cuboid = length × width × height

EXAMPLE

Calculate the volume of each shape. State the units in your answers.

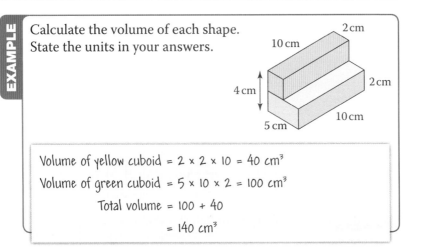

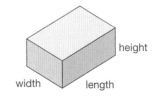

Volume of yellow cuboid = 2 × 2 × 10 = 40 cm^3
Volume of green cuboid = 5 × 10 × 2 = 100 cm^3
Total volume = 100 + 40
= 140 cm^3

A **prism** is a 3D shape with the same **cross-section** throughout its length.

cross section

length

● Volume of a prism = area of cross-section × length

EXAMPLE

Calculate the volume of this triangular prism. State the units of your answer.

Area of triangle = $\frac{1}{2}$ × 4 × 2 = 4 m^2
Volume of prism = area of triangle × 8
= 4 × 8 = 32 m^3

Area of triangle = $\frac{1}{2}bh$

● A cylinder is a prism with circular cross-section.
● Volume of a **cylinder** = area of circle × height

EXAMPLE

Find the volume and surface area of this cylinder. Give your answers to 3 significant figures.

Area of circle = π × 3^2 = 28.274 ... cm^2
Volume = 28.274 ... × 7 = 198 cm^3 (3 sf)
Surface area = 2 × π × 3^2 + 2 × π × 3 × 7
= 188.495... cm^2 = 188 cm^2 (3 sf)

> Do not round intermediate steps of the calculation.

> Give answers to a sensible degree of accuracy. Here, you are asked for 3 sf.

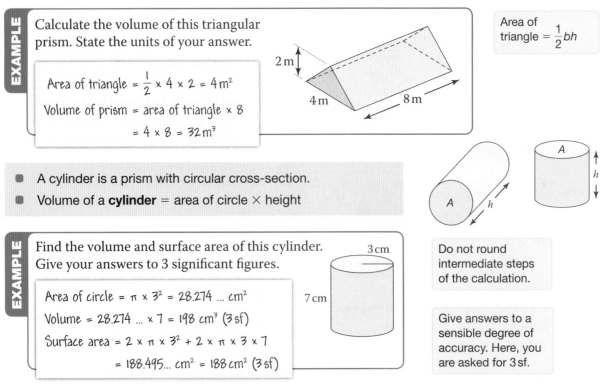

Exercise 15.2S

1 Calculate the volume, in cm³, of each cuboid made of cm³ blocks.

a

b

2 Calculate the volume of each cuboid. State the units of your answers

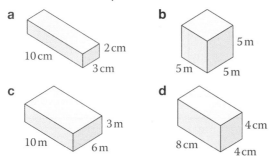

a 10 cm, 2 cm, 3 cm

b 5 m, 5 m, 5 m

c 3 m, 10 m, 6 m

d 4 cm, 8 cm, 4 cm

3 Calculate the volume of each shape. State the units of your answers.

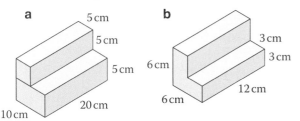

a 5 cm, 5 cm, 5 cm, 20 cm, 10 cm

b 3 cm, 3 cm, 6 cm, 12 cm, 6 cm

4 Calculate the area of cross-section and the volume for each prism.

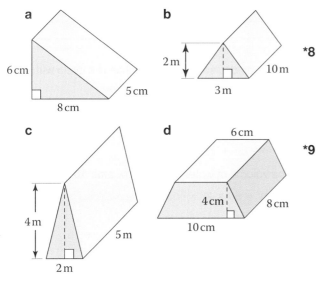

a 6 cm, 5 cm, 8 cm

b 2 m, 10 m, 3 m

c 4 m, 5 m, 2 m

d 6 cm, 4 cm, 8 cm, 10 cm

5 Find the volume of each cylinder. Give your answers to 3 significant figures.

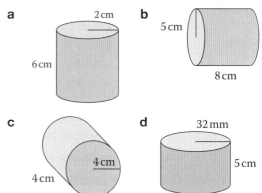

a 2 cm, 6 cm

b 5 cm, 8 cm

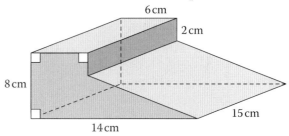

c 4 cm, 4 cm

d 32 mm, 5 cm

6 Find the volume of this shape.

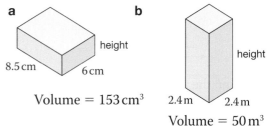

6 cm, 2 cm, 8 cm, 14 cm, 15 cm

7 Calculate each unknown length. Give your answers to a suitable degree of accuracy.

a 8.5 cm, 6 cm, height

Volume = 153 cm³

b 2.4 m, 2.4 m, height

Volume = 50 m³

*8 **a** The volume of a cylinder is 510 cm³. The radius is 4.5 cm. Calculate the height.

b A cylindrical can has a capacity of 440 ml. The height is 10 cm. Calculate the diameter.

*9 The diagram shows a planter. How many litres of compost does it take to fill it?

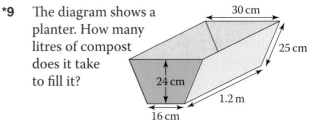

30 cm, 25 cm, 24 cm, 1.2 m, 16 cm

15.2 Volume of a prism

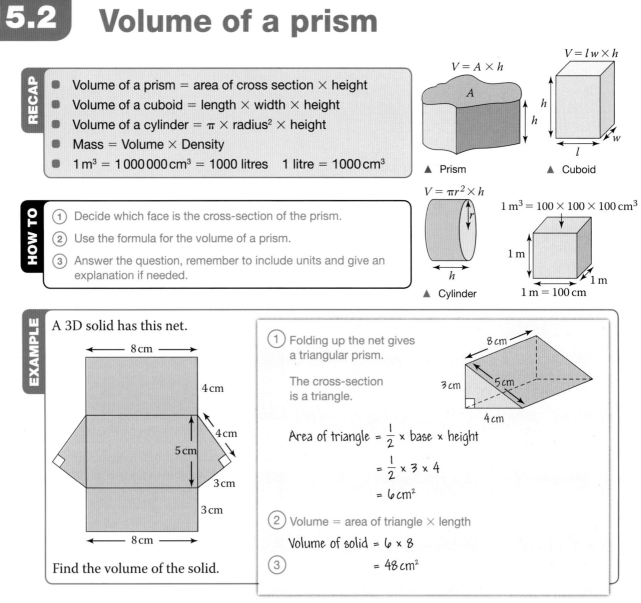

RECAP

- Volume of a prism = area of cross section × height
- Volume of a cuboid = length × width × height
- Volume of a cylinder = π × radius^2 × height
- Mass = Volume × Density
- $1\,m^3 = 1\,000\,000\,cm^3 = 1000$ litres   1 litre = $1000\,cm^3$

$V = A \times h$

▲ Prism

$V = l\,w \times h$

▲ Cuboid

HOW TO

① Decide which face is the cross-section of the prism.

② Use the formula for the volume of a prism.

③ Answer the question, remember to include units and give an explanation if needed.

$V = \pi r^2 \times h$

▲ Cylinder

$1\,m^3 = 100 \times 100 \times 100\,cm^3$

$1\,m = 100\,cm$

EXAMPLE

A 3D solid has this net.

8 cm

4 cm

4 cm

5 cm

3 cm

3 cm

8 cm

Find the volume of the solid.

① Folding up the net gives a triangular prism.

The cross-section is a triangle.

8 cm

3 cm   5 cm

4 cm

Area of triangle = $\frac{1}{2}$ × base × height

= $\frac{1}{2}$ × 3 × 4

= 6 cm²

② Volume = area of triangle × length

Volume of solid = 6 × 8

③   = 48 cm²

EXAMPLE

Jessica has a cylindrical vase.

The vase has a radius of 4 cm and a height of 15 cm.

Jessica fills the vase from the tap.

Water flows from the tap at a rate of 80 cm³ per second.

Jessica turns on the tap and fills the vase for 10 seconds.

Will the vase overflow?

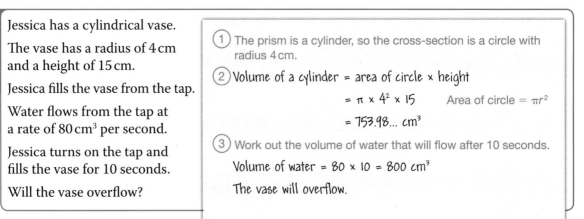

① The prism is a cylinder, so the cross-section is a circle with radius 4 cm.

② Volume of a cylinder = area of circle × height

= π × 4² × 15   Area of circle = πr^2

= 753.98... cm³

③ Work out the volume of water that will flow after 10 seconds.

Volume of water = 80 × 10 = 800 cm³

The vase will overflow.

Geometry Working in 3D

Exercise 15.2A

1 A 3D solid has this net.

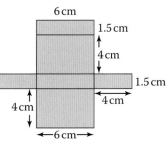

Find the volume of the solid.

2 The diagrams show the plan and the front elevation of different solids.

Find the volume of each solid.

a

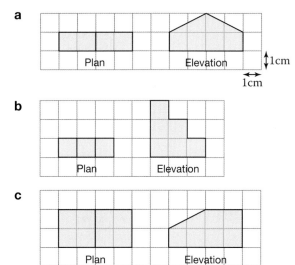

b

c

3 Marcus is planning to go camping, but requires a tent with a capacity greater than 5 m³. He can borrow the tent shown.
Does it fulfil Marcus' needs?

4 A cuboid has volume 200 cm³.

Write down three different possible dimensions of the cuboid.

5 Draw and label a sketch of a triangular prism with volume 120 cm³.

6 a Calculate the volume of this cuboid. State the units of your answer.

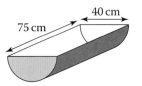

6 The cuboid is enlarged by scale factor 3.

b Calculate the dimensions of the new cuboid.

c Calculate the volume of the new cuboid.

d Copy and complete this sentence:

> For an enlargement of scale factor 3, the volume increases by multiplying by ____.

7 A block has a length and width of 22.50 mm, and a height of 3.15 mm. It has a mass of 9.50 g.

$$\text{Density} = \frac{\text{mass}}{\text{volume}}$$

a Find the density of the metal from which the block is made, giving your answer in g/cm³.

b How many blocks can be made from 1 kg of the material?

8 A steel cable weighs 2450 kg.

The cable has a uniform circular cross-section of radius 0.85 cm.

The steel from which the cable is made has a density of 7950 kg/m³.

Find the length of the cable.

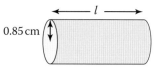

9 Water is poured into this trough at a rate of 2 litres per minute. How long will it take to fill the trough? Give your answer to a sensible degree of accuracy.

*10 Each can of beans has diameter 76 mm and height 105 mm.

a What are the dimensions of the box? State any assumptions you make. Give your answer in cm.

b Find the percentage of the volume of the box that is filled by the cans.

Q 1137, 1138, 1139, 1246   SEARCH

15.3 Volume and surface area

Pyramids and **cones** are 3D shapes with sides that taper to a point.

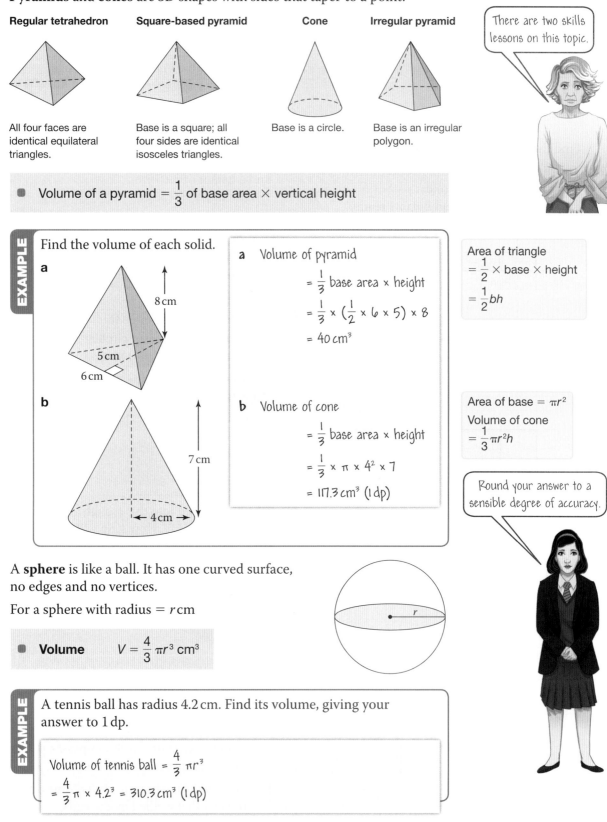

Regular tetrahedron	Square-based pyramid	Cone	Irregular pyramid

All four faces are identical equilateral triangles.

Base is a square; all four sides are identical isosceles triangles.

Base is a circle.

Base is an irregular polygon.

There are two skills lessons on this topic.

● Volume of a pyramid $= \frac{1}{3}$ of base area $\times$ vertical height

EXAMPLE

Find the volume of each solid.

a

8 cm

5 cm

6 cm

b

7 cm

4 cm

a Volume of pyramid

$= \frac{1}{3}$ base area $\times$ height

$= \frac{1}{3} \times (\frac{1}{2} \times 6 \times 5) \times 8$

$= 40 \, cm^3$

b Volume of cone

$= \frac{1}{3}$ base area $\times$ height

$= \frac{1}{3} \times \pi \times 4^2 \times 7$

$= 117.3 \, cm^3$ (1 dp)

Area of triangle
$= \frac{1}{2} \times$ base $\times$ height
$= \frac{1}{2} bh$

Area of base $= \pi r^2$

Volume of cone
$= \frac{1}{3} \pi r^2 h$

Round your answer to a sensible degree of accuracy.

A **sphere** is like a ball. It has one curved surface, no edges and no vertices.

For a sphere with radius $= r$ cm

r

● **Volume** $V = \frac{4}{3} \pi r^3 \, cm^3$

EXAMPLE

A tennis ball has radius 4.2 cm. Find its volume, giving your answer to 1 dp.

Volume of tennis ball $= \frac{4}{3} \pi r^3$

$= \frac{4}{3} \pi \times 4.2^3 = 310.3 \, cm^3$ (1 dp)

Exercise 15.3S1

1 Work out the value for each expression when $a = 2$, $b = 6$ and $c = 15$.

    a $a \times b \div 3$ **b** $\frac{1}{3} \times a \times c$

    c $4 \times c \div 3$ **d** $\frac{4}{3} \times b^2$

2 Calculate the areas of these rectangles.

 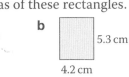

    a   4 cm, 7 cm   **b**   5.3 cm, 4.2 cm

3 Find the areas of these triangles.

 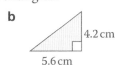

    a   2 cm, 7.5 cm   **b**   4.2 cm, 5.6 cm

4 Find the area of the parallelogram.

    2.5 cm, 6 cm

5 Find the area of the trapezium.

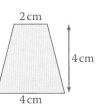

    2 cm, 4 cm, 4 cm

6 Calculate the areas of these circles.

    a   radius = 7 cm   **b**   radius = 5 cm

7 Use the formula:

> volume of a pyramid
> $= \frac{1}{3}$ of base area $\times$ vertical height

to calculate the volumes of pyramids with these base areas and vertical heights.

Include the correct units in your answer.

	base area	vertical height
a	18 cm²	8 cm
b	25 m²	9 m
c	14 cm²	12 cm
d	18 mm²	8 mm
e	2.4 cm²	1.6 cm
f	3.5 m²	1.44 m

8 Find the volume of each solid.

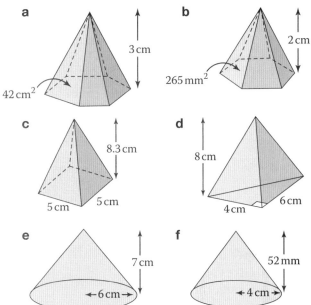

    a   3 cm, 42 cm²

    b   2 cm, 265 mm²

    c   8.3 cm, 5 cm, 5 cm

    d   8 cm, 4 cm, 6 cm

    e   7 cm, 6 cm

    f   52 mm, 4 cm

9 Use your calculator to work out each calculation.

Give your answer to 1 decimal place.

    a $\pi \times 4$ **b** $\pi \times 5^2$

    c $\frac{4}{3} \times \pi \times 9$ **d** $\frac{4}{3} \times \pi \times 49$

    e $\frac{4}{3} \times \pi \times 10^2$ **f** $\frac{4}{3} \times \pi \times 12^2$

    g $\frac{4}{3} \times \pi \times 2.6^2$ **h** $\frac{4}{3} \times \pi \times 4.7^2$

10 Use the formula:

> volume of a sphere $= \frac{4}{3}\pi r^3$

to calculate the volume of each sphere.

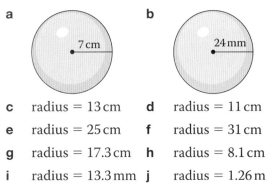

    a   7 cm   **b**   24 mm

    c   radius = 13 cm   **d**   radius = 11 cm

    e   radius = 25 cm   **f**   radius = 31 cm

    g   radius = 17.3 cm   **h**   radius = 8.1 cm

    i   radius = 13.3 mm   **j**   radius = 1.26 m

 Q 1107, 1122, 1136   SEARCH

15.3 Volume and surface area

Surface area is the total area of all the surfaces of a 3D solid.

- For 3D solids with flat surfaces, you work out the area of each surface and add them together.

3D solids such as cylinders and cones have curved surfaces.

The curved face of a **cylinder** opens out to make a rectangle.
For a cylinder with radius $= r$ and height $= h$

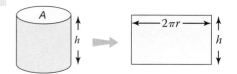

- Surface area of cylinder $= 2 \times$ area of circle $+$ area of curved face
$$= 2\pi r^2 + 2\pi rh.$$

Find the volume and surface area of this cylinder.
Give your answers to 3 significant figures.

3 cm

7 cm

Area of circle $= \pi \times 3^2 = 28.274 \ldots \text{cm}^2$
Volume $= 28.274 \ldots \times 7 = 198 \text{ cm}^3$ (3 sf)
Surface area $= 2 \times \pi \times 3^2 + 2 \times \pi \times 3 \times 7$
$= 188.495 \ldots \text{cm}^2 = 188 \text{ cm}^2$ (3 sf)

A **cone** is a solid with a circular base.

For a cone with radius $= r$ and slant height $= l$

- Curved surface area of cone $= \pi rl$

- Surface area of cone $=$ curved surface area $+$ area of base
$$= \pi rl + \pi r^2$$

Find the total surface area of the cone.
Give your answer to 3 significant figures.

Surface area $= \pi \times 5 \times 8 + \pi \times 5^2$
$= 204.203 \ldots \text{cm}^2 = 204 \text{ cm}^2$

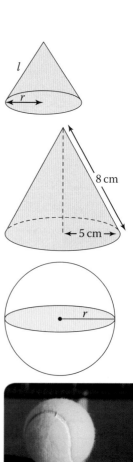

8 cm

5 cm

For a sphere with radius $= r$ cm

- Surface area of a sphere $= 4\pi r^2 \text{ cm}^2$

A tennis ball has radius 4.2 cm. Find its surface area, giving your answer to 1 dp.

Surface area $= 4\pi r^2$
$= 4\pi \times 4.2^2 = 221.7 \text{cm}^2$ (1 dp)

Geometry Working in 3D

Exercise 15.3S2

1 Calculate the surface areas of these cuboids.

a

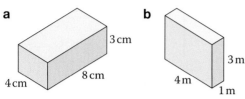

3 cm

8 cm

4 cm

b

3 m

4 m

1 m

2 Use the formula:

> Curved surface area of a cylinder = $2\pi rh$

to find the curved surface area of cylinders with radius (r) and height (h).

Give your answers to 1 decimal place and include the correct units.

a radius = 4 cm, height = 9 cm

b radius = 5 cm, height = 7.5 cm

c $r = 16$ cm, $h = 8$ cm

d $r = 11.3$ cm, $h = 26$ cm

e $r = 13.9$ cm, $h = 2.9$ cm

f $r = 13$ mm, $h = 2.36$ mm

3 Use the formula:

> Total surface area of a cylinder = $2\pi r^2 + 2\pi rh$

to find the total surface area of each cylinder in question **2**.

Give your answers to 1 decimal place and include the correct units.

4 Find the surface area of each cylinder. Give your answers to 3 significant figures.

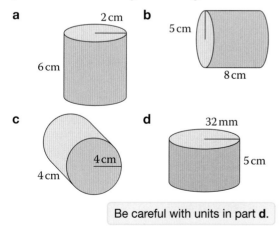

a 2 cm

6 cm

b 5 cm

8 cm

c 4 cm

4 cm

d 32 mm

5 cm

> Be careful with units in part **d**.

5 Use the formula:

> Curved surface area of cone = πrl

to find the curved surface area of cones with radius (r) and slant height (l).

Give your answers to 1 decimal place and include the correct units.

a $r = 5$ cm, $l = 8.2$ cm

b $r = 3$ cm, $l = 6$ cm

c $r = 45$ mm, $l = 30$ mm

d $r = 2.5$ cm, $l = 30$ mm

e $r = 6.7$ cm, $l = 10.5$ cm

f $r = 135$ mm, $l = 18.5$ cm

6 Use the formula:

> Total surface area of cone = $\pi rl + \pi r^2$

to find the total surface area of each cone in question **5**.

Give your answers to 1 decimal place and include the correct units.

7 Find the total surface area of each cone.

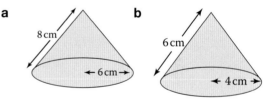

a 8 cm

← 6 cm →

b 6 cm

← 4 cm →

8 Use the formula:

> Surface area of a sphere = $4\pi r^2$

to find the surface areas of these spheres.

Give your answers to 1 decimal place and include the correct units.

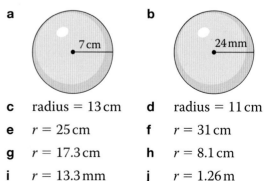

a 7 cm

b 24 mm

c radius = 13 cm

d radius = 11 cm

e $r = 25$ cm

f $r = 31$ cm

g $r = 17.3$ cm

h $r = 8.1$ cm

i $r = 13.3$ mm

j $r = 1.26$ m

15.3 Volume and surface area

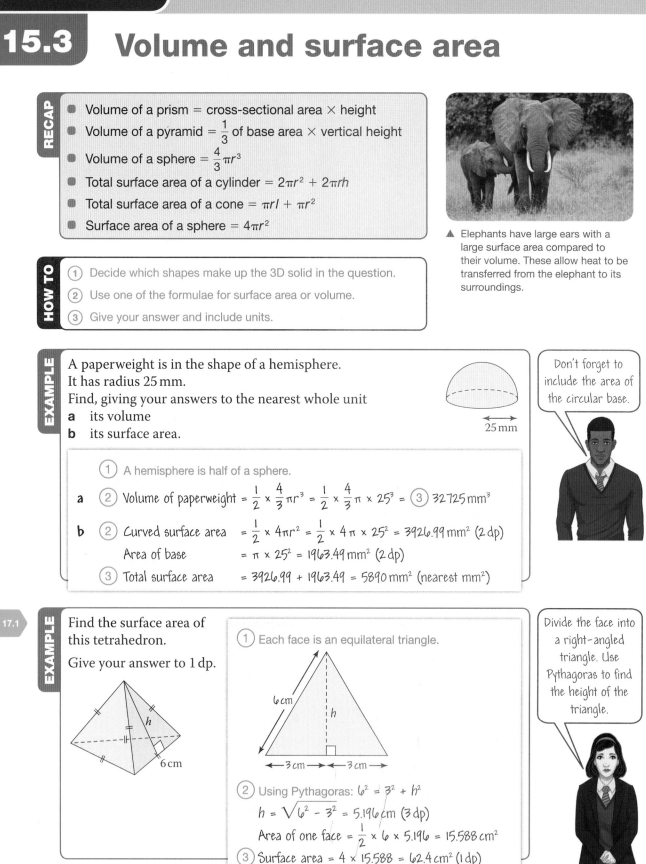

RECAP

- Volume of a prism = cross-sectional area × height
- Volume of a pyramid = $\frac{1}{3}$ of base area × vertical height
- Volume of a sphere = $\frac{4}{3}\pi r^3$
- Total surface area of a cylinder = $2\pi r^2 + 2\pi rh$
- Total surface area of a cone = $\pi rl + \pi r^2$
- Surface area of a sphere = $4\pi r^2$

▲ Elephants have large ears with a large surface area compared to their volume. These allow heat to be transferred from the elephant to its surroundings.

HOW TO

(1) Decide which shapes make up the 3D solid in the question.

(2) Use one of the formulae for surface area or volume.

(3) Give your answer and include units.

EXAMPLE

A paperweight is in the shape of a hemisphere.
It has radius 25 mm.
Find, giving your answers to the nearest whole unit
a its volume
b its surface area.

25 mm

Don't forget to include the area of the circular base.

(1) A hemisphere is half of a sphere.

a (2) Volume of paperweight = $\frac{1}{2} \times \frac{4}{3}\pi r^3 = \frac{1}{2} \times \frac{4}{3}\pi \times 25^3 =$ (3) $32\,725\,\text{mm}^3$

b (2) Curved surface area = $\frac{1}{2} \times 4\pi r^2 = \frac{1}{2} \times 4\pi \times 25^2 = 3926.99\,\text{mm}^2$ (2 dp)

Area of base = $\pi \times 25^2 = 1963.49\,\text{mm}^2$ (2 dp)

(3) Total surface area = $3926.99 + 1963.49 = 5890\,\text{mm}^2$ (nearest mm²)

17.1

EXAMPLE

Find the surface area of this tetrahedron.

Give your answer to 1 dp.

6 cm

h

6 cm

(1) Each face is an equilateral triangle.

6 cm

h

←3 cm→←3 cm→

Divide the face into a right-angled triangle. Use Pythagoras to find the height of the triangle.

(2) Using Pythagoras: $6^2 = 3^2 + h^2$

$h = \sqrt{6^2 - 3^2} = 5.196\,\text{cm}$ (3 dp)

Area of one face = $\frac{1}{2} \times 6 \times 5.196 = 15.588\,\text{cm}^2$

(3) Surface area = $4 \times 15.588 = 62.4\,\text{cm}^2$ (1 dp)

Exercise 15.3A

1 Calculate the surface areas of these shapes.

a

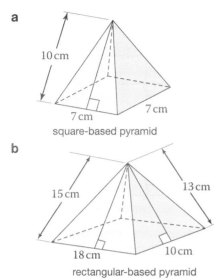

square-based pyramid

b

15 cm 13 cm

18 cm 10 cm

rectangular-based pyramid

2 James is going to fill a paper cone with sweets. He can choose between cone X and cone Y. Cone X has top radius 4 cm and height 8 cm. Cone Y has top radius 8 cm and height 4 cm.

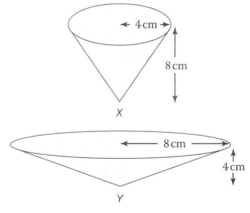

a Which cone has the greater volume?

b What is the difference in volume between cone X and cone Y?

3 Laura made a model rocket from a cylinder with height 6 cm and a cone with vertical height 2.4 cm.

The radius of the cylinder and the cone is 1.8 cm.

Find the volume of the rocket.

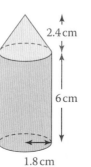

4 Find the surface area of each 3D solid.

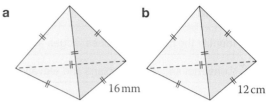

a

16 mm

b

12 cm

5 Rachel bought a cylindrical tube containing three power balls. Each ball is a sphere of radius 5 cm. The balls touch the sides of the tube. The balls touch the top and bottom of the tube.

Find the percentage volume of the tube that is filled by the power balls.

5 cm

6 **a** Which has the greater volume, a sphere with diameter 3 cm or a cube with side length 3 cm?

b Which has the greater surface area, a sphere with diameter 3 cm or a cube with side length 3 cm?

7 **a** A cube has sides 4 cm long. Find the surface area and the volume of the cube.

b A second cube has sides twice as long. Find the surface area and volume of this cube.

c Compare the surface areas and volumes of the two cubes. How many times larger is

i the surface area

ii the volume of the second cube?

8 **a** Sketch a net of this pyramid and work out its surface area.

b Find the dimensions of the smallest rectangular card that you could use for your net.

c Work out the percentage of the card that is wasted.

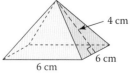

4 cm

6 cm 6 cm

Q 1107, 1122, 1136   SEARCH

Summary

Checkout

You should now be able to...

✔ Identify the numbers of faces, edges and vertices of 3D shapes.	1 – 2
✔ Construct and interpret plans and elevations of 3D shapes.	3 – 4
✔ Calculate the volume and surface area of cuboids, cylinders and other prisms.	5 – 6
✔ Calculate the volume and surface area of spheres, pyramids, cones and composite solids.	7 – 8

Language | Meaning | Example

Language	Meaning	Example
Face	A flat surface of a solid.	vertex, face, edge
Edge (solid)	A line along which two faces meet.	
Vertex **Vertices (plural)**	A point at which two or more edges meet.	
Cube	A 3D solid with 6 identical square faces, 12 equal edges and 8 vertices.	Cube, Cuboid
Cuboid	A 3D solid with 6 rectangular faces. A cuboid is a rectangular prism.	
Prism	A 3D solid with a constant cross-section.	Cylinder, triangular prism.
Pyramid	A 3D solid with a polygon as its base. All the other faces are triangular in shape and meet at a single vertex.	Tetrahedron, square-based pyramid, irregular pyramid.
Net	A 2D shape that can be folded to make a 3D solid.	
Plan	A 2D representation of an object as seen from above.	Plan
Elevation	A 2D representation of an object as seen from the front or side.	Front Side
Surface area	The total area of all the faces of a 3D solid.	Surface area $= 6 \times 4^2 = 96\,\text{cm}^2$
Volume	The amount of space occupied by, or inside, a 3D shape.	Volume $= 4^3 = 64\,\text{cm}^3$ 4 cm, 4 cm, 4 cm
Cylinder	A prism with a circular cross-section.	Cylinder Cone Sphere
Cone	A solid with a circular base and one vertex.	
Sphere	A solid with every point on its surface the same distance from the centre.	

Review

1 How many faces, edges and vertices are there on

   a a cuboid

   b a square-based pyramid?

2 **i** What are the names of these 3D shapes?

   ii How many faces, edges and vertices does each have?

   a   **b**

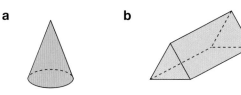

3

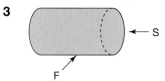

Draw

   a the front elevation (*F*),

   b the side elevation (*S*) of the cylinder.

4 Here are three elevations of a 3D solid made from cubes.

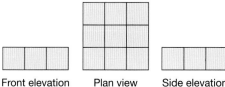

Front elevation   Plan view   Side elevation

Draw the 3D solid these views come from.

5 A cuboid has dimensions 4 m, 2 m and 1.5 m.

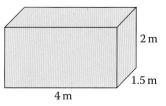

Calculate

   a the volume

   b the surface area of the cuboid.

The cuboid is made from a material with a density of 2.5 kg/m³.

   c What is the mass of the cuboid?

6 A cylinder has a radius of 11 cm and a length of 20 cm. Give your answers to 3 significant figures. Calculate

   a the volume

   b the curved surface area

   c the total surface area assuming the cylinder is closed at both ends.

7 Use the formula $SA = \pi rl + \pi r^2$ to find the surface area of this cone. Give your answer to 3 significant figures.

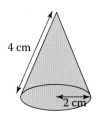

8 Use the formula $V = \frac{4}{3}\pi r^3$ to find the volume of this sphere. Give your answer to 3 significant figures.

What next?

Score			
	0 – 3		Your knowledge of this topic is still developing. To improve look at MyMaths: 1078, 1098, 1106, 1107, 1122, 1136, 1137, 1138, 1139, 1246
	4 – 7		You are gaining a secure knowledge of this topic. To improve look at InvisiPens: 15Sa – i
	8		You have mastered these skills. Well done you are ready to progress! To develop your exam technique look at InvisiPens: 15Aa – g

Assessment 15

1   Jemima matches these mathematical names to the solids shown.
    Correct any mistakes Jemima has made.   [3]

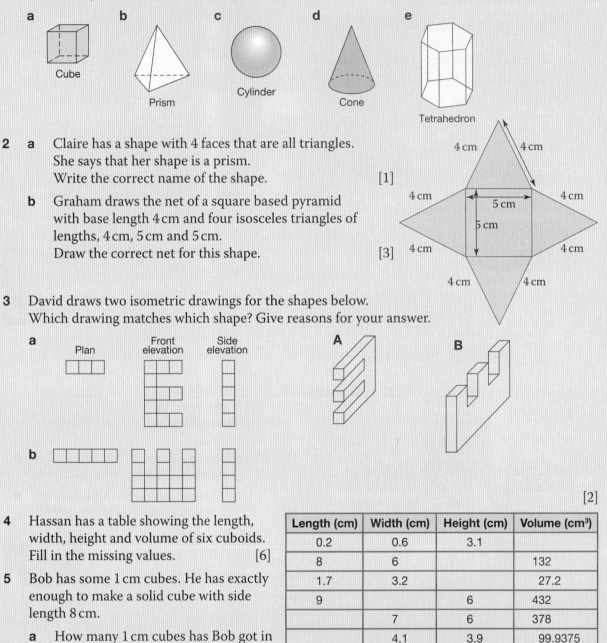

a Cube   **b** Prism   **c** Cylinder   **d** Cone   **e** Tetrahedron

2   **a** Claire has a shape with 4 faces that are all triangles.
          She says that her shape is a prism.
          Write the correct name of the shape.   [1]

    b Graham draws the net of a square based pyramid
          with base length 4 cm and four isosceles triangles of
          lengths, 4 cm, 5 cm and 5 cm.
          Draw the correct net for this shape.   [3]

3   David draws two isometric drawings for the shapes below.
    Which drawing matches which shape? Give reasons for your answer.

    a Plan   Front elevation   Side elevation   **A**   **B**

    b   [2]

4   Hassan has a table showing the length,
    width, height and volume of six cuboids.
    Fill in the missing values.   [6]

Length (cm)	Width (cm)	Height (cm)	Volume (cm³)
0.2	0.6	3.1	
8	6		132
1.7	3.2		27.2
9		6	432
	7	6	378
	4.1	3.9	99.9375

5   Bob has some 1 cm cubes. He has exactly
    enough to make a solid cube with side
    length 8 cm.

    a How many 1 cm cubes has Bob got in
          total?   [1]

    b How many cubes with side length 2 cm can he make?   [2]

    c How many cubes with side length 4 cm can he make?   [2]

    d **i**   How many cubes with side length 3 cm can he make?   [2]

        ii How many 1 cm cubes are left over?   [1]

6 A cuboid is made from 60 interlocking 1 cm cubes. Write down the dimensions of all possible cuboids that can be constructed from all of these cubes. [6]

7 Freya has a cubic metre. Each side is 1 m long. For each of Freya's statement, decide if she is correct. Give reasons for your answers and correct any of her mistakes.

   a The volume of the cube is 1 cubic metre. [1]

   b The volume of the cube is 1000 cubic cm. [1]

   c The volume of the cube is 10 000 cubic mm. [1]

8 Liz and Sue run a jewellery shop. Sue has cuboid storage boxes, 24 cm × 30 cm × 36 cm.

   a Liz has some earrings packed in small boxes with dimensions 6 cm × 5 cm × 2 cm. How many of Liz's boxes will fit into Sue's? [3]

   b Sue says that the ratio of her surface area to Liz's surface area is the same as the ratio of the volumes. Is Sue correct? Give reasons for your answer. [7]

9 A waste paper basket is a cylinder of radius 14 cm and height 36 cm.

   a Calculate its volume. [2]

   b Find the total surface area of the basket in m². [4]

10 A slab of cheese is in the shape of a prism with a right angled triangular cross section. The triangle has sides of 1.5, 3.6 and 3.9 inches. The length of the prism is 5 inches.

   Calculate its   **a** total surface area   [4]   **b** volume. [2]

11 A child's ice cream cornet is in the shape of a cone of height 12 cm with a circular top of outer radius 4 cm. The inner radius of the top is 3.9 cm. Work out the volume of the biscuit that makes up the cone. [6]

12 A pyramid made of gold has a rectangular base. The dimensions of the base are 55 cm × 40 cm. The perpendicular height of the pyramid is 65 cm.

   a Calculate its volume. [3]

   b The gold is melted down and recast as a number of spheres with diameter 10 cm. How many complete spheres can be made? [4]

13 A tube is 8 cm long, with an outer radius of 0.5 cm and a central hole of diameter 0.5 cm.

   a Calculate the volume of the tube. [3]

   b Find the mass of the tube if its density is 2.6 g/cm³. [2]

14 'Our Plaice' fish and chip shop sells chips in two sizes.
'Slim Jims' are 13.5 cm long with a square cross section of side 3 mm.
'Chunky Dunkies' are 6 cm long with a square cross section of side 4.5 mm.

   a Does each chip have the same volume? [2]

   b Healthier chips have a smaller surface area so they absorb less fat. Which chips are healthier to eat? [4]

15 A child's toy consists of a hemisphere and a cone with the same radius, as shown. Calculate

   a the volume [6]

   b the surface area of the toy. [6]

5 cm   4 cm   6 cm

Life skills 3: Getting ready

Abigail, Mike, Juliet and Raheem are now ready to start making the building suitable for their restaurant. They have also chosen some suppliers to work with, and are considering what stocks of non-perishable food and drink they will need. They analyse the data that they have gathered from their market research to help set their menu and price list.

Task 1 – The ingredients

One of the starters on the menu is an old family recipe that Mike borrowed from his grandmother: *Artichokes with anchovies*.

The friends are keen to avoid waste, and want to use ingredients efficiently. Each tin of anchovy fillets contains 24 anchovy fillets.

Each jar of artichoke hearts contains 20 artichoke hearts.

How many portions of *Artichokes with anchovies* will they have to make in order to use up complete tins of anchovy fillets and complete jars of artichoke hearts with no leftovers?

Artichokes with anchovies – serves 1

2 artichoke hearts
3 anchovy fillets
1 tablespoon of olive oil
8 tablespoons of butter, softened
2 garlic cloves
1 pinch each, salt and pepper
Juice of 1 lemon

Task 2 – Price list

p.116

They decide to look again at the data they collected from their small pilot survey (*Life skills 1: The business plan*).

a On graph paper, plot separate male and female scatter graphs of amount prepared to spend against age.

b For each of these graphs, describe the correlation and explain what it suggests.

c Should they use these graphs alone to decide a specific age group to target for their restaurant? Why?

Task 3 – Supplier meeting

Raheem drives from the restaurant to the vegetable suppliers, 9 miles away. Each friend keeps a log of all business-related travel and time spent in meetings so that they can claim back costs on expenses and keep a record of how much time they are investing.

a Draw a distance-time graph showing Raheem's journey.

b At what time does he arrive back at the restaurant?

Veggie-r-us supplier meeting 3, journey log

Total distance to supplier, 9 miles.

1. Departed restaurant at 1 pm.

2. 6 minute drive down high street (2 miles).

3. 7 miles at 60 mph (miles per hour) on motorway.

4. Reach supplier, stay there for 10 minutes.

5. 3 miles on motorway at 60 mph.

6. Stop at motorway services (5 minutes).

7. 4 miles on motorway at 40 mph (bad traffic).

8. 2 miles through town (4 minutes).

Task 4 – Maintenance

They need to hire a plumber to fix some pipes in the restaurant.

There are two local plumbers who have been recommended by friends.

a On the same axes, draw graphs showing the cost (excluding parts) of hiring each plumber against the time taken.

b For what range of times is Bill cheaper than Alan?

Both quote that the job will take about 6.5 hours.

c Who should they get to do the job? Explain your reasons.

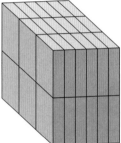

4.4 cm

14.2 cm

Carton

7.3 cm

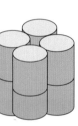

3.7 cm

Can

▲ Top: dimensions of one carton and one can.
Bottom: stacked arrangement of cartons and an example of how cans are able to stack.

Task 5 – Stock keeping

The friends need to decide between buying cartons or cans of tomatoes.

The cartons are stacked in 2 layers in a box. Each layer is 6 cartons deep and 3 cartons across.

a Find the volume of a box. Add an extra 5% to the volume for packaging.

The boxes are stored in a cuboid cupboard with dimensions:

70 cm deep, 115 cm wide and 180 cm high.

b Calculate the volume of the cupboard in cm^3 and m^3.

c What percentage of the space in the cupboard do 30 cardboard boxes take up?

d Find the surface area of one carton.

The cans are cylindrical with a radius of 3.7 cm.

e Find the area of the circular base of the can.

f One can has the same volume as a carton. Find the height of a can.

g The cans can be stacked as shown in the diagram. Work out how many cans could fit in the same space taken up by one cardboard box.

h If the cans and cartons cost the same which would you buy? Give your reasons.

Annual amount spent eating out (£)	Number of people
0–400	36
Over 400–800	59
Over 800–1200	61
Over 1200–1600	24
Over 1600–2000	20

▲ Annual spend on eating out in Newton-Maxwell.

Task 6 – Repeat business

The table shows some results from the larger survey.

a Calculate an estimate of the mean annual amount spent eating out.

b Draw a histogram showing the data.

c Does you answer for the mean agree with what your histogram shows? Give your reasons.

Juliet assumes that a given person spend the same amount at each of the four local restaurants.

d If Juliet is right, how much should the friends expect the average person to spend at their restaurant in a year?

16 Handling data 2

Introduction

In the UK, literacy is almost taken for granted. By the time people reach adulthood, they can generally read and write. In many other countries, this is not the case. Statistics show that there is a correlation, or link, between the wealth per person of a country (as measured in 'GDP per capita') and the adult literacy rate. In 2013, the UK GDP and literacy rate $36 000 and 99% respectively; by comparison, in Sierra Leone they were $2000 and 35%.

The branch of statistics that deals with the relationship between variables is called correlation.

What's the point?

Understanding the relationship between quantities helps us to make informed decisions on a global scale. Literacy problems will not be resolved effectively unless poverty is also tackled.

Objectives

By the end of this chapter you will have learned how to …

- Interpret and construct tables, graphs and charts for discrete, continuous and grouped data.
- Use the median, mean, modal class and range to interpret and compare distributions.
- Use correlation to describe scatter graphs but know that it does not imply causation.
- Draw estimated lines of best fit and make predictions but understand their limitations.
- Interpret and construct line graphs for time series data.

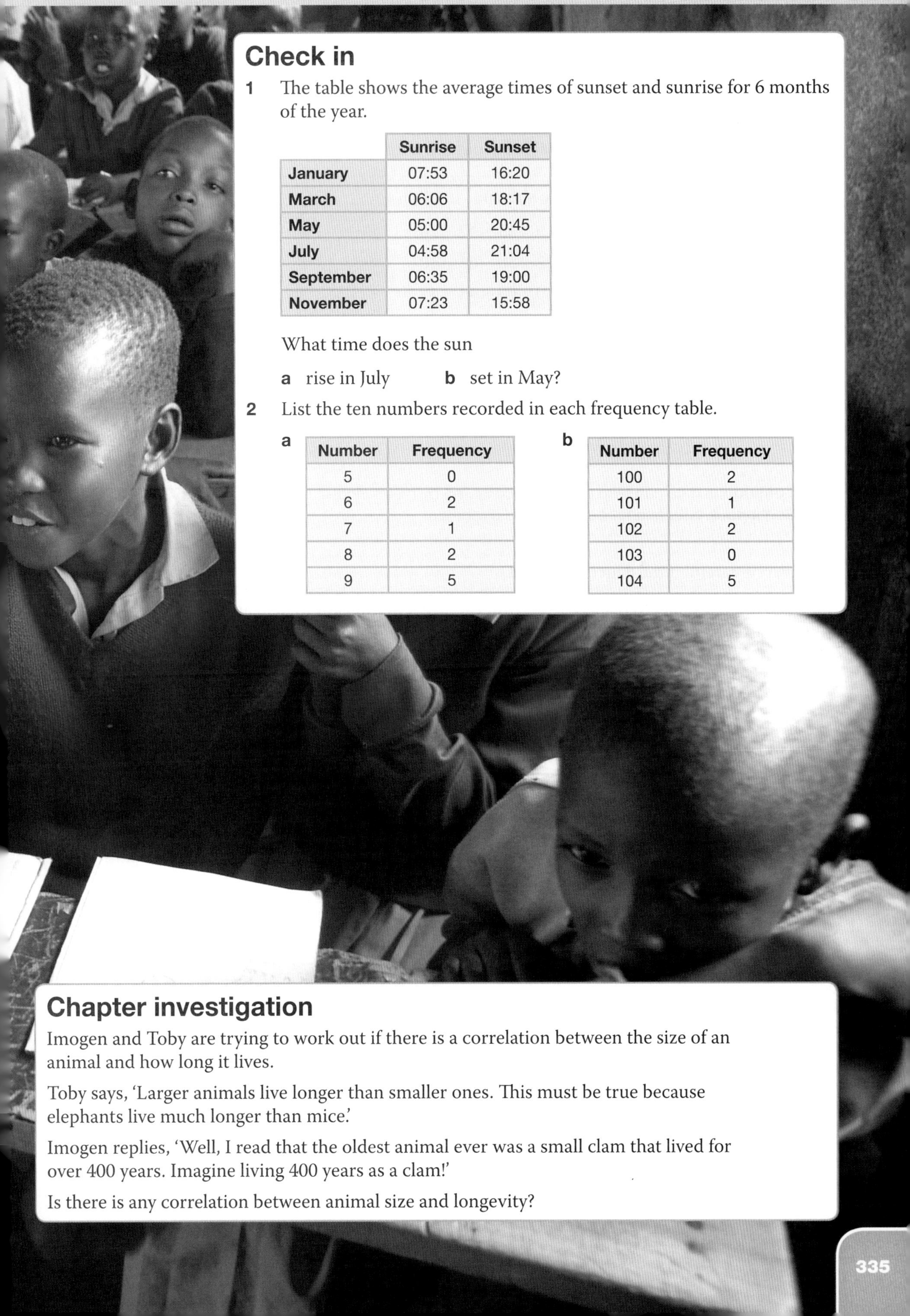

Check in

1 The table shows the average times of sunset and sunrise for 6 months of the year.

	Sunrise	Sunset
January	07:53	16:20
March	06:06	18:17
May	05:00	20:45
July	04:58	21:04
September	06:35	19:00
November	07:23	15:58

What time does the sun

a rise in July   b set in May?

2 List the ten numbers recorded in each frequency table.

a

Number	Frequency
5	0
6	2
7	1
8	2
9	5

b

Number	Frequency
100	2
101	1
102	2
103	0
104	5

Chapter investigation

Imogen and Toby are trying to work out if there is a correlation between the size of an animal and how long it lives.

Toby says, 'Larger animals live longer than smaller ones. This must be true because elephants live much longer than mice.'

Imogen replies, 'Well, I read that the oldest animal ever was a small clam that lived for over 400 years. Imagine living 400 years as a clam!'

Is there is any correlation between animal size and longevity?

Some surveys produce data with many different values.

You can **group** data into **class intervals** to avoid too many categories.

EXAMPLE

The exam marks of class 10A are shown:

35	47	63	25	31	8	19	55	47	14
24	36	56	61	15	43	22	50	66	10
36	45	18	20	53	31	40	60	44	47

Complete a **grouped frequency table**.

Mark	Tally	Frequency
1–20	ⅢⅡ	7
21–40	Ⅲ Ⅲ	9
41–60	Ⅲ Ⅲ Ⅰ	11
61–80	Ⅲ	3

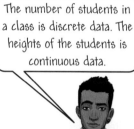

Ⅲ = 5

Check that the frequencies add to 30.

- **Discrete** data can only take exact values (usually collected by counting).

- **Continuous** data can take any value (collected by measuring). Continuous data cannot be measured exactly.

> The number of students in a class is discrete data. The heights of the students is continuous data.

You must be careful if the data is **grouped**.

You can use a **bar chart** to display grouped discrete data.

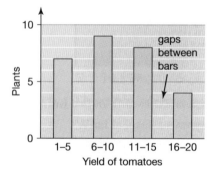

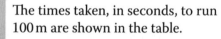

You can use **a histogram** to display grouped continuous data.

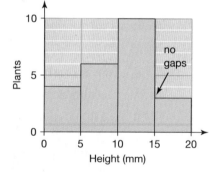

EXAMPLE

The times taken, in seconds, to run 100 m are shown in the table.

Time (seconds)	Number of people
$0 < t \leqslant 10$	0
$10 < t \leqslant 20$	4
$20 < t \leqslant 30$	6
$30 < t \leqslant 40$	3

Draw a frequency diagram to illustrate this information.

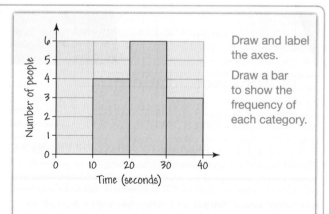

Draw and label the axes.

Draw a bar to show the frequency of each category.

Exercise 16.1S

1 a Copy and complete the frequency table using these test marks.

8 14 21   4 15 22 25 24 15 11

10 17 24 20 13 16 12   9   3 14

20 10 16 15   7 23 23 14 15 16

8   2   9 19 12 10 10 20 13 13

15 17 11 14 19 20 23 23 24   5

Test mark	Tally	Frequency
1–5		
6–10		
11–15		
16–20		
21–25		

b Calculate the number of people who took the test.

2 a Copy and complete the frequency table using these weights of people, given to the nearest kilogram.

48 63 73 55 59 61 70 63 58 67

46 45 57 58 63 71 60 47 49 51

53 61 68 65 70 60 52 59 50 49

48 47 63 61 58 71 53 51 60 70

Weight (kg)	Tally	Number of people
$45 \leqslant w < 50$		
$50 \leqslant w < 55$		
$55 \leqslant w < 60$		
$60 \leqslant w < 65$		
$65 \leqslant w < 70$		
$70 \leqslant w < 75$		

b Calculate the number of people in the sample.

c Draw a histogram to show the data.

d Copy and complete the frequency table.

Weight (kg)	Tally	Number of people
$45 \leqslant w < 55$		
$55 \leqslant w < 65$		
$65 \leqslant w < 75$		

e Draw a grouped bar chart to show the data.

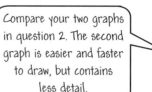
Compare your two graphs in question 2. The second graph is easier and faster to draw, but contains less detail.

3 a Copy and complete the frequency table using these heights of people.

153 134 155 142 140 163 150 135

170 156 171 161 141 153 144 163

140 160 172 157 136 160 134 154

176 154 173 179 160 152 170 148

151 165 138 143 147 144 156 139

Height (cm)	Tally	Number of people
$130 < h \leqslant 140$		
$140 < h \leqslant 150$		
$150 < h \leqslant 160$		
$160 < h \leqslant 170$		
$170 < h \leqslant 180$		

b Draw a histogram to show the data.

4 The depth, in centimetres, of a reservoir is measured daily throughout April.

Draw a histogram to show the depths.

Depth (cm)	Number of days
$0 < d \leqslant 5$	1
$5 < d \leqslant 10$	5
$10 < d \leqslant 15$	14
$15 < d \leqslant 20$	8
$20 < d \leqslant 25$	2

5 The exam marks of 36 students are shown in the frequency table.

Draw a bar chart to show the exam marks.

Exam mark (%)	Number of students
1 to 20	4
21 to 40	8
41 to 60	13
61 to 80	6
81 to 100	5

Q 1193, 1196   SEARCH

16.1 Frequency diagrams

- You can group data into class intervals when there are many values.
- Discrete data can only take exact values (usually collected by counting), for example the number of students in a class.
- Continuous data can take any value (collected by measuring), for example the heights of the students in a class.
- You can use a bar chart to display grouped discrete data.
- You can use a grouped bar chart to display continuous data.

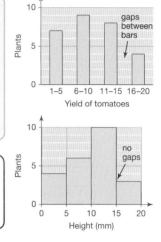

HOW TO

① Read information from the diagram.
Make sure that you read the scale carefully.

② Use the information from the diagram to answer the question.

EXAMPLE

The table shows some information about the times taken by a bus to travel from Cookham to Marlow.

Time (minutes)	Frequency
$6 \leqslant x < 8$	
$8 \leqslant x < 10$	8
$10 \leqslant x < 12$	
$12 \leqslant x < 14$	7
$14 \leqslant x < 16$	2

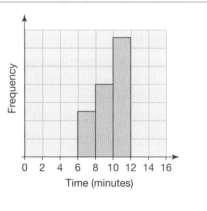

a Copy and complete the histogram and the table.
Include a scale on the vertical axis of the histogram.

b The bus company claims that

Half of our journeys from Cookham to Marlow take less than 10 minutes!

Is the bus company correct?

① The $8 \leqslant x < 10$ category has frequency 8, so the scale on the axis goes up in twos.

② Complete the table and the graph.

a

Time (minutes)	Frequency
$6 \leqslant x < 8$	5
$8 \leqslant x < 10$	8
$10 \leqslant x < 12$	13
$12 \leqslant x < 14$	7
$14 \leqslant x < 16$	2
Total	35

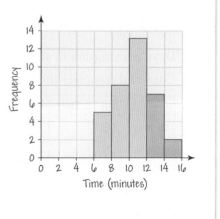

b Add up the frequencies to find the total number of buses.

Less than 10 minutes $= \dfrac{13}{35} < \dfrac{1}{2}$

The bus company is wrong.

Exercise 16.1A

1 The histogram shows the best distances, in metres, that athletes threw a javelin in a competition.

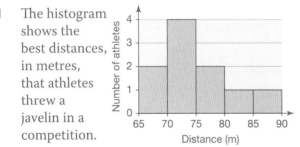

a State the number of athletes who threw the javelin

  i between 65 and 70 metres

  ii between 80 and 85 metres.

b In which class interval was the winner?

c Calculate the total number of athletes who threw a javelin.

2 A garden centre gave a discount to its staff. It kept a record of how much each staff member spent during one month. The values are recorded on the histogram.

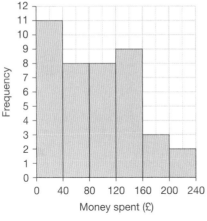

a Is it possible for the actual range of money spent to be

  i £159

  ii £160

  iii £161?

  Explain how you worked out your answers.

b How many people work at the garden centre in total?

c A person at the garden centre is chosen at random. What is the probability that they spent more than £120 in one month?

3 Ali surveyed 40 students to find out how much time they spent doing homework in the past two weeks.

Find the value of x and complete the frequency table and the histogram.

Time spent doing homework, t, hours	Frequency
$0 \leq t < 5$	
$5 \leq t < 10$	x
$10 \leq t < 15$	
$15 \leq t < 20$	13
$20 \leq t < 25$	$2x$

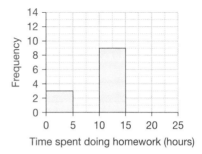

4 David carries out a survey to find the distance travelled to work by the staff at a school. His results are shown on the histogram.

Distance travelled to work for all school staff

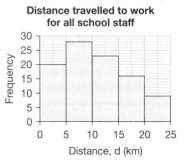

David kept his data for the teachers, but lost the data for the non-teachers he surveyed.

Distance, d (km)	Teachers	Non-teachers
$0 \leq d < 5$	13	
$5 \leq d < 10$	17	
$10 \leq d < 15$	14	
$15 \leq d < 20$	11	
$20 \leq d < 25$	5	

Complete the frequency table to show the data for the non-teachers at the school.

 1193, 1196 SEARCH

16.2 Averages and spread 2

- You can put large amounts of continuous data into a **grouped frequency table**.

A grouped frequency table does not tell you the actual data values so you can only find estimates of the averages.

- You use **estimates** of averages to summarise the data.

> - For grouped data in a frequency table, you can calculate
>   - the **estimated mean**
>   - the **modal class**
>   - the **class interval** in which the median lies.

> The modal class is the class with the greatest frequency.

EXAMPLE

The table shows the time taken, to the nearest minute, by a group of students to solve a crossword puzzle.

Time, t, minutes	Frequency
$5 < t \leqslant 10$	2
$10 < t \leqslant 15$	14
$15 < t \leqslant 20$	13
$20 < t \leqslant 25$	6
$25 < t \leqslant 30$	1

For these data, work out

a the **modal class**

b the class containing the **median**

c an estimate for the **mean**.

a Modal class is $10 < t \leqslant 15$

b Class containing median is $15 < t \leqslant 20$

c

Time t minutes	Frequency	Midpoint	Midpoint × frequency
$5 < t \leqslant 10$	2	7.5	7.5 × 2 = 15
$10 < t \leqslant 15$	14	12.5	12.5 × 14 = 175
$15 < t \leqslant 20$	13	17.5	17.5 × 13 = 227.5
$20 < t \leqslant 25$	6	22.5	22.5 × 6 = 135
$25 < t \leqslant 30$	1	27.5	27.5 × 1 = 27.5
Total	36		580

Total number of students

Total time

Estimated mean = $\dfrac{\text{Estimated total time}}{\text{Total number of students}}$

Estimated mean = $\dfrac{580}{36}$ = 16.1 (1 dp)

Put two extra columns in the table.

Find the midpoint of each interval.

Find the totals of the Frequency and Midpoint × frequency columns.

Exercise 16.2S

1 The weights, to the nearest kilogram, of 25 men are shown.

69	82	75	66	72
73	79	70	74	68
84	63	69	88	81
73	86	71	74	67
80	86	68	71	75

Note:
You can use a scientific calculator to work out the mean of grouped data. You should find out how to do this on your calculator.

a Copy and complete the frequency table.

Weight (kg)	Tally	Number of men
60 to 64		
65 to 69		
70 to 74		
75 to 79		
80 to 84		
85 to 89		

b State the modal class.

c Find the class interval in which the median lies.

2 The speeds of 10 cars in a 20 mph zone are shown in the frequency table.

Speed (mph)	Mid-value	Number of cars	Mid-value × Number of cars
11 to 15		1	
16 to 20		6	
21 to 25		2	
26 to 30		1	

a Calculate the number of cars that are breaking the speed limit.

b Copy the frequency table and calculate the mid-values for each class interval.

c Complete the last column of your table and find an estimate of the mean speed.

3 The grouped frequency tables give information about the time taken to solve four different crosswords.

For each table, copy the table, add extra working columns and find

i the modal class

ii the class containing the median

3 **iii** an estimate of the mean.

a

Time, t, minutes	Frequency
$5 < t \leqslant 10$	2
$10 < t \leqslant 15$	14
$15 < t \leqslant 20$	13
$20 < t \leqslant 25$	6
$25 < t \leqslant 30$	1

b

Time, t, minutes	Frequency
$0 < t \leqslant 10$	3
$10 < t \leqslant 20$	6
$20 < t \leqslant 30$	4
$30 < t \leqslant 40$	5
$40 < t \leqslant 50$	2

c

Time, t, minutes	Frequency
$5 < t \leqslant 10$	8
$10 < t \leqslant 15$	5
$15 < t \leqslant 20$	7
$20 < t \leqslant 25$	4
$25 < t \leqslant 30$	0
$30 < t \leqslant 35$	1

d

Time, t, minutes	Frequency
$5 < t \leqslant 15$	3
$15 < t \leqslant 25$	9
$25 < t \leqslant 35$	7
$35 < t \leqslant 45$	8
$45 < t \leqslant 55$	2
$55 < t \leqslant 65$	1

4 The heights of 50 Year 10 students were measured. The results are shown in the table.

Height, h, cm	Number of students
$150 \leqslant h < 155$	3
$155 \leqslant h < 160$	5
$160 \leqslant h < 165$	15
$165 \leqslant h < 170$	25
$170 \leqslant h < 175$	2

a What is the modal group?

b Estimate the mean height.

c Which class interval contains the median?

Q 1201, 1202   SEARCH

16.2 Averages and spread 2

- For large amounts of data presented as grouped data in a frequency table, you cannot calculate the exact mean, mode or median. Instead, you can calculate:
  - the estimated mean
  - the modal class
  - the class interval in which the median lies.

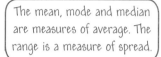

The mean, mode and median are measures of average. The range is a measure of spread.

HOW TO

Compare grouped data

1. Compare a measure of average such as the modal class, median class or estimated mean.
2. Compare the ranges of the data sets.

EXAMPLE

The tables show the ages of people attending concerts to see the bands Badness and Cloudplay.

Badness	
Age, *a*, years	Frequency
$20 \leq a < 30$	1600
$30 \leq a < 40$	4300
$40 \leq a < 50$	2100
$50 \leq a < 60$	1000

Cloudplay	
Age, *a*, years	Frequency
$10 \leq a < 20$	2800
$20 \leq a < 30$	4600
$30 \leq a < 40$	3300
$40 \leq a < 50$	700
$50 \leq a < 60$	500

You could also compare the median class or the estimated mean.

Compare the ages of the people attending the two concerts.

1. Compare a measure of average.

The modal age is greater at Badness concerts then Cloudplay concerts.

The highest frequency for Badness is in the class interval 30–40 years old, whereas the highest frequency for Cloudplay is in the interval 20–30 years old.

2. Compare the range in ages.

There is more variation in the ages of the people at Cloudplay concerts than Badness concerts.

The range for Cloudplay is 60 – 10 = 50 years, whereas for Badness it is 60 – 20 = 40 years.

Exercise 16.2A

1 Jayne kept a daily record of the number of miles she travelled in her car during two months.

	December	January
Miles travelled, m	Frequency	Frequency
$0 < m \leqslant 20$	3	0
$20 < m \leqslant 40$	8	5
$40 < m \leqslant 60$	10	12
$60 < m \leqslant 80$	6	8
$80 < m \leqslant 100$	4	6

a Estimate the mean number of miles for each month.

b Find the modal class and median class for each month.

c Compare the number of miles Jayne travelled in December and January.

2 David carried out a survey to find the time taken by 120 teachers and 120 office workers to travel home from work.

Teachers

Time taken, t, minutes	Frequency
$0 < t \leqslant 10$	12
$10 < t \leqslant 20$	33
$20 < t \leqslant 30$	48
$30 < t \leqslant 40$	20
$40 < t \leqslant 50$	7

Office workers

Time taken, t, minutes	Frequency
$10 < t \leqslant 20$	2
$20 < t \leqslant 30$	21
$30 < t \leqslant 40$	51
$40 < t \leqslant 50$	28
$50 < t \leqslant 60$	18

a Estimate the mean number of minutes for the teachers and office workers.

b Find the modal class and median class for the teachers and office workers.

c Make comparisons between the time taken by the teachers and office workers to travel home from work.

3 The weights of some apples are shown in the table.

Weight, w (g)	Frequency
$30 \leqslant w < 40$	6
$40 \leqslant w < 50$	28
$50 \leqslant w < 60$	21
$60 \leqslant w < 70$	25

Granny Smith apples have a mean weight of 45 g and a range of 39 g.

Is the data in the table about Granny Smith apples?

4 A company produces three million packets of crisps each day. It states on each packet that the bag contains 25 g of crisps. To test this, a sample of 1000 bags are weighed.

The table shows the results.

Weight, w (g)	Frequency
$23.5 \leqslant w < 24.5$	20
$24.5 \leqslant w < 25.5$	733
$25.5 \leqslant w < 26.5$	194
$26.5 \leqslant w < 27.5$	53

Is the company justified in stating that each bag contains 25 g of crisps?

Show your workings and justify your answer.

5 Two machines are each designed to produce paper 0.3 mm thick.

The tables show the actual output of a sample from each machine.

	Machine A	Machine B
Thickness, t (mm)	Frequency	Frequency
$0.27 \leqslant t < 0.28$	2	1
$0.28 \leqslant t < 0.29$	7	50
$0.29 \leqslant t < 0.30$	32	42
$0.30 \leqslant t < 0.31$	50	5
$0.31 \leqslant t < 0.32$	9	2

Compare the output of the two machines using suitable calculations.

Which machine is producing paper closer to the required thickness?

Q 1201, 1202   SEARCH

16.3 Scatter graphs and correlation

You can use a **scatter graph** to compare two sets of data, for example, height and weight.

● The data is collected in pairs and plotted as coordinates.

● If the points lie roughly in a straight line, there is a **linear relationship** or **correlation** between the two **variables**.

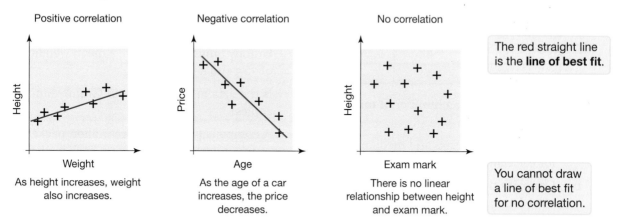

Positive correlation	Negative correlation	No correlation

As height increases, weight also increases.

As the age of a car increases, the price decreases.

There is no linear relationship between height and exam mark.

The red straight line is the **line of best fit**.

You cannot draw a line of best fit for no correlation.

If the points lie close to the line of best fit, the correlation is strong.

EXAMPLE

The exam results (%) for Paper 1 and Paper 2 for 10 students are shown.

Paper 1	56	72	50	24	44	80	68	48	60	36
Paper 2	44	64	40	20	36	64	56	36	50	24

a Draw a scatter graph and line of best fit.

b Describe the relationship between the Paper 1 results and Paper 2 results.

a Plot the exam marks as coordinates. The line of best fit should be close to all the points, with approximately the same number of crosses on either side of the line.

b There is a positive correlation. Students who did well on Paper 1 did well on Paper 2. Students who did not do well on Paper 1 did not do well on Paper 2 either.

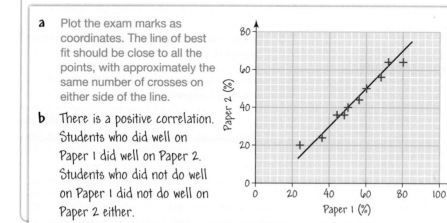

Points that are an exception to the general pattern of the data are called **outliers**.

Performing well on Paper 1 does not cause you to perform well on Paper 2. The tests could be on completely different subjects.

● If two variables are correlated it does not always mean that one **causes** the other.

The events could both be a result of a common cause, or there could be no connection at all between the events.

Exercise 16.3S

1 Match each scatter graph to its type of correlation.

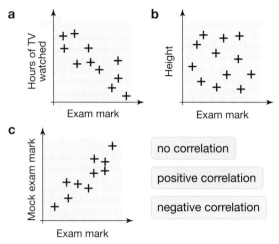

a

b

c

no correlation

positive correlation

negative correlation

2 Describe the points A, B, C, D and E on each scatter graph.

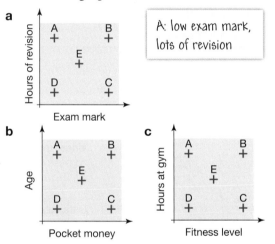

a

A: low exam mark, lots of revision

b

c

3 Match a type of correlation to each scatter graph.

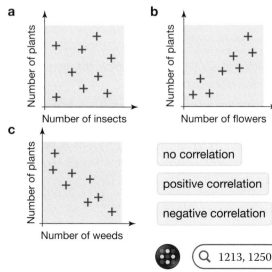

a

b

c

no correlation

positive correlation

negative correlation

4 The table shows the amount of water used to water plants and the daily maximum temperature.

Water (litres)	25	26	31	24	45	40	5	13	18	28
Maximum temperature (°C)	24	21	25	19	30	28	15	18	20	27

a Copy and complete the scatter graph for this information.

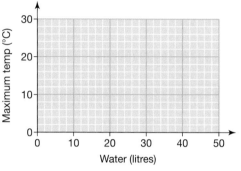

b State the type of correlation shown in the scatter graph.

c Copy and complete these sentences:

   i As the temperature increases, the amount of water used _____.

   ii As the temperature decreases, the amount of water used _____.

5 The times taken, in minutes, to run a mile and the shoe sizes of ten athletes are shown in the table.

Shoe size	10	$7\frac{1}{2}$	5	9	6	$8\frac{1}{2}$	$7\frac{1}{2}$	$6\frac{1}{2}$	8	7
Time (mins)	9	8	8	7	5	13	15	12	5	6

a Draw a scatter graph to show this information.

   Use 2 cm to represent 1 shoe size on the horizontal axis.

   Use 2 cm to represent 5 minutes on the vertical axis.

b State the type of correlation shown in the scatter graph.

c Describe, in words, any relationship that the graph shows.

Q 1213, 1250   SEARCH

16.3 Scatter graphs and correlation

RECAP

- You can compare two sets of data on a scatter graph.
- The data is collected in pairs and plotted as coordinates.
- If the points lie roughly in a straight line, there is a correlation between the two variables.

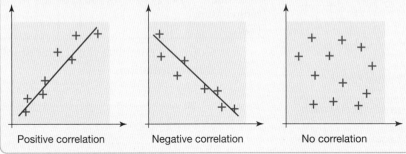

Positive correlation   Negative correlation   No correlation

You can use the line of best fit to make predictions for a value that falls in the range of the data - this is called **interpolation**. Interpolation is reliable, especially if the correlation is strong.

You can use the line of best fit to make predictions for a value that falls outside the range of the data - this is called **extrapolation**. Extrapolation is not always reliable as you cannot be sure that pattern holds for values outside of the data collected.

HOW TO

① Describe and interpret the relationship shown by the diagram.

② Make predictions based on the correlation shown.

EXAMPLE

The scatter graph shows the number of goals scored by 21 football teams in a season plotted against the number of points gained.

a Describe the relationship between the goals scored and the number of points.

b Describe the goals and points for team A.

c If a team scored 45 goals, how many points would you expect it to have?

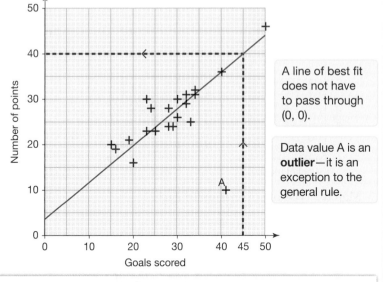

A line of best fit does not have to pass through (0, 0).

Data value A is an **outlier**—it is an exception to the general rule.

① Describe and interpret the relationship shown by the diagram.

a The graph shows a positive correlation: the more goals scored, the more points gained.

b Team A has scored lots of goals but has gained very few points.

② Make predictions based on the correlation shown.

c Reading from the graph, and using the line of best fit as a guide, you could expect a team that scored 45 goals to gain 40 points.

Exercise 16.3A

1 The scatter graph shows the exam results (%) for Paper 1 and Paper 2 for 11 students.

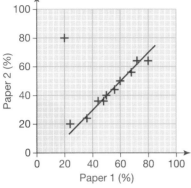

a Describe the relationship between Paper 1 and Paper 2 results.

b Identify an outlier and describe how this student performed in the papers.

c If a student scored 40 in Paper 1, what score would you expect them to gain in Paper 2?

d Jenna extends the line of best fit on the graph and tries to predict the score that a student who scored 100 on Paper 2 will score on Paper 1.

Will her estimate make sense?

2 The graph shows the marks in two papers achieved by ten students.

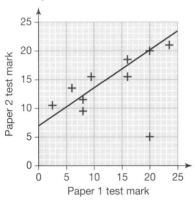

Use the line of best fit to estimate

a the Paper 2 mark for a student who scored 13 in Paper 1

b the Paper 1 mark for a student who scored 23 in Paper 2.

c Describe the outlier in the scatter graph. How well did this student perform on the two papers?

3 The table shows the age and diameter, in centimetres, of trees in a forest.

Age (years)	10	27	6	22	15	25	11	16	21	19
Diameter (cm)	20	78	9	65	38	74	25	44	59	50

a Draw a scatter diagram to show the information.

b State the type of correlation between the age and diameter of the trees.

> Use 2 cm to represent 10 centimetres on the horizontal axis, numbered 0 to 80.
>
> Use 2 cm to represent 5 years on the vertical axis, numbered 0 to 30.

c Draw a line of best fit.

d If the diameter of a tree is 55 cm, estimate the age of the tree.

e Use your graph to estimate the diameter of a tree that is one year old.

f Explain why the estimate in part **d** is more reliable than the estimate in part **e**.

4 The table gives the marks earned in two exams by 10 students.

Maths %	70	76	61	70	89	65	59	58	73	82
Statistics %	78	82	74	75	93	70	66	62	77	89

a Draw a scatter graph for the data.

> Choose a sensible scale for the graph. Try to make your graph as large as possible.

b Describe the correlation and the relationship shown.

c Draw a line of best fit.

d Predict the Statistics mark for a student who scored 62% in Maths.

e Could you use your graph to predict the Maths mark for a student who scored 32% in Statistics? Give your reasons.

16.4 Time series

You can use a **line graph** to show how data changes as time passes.

The data can be discrete or continuous.

The temperature of a liquid is measured every minute.

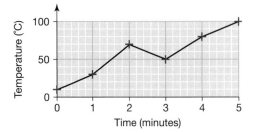

Time is always the **horizontal** axis. Time could be seconds, minutes, hours, days, weeks, months or years.

This is an example of a **time series graph**.

- A time series graph shows
  - how the data changes over time, or the **trend**
  - each individual value of the data.

EXAMPLE

The table shows the average monthly rainfall, in centimetres, in Sheffield over the last 30 years.

Month	Jan	Feb	Mar	Apr	May	Jun	Jul	Aug	Sep	Oct	Nov	Dec
Rainfall (cm)	8.7	6.3	6.8	6.3	5.6	6.7	5.1	6.4	6.4	7.4	7.8	9.2

Draw a time series graph to show this information.

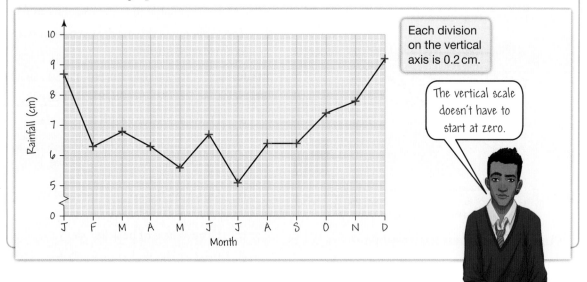

Each division on the vertical axis is 0.2 cm.

The vertical scale doesn't have to start at zero.

Exercise 16.4S

1 The number of photographs taken each day during a 7-day holiday is given.

Sunday	Monday	Tuesday	Wednesday	Thursday	Friday	Saturday
8	12	11	16	19	2	13

Copy and complete the line graph to show this information.

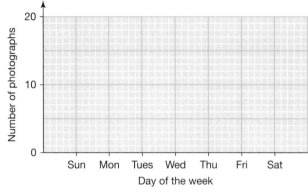

2 The numbers of DVDs rented from a shop during a week are shown.

Sunday	Monday	Tuesday	Wednesday	Thursday	Friday	Saturday
18	9	7	11	15	35	36

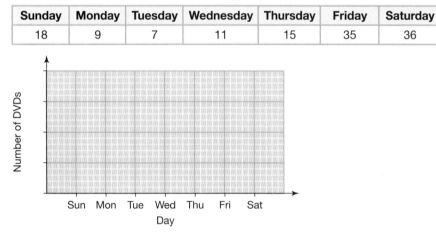

Copy and complete the line graph, choosing a suitable vertical scale.

3 Every year on his birthday, Peter's weight in kilograms is measured.

Age	2	3	4	5	6	7	8	9	10	11	12	13
Weight (kg)	14	16	18	20	22	25	28	31	34	38.5	40	45

Draw a line graph to show the weights.

4 The hours of sunshine each month are shown in the table.

Jan	Feb	Mar	Apr	May	Jun	Jul	Aug	Sep	Oct	Nov	Dec
43	57	105	131	185	176	194	183	131	87	53	35

Draw a line graph to show the hours of sunshine.

5 The daily viewing figures, in millions, for a reality TV show are shown.

Day	Sat	Sun	Mon	Tue	Wed	Thu	Fri
Viewers (in millions)	3.2	3.8	4.3	4.5	3.1	5.2	7.1

Draw a line graph to show the viewing figures.

 1198, 1939   SEARCH

16.4 Time series

- You can use a line graph to show how data changes over time.
- This is sometimes called a time series graph. It shows:
  - how the data changes over time, or the trend
  - each individual value of the data.
- This graph shows the how the temperature changes over time.

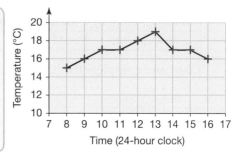

① Draw axes on graph paper and label with suitable scales.

② Plot the data from the table as coordinate pairs. Join the plotted points with straight lines.

③ Discuss any short term trends, seasonal variation and any longer term trends.

Jenny's quarterly gas bills over a period of two years are shown in the table.

	Jan–March	April–June	July–Sept	Oct–Dec
2003	£65	£38	£24	£60
2004	£68	£42	£30	£68

Plot the data on a graph and comment on any pattern in the data.

The graph shows seasonal variation.

① Draw axes on graph paper with time on the horizontal axis. Plot time on the horizontal axis, J–M means Jan–March.

② Plot the coordinates as crosses on the grid. Join them up with straight lines.

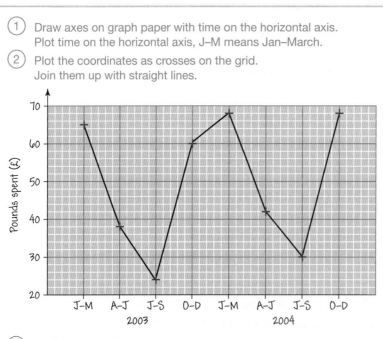

③ Gas bills are highest in the winter months and lowest in the summer months. This annual pattern appears to repeat itself.

There is a slight trend for the bills to rise from year to year.

Exercise 16.4A

For each of questions **1–4**

   a   Plot the data on a graph

   b   Comment on any patterns in the data.

1   The table shows Ken's monthly mobile phone bills.

Jan	Feb	Mar	April	May	June	July	Aug	Sept	Oct	Nov	Dec
£16	£12	£15	£18	£16	£18	£12	£10	£12	£15	£16	£20

2   The table shows Mary's quarterly electricity bills over a two-year period.

	Jan–March	April–June	July–Sept	Oct–Dec
2004	£45	£20	£15	£48
2005	£54	£24	£18	£50

3   The table shows monthly ice-cream sales at Angelo's shop during one year.

Jan	Feb	Mar	April	May	June	July	Aug	Sept	Oct	Nov	Dec
£16	£12	£15	£18	£38	£48	£52	£58	£18	£15	£16	£40

4   A town council carried out a survey over a number of years to
find the percentage of local teenagers who used the town's library.
The table shows the results.

year	1998	1999	2000	2001	2002	2003	2004	2005
%	14	18	24	28	25	20	18	22

5   This news report was written about sales representatives of a small firm.

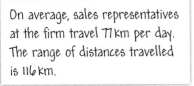

On average, sales representatives at the firm travel 77 km per day. The range of distances travelled is 116 km.

These two graphs were drawn to summarise the distances travelled daily by
representatives at two different firms.

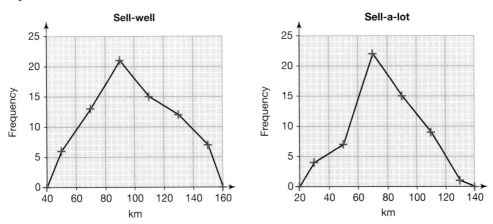

Use the graphs to identify which firm was being reported on.
Give a reason for your choice.

Q 1198, 1939   SEARCH

Summary

Checkout

You should now be able to...

	Test it Questions
✓ Interpret and construct tables, graphs and charts for continuous and grouped data.	1
✓ Use the median, mean, modal class and range to interpret and compare distributions.	2
✓ Use correlation to describe scatter graphs but know that it does not imply causation.	3
✓ Draw estimated lines of best fit and make predictions but understand their limitations.	4
✓ Interpret and construct line graphs for time series data.	5

Language   Meaning   Example

Language	Meaning	Example
Modal class	The most commonly occurring class in a set of grouped data	(see table below)
Estimated mean	An estimate for the mean of the data using an approximation for the total of the values. This approximation is found by multiplying the midpoint of each group by the frequency.	

Time, t, minutes	Frequency	Midpoint	Mid-point · Frequency
$5 < t \leqslant 10$	2	7.5	15
$10 < t \leqslant 15$	3	12.5	37.5
$15 < t \leqslant 20$	6	17.5	105
$20 < t \leqslant 25$	1	22.5	22.5

Modal class = $15 < t \leqslant 20$

$$\text{Estimated mean} = \frac{15 + 37.5 + 105 + 22.5}{2 + 3 + 6 + 1}$$

$$= \frac{180}{12} = 15 \text{ minutes}$$

Language	Meaning
Scatter graph	A graph that shows how two sets of numerical data are related.
Line of best fit	The single line that best represents the general direction of a set of points
Correlation (Linear relationship)	If the points lie roughly on a straight line there is a correlation between the two variables. It is a measure of how strongly they appear to be related.

Positive correlation   Negative correlation   No correlation

Language	Meaning
Time series graph	A graph that shows how a measurement changes with time.
Line graph	A graph where points are joined with straight lines.
Trend	The direction in which data appears to head as it changes over time.

Review

1 Chloe measures the length of 16 worms (in cm).

8.5, 10.7, 12.4, 12.8, 9.9,   11.5, 10,   8.9, 9.3, 12.2, 11.8, 12.5, 14.4, 10.2, 13.1, 12.2

Copy and complete the frequency table for this data.

Length, L (cm)	Frequency
$8 \leqslant L < 10$	
$10 \leqslant L < 12$	
$12 \leqslant L < 14$	
$14 \leqslant L < 16$	

2 For the data in question **1**

a what is the modal class?

b In which class does the median lie?

c Estimate the mean length.

3

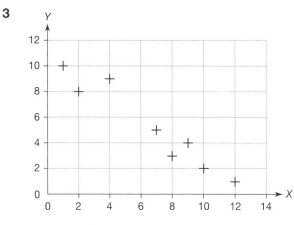

a Describe the type of correlation shown in this scatter diagram.

b Evie says 'this diagram proves that a lower value of X causes a higher value of Y'. Is Evie correct?

4 The table shows the results of a group of students in two different maths tests.

Name	Test A	Test B
Aaron	12	15
Brooke	13	16
Colton	8	8
Daisy	5	8
Ellis	5	9
Frankie	8	10
Glen	10	14
Harriet	7	11
Isaac	8	14
Jasmine	6	9

a Plot this data on a scatter diagram.

b What type of correlation does the diagram show?

c Draw a line of best fit on the graph.

d Estimate the result in Test B of a student who scored 9 on Test A.

e Kevin has only sat Test B and scored 4.

   i Estimate his result in Test A.

   ii What could be wrong with this estimate and why?

5 Draw a line graph for this time series data.

Time	Temperature (°C)
06:00	15
09:00	17
12:00	20
15:00	22
18:00	24
21:00	21

What next?

Score			
	0 – 2		Your knowledge of this topic is still developing. To improve look at MyMaths: 1193, 1196, 1198, 1201, 1202, 1213, 1250
	3 – 4		You are gaining a secure knowledge of this topic. To improve look at InvisiPens: 16Sa – h
	5		You have mastered these skills. Well done you are ready to progress! To develop your exam technique look at InvisiPens: 16Aa – d

Assessment 16

1 The table shows the age distribution of the members of an athletics club.

Age (Y) (yrs)	$10 \leqslant Y < 15$	$15 \leqslant Y < 20$	$20 \leqslant Y < 25$	$25 \leqslant Y < 30$	$30 \leqslant Y < 35$	$35 \leqslant Y < 40$
People	16	19	27	34	29	25

    a Draw a grouped bar chart to illustrate this data. [4]

    b The club's rules only allow athletes between 10 and 20 years old to represent the club in junior competitions. How many athletes do they have to choose from? [1]

2 A police officer records the speeds of 250 vehicles along the M6 in a grouped frequency table.
The national speed limit is 70 mph.
The police officer wants to work how many vehicles were breaking the speed limit.
He also knows that there were no vehicles travelling less than 10 mph. Give two criticisms of the table. [2]

Speed (mph)	No. of Vehicles
$0 < V \leqslant 20$	6
$20 < V \leqslant 40$	42
$40 < V \leqslant 60$	104
$60 < V \leqslant 80$	87
$80 < V \leqslant 100$	11

3 Malia made a grouped bar chart to show the ages of all the people staying in a hotel but forgot to include a scale on the vertical axis.

    a Write down the modal class. [1]

    b Explain how you can work out the modal class without a scale. [1]

Malia knows that 24 guests were under 20 years old.

    c How many people were in the hotel altogether? [3]

    d Which class contains the median? [1]

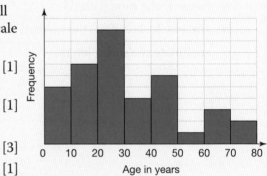

4 Olivia counted the number of tea bags in 30 boxes. The boxes are supposed to contain 400 tea bags. The table shows the difference between actual number of tea bags and 400.

Difference (d)	$0 \leqslant d < 5$	$5 \leqslant d < 10$	$10 \leqslant d < 15$	$15 \leqslant d < 20$
Frequency	13	11	5	1

    a Draw a grouped bar chart to show this information. [4]

    b Estimate the mean difference. [5]

    c Olivia counted the number of tea bags in the 30 boxes and divided by 30 to find the mean. Olivia's value for the mean is 400 tea bags. Explain how this could be possible even though your answer to part **b** is not zero. [1]

5 The table shows the extension of a spring when various masses are added to it.

Mass (g)	50	100	150	200	250	300	350	400	450
Extension (cm)	3.1	6.0	9.2	12.5	15.3	18.8	21.7	25.0	28.0

    a Plot a scatter diagram of the data. [4]

    b Draw the line of best fit and comment on the type of correlation. [3]

    c Use your line of best fit to estimate

        i The extension when a mass of 325 g is added to the spring [1]

        ii The mass added when an extension of 20 cm is produced [1]

    d Lewis wants to estimate the extension when a mass of 600 g is added to the spring. Should Lewis use the line of best fit? Give reasons for your answer. [2]

6   The table shows the results of a group of key Stage 2 children's English and Science tests.

Science	25	31	33	37	48	54	56	61	75	80	89	90	93	96	98
English	86	47	58	12	76	55	82	60	59	52	66	80	15	72	68

a   Plot a scatter diagram for the data.   [4]

b   Draw the line of best fit and comment on the correlation.   [3]

c   Mr Patel thinks that the results for the English test for two students are mixed up.
     Which two results do not fit the pattern of the data?   [1]

d   Use your line to estimate

    i   a Science Mark corresponding to an English mark of 52   [1]

    ii   an English Mark corresponding to a Science mark of 52.   [1]

7   The number of patients in the waiting room of a Health Centre was recorded at hourly intervals
     and the results recorded on this time series graph.

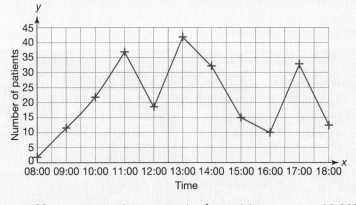

a   How many patients were in the waiting room at 10:00?   [1]

b   How many patients were in the waiting room at 12:00?   [1]

c   How many patients were in the waiting room at 14:30?   [1]

d   Why is your answer to c is only an approximation?   [1]

e   What is the maximum number of patients recorded in the waiting room?   [1]

f   Give a possible reason for the three peaks in the time series.   [1]

8   Scream4IceCream issued their sales figures for the past two years.
     The figures are to the nearest £1000.

	Jan	Feb	Mar	Apr	May	Jun	Jul	Aug	Sep	Oct	Nov	Dec
Year 1	100	95	125	150	176	203	251	266	204	131	101	88
Year 2	70	70	105	135	165	176	215	189	217	145	133	110

a   Draw a time series diagram for both sets of data on the same axes.   [6]

b   Compare the summer and winter sales for these periods.   [4]

17 Calculations 2

Introduction

An electron microscope is much more powerful than a normal microscope, and is used to look at very small objects. Biologists use electron microscopes to look at micro-organisms such as viruses. Materials scientists might use one to analyse crystalline structures. Nowadays there are electron microscopes that can detect objects that are smaller than a nanometre (0.000000001 m), so they can 'see' molecules and even individual atoms.

What's the point?

Mathematics needs to be able to describe very small quantities, such as lengths measured in nanometres. Without this ability, researchers could not analyse microscopic organisms like viruses.

Objectives

By the end of this chapter you will have learned how to …

- Calculate with roots and with integer indices.
- Calculate exactly with fractions and multiples of π.
- Calculate with and interpret numbers written in standard form.

1 Work out the value of each of these products.

    a $3^2 \times 5$ b $2^2 \times 5^2$

2 Convert between these units.

    a   23 cm into metres   b   2.4 km into centimetres.

Chapter investigation

A picometre is a unit of length that is used to describe atomic distances.
It is 0.000000000001 m in length.

See if you can find out about the smallest metric unit of length that has a name, and if it has any possible real-world uses. What about the largest unit of length?

17.1 Calculating with roots and indices

A **square root** is a number that when multiplied by itself gives a result equal to a given number. Square roots are written using $\sqrt{}$ notation.

A **cube root** is a number that when multiplied by itself and then multiplied by itself again gives a result equal to a given number. Cube roots are written using $\sqrt[3]{}$ notation.

Powers of the same number can be multiplied and divided.

> Use a calculator to find a square root using the $\boxed{\sqrt{x}}$ function key.

- When multiplying, you add the indices.

$$5^3 \times 5^4 = (5 \times 5 \times 5) \times (5 \times 5 \times 5 \times 5) = 5^7 \Rightarrow 5^{3+4} = 5^7$$

- When dividing, you subtract the indices.

$$3^5 \div 3^2 = \frac{3 \times 3 \times 3 \times 3 \times 3}{3 \times 3} = 3 \times 3 \times 3 = 3^3 \Rightarrow 3^{5-2} = 3^3$$

> Use a calculator to find a cube root using the $\boxed{\sqrt[3]{x}}$ function key.

- To raise a power, multiply the indices.

$$(9^2)^3 = (9 \times 9) \times (9 \times 9) \times (9 \times 9) = 9^{2 \times 3} = 9^6$$

> **EXAMPLE**
>
> Write the answers to these in index form.
> a $5^5 \times 5^3$
> b $4^7 \div 4^3$
> c $(6^3)^4$
>
> a $5^5 \times 5^3 = 5^{5+3} = 5^8$
> b $4^7 \div 4^3 = 4^{7-3} = 4^4$
> c $(6^3)^4 = 6^{3 \times 4} = 6^{12}$

Using the rule for multiplication, $7^3 \times 7^0 = 7^3$. Since multiplying by 7^0 leaves the 7^3 unchanged, 7^0 must be equal to 1.

> These rules apply to numbers and algebraic powers.

- Any number (except 0) to the power 0 is 1: $x^0 = 1$ for any value of x, if $x \neq 0$.

You know that $3^2 \times 3 = (3 \times 3) \times 3 = 3^3$.
You can write this as $3^2 \times 3^1 = 3^{(2+1)} = 3^3$, so $3 = 3^1$.

- Any number to the power 1 is just the number itself: $x^1 = x$ for any value of x.

> **EXAMPLE**
>
> Write the value of
> a 19^0
> b 8^1
> c $(16^3 - 81 \times 17)^0$
> d $(4.8)^1$
>
> a $19^0 = 1$
> b $8^1 = 8$
> c There is no need to evaluate the expression in the brackets.
> $(16^3 - 81 \times 17)^0 = 1$
> d $(4.8)^1 = 4.8$

Number Calculations 2

Exercise 17.1S

1 Find these numbers.

  a $\sqrt{25}$ b $\sqrt{9}$ c $\sqrt{16}$

  d $\sqrt{1}$ e $\sqrt{4}$ f $\sqrt{36}$

2 Calculate these using a calculator, giving your answers to 2 dp as appropriate.

  a $\sqrt{40}$ b $\sqrt{61}$ c $\sqrt{180}$

  d $\sqrt{249}$ e $\sqrt{676}$ f $\sqrt{1234}$

3 Calculate these using the $\sqrt[3]{}$ key on your calculator. Give your answers to 2 dp where appropriate.

  a $\sqrt[3]{27}$ b $\sqrt[3]{512}$ c $\sqrt[3]{3375}$

  d $\sqrt[3]{100}$ e $\sqrt[3]{24389}$ f $\sqrt[3]{16129}$

4 Find the value of

  a 5^2 b 2^3 c 3^3

  d 8^2 e 12^2 f 13^2

5 Find the value of

  a 3^4 b 1^5 c 2^7

  d 3^6 e 10^6 f 4^4

6 Use the y^x function key on your calculator to work these out.

  a 12^3 b 6^6 c 21^3

  d 16^5 e 13^3 f 14^6

7 Evaluate

  a 5^1 b 6^1

  c 6^0 d 7^0

  e $(4 + 88^2)^0$ f $(4^2 + 5^2)^1$

  g $(92.5)^0$ h 0^1

8 Write the answers to these multiplications in index form.

  a $7 \times 7 \times 7$ b 3×3^2

  c 5×5^2 d $6 \times 6 \times 6^2$

  e $5^3 \div 5$ f $8^4 \times 8$

  g $9^3 \times 9^2 \times 9$ h $8^7 \times 8$

9 Simplify these expressions, giving your answers in index form.

  a $6^2 \times 6^3$ b $4^5 \times 4^4$

  c $2^6 \times 2^7$ d $11^5 \times 11^2$

  e $1^{17} \times 1^{13}$ f $7^8 \times 7^4$

  g $3^6 \times 3^6$ h $9^9 \times 9^1$

10 Simplify these expressions, giving your answers in index form where appropriate.

  a $7^8 \div 7^6$ b $8^6 \div 8^2$

  c $3^3 \div 3^2$ d $9^{11} \div 9^8$

  e $4^7 \div 4^1$ f $2^9 \div 2^9$

  g $12^8 \div 12^6$ h $6^{13} \div 6^{13}$

11 Use the relationship $(x^m)^n = x^{mn}$ to simplify these expressions, giving your answers in index form.

  a $(2^3)^2$ b $(4^2)^5$ c $(7^2)^2$

  d $(5^5)^3$ e $(3^4)^4$ f $(6^2)^2$

  g $(5^7)^3$ h $(10^4)^4$

12 Simplify these expressions, giving your answers in index form.

  a $3^4 \times 3^2 \div (3^3 \times 3^2)$

  b $(5^6 \div 5^2) \times 5^4 \times 5^2$

  c $(4^5 \div 4^2) \div (4^6 \div 4^5)$

  d $(7^9 \div 7^2) \div (7^2 \times 7^3)$

  e $(8^7 \div 8^4) \times 8^5 \times 8^3$

  f $9^3 \times (9^5 \div 9^2) \times 9^4$

13 Simplify these expressions, giving your answers in index form.

  a $\dfrac{4^2 \times 4^2}{4^2}$ b $\dfrac{6^3 \times 6^4}{6^5}$

  c $\dfrac{9^8}{9^2 \times 9^4}$ d $\dfrac{8^6 \div 8^3}{8^2}$

  e $\dfrac{5^9 \times 5^4}{5^3 \times 5^7}$ f $\dfrac{6^3 \times 6^4}{6^5 \div 6^3}$

  g $\dfrac{8^9 \div 8^2}{8^7 \div 8^2}$ h $\dfrac{10^6 \div 10^2}{10^2 \times 10^2}$

14 Simplify these expressions, giving your answers in index form.

  a $\left(\dfrac{4^4}{4^2}\right)^2$ b $\dfrac{(3^3)^2 \times 5^5}{3^7}$

  c $\dfrac{6^3 \times (6^2)^4}{(6^5 \div 6^3)^2}$ d $\left(\dfrac{5^8 \div 5^2}{5^4 \div 5^2}\right)^3$

  e $2^8 \times \left(\dfrac{2^4 \div 2^2}{2^9 \div 2^8}\right)^2$ f $\left(\dfrac{7^{15} \div 7^2}{7^8 \times 7^2}\right)^4$

  g $\dfrac{(3^3)^5}{3^4 \times 3^2} \div (3^2)^3$ h $9^2 \times \left(\dfrac{9^7 \div 9}{9^2 \times 9^3}\right)^3$

1033, 1924, 1951   SEARCH

17.1 Calculating with roots and indices

RECAP

- Add indices when multiplying powers of the same number.
- Subtract indices when dividing powers of the same number.
- Multiply indices when finding the power of a power.
- Any number (except 0) to the power of 0 is 1.

$3^2 \times 3^4 = 3^{2+4} = 3^6$

$5^7 \div 5^3 = 5^{7-3} = 5^4$

$(7^2)^5 = 7^{2 \times 5} = 7^{10}$

$(4.86)^0 = 1$

HOW TO

(1) Read the question carefully; decide how to use your knowledge of powers and roots.

(2) Apply the index laws and look out for chances to use known facts about squares and cubes.

(3) Answer the question; make sure that you leave your answer in index form if the question asks for it.

EXAMPLE

For the calculation $2^3 \times 5^4$, Jack wrote:

$2 \times 5 = 10$, and $3 + 4 = 7$.

So, the answer is 10^7.

Explain why Jack is incorrect. Find the correct answer.

> The base is the number you are multiplying.

(1) The index laws only apply when both numbers have the same base.

(2) For this calculation you need to evaluate each term and then multiply them together:

(3) $2^3 = 8$ and $5^4 = 625$ $\quad 2^3 \times 5^4 = 8 \times 625 = 5000$

EXAMPLE

$2^{12} = 4^a$. Find the value of a.

(1) Rewrite 2^{12} as a power of 4. This will give you the value of a.

(2) Use the index law for the power of a power and the fact that $4 = 2^2$.

$2^{12} = (2^2)^6 = 4^6$

(3) $a = 6$

EXAMPLE

Explain why the square of a prime number has exactly three factors.

> For example, 169 has three factors 1, 13 and 169.

(1) A prime number, p, has two factors, 1 and p (itself).

(2) List the factors of the square of a prime number.

$p^2 = 1 \times p^2$ and $p^2 = p \times p$

(3) A prime number squared has exactly three factors because the only factors are: 1, the prime number p and the number itself p^2.

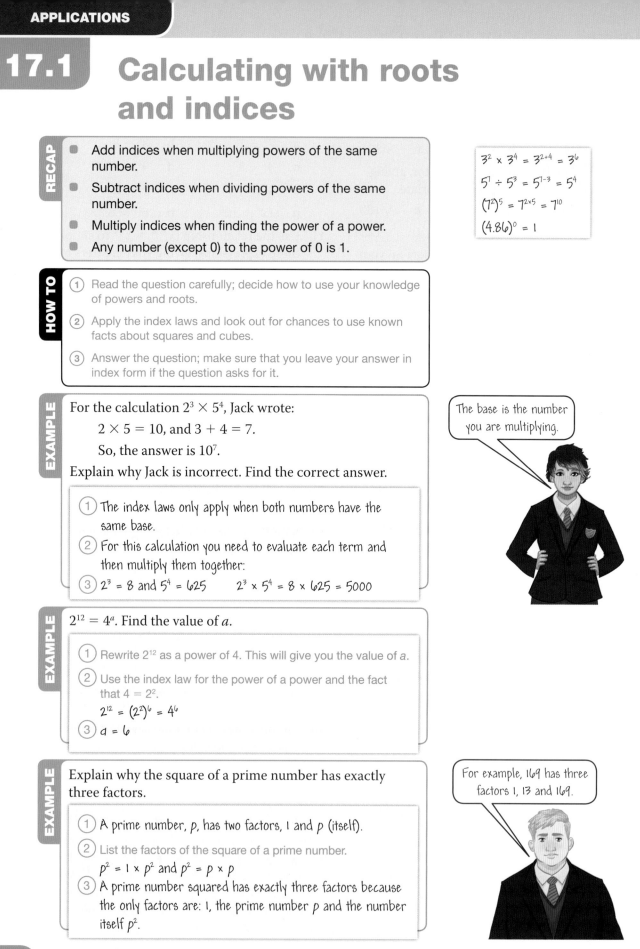

Exercise 17.1A

1 For the calculation $3^2 + 3^2 + 3^2$, Kira wrote:

$$3^2 + 3^2 + 3^2 = (3 + 3 + 3)^2 = 9^2$$

Explain why Kira is incorrect.
Find the correct answer.

2

$6^0 = 1$	$6^1 = 6$
$6^2 = 36$	$6^3 = 216$
$6^4 = 1296$	$6^5 = 7776$
$6^6 = 46656$	

Use the table to:

a Explain why $36 \times 216 = 7776$

b Work out $46656 \div 36$

3 Find the value of the letter in each equation.

a $3^4 \times 9 = 3^a$ **b** $5^b = 5^8 \div 25$

c $9^5 = 3^c$ **d** $4^{10} = 16^d$

4 Explain why the cube of a prime number has exactly four factors.

5 Which of the following are correct?
Change one symbol or number to correct any mistakes.

a $2^3 + 2^5 = 2^8$ **b** $3^2 \times 3^3 = 3^5$

c $(5^2)^3 = 5^6$ **d** $(7^3)^4 = 7^7$

e $4^5 \div 4^3 = 4^8$ **f** $6^4 \div 6^3 = 6$

6 Arrange these numbers in order of size, starting with the smallest first

8^0 7^1 6^2 5^3 4^4 3^5 2^6 1^7 0^8

7 Myia runs a secret maths club.
To get in to the club, you have to give her two numbers. Myia squares the numbers, adds these together, then square roots the answer. If the answer is a whole number you can join the club.

Which of these pairs of numbers would get you into her club?

a 3 and 4   **b** 4 and 5

c 6 and 12   **d** 8 and 15

e Can you find two more pairs that would get you into Myia's club?

8 The Richter scale measures how forceful an earthquake is. Each level is ten times more powerful than the previous value – so level 4 is 10 times as powerful as level 3.

a How many times more powerful is a level 4 earthquake compared to a level 1?

b How many times more powerful is one described as 'Complete devastation' (level 9) compared to 'Buildings shake' (level 5)?

c What level is an earthquake 10^5 times more powerful than level 3?

*9 Here are twelve number cards:

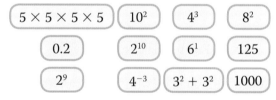

$5 \times 5 \times 5 \times 5$	10^2	4^3	8^2
0.2	2^{10}	6^1	125
2^9	4^{-3}	$3^2 + 3^2$	1000

a Find a pair of cards where

   i they are both the same value

   ii one is twice the other

   iii one is ten times as big as the other

   iv they are both odd numbers

   v they multiply together to make 1

   vi they are reciprocals

   vii they multiply together to make 5^7

   viii they multiply together to make 5^2.

b Which card

   i is 5^{-1}

   ii is three times the cube root of 216

   iii has the largest value

   iv has the smallest value?

   🔍 1033, 1924, 1951   SEARCH

17.2 Exact calculations

- A fraction will have a **terminating** decimal equivalent if the only prime factors of the denominator are 2 or 5.
- All other fractions give **recurring** decimals.

$$\frac{3}{20} = \frac{3}{(2 \times 2 \times 5)} = 0.15$$

$$\frac{1}{18} = \frac{1}{(2 \times 3 \times 3)} = 0.05555... = 0.0\dot{5}$$

When you calculate with fractions, you should work with the numbers in fraction form as far as possible.

EXAMPLE

Give the exact answer to each calculation.

a $\dfrac{1}{2} + \dfrac{2}{3}$

b $\dfrac{3}{4} \times \dfrac{2}{9}$

c $\dfrac{4}{5} \div \dfrac{3}{10}$

a $\dfrac{1}{2} + \dfrac{2}{3} = \dfrac{3}{6} + \dfrac{4}{6} = \dfrac{7}{6} = 1\dfrac{1}{6}$

b $\dfrac{\cancel{3}^1}{\cancel{4}_2} \times \dfrac{\cancel{2}^1}{\cancel{9}_3} = \dfrac{1}{2} \times \dfrac{1}{3} = \dfrac{1}{6}$

c $\dfrac{4}{5} \div \dfrac{3}{10} = \dfrac{4}{\cancel{5}_1} \times \dfrac{\cancel{10}^2}{3} = \dfrac{4}{1} \times \dfrac{2}{3} = \dfrac{8}{3} = 2\dfrac{2}{3}$

> If the question asks for an 'exact answer', avoid using your calculator.

5.3

EXAMPLE

Which of these calculations gives a different answer to the others?

a $1\dfrac{3}{4} - 1\dfrac{1}{2}$ **b** $\dfrac{1}{2} \times 1\dfrac{1}{2}$ **c** $3\dfrac{1}{8} - 2\dfrac{7}{8}$ **d** $2\dfrac{1}{2} \div 10$

You need to work out all four parts to see which is different.

a $1\dfrac{3}{4} - 1\dfrac{1}{2} = \dfrac{1}{4}$

b $\dfrac{1}{2} \times 1\dfrac{1}{2} = \dfrac{1}{2} \times \dfrac{3}{2}$
$= \dfrac{3}{4}$

c $3\dfrac{1}{8} - 2\dfrac{7}{8} = \dfrac{2}{8}$
$= \dfrac{1}{4}$

d $2\dfrac{1}{2} \div 10 = \dfrac{5}{2} \times \dfrac{1}{10}$
$= \dfrac{5}{20}$
$= \dfrac{1}{4}$

Part **b** is different to the others.

When you calculate with π, you can leave your answers in terms of π to give an exact answer.

EXAMPLE

Evaluate $2\pi(4^2 + 4 \times 6)$

First simplify, leaving π in the expression.

$2\pi(4^2 + 4 \times 6) = 2\pi(16 + 24) = 2\pi \times 40 = 80\pi$

> Using π you can give exact answers to problems involving circles, cylinders and spheres.

11.1

EXAMPLE

Jennifer has a circular piece of fabric with radius 6 cm.
She cuts a square with sides 3 cm from the circle.
Find the exact area of the remaining fabric.

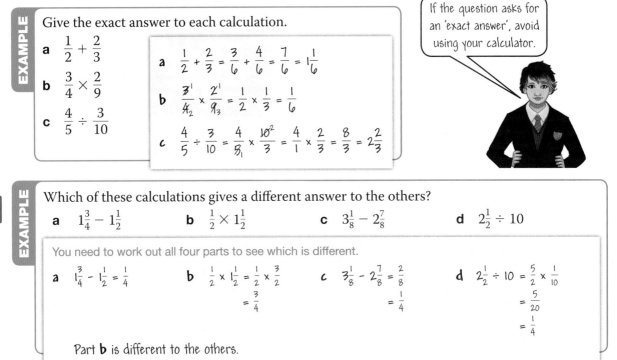

Area of a circle = πr^2

Area of the circle = $\pi \times 6^2 = \pi \times 36 = 36\pi \text{ cm}^2$

Area of the square = $3 \times 3 = 9 \text{ cm}^2$

Remaining area = $36\pi - 9 \text{ cm}^2$ Leave π in the answers.

Number Calculations 2

Exercise 17.2S

Give exact answer to all questions.

1   Evaluate exactly

   a $\dfrac{1}{3} + \dfrac{1}{5}$ **b** $\dfrac{2}{5} + \dfrac{3}{7}$

   c $\dfrac{3}{8} - \dfrac{2}{7}$ **d** $\dfrac{8}{15} + \dfrac{4}{9}$

   e $6\dfrac{1}{2} - 1\dfrac{5}{8}$ **f** $\dfrac{2}{5} + \dfrac{1}{3} + \dfrac{1}{4}$

   g $\dfrac{3}{8} + \dfrac{1}{2} - \dfrac{2}{5}$ **h** $7\dfrac{2}{9} + 2\dfrac{1}{4}$

2   Evaluate exactly

   a $\dfrac{2}{3} \times \dfrac{3}{4}$ **b** $\dfrac{5}{9} \div \dfrac{1}{3}$

   c $2\dfrac{1}{2} \times \dfrac{5}{8}$ **d** $\dfrac{8}{9} \div 1\dfrac{2}{3}$

   e $3\dfrac{1}{2} \div 2\dfrac{1}{4}$ **f** $5\dfrac{1}{5} \times 2\dfrac{3}{4}$

   g $8\dfrac{2}{5} \div 3\dfrac{1}{7}$ **h** $3\dfrac{1}{2} \times 7\dfrac{5}{9}$

3   For each answer from questions **1** and **2**, either give an exact decimal equivalent of the answer, or explain why it is not possible to do so.

4   Evaluate exactly

   a $\dfrac{2}{5} \times \left(\dfrac{3}{4} + \dfrac{2}{3}\right)$

   b $\left(\dfrac{2}{3} + \dfrac{4}{5}\right) \div \left(\dfrac{2}{5} + \dfrac{3}{7}\right)$

   c $\dfrac{5}{6} \div \dfrac{3^2 + 4^2}{10}$

   d $\left(\dfrac{3}{7}\right)^2 \times \left(\dfrac{4}{5} - \dfrac{1}{7}\right)$

   e $\left(\dfrac{1}{2} + \dfrac{5}{9}\right)^2 + \left(\dfrac{2}{3}\right)^3$

   f $\left(\dfrac{5}{6} + \dfrac{1}{2}\right) - \left(\dfrac{4}{7} \times 3\right)$

5   Julia says that all these four calculations give the same answer. Is she correct? (Show your working.)

   a $\dfrac{3}{4} + \dfrac{1}{2}$ **b** $\dfrac{1}{2} \times 2\dfrac{1}{2}$

   c $2\dfrac{3}{4} - 1\dfrac{1}{2}$ **d** $3 \div 2\dfrac{2}{5}$

6   Simplify these expressions

   a $\pi + \pi$ **b** $2\pi + \pi$

   c $4 + \pi - 2 + \pi$ **d** $\pi(4^2 - 6)$

   e $2\pi(4^2 - 7)$ **f** $4\pi(2 + \sqrt{4})$

   g $\dfrac{42\pi}{4}$ **h** $3 \times (7 - \sqrt{9})\pi$

7   Simplify these expressions

   a $\pi(6^2 - 4)$ **b** $\sqrt{49} + \pi(7 - 5)$

   c $4(7 + \sqrt{2})$ **d** $\pi(8^2 - \sqrt{4})$

   e $2\pi(3 - 2\pi) + (2\pi)^2$

   f $\pi^2 - \pi(\pi - 4)$

8   Elisa has a square piece of card with sides 8 cm.
   She cuts a circle with radius 2 cm from the card.
   Find the exact area of card left over.

*9   Simplify each algebraic expression by collecting like terms.
   Match each expression to one of these cards.

   $\dfrac{3}{4}x - 1$ $\dfrac{1}{4}x + 1$

   $\dfrac{3}{4}x + 1$ $2x + 5\pi$

   $5x + 8\pi$

   $8x + 5\pi$

   a $\sqrt{9} - \sqrt{4} + \dfrac{1}{2}x + \dfrac{1}{4}x$

   b $\pi + \sqrt{64}x + 4\pi - \sqrt{36}x$

   c $\sqrt{25}x - 6\pi + 11\pi + \sqrt{9}x$

   d $\left(\dfrac{3}{8} \div \dfrac{3}{7} - \dfrac{1}{8}\right)x - \dfrac{6^2 + 8^2}{10^2}$

   e $\dfrac{1}{2}x + 3\dfrac{1}{4} - \dfrac{1}{4}x - \left(\dfrac{3}{2}\right)^2$

   f $2\pi(\sqrt{81} - \sqrt{49}) + \left(\dfrac{7}{4} \times 3\right)x + 4\pi - \dfrac{1}{4}x$

10   Geoff is laying a rectangular lawn with semicircular flowerbeds at each end.

   11.1

   The rectangle is 6 m wide by $8\dfrac{1}{2}$ m long.

   a What is the perimeter of the rectangle?
   b What is the perimeter of the entire shape?
   c What is the area of the lawn?
   d What is the area of a flowerbed?
   e What is the area of the entire shape?

Q 1017, 1040, 1047   SEARCH

17.2   Exact calculations

HOW TO

1. When using fractions, work with exact fractions as long as possible.
2. You may need to use your knowledge of length, area and volume formulae.
3. Simplify expressions and collect terms with π together.

EXAMPLE

A bicycle wheel rotates 3000 times. The diameter of the wheel is 50 cm.

Write an exact expression for how far the bicycle travels in km.

50 cm

1. The circumference is equal to one rotation of the wheel.

$C = 50 \times \pi \, cm = \frac{1}{2}\pi \, m$

2. $50 \, cm = \frac{1}{2} m$. Use exact fractions.

Distance $= \frac{1}{2}\pi \times 3000$ The wheel rotates 3000 times.

3. $= 15000\pi \, m$ travelled.

Convert to km.

$15000\pi \div 1000 = 15\pi \, km$

EXAMPLE

15.2

Find an exact expression for the total volume of these three shapes.

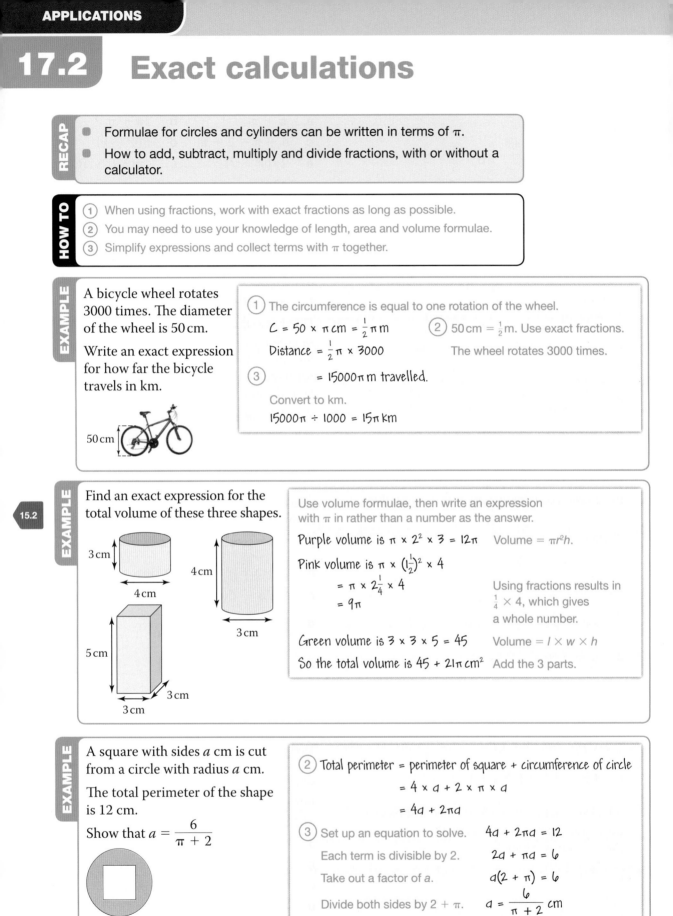

3 cm

4 cm

4 cm

5 cm

3 cm

3 cm

3 cm

Use volume formulae, then write an expression with π in rather than a number as the answer.

Purple volume is $\pi \times 2^2 \times 3 = 12\pi$ Volume $= \pi r^2 h$.

Pink volume is $\pi \times (1\frac{1}{2})^2 \times 4$

$= \pi \times 2\frac{1}{4} \times 4$ Using fractions results in

$= 9\pi$ $\frac{1}{4} \times 4$, which gives a whole number.

Green volume is $3 \times 3 \times 5 = 45$ Volume $= l \times w \times h$

So the total volume is $45 + 21\pi \, cm^2$ Add the 3 parts.

EXAMPLE

A square with sides a cm is cut from a circle with radius a cm.

The total perimeter of the shape is 12 cm.

Show that $a = \dfrac{6}{\pi + 2}$

2. Total perimeter = perimeter of square + circumference of circle

$= 4 \times a + 2 \times \pi \times a$

$= 4a + 2\pi a$

3. Set up an equation to solve. $4a + 2\pi a = 12$

Each term is divisible by 2. $2a + \pi a = 6$

Take out a factor of a. $a(2 + \pi) = 6$

Divide both sides by $2 + \pi$. $a = \dfrac{6}{\pi + 2}$ cm

Exercise 17.2A

1 Arrange these cards so they make a true statement.

$\sqrt{100}$ 18 6^2 $\sqrt{64}$ = ÷ −

2 In an addition pyramid, you add the numbers directly below to get the number above.

Complete the addition pyramid.

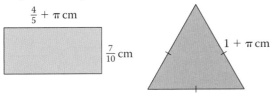

12

$2 + \pi$ $8 - 2\pi$

3 Which shape has the largest perimeter? Explain how you decide.

$\frac{4}{5} + \pi$ cm

$\frac{7}{10}$ cm

$1 + \pi$ cm

4 Joshua's bicycle wheel has a diameter of 0.4 m.

On his journey to work, the wheel rotates 5500 times.

Write an exact expression for how far Joshua travels to work in km.

5 A square tile has area 4 cm². A quadrant has area π cm².

π 4

a Choose four of the six expressions to describe the total area of these shapes.

i

ii

iii

iv

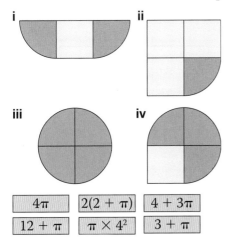

| 4π | $2(2 + \pi)$ | $4 + 3\pi$ |
| $12 + \pi$ | $\pi \times 4^2$ | $3 + \pi$ |

*b Explain why the quadrant and square fit together exactly.

*6 A tin can is 3 inches in diameter and $4\frac{1}{4}$ inches high. It has a base but no lid. It is made from metal which is $\frac{1}{32}$ of an inch thick.

Work out

3 inches

$4\frac{1}{4}$ inches

15.2

a the area of the base

b the circumference of the can

c the curved surface area

*d the volume of metal used

*e the volume of the can.

7 Anna is making a three-tiered wedding cake.

Each layer of the cake is a cylinder.
The top tier has radius 8 cm and height 8 cm.
The middle tier has radius 12 cm and height 12 cm.
The bottom tier has radius 16 cm and height 16 cm.

a Find an exact expression for the total surface area of the cake.

b Anna can buy circular sheets of icing with diameter 40 cm.
She buys five sheets to cover the surface of the cake, except the base. What percentage of the icing will she have left?

8 A circle with radius b cm is cut from a square with sides $2b$ cm.

The total perimeter of the shape is 20 cm.
Show that $b = \frac{10}{\pi + 4}$ cm

Q 1017, 1040, 1047   SEARCH

17.3 Standard form

You can use **standard form** to represent large and small numbers.

● In standard form, a number is written as $A \times 10^n$, where $1 \leqslant A < 10$ and n is an integer.

EXAMPLE

Write these numbers in standard form.

a 235 **b** 0.23×10^6

c 0.45 **d** 0.000 000 416

a $235 = 2.35 \times 10^2$ **b** $0.23 \times 10^6 = 2.3 \times 10^5$

c 4.5×10^{-1} **d** 4.16×10^{-7}

EXAMPLE

Write these numbers in order, starting with the smallest.

$6.35 \times 10^4, \quad 5.44 \times 10^4, \quad 6.95 \times 10^3, \quad 7.075 \times 10^2, \quad 9.9 \times 10^{-1}$

The correct order is

$9.9 \times 10^{-1}, \quad 7.075 \times 10^2, \quad 6.95 \times 10^3, \quad 5.44 \times 10^4, \quad 6.35 \times 10^4$

Then compare numbers with the same powers of 10: $5.44 < 6.35$.

First order the powers of 10: $10^{-1} < 10^2 < 10^3 < 10^4$.

● You can multiply and divide in standard form. Multiply or divide the numbers and then multiply or divide the powers.

EXAMPLE

Calculate

a $(3.6 \times 10^5) \div (1.2 \times 10^3)$ **b** $(5.4 \times 10^4) \times (2 \times 10^{-3})$

a $(3.6 \times 10^5) \div (1.2 \times 10^3) = (3.6 \div 1.2) \times (10^5 \div 10^3)$
$= 3 \times 10^{(5-3)} = 3 \times 10^2$

b $(5.4 \times 10^4) \times (2 \times 10^{-3}) = (5.4 \times 2) \times (10^4 \times 10^{-3})$
$= 10.8 \times 10^{4-3} = 1.08 \times 10^2$

● You can add and subtract in standard form. Write the numbers as ordinary numbers, and then add or subtract the numbers. Give your answer in standard form.

EXAMPLE

Calculate $3.2 \times 10^5 + 7.1 \times 10^4$.
Give your answer in standard form.

$3.2 \times 10^5 + 7.1 \times 10^4 = 320\,000 + 71\,000$
$= 391\,000 = 3.91 \times 10^5$

● You need to know how to enter standard form calculations into your calculator and how to interpret the display.

EXAMPLE

Use your calculator to work out $(6.43 \times 10^6) \div (4.21 \times 10^{-2})$.

| 1.5273 159 14 08 |

$(6.43 \times 10^6) \div (4.21 \times 10^{-2}) = 1.53 \times 10^8$ (to 3 sf)

Exercise 17.3S

1 Write these numbers as powers of 10.

   a 100   **b** 10

   c 1000   **d** 1

   e 10 000   **f** 1 000 000

   g 100 000   **h** 100 000 000

2 Write these numbers as powers of 10.

   a 0.01   **b** 0.1

   c 0.001   **d** 0.00001

   e 0.0001   **f** 0.000 000 1

   g 0.000 001   **h** 1.0

3 Convert to ordinary numbers.

   a 10^3 **b** 10^6

   c 10^5 **d** 10^9

   e 10^4 **f** 10^1

   g 10^2 **h** 10^7

4 Convert to ordinary numbers.

   a 10^0 **b** 10^{-2}

   c 10^{-5} **d** 10^{-3}

   e 10^{-7} **f** 10^{-1}

   g 10^{-4} **h** 10^{-6}

5 Work out these calculations and give your answers in index form.

   a $10^2 \times 10^3$ **b** $10^4 \times 10^5$

   c $10^5 \times 10^3$ **d** $10^6 \div 10^3$

   e $10^8 \div 10^4$ **f** $10^6 \div 10^2$

6 Work out these calculations and give your answers in index form.

   a $10^4 \div 10^6$ **b** $10^3 \div 10^7$

   c $10^2 \div 10^{10}$ **d** $10 \div 10^9$

   e $1 \div 10^8$ **f** $10^7 \div 10$

7 Write these numbers in standard form.

   a 200   **b** 800

   c 9000   **d** 650

   e 6500   **f** 952

   g 23.58   **h** 255.85

   i 0.3   **j** 0.0047

   k 0.000 078   **l** 0.4485

8 These numbers are in standard form. Write each of them as an 'ordinary' number.

   a 5×10^2 **b** 3×10^3

   c 1×10^5 **d** 2.5×10^2

   e 4.9×10^3 **f** 3.8×10^6

   g 7.5×10^{11} **h** 8.1×10^{18}

9 Although they are written as multiples of powers of 10, these numbers are not in standard form. Rewrite each of them correctly in standard form.

   a 60×10^1 **b** 45×10^3

   c 0.65×10^1 **d** 0.05×10^8

   e 28×10^{-2} **f** 0.4×10^{-1}

   g 13.5×10^{-4} **h** 12×10^{-8}

10 Evaluate these calculations, giving your answers in standard form. Do not use a calculator.

   a $(2 \times 10^2) \times (2 \times 10^3)$

   b $(3 \times 10^4) \times (3 \times 10^3)$

   c $(5 \times 10^3) \times (5 \times 10^4)$

   d $(8 \times 10^7) \times (3 \times 10^5)$

   e $(2.5 \times 10^{-3}) \times (2 \times 10^2)$

   f $(4.6 \times 10^{-6}) \times (2 \times 10^{-2})$

11 Evaluate these calculations, showing your working. Do not use a calculator; give your answers in standard form.

   a $(4 \times 10^4) \div (2 \times 10^2)$

   b $(8.4 \times 10^9) \div (4.2 \times 10^5)$

   c $(2 \times 10^6) \div (4 \times 10^4)$

   d $(3 \times 10^5) \div (4 \times 10^2)$

12 Use a calculator to find the value of these in standard form.

   a $(6.4 \times 10^{-4}) + (7.1 \times 10^{-3})$

   b $(9.9 \times 10^5) - (2.7 \times 10^4)$

   c $(4.8 \times 10^{-6}) + (3.9 \times 10^{-5})$

   d $(3.3 \times 10^2) - (7.5 \times 10^1)$

   e $(9.8 \times 10^5) - (6.4 \times 10^5)$

   f $(3.5 \times 10^{-2}) + (9.7 \times 10^{-3})$

17.3 Standard form

- You can write a number in standard form as $A \times 10^n$, where n is a positive or negative integer and $1 \leqslant A < 10$.

$1570000 = 1.57 \times 10^6$

$0.0000204 = 2.04 \times 10^{-5}$

HOW TO

1. Read the question carefully and decide which calculation you need to carry out.

2. Calculate using your knowledge of standard form.

3. Give your answer in standard form and check that your answer is a sensible order of magnitude.

EXAMPLE

The mass of a carbon atom is 2×10^{-23} g.

How many atoms are there in one gram of carbon?

You would expect there to be a very large number of atoms, so the answer is sensible.

1. Divide 1 g by the mass of one carbon atom.

2. Write both numbers in standard form.

$1 \div (2 \times 10^{-23}) = 1 \times 10^0 \div (2 \times 10^{-23})$

Divide the numbers and then divide the powers of 10.

$= (1 \div 2) \times (10^0 \div 10^{-23})$

$= 0.5 \times 10^{0 - -3}$

$= 0.5 \times 10^{23}$

3. Give your answer in standard form.

There are 5×10^{22} atoms in one gram of carbon.

EXAMPLE

A butterfly weighs 4.5×10^{-4} kg. A ladybird weighs 2.1×10^{-5} kg.

An ant weighs 3.4×10^{-5} kg. An ant can carry 20 times its own weight.

Could an ant carry a butterfly and a ladybird?

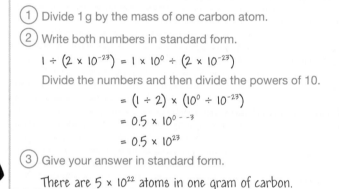

1. Compare the combined weight of a butterfly and a ladybird to 20 times the weight of an ant.

Weight of butterfly and ladybird $= 4.5 \times 10^{-4} + 2.1 \times 10^{-5}$

$= 0.00045 + 0.000021$

$= 0.000471$

$= 4.71 \times 10^{-4}$ kg

Weight an ant can carry $= 20 \times 3.4 \times 10^{-5}$

$= (2 \times 3.4) \times (10 \times 10^{-5})$

$= 6.8 \times 10^{-4}$ kg

2. Writing the answers in standard form makes it easy to compare the weights.

$6.8 \times 10^{-4} > 4.71 \times 10^{-4}$

An ant could carry a butterfly and a ladybird.

Exercise 17.3A

1 Write these measurements using standard form.

   a One hundredth of a kilometre

   b Two thousandths of a gram

   c Five millionths of a metre

   d 11 thousandths of a litre

2 Orla thinks she has answered her maths homework correctly in standard form. Check her answers and correct any mistakes.

   a $3.2 \times 10^5 \times 2.5 \times 10^6 = 8 \times 10^{11}$

   b $5.2 \times 10^8 \times 6.3 \times 10^4 = 32.8 \times 10^{12}$

   c $4 \times 10^5 + 3 \times 10^4 = 7 \times 10^9$

   d $8 \times 10^3 \div 5 \times 10^6 = 1.6 \times 10^{-3}$

3 The American value for a billion is 1×10^9. The British value for a billion used to be 1×10^{12}.
How many times smaller was an American billion than a British billion?

4 The mass of the Sun is 2×10^{30} kg. The mass of the Earth is 6×10^{24} kg. How many times heavier than the Earth is the Sun?

5 The mass of one atom of the element mercury is 3.3×10^{-22} g. The mass of the planet Mercury is 3.3×10^{23} kg.
How many mercury atoms are there in 3.3×10^{23} kg?

6 The width of a plant cell is 60 micrometres. A micrometre is 1×10^{-6} m (one millionth of a metre).
The diagram of a plant cell in a science textbook has width 3 cm.
How many times bigger is the diagram than the real plant cell?

7 Light travels about 3×10^8 metres per second.

   a Find the time it takes for light to travel 1 metre.

   b Find the distance light travels in 1 year. Give your answers in standard form.

8 A mass of 12 grams of carbon contains about 6.0×10^{23} carbon atoms.

   a Write 12 grams as a mass in kilograms using standard form.

   b Estimate the mass of one carbon atom, giving your answer in kilograms in standard form, correct to 2 significant figures.

9 A bumblebee weighs 5.2×10^{-5} kg. An adult man weighs 70 kg. A bumblebee can carry 75% of its weight.

How many bumblebees would it take to lift a man?
Give your answer in standard form to 3 sf.

10 As the moon orbits Earth the distance between them varies between 4.06×10^5 km and 3.63×10^5 km. Find the difference between these two distances.

11 The masses of the eight planets in our solar system are listed in the table.

Mercury	3.30×10^{23}
Venus	4.87×10^{24}
Earth	5.97×10^{24}
Mars	6.42×10^{23}
Jupiter	1.90×10^{27}
Saturn	5.68×10^{26}
Uranus	8.68×10^{25}
Neptune	1.02×10^{26}

The masses are given in kg.

Carrie calculated the total mass of the planets.
Her answer is $7.686\,12 \times 10^{26}$ kg.

   a Which planet did Carrie forget to include in her total?

   b Use your calculator to work out the correct total mass.

1049, 1050, 1051   SEARCH

Summary

Checkout

You should now be able to...

	Test it Questions
✓ Calculate with roots and with integer indices.	**1, 2**
✓ Calculate exactly with fractions and multiples of π.	**3 – 6**
✓ Calculate with and interpret numbers written in standard form.	**7 – 9**

Language

Language	Meaning	Example
Index **Base** **Power**	In index notation, the index or power shows how many times the base has to be multiplied by itself. The plural of index is **indices**.	Index $2^3 = 2 \times 2 \times 2 = 8$ Base
Index laws	The rules for how to multiply, divide or raise to a power numbers written as powers of the *same* base.	$5^3 \times 5^2 = 5^{3+2} = 5^5$ $5^3 \div 5^2 = 5^{3-2} = 5^1 = 5 \qquad 5^0 = 1$ $(5^3)^2 = 5^{3 \times 2} = 5^6$
Square root	A number that when multiplied by itself is equal to the number underneath the square root symbol.	$4 \times 4 = 16$ is the square of 4 $\sqrt{16} = 4$ is the square root of 16
Cube root	A number that when multiplied by itself and then by itself again is equal to the number underneath the cube root symbol.	$4 \times 4 \times 4 = 64$ is the cube of 4 $\sqrt[3]{16} = 4$ is the cube root of 64
Terminating	A decimal with a definite number of digits.	$\frac{1}{8} = 0.125$
Recurring	A decimal with a repeating pattern that goes on forever.	$\frac{1}{3} = 0.333... = 0.\dot{3}$ $\frac{9}{11} = 0.818181... = 0.\dot{8}\dot{1}$
Pi, π	The ratio of a circle's circumference to its diameter.	Circumference $= \pi \times$ diameter $\pi = 3.141\,592\,653\,589\,793$ (15 dp)
Exact calculation	A calculation that does not involve decimals that have been rounded or other approximations. Exact answers are given in terms of integers, fractions and π.	$2 \times \pi \times \left(\frac{1}{3}\right)^2 + 2\pi \times \frac{1}{3}$ $= 2\pi \times \frac{1}{9} + \frac{2}{3}\pi - \left(\frac{2}{9} + \frac{2}{3}\right)\pi$ $= \frac{8}{9}\pi$
Standard form	A number written as a decimal between 1 and 10 multiplied by a power of 10.	$498000 = 4.98 \times 10^5$ $0.0056 = 5.6 \times 10^{-3}$

Review

1 Simplify these expressions.

a $5^3 \times 5^7$ **b** $3^{12} \div 3^8$

c $(7^4)^3$ **d** $2^4 \times 2^3 \div 2^7$

e $\dfrac{3^2 \times 3^7}{3^3 \times 3^3}$ **f** $\left(\dfrac{4^8 \div 4^3}{4^3}\right)^2$

2 Write down the numerical value of these expressions.

a $2^3 \times 2^4$ **b** $4^3 \div 4$

c 5^1 **d** $(2^3)^2$

e 8^0 **f** $3^{12} \div 3^9$

g $\sqrt{36}$ **h** $\sqrt[3]{8}$

3 Find the exact answers to these calculations.

a $\dfrac{3}{10} \times 2$ **b** $\dfrac{2}{7} \times \dfrac{3}{4}$

c $\dfrac{2}{5} + \dfrac{1}{5}$ **d** $\dfrac{7}{8} - \dfrac{1}{3}$

e $\dfrac{9}{8} + \dfrac{1}{6}$ **f** $3\dfrac{1}{4} - 1\dfrac{1}{5}$

g $\dfrac{11}{3} \times \dfrac{2}{3}$ **h** $3\dfrac{1}{7} \times \dfrac{1}{4}$

i $\dfrac{6}{7} \div 3$ **j** $\dfrac{1}{9} \div \dfrac{2}{3}$

4 Calculate the areas of these shapes.

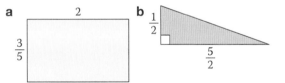

5 Simplify these expressions.

a $5 + 2\pi - 3 + \pi$ **b** $\pi(3^2 - \sqrt{16})$

c $\dfrac{1}{2} \times \dfrac{\pi}{3} + \dfrac{\pi}{4}$ **d** $\dfrac{4}{3}\pi\left(\dfrac{3}{2}\right)^3$

6 a Calculate the exact areas of the circle and the semi-circle.

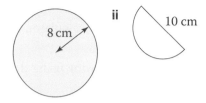

b Calculate the exact circumference of the circle and the perimeter of the semi-circle.

7 Write these approximations in standard form.

a The population of China in 2014: 1 370 000 000

b The closest Earth comes to Mars: 54 600 000 miles

c The amount of vitamin C in an orange: 0.0697 g

d The size of a grain of sand: 0.0000625 m

8 Write these as ordinary numbers.

a 3.5×10^5 **b** 8.21×10^8

c 2.7×10^{-3} **d** 2.07×10^{-7}

9 Calculate these and leave your answer in standard form.

a $(2 \times 10^7) \times (3 \times 10^3)$

b $(3.5 \times 10^6) \times (2 \times 10^{-2})$

c $(8 \times 10^9) \div (2 \times 10^4)$

d $(1.2 \times 10^{-3}) \div (2 \times 10^{-2})$

e $2.4 \times 10^5 + 2.4 \times 10^3$

f $3.6 \times 10^3 - 1.8 \times 10^2$

What next?

Score			
	0 – 4		Your knowledge of this topic is still developing. To improve look at MyMaths: 1017, 1033, 1040, 1047, 1049, 1050, 1051, 1924
	5 – 8		You are gaining a secure knowledge of this topic. To improve look at InvisiPens: 17Sa – c
	9		You have mastered these skills. Well done you are ready to progress! To develop your exam technique look at InvisiPens: 17Aa – c

Assessment 17

1 Peter says that   **a** $0^2 = 1$ **b** $49^{\frac{1}{2}} = 7$ **c** $(-3)^2 = -9$

Soraya says that   **a** $0^2 = 0$ [1]   **b** $49^{\frac{1}{2}} = \frac{1}{49}$ [1]   **c** $(-3)^2 = 9$ [1]

Without using your calculator, for each value who is correct?
Show your working.

2 Work out the following using the appropriate keys on a calculator.

   a 2.1^3 [1]   **b** 7.2^4 [1]   **c** 1.99^{10} [1]

   d $\sqrt{841}$ [1]   **e** $\sqrt{77}$ [1]   **f** $\sqrt[3]{654.321}$ [1]

3 **a** Paige says that to work out $15^7 \times 15^5$ you multiply the indices. Is she correct?
Work out the expression and leave your answer as a single power. [1]

   b Hayley says that to work out $(3^4)^5$ you multiply the indices. Is she correct?
Work out the expression and leave your answer as a single power. [1]

   c Matt says that $(3^4)^0 = 3^4$. Eliza says that $(3^4)^0 = 1$. Lois says that $(3^4)^0 = 3$.
Who is correct? Give reasons for your answer. [2]

   d Mike says that $7^7 \times 7^2 \div 7^6 = 7^8$. Show that Mike is incorrect. [1]

4 A dice is a cube of side 2.3 cm. Calculate the volume of the dice. [2]

5 **a** Joe is batting in a cricket tournament. He scores 6 sixes in each of 6 games
every day for 6 days. How many sixes does Joe score? [2]

   b Joss scores x sixes in each of x games every day for x days, getting 125 sixes overall.
Find x. [2]

6 Find the values of a, b and c that make these expressions correct.

   a $4^2 \times 64 = 4^a$ [2]   **b** $8^b = 4^6$ [2]   **c** $(3^2)^4 = 9^c$ [2]

7 Use four different digits to make this sum correct. $\dfrac{\square}{\square} + \dfrac{\square}{\square} = \dfrac{1}{2}$

Write down four different answers. [4]

8 240 pupils sat an exam and only 48 answered the last question correctly.

   a What fraction of the pupils got the question right? [2]

   b How many pupils got the question wrong? [1]

9 The Chiefs played the Saints at rugby. There were 55 points scored in the match.

The Chiefs scored $\frac{8}{11}$ of those points.

   a How many points did the Chiefs score? [2]

   b How many points did the Saints score? [1]

10 A vegetable box has potatoes, carrots and leeks in it.

$\frac{3}{5}$ of the box is potatoes and $\frac{1}{3}$ is carrots. There are 4 leeks.

   a What fraction of the box is made up of potatoes and carrots together? [2]

   b What fraction of the box is leeks? [1]

   c How many potatoes are there in the box? [4]

11 Romeo and Juliet are saving up to get married.

Romeo saves $\frac{7}{25}$ of his salary and Juliet saves $\frac{7}{20}$ of hers.

Romeo's annual salary is £15 500 and Juliet's is £13 400.

    a   Who saves most annually and by how much?   [4]

    b   How much do they save in total during the year?   [2]

12 An athlete is $13\frac{3}{4}$ stone when he starts weight training.

After six months, he weighs $9\frac{5}{8}$ stone.

What fraction of his original weight is his new weight?   [3]

13 **a**   The areas of some of our oceans and seas in mi^2 are shown.
       Convert them to ordinary numbers and write them in increasing order of size.

Malay Sea: $3.14 \times 10^6\,\text{mi}^2$	Indian Ocean: $2.84 \times 10^7\,\text{mi}^2$
Bering Sea: $8.76 \times 10^5\,\text{mi}^2$	Caribbean Sea: $1.06 \times 10^6\,\text{mi}^2$
English Channel: $2.9 \times 10^4\,\text{mi}^2$	Baltic Sea: $1.46 \times 10^5\,\text{mi}^2$

                                                        [4]

    b   Which is bigger: 1×10^9 or 999 999 999 and by how much?   [2]

    c   Find the value of n in each of the following equations.

         i $4.7 \times 10^n = 47\,000$ [1]   **ii** $6.81 \times 10^n = 681$ [1]

         iii $3.467 \times 10^n = 3\,467\,000$ [1]   **iv** $27.5 \times 10^n = 0.0275$ [2]

    d   Rewrite each of these sentences using ordinary numbers.

         i   The energy released per second by the wingbeat of a bee is 8×10^{-4} joules/s.   [1]

         ii   The distance from Mexico City to Moscow is $1.0763 \times 10^4\,\text{km}$.   [1]

    e   Rewrite this sentence using standard form.

        The average length of a bedbug is $\dfrac{4}{1\,000\,000}$ km thick.   [1]

    f   What is this length in mm?   [1]

14 **a**   The Wright Brothers 'Flyer 1', the world's first aircraft, had a mass of $3.4 \times 10^2\,\text{kg}$.
       The Saturn V Rocket had a mass of $2.96 \times 10^6\,\text{kg}$.
       How many Flyers have the same mass as a Saturn?   [2]

    b   'Flyer 1' attained a speed of $3.04 \times 10^0\,\text{ms}^{-1}$ on its first flight and the supersonic airliner, Concorde, attained a speed of $2.179 \times 10^3\,\text{kmh}^{-1}$.
       How many times faster was the Concorde than 'Flyer 1'?   [4]

15 There are approximately 4.336×10^9 stars in our galaxy and about 5.776×10^3 stars visible to the naked eye.

    a   What fraction of the galaxy can we see?   [2]

    b   Write your fraction in the form $\frac{1}{x}$.   [1]

16 A triathlon has 3 stages, the largest triathlon has a $3.8 \times 10^0\,\text{km}$ swim, $1.8 \times 10^2\,\text{km}$ cycle ride and $0.42195 \times 10^2\,\text{km}$ run.

How far is the race in full? Give your answer to 3 sf

    a   as an ordinary number   [3]

    b   in standard form.   [1]

18 Graphs 2

Introduction

When you look at a ball in flight, you already know certain things about its path, or trajectory. It will travel in a smooth curve. It will reach a maximum point. Its downward path will tend to be a mirror image of the upwards path. In football, a goalkeeper knows this as well, so a striker might put spin on the ball to make its flight less predictable.

The path of the ball can be modelled mathematically by a type of equation called a quadratic equation, which is described in this chapter.

What's the point?

An appreciation of quadratic equations and their graphs enables us to understand how an object moves under gravity, and tells us where it's likely to land!

Objectives

By the end of this chapter you will have learned how to …

- Draw graphs to identify and interpret roots, intercepts and turning points of quadratic functions.
- Solve a quadratic equation by finding approximate solutions using a graph.
- Recognise, sketch and interpret graphs of linear, quadratic and simple cubic functions.
- Recognise, sketch and interpret the reciprocal function $y = \frac{1}{x}$.
- Plot and interpret real-life graphs.

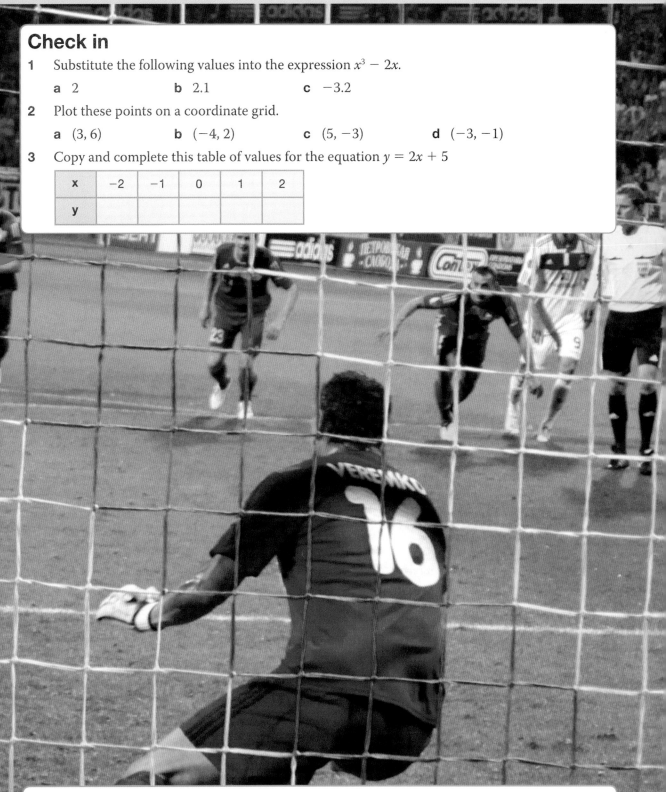

Check in

1 Substitute the following values into the expression $x^3 - 2x$.

   a 2   **b** 2.1   **c** -3.2

2 Plot these points on a coordinate grid.

   a $(3, 6)$ **b** $(-4, 2)$ **c** $(5, -3)$ **d** $(-3, -1)$

3 Copy and complete this table of values for the equation $y = 2x + 5$

x	−2	−1	0	1	2
y					

Chapter investigation

It takes 20 people 18 days to build an extension to a sports centre.

How long would it take one person? State any assumptions that you have made.

Find out how long the job would take for different numbers of people.
Can you draw a graph to show this information?

18.1 Properties of quadratic functions

A **quadratic** function includes a 'squared' term, for example x^2.
These are all quadratic functions:

x^2 $\qquad\qquad x^2 + 3$ $\qquad\qquad x^2 + 3x - 1$ $\qquad\qquad 2x - x^2$

- To draw a graph of a quadratic function:
  - Draw a table of values
  - Calculate the value of y for each value of x
  - Draw a suitable grid
  - Plot the (x, y) pairs and join them with a smooth curve.

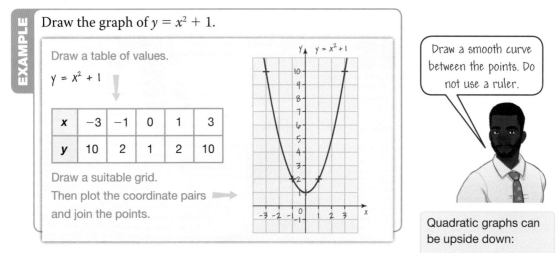

EXAMPLE

Draw the graph of $y = x^2 + 1$.

Draw a table of values.

$y = x^2 + 1$

x	−3	−1	0	1	3
y	10	2	1	2	10

Draw a suitable grid.
Then plot the coordinate pairs ➡
and join the points.

Draw a smooth curve between the points. Do not use a ruler.

Quadratic graphs can be upside down:

Maximum point

Graphs of quadratic functions have a distinctive shape.

- The graph of a quadratic function
  - is always a U-shaped curve
  - is symmetrical about a vertical line
  - always has a maximum point or a minimum point.

EXAMPLE

Plot the curve $y = x^2 - x$ for $-2 \leqslant x \leqslant 2$ and find the coordinates of its minimum point.

Make a table of values.

x	−2	−1	0	1	2
x^2	4	1	0	1	4
$-x$	2	1	0	−1	−2
$y = x^2 - x$	6	2	0	0	2

Draw the x-axis from −2 to 2 and the y-axis from −1 to 7.

Plot the points (−2, 6), (−1, 2), ..., (2, 2).

Join the points in a smooth curve.

The **minimum** point is $\left(\frac{1}{2}, -\frac{1}{4}\right)$.

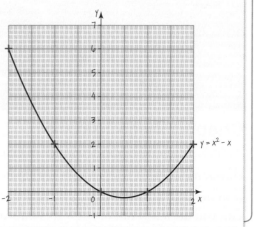

Exercise 18.1S

1 **a** Copy and complete the table of values for $y = x^2$.

x	−3	−1	0	1	3
y					

b Draw a pair of axes from 0 to 10 on the y-axis and from −5 to +5 on the x-axis.

c Plot the coordinate pairs on the grid.

d Join the points with a smooth curve.

2 Draw the graphs of $y = x^2 + 2$ and $y = x^2 + 3$ on the same axes as your graph from question **1**. What do you notice?

3 Draw the graphs of $y = x^2 − 1$ and $y = x^2 − 3$ on the same pair of axes. What do you notice?

4 Match these graphs to their equations.

a $y = x^2 + 1$

b $y = x^2 − 2$

c $y = 2 + x^2$

d $y = x^2 − 1$

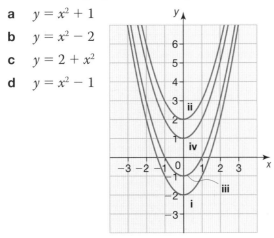

5 **a** Copy and complete this table to show if each graph will be a straight line or a parabola.

b Add an equation of your own in each column.

Straight line	Parabola

$y = 3x − 2$

$y = x^2 − 2$

$3x + 2y = 8$

$y = 10 + x^2$ $y = x^2 + 2x + 1$ $y = x$

6 **a** Draw axes labelled from −4 to 4 on the x-axis and −5 to +15 on the y-axis.

b Copy and complete this table for $y = x^2 − 2$.

x	−4	−3	−2	−1	0	1	2	3	4
x^2	16							9	
$y = x^2 − 2$	14							7	

c Plot the points that you have found in part **b** on your axes from part **a**. Join them to form a smooth parabola.

7 For each equation

i make a table with x-values from −4 to 4 and find the corresponding y-values

ii draw an x-axis from −4 to 4 and a suitable y-axis

iii plot the points and join them to form a parabola

iv write the coordinates of the minimum point of each parabola.

a $y = x^2 + 3$

b $y = 2x^2$

c $y = 3x^2 − 1$

d $y = x^2 + x$

> In part **b**, square before you multiply by 2.

8 **a** Copy and complete the table of values for the graph $y = x^2 + x + 1$.

x	−3	−2	−1	0	1	2	3
x^2	9						
y	7				3		

b Plot the points for x and y and join them to form a smooth parabola.

c What is the approximate minimum value of $x^2 + x + 1$ and for what value of x does it occur?

9 Imagine that these quadratic graphs are drawn.

A $y = 4 + 3x − x^2$ **B** $y = x^2 + 2x − 4$

C $y = 2x^2 − 4$ **D** $y = x^2 − 3x − 4$

Which graphs have the same y-intercept?

18.1 Properties of quadratic functions

- The graph of a quadratic function
  - is always a U-shaped curve
  - is symmetrical about a vertical line
  - always has a maximum point or a minimum point.

Quadratic graphs can be upside down:

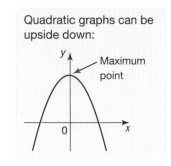

HOW TO

To use a graph to solve a quadratic equation or real-life problem

① Draw up a table of values. Calculate the values of y for at least three x-values. You may decide to write a formula for the situation in the question first.

② Draw a suitable grid, label the axes and plot the graph.

③ Use your graph to give the answer in the context of the question.

EXAMPLE

The parabola $y = 1.2x - 0.02x^2$ models a javelin throw.

y = the height of a javelin

x = the horizontal distance.

a How far was the javelin thrown?

b Find the maximum height of the javelin.

① Make a table of values.

x	0	10	20	30	40	50	60
y	0	10	16	18	16	10	0

② Plot the points and draw a smooth curve to join the points.

a ③ The javelin lands when $y = 0$.

The javelin was thrown 60 m.

b The maximum height was 18 m.

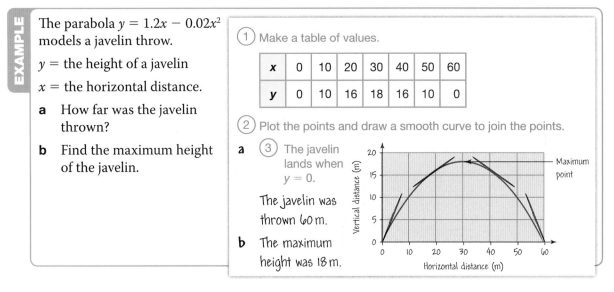

Quadratic equations can be solved using a graph.

EXAMPLE

Solve $x^2 + 5x + 6 = 2$

① Make a table of values.

x	−5	−4	−3	−2	−1	0
x²	25	16	9	4	1	0
5x	−25	−20	−15	−10	−5	0
y	6	2	0	0	2	6

② Plot the points and draw a smooth curve to join the points.

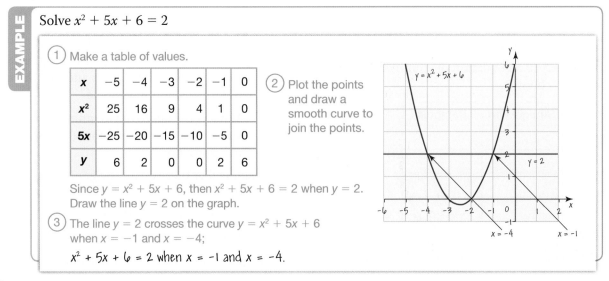

Since $y = x^2 + 5x + 6$, then $x^2 + 5x + 6 = 2$ when $y = 2$. Draw the line $y = 2$ on the graph.

③ The line $y = 2$ crosses the curve $y = x^2 + 5x + 6$ when $x = -1$ and $x = -4$;

$x^2 + 5x + 6 = 2$ when $x = -1$ and $x = -4$.

Algebra Graphs 2

Exercise 18.1A

1 Sketch graphs of these functions showing clearly all their main features.

   a $y = 10 + 3x - x^2$

   b $y = x^2 - 6x + 9$

  *c $y = 2x^2 - 3x - 5$

2 A ball is thrown into the air. The formula $y = 20x - 4x^2$ shows its height, y metres, above the ground x seconds after it is thrown.

   a Copy and complete the table of values to show the height of the ball during its first five seconds.

Time (x)	0	1	2	3	4	5
20x						
4x^2						
Height (y)						

   b Use the table to plot a graph to show the ball's height against time.

   c Use your graph to find

     i the maximum height reached by the ball and the time at which it reaches this height

    ii two times when the ball is 12 metres above the ground

    iii the interval of time when the ball is above 15 metres.

3 The height above ground of a javelin is modelled by the function $h = 100 + 48t - t^2$ where h = height in centimetres and t = length of throw in metres.

   a Complete the table and plot the graph of the function.

t	0	10	20	30	40	50	60
h							

   b Find the maximum height of the javelin.

   c State the length of the throw.

4 Some graphs are drawn on these axes.

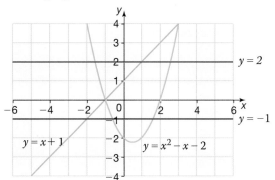

Use the graphs to find the approximate solutions of

   a $x^2 - x - 2 = 2$

   b $x^2 - x - 2 = -1$

   c $x^2 - x - 2 = x + 1$

5 The graph shows $y = x^2 - 2x - 3$.

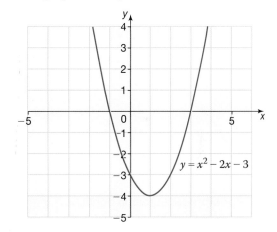

Copy the graph and, by adding lines, use it to find the approximate solutions of

   a $x^2 - 2x - 3 = 1$

   b $x^2 - 2x - 3 = -3$

   c $x^2 - 2x - 3 = -4$

   d $x^2 - 2x - 3 = x - 2$

   e $x^2 - 2x - 3 = 1 - x$

   f $x^2 - 2x - 3 = 0$

*6 Draw appropriate graphs to find the approximate solutions of

   a $x^2 - 2 = 5$ **b** $x^3 + x = 2x - 1$

   c $2x^2 - x = 0$ **d** $x^3 - x^2 = 2$

Q 1168, 1169, 1959   SEARCH

18.2 Sketching functions

- A **cubic** equation contains a term in x^3. It has a distinctive **S-shaped** graph.

Draw the graph of $y = x^3 - 1$ and use it to estimate the value of y when $x = 2.5$.

First make a table of x and y values.

x	-3	-2	-1	0	1	2	3
x^3	-27	-8	-1	0	1	8	27
$y = x^3 - 1$	-28	-9	-2	-1	0	7	26

$(-3)^3 = -27$

Draw the x-axis from -3 to 3 and the y-axis from -30 to 30.
Plot the points $(-3, -28)$, $(-2, -9)$ and so on, and join them in a smooth curve.

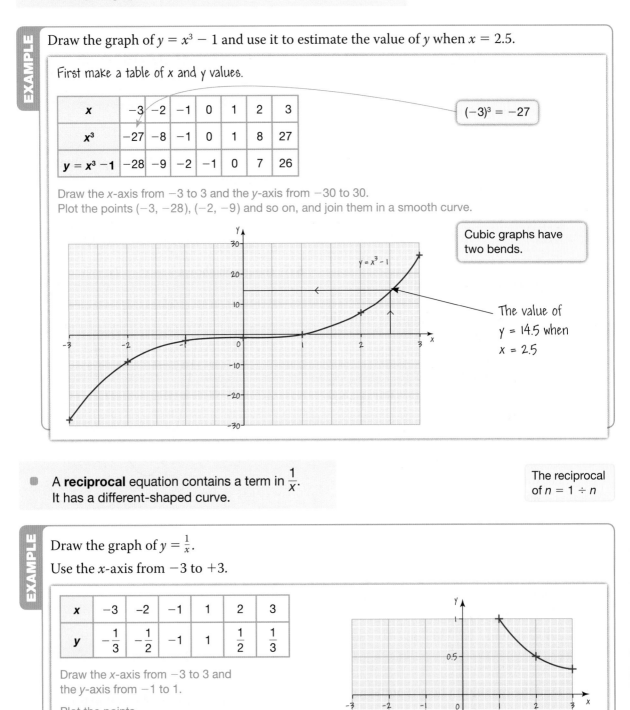

Cubic graphs have two bends.

The value of $y = 14.5$ when $x = 2.5$

- A **reciprocal** equation contains a term in $\frac{1}{x}$.
  It has a different-shaped curve.

The reciprocal of $n = 1 \div n$

Draw the graph of $y = \frac{1}{x}$.

Use the x-axis from -3 to $+3$.

x	-3	-2	-1	1	2	3
y	$-\frac{1}{3}$	$-\frac{1}{2}$	-1	1	$\frac{1}{2}$	$\frac{1}{3}$

Draw the x-axis from -3 to 3 and the y-axis from -1 to 1.

Plot the points.

Note that you cannot plot $x = 0$ as $\frac{1}{0}$ is not a defined value.

Algebra Graphs 2

Exercise 18.2S

1 a Which equations are cubic?

b Which equations are reciprocal?

$$x = 4 \qquad y = x^3 - x - 6 \qquad y = 5x$$

$$y = x^3 \qquad y = 3 - \frac{2}{x} \qquad y = \frac{4}{x}$$

2 a Draw an x-axis from -3 to 3 and a y-axis from -30 to $+30$.

b Copy and complete the table for $y = x^3 + 1$.

x	−3	−2	−1	0	1	2	3
y	−26						28

when $x = -3$, $y = (-3)^3 + 1 = -27 + 1$

c Plot the coordinates that you have found in part **b** on your axes from part **a**. Join them to form a smooth, S-shaped curve.

> When using your calculator, remember brackets.

d Use your graph to estimate the value of y when

i $x = 1.5$ **ii** $x = 0.5^3 + 1$

3 For each equation, copy and complete the table of values. Plot these points on suitable axes and join them to form a smooth curve.

a $y = x^3 - 4$

x	−2	−1	0	1	2	3
x³					8	
x³ − 4					4	

b $y = \frac{2}{x}$

x	−2	−1	1	2	3
$\frac{1}{x}$				$\frac{1}{2}$	
y				1	

3 c $y = x^3 + x + 1$

x	−2	−1	0	1	2	3
x³					8	
x + 1					3	
y					11	

d $y = \frac{3}{x} + 1$

x	−2	−1	1	2	3
$\frac{3}{x}$	$-\frac{3}{2}$				
y	$-\frac{1}{2}$				

4 Draw graphs of these functions for the range of x-values given.

a $y = x^3 + 3x$, for $-3 \leqslant x \leqslant 3$

b $y = x^3 + x - 2$, for $-3 \leqslant x \leqslant 3$

c $y = x^3 + x - 4$, for $-3 \leqslant x \leqslant 3$

d $f(x) = x^3 - x^2 + 3x$, for $-3 \leqslant x \leqslant 3$

5 Draw graphs of these functions for the range of x-values given.

a $y = -x^3 - 3x$, for $-3 \leqslant x \leqslant 3$

b $y = 2 - x^3 - x$, for $-3 \leqslant x \leqslant 3$

c $y = 4 - x^3 - x$, for $-3 \leqslant x \leqslant 3$

d $f(x) = -3x - x^3 + x^2$, for $-3 \leqslant x \leqslant 3$

e Compare your graphs with the graphs in question **4**. What do you notice?

6 a Plot the graph $f(x) = x^3 - 2x^2 + x + 4$, for $-3 \leqslant x \leqslant 3$.

b Use your graph to find

i the value of x when $y = -20$

ii the value of y when $x = 1.7$.

7 Plot these functions for the range given.

a $y = \frac{12}{x - 2}$, for $-2 \leqslant x \leqslant 6$

b $y = \frac{x}{x + 4}$, for $-4 \leqslant x \leqslant 4$

c $f(x) = \frac{6}{x + x} - 2$ for $-3 \leqslant x \leqslant 3$

d $f(x) = 2 + \frac{1}{1 - x}$, for $-3 \leqslant x \leqslant 3$

Q 1071, 1172, 1958, 1960   SEARCH

18.2 Sketching functions

- A linear function is of the form $y = mx + c$. The graph is a straight line with gradient $= m$ and y-intercept $= c$.
- A quadratic function is of the form $y = ax^2 + bx + c$.
- A cubic function is of the form $y = ax^3 + bx^2 + cx + d$.
- A reciprocal function is of the form $y = \dfrac{a}{x}$.

▲ Linear   ▲ Quadratic

▲ Cubic   ▲ Reciprocal

HOW TO

Here is a general strategy for solving problems involving quadratic graphs. It can be adapted to other types of function.

1. Identify the roots and y-intercept of the function.
2. Sketch a graph.
3. Identify whether the roots or turning point help solve the problem.
4. Interpret the solution.

EXAMPLE

A company manufactures and sells packs of batteries. If the selling price of their batteries is too low they make no profit. If the selling price is too high they do not sell enough to make a profit. The company works out that their profit is a function of price:

$$P = (2 - s)(s - 5)$$

where P = profit (p) and s = selling price (£).

a At what selling price will the company start losing money?

b What selling price will maximise their profit? How much profit will they make in this case?

1. The roots of the function are $s = 2$ and $s = 5$.

   $x = 0 \Rightarrow y = 2 \times -5 = -10$

   The P-intercept $= -10$.

2. (diagram)

a If the company charges more than £5 they will make a loss. (4)

b 3. The maximum point is needed. Since a parabola is symmetrical the maximum occurs halfway between the roots: $(5 + 2) \div 2 = 3.5$.

   4. A selling price of £3.50 will maximise their profits. By substitution,
   $P = -3.5^2 + 7 \times 3.5 - 10 = 2.25$. Their profit is $2.25\,p$.

EXAMPLE

18.1

Oona is asked to plot the graph of $y = (x + 3)(14 - x)$.
The diagram shows her correct solution.
Oona is then asked to write the function in the form $y = ax^2 + bx + c$.
She multiplies out the brackets in the function and writes

$y = (x + 3)(14 - x) \Rightarrow y = x^2 + 11x + 42$

How does the graph show you that Oona's solution is *wrong*?

When Oona multiples out the brackets she has written an x^2 term that is positive.

When a parabola has a maximum it has an x^2 term that is negative.

The correct answer is $y = -x^2 + 11x + 42$. So $a = -1$, $b = 11$ and $c = 42$.

Algebra Graphs 2

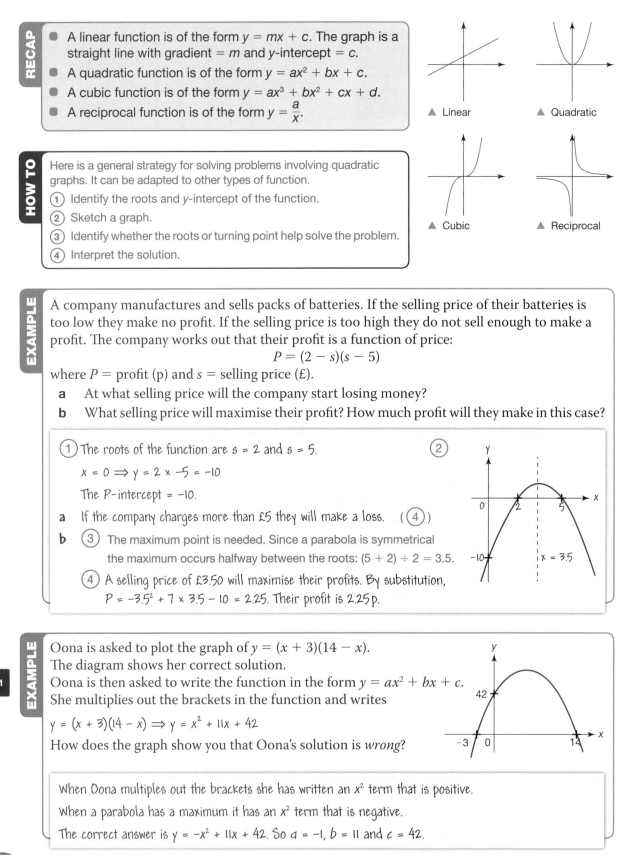

Exercise 18.2A

1 a Sketch the graph of the function
$y = x(8 + x)$.

b Write the function in the form
$y = ax^2 + bx + c$.
Write the values of a, b and c.

2 A company establishes a profit function of
$P = (1 - s)(s - 5)$ where P = profit (p) and
s = selling price (£).

a Sketch the graph of the profit function.

b At what prices would the profit be zero?

c What selling price will maximise their
profit? How much profit will they make
in this case?

3 A company uses a profit function of
$P = (50 - s)(s - 120)$ where P = profit (£)
and s = selling price (£).

a Sketch the graph of the profit function.

b At what prices would the profit be zero?

c What selling price will maximise their
profit? How much profit will they make
in this case?

4 A company uses a profit function of
$P = (15 - s)(s - 60)$ where P = profit (£)
and s = selling price (£).

a What is the maximum profit they expect
to make?

b What other information is provided by
the profit function?

5 An arrow is fired. Its height above ground is
modelled by the function $h = (39 - t)(t + 1)$
where h = height in centimetres and
t = length of shot in metres.

a Sketch the graph of the function.

b Find the maximum height of the arrow.

c State the length of the shot.

*6 The jet of water in a
fountain is modelled
by the function
$y = 10x(0.5 - x)$
where x = distance
from source (m) and
y = height (m).

*6 a At what distance from the source does
the jet enter the water again?

b What is the maximum height reached
by the jet?

c Write the function in the form
$y = ax^2 + bx + c$.

7 Match each of the functions **a** to **d** to their
graphs and their factorised forms.

a $y = 7x - x^2$ **b** $y = x^2 - 8x + 16$

c $y = -x^2 + 3x + 4$ **d** $y = x^2 - 4$

A

B

C

D

1 $y = (4 - x)(1 + x)$

2 $y = (x - 4)(x - 4)$

3 $y = x(7 - x)$

4 $y = (x + 2)(x - 2)$

8 Sketch the graph of the function
$y = x (x + 2)(x + 1)$.

9 Another cubic function is written as
$y = (x + 1)(x + 4)(x - 2)$. Sketch the graph
of this function.

Use graphing software for questions **10** and **11**.

10 a Plot the graph of $y = ax + a$.

b What do you notice? Explain why this
happens.

11 a Plot the graph of $y = x^3 + a$. Use the
constant controller to vary the value
of a. What do you notice?

b Repeat for each of the following

i $y = (x + a)^3$ **ii** $y = \dfrac{a}{x}$

iii $y = \dfrac{1}{x} + a$

 Q 1071, 1172, 1958, 1960 SEARCH

18.3 Real-life graphs

● The shape of a graph shows the **trend**.

The graph shows the numbers of video recorders sold over a 10 year period.

The trend is that the number of video recorders sold is **decreasing**.

You can read information from a graph, but read the axis labels and scale carefully.

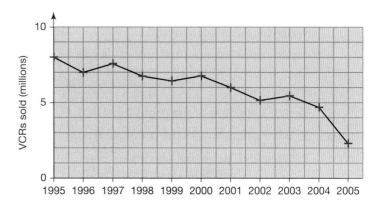

EXAMPLE

The graph shows the amount of rainwater in a barrel over a few days.

a   On day 1 it rained heavily. What happened to the amount of water in the barrel?

b   On which day was 25 litres poured out of the barrel?

c   What happened to the amount of water on day 2? Suggest a reason for this.

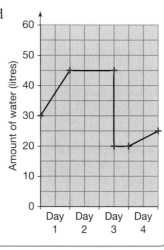

a   The amount of water increased.

b   Day 3, as the amount suddenly reduced by 25 litres.

c   Amount of water stayed the same. It probably did not rain on day 2 and no water was poured out.

Think what could affect the amount of water.

● A straight line shows that a quantity is changing at a steady rate.

The steeper the slope, the faster the change.

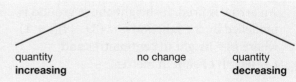

quantity **increasing**   no change   quantity **decreasing**

The graphs show the water level as two tanks fill with water. Water is poured into both tanks at a steady rate.

The first tank fills more slowly, as it is wider. The second tank fills more quickly, as it is narrower.

The steeper the slope, the faster the change in water level.

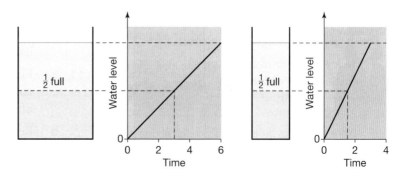

Algebra   Graphs 2

Exercise 18.3S

1 The graph shows sales of 'Time 2 Chat' mobile phones.

a How many phones were sold in March?

b How many phones were sold in June?

c How many more phones were sold in June than in January?

d Here are the sales figures for the next three months.

Month	October	November	December
Number of phones sold	400	325	400

Copy the graph and complete it for this information.

e What happened to sales in November and December?

Suggest a reason for this.

f What overall trend does the graph show?

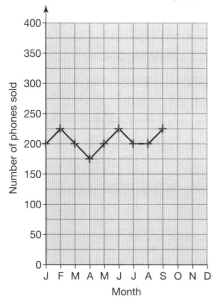

2 The graph shows how a bean plant grew from a seed over several weeks.

a How tall was the plant after 6 weeks?

b How much did the plant grow between weeks 8 and 10?

c How tall did the plant grow in total?

d How much did the plant grow between 15 and 20 weeks?

2 e Is the plant likely to reach a height of 3 metres? Explain your answer.

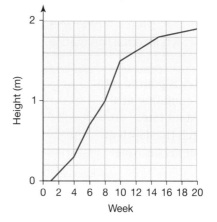

3 Match the four sketch graphs with the containers.

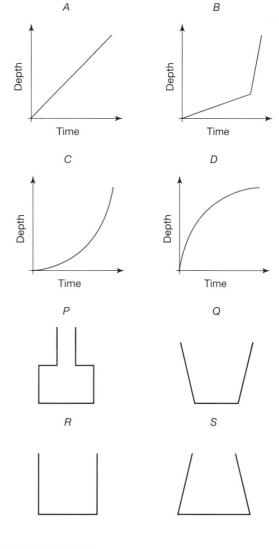

Q 1184, 1322   SEARCH

18.3 Real-life graphs

- The gradient of a line shows how fast a quantity is changing.
  - A straight line implies a constant rate of change.
  - A horizontal line means there is no change.

EXAMPLE

Water is poured into this container at a constant rate.
Sketch the depth of water versus time.

Container widens – rate of filling slows.

Container narrow – fills quickly.

Container wide – fills slowly.

HOW TO

To identify the type of function that can be used to model some data
1. Plot the points on a graph and draw a smooth curve through them.
2. Compare the shape of the data with the common types of function. You may need to stretch or translate the curve.
3. ATQ

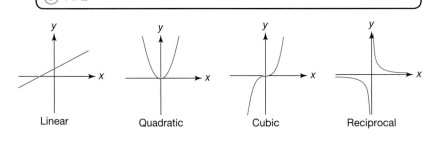

Linear   Quadratic   Cubic   Reciprocal

EXAMPLE

Beth is investigating the features of an engine.
The table shows the power generated by different amounts of torque.

Torque (N/m)	1	2	3	4	5	6	7
Power (W)	22	42	53	63	62	55	40

a Plot the data on a graph.

b What type of function could be used to model the data?

c Estimate the power when a torque of 5.5 N/m is applied.

Torque is a turning force. You can still complete the question without knowing that detail though.

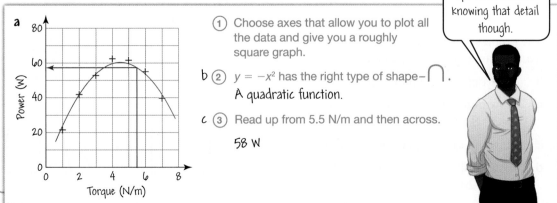

a
① Choose axes that allow you to plot all the data and give you a roughly square graph.

b ② $y = -x^2$ has the right type of shape– ∩.
A quadratic function.

c ③ Read up from 5.5 N/m and then across.

58 W

Algebra Graphs 2

Exercise 18.3A

1 Georgina is playing a computer game.

Describe the type of function that created the path of the bird.

2 The diagrams show four different types of beaker.

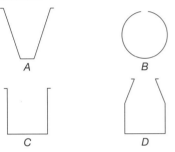

Liquid is poured into a beaker at a constant rate. A graph is plotted to show 'time' on the horizontal axis and 'depth of liquid' on the vertical axis.

a Sketch the graph for each beaker.

b Match each beaker to the type of function that best describes the graph.

Cubic	Quadratic
No standard function	Linear

3 A lottery offers a jackpot of £1000. Entrants choose any three numbers from a set of seven numbers. It is likely that the jackpot will have to be shared.

a Copy and complete the table to show the prizes available in different cases.

Winners	1	2	3	4	5	6
Prize (£)						

b Plot a graph to show this information.

c What type of function describes the relationship?

4 Steve is a scientist developing energy-efficient LCD displays. Power is required to update his display. When this happens, the 'pulse size' (in volts) is connected to the temperature. Steve has this set of data

°C	−10	−5	0	10	23	40
Volts	20.5	17.7	16	13.6	12.5	12.1

a Plot the data on a graph.

b What type of function best describes the data?

c Steve has worked out a function that connects temperature and pulse size. He uses his formula to work out that if temperature = −20°C, then pulse size = 30 V. He also knows that the pulse size cannot be less than 11.8 V. Do these facts confirm your reasoning in **b**? Explain why.

Use graph plotting software for question 5.

*5 During the 1960s British mathematicians developed radar that could track the position of artillery fire. From this it could be worked out where the fire had been launched.

Artillery fire follows a parabolic path. The diagram shows a simplified example.

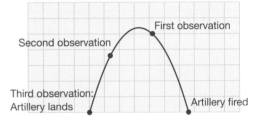

a Plot the points (6, 12), (2, 8) and (0, 0). Use these as three observations. The artillery lands at (0, 0).

b Plot the graph of $y = -a(x + b)^2 + c$.

c Vary the values of a, b and c. Find a quadratic curve that passes through all three points.

d Write the coordinates of the point where the artillery is fired from.

e Work with a partner. Choose three points, including (0, 0). Challenge them to find a quadratic curve that passes through those points.

Summary

Checkout

You should now be able to...

✔	Draw graphs to identify and interpret roots, intercepts and turning points of quadratic functions.	**1, 2**
✔	Solve a quadratic equation by finding approximate solutions using a graph.	**3**
✔	Recognise, sketch and interpret graphs of linear, quadratic, cubic and reciprocal functions.	**4, 5**
✔	Plot and interpret real-life graphs.	**6, 7**

Language   Meaning   Example

Language	Meaning	Example
Quadratic function	A function of the form $ax^2 + bx + c$. They have a characteristic $\cap$- or $\cup$-shape.	
Cubic function	A function of the form $ax^3 + bx^2 + cx + d$. They have a characteristic S-shape.	
Reciprocal function	A function of the form $\dfrac{c}{x}$. They have two parts for negative and positive x.	
Turning point	A point on a curve where the curve changes from rising to falling, $\cap$ or falling to rising, $\cup$	
Root	The points where curve crosses the x-axis.	
y-intercept	The point where a curve crosses the y-axis.	
Solve solution	Find a value for the unknown variable that will make the equation true.	$4x - 3 = 9$ is true when $x = 3$. $x = 3$ is the solution to the equation.

Review

1 Here is the graph of $y = x^2 - 4x - 5$.

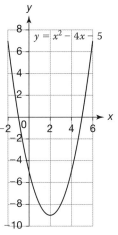

Write down

  a the y-intercept of the graph

  b the x-intercepts of the graph

  c the roots of the equation $x^2 + 4x - 5 = 0$

  d the coordinates of the turning point of the curve.

2 **a** Copy and complete this table of values for $y = x^2$.

x	−4	−3	−2	−1	0	1	2	3	4
y									

  b Draw the graph of $y = x^2$ for values of x between −4 and 4.

  c State the coordinates of the turning point on the curve.

3 **a** Use a table of values, with x from −4 to 4, to draw the graph of $y = x^2 - 3x - 6$.

  b By adding lines, use your graph to find approximate solutions to

    i $x^2 - 3x - 6 = 0$

    ii $x^2 - 3x - 6 = -4$

4 Sketch the graph of $y = \frac{1}{x}$ for $x \neq 0$.

5 The graph shows these four functions.

$$y = x^3 - 4x \qquad y = \frac{2}{x} \qquad y = 2x + 3 \qquad y = x^2 + 3x$$

5 Match each graph to its mathematical name and equation.

> Linear   Reciprocal   Cubic   Quadratic

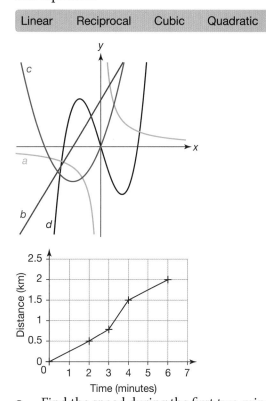

6

  a Find the speed during the first two mins.

  b What was the average speed, in km/h, during the 6 minute journey?

7 **a** Plot a speed–time graph for the information given in this table.

Time (s)	0	4	7	10
Speed (m/s)	5	3	3	0

  b What is happening to the speed during the first 4 seconds?

  c What is happening between 4 s and 7 s?

  d What is the rate of change of speed during the final 3 seconds?

What next?

Score	0 – 3		Your knowledge of this topic is still developing.
			To improve look at MyMaths: 1168, 1169, 1071, 1172, 1180, 1184, 1316, 1322
	4 – 6		You are gaining a secure knowledge of this topic.
			To improve look at InvisiPens: 18Sa – d
	7		You have mastered these skills. Well done you are ready to progress!
			To develop your exam technique look at InvisiPens: 18Aa – c

Assessment 18

1 **a** Draw the graph of $y = x^2 - x - 6$
and $y = x$ on a copy of this grid. [6]

b Estimate the coordinates of the
points where the graphs intersect. [1]

c Estimate the coordinates of the minimum
point on the quadratic curve. [1]

d Use your graph to estimate the solutions of
the equation $x^2 - x - 6 = 0$ [2]

e By factorising, solve the equation $x^2 - x - 6 = 0$.
Compare your answer with your estimate in part **d**. [2]

2 A metal spring stretches when a mass of m grams hung on the end.

The distance stretched, d cm, is given by the formula $d = 8 + \dfrac{m}{15}$.

a Complete the table of values for m and d.
Write your answers to 1 dp where appropriate. [2]

m	10	20	30	40	50	60	70	80	90	100
d										

b Plot the graph of d against m for values of m from 0 to 100. [2]

c Use your graph to

 i find the stretch, d, when a mass of 75 g is hung on the spring [1]

 ii find the mass hung on the spring when $d = 9$ cm. [1]

d Complete this sentence.
If the mass of the weight hung on the spring increases by 1 g,
the distance the spring stretches increases by _____ cm. [1]

e What is the length of the spring when no weights are hung on it? [1]

3 The average safe stopping distance for cars consists of two parts:

a thinking distance $\dfrac{v}{2.7}$ and a braking distance $\dfrac{3v^2}{125}$.

The formula for the safe stopping distance, d metres, is given by the equation

$d = \dfrac{3v^2}{125} + \dfrac{v}{2.7}$, where v is the speed of the car in km/h.

a Complete the table of values for d and v. Write your distances to the nearest metre. [3]

v	0	10	20	30	40	50	60	70	80	90	100
d											

b Draw the graph of $d = \dfrac{3v^2}{125} + \dfrac{v}{2.7}$. [3]

c Use your graph to find the safe stopping distance when a car travels at

 i 25 km/h [1] **ii** 42 km/h [1] **iii** 77 km/h. [1]

4   A crystal glass making firm makes £y profit from the sale of x glasses.

The formula they use to estimate their profits is $y = 5x - 2000 - \dfrac{x^2}{1000}$

a   Complete the table of values.   [3]

x	0	500	1000	1500	2000	2500	3000	3500	4000	4500	5000
y											

b   Use appropriate axes and scales, plot the graph of this function.   [3]

c   Use your graph to estimate

   i   the profit when 900 glasses are made   [1]

   ii   the profit when 3750 glasses are made   [1]

   iii the number of glasses made when the profit is £3750   [1]

   iv   the maximum profit the firm can make   [1]

   v   how many glasses they must sell to make the maximum profit   [1]

   vi   the smallest number of glasses the firm must make to avoid making a loss   [1]

   vii the greatest number of glasses the firm must make to avoid making a loss   [1]

   viii the range of values of x for which the profit is more than £2500.   [1]

5   Match the graphs **A – E** to one of the five statements **i–v**.   [4]

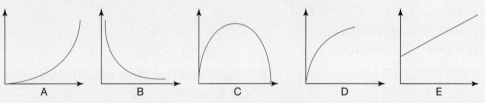

   A   B   C   D   E

a   The cost of gas is a set amount added to a price per unit.

b   The motion of a cricket ball after being thrown.

c   The time taken to drive a set distance compared to the average speed.

d   The height of water in a conical flask being filled at a steady rate.

e   The height of water in a hemispherical flask being filled at a steady rate.

6   a   Taking values of x from -3 to 3, draw the graph of $y = x^3 + x^2 - 12x$.   [6]

   b   On the same grid, draw the graph of $y = 4 - 8x$ [2]

   c   Use your graph to solve the equation $4 - 8x = 0$ [1]

   d   Use your graph to solve the equation $x^3 + x^2 - 12x = 0$ [3]

   e   Find and record the coordinates of the points where the two graphs intersect.   [3]

   f   Complete the sentence: The solutions to $x^3 + x^2 -$ _____ $x -$ _____ $= 0$ are the points
       of intersection of the graphs of $y = x^3 + x^2 - 12x$ and $y = 4 - 8x$ [2]

7   The product of x and y is 36.

   a   Write down a formula for y in the form $y = \dots$ [1]

   b   Draw a suitable graph the equation in part **a** for x values between -12 and 12.   [6]

   c   Use your graph to find a pair of numbers that multiply to give 36 and add
       together to give 13. Why is there only one solution but two points of intersection?   [4]

Revision 3

1 **a** An orchard owner packs apples in boxes. Each box contains the same number of apples. She delivers 240 apples to one shop and 144 to another. What is the largest number of apples in any box? [4]

b Kevin says that the LCM of the numbers 180 and 210 is 36 and the HCF is 1260. Correct his mistakes. [4]

c Show that the least positive whole number that 180 must be multiplied by to make the result a perfect square is 5. [1]

2 Find the value of these expressions.

a 11^2 [1] **b** 2.6^2 [1]

c $\sqrt{441}$ [1] **d** $\sqrt{2.25}$ [1]

e 1^7 [1] **f** $(-3)^3$ [1]

g 3.2^3 [1] **h** $\sqrt[3]{46.656}$ [1]

i $\sqrt[3]{(-592.704)}$ [1] **j** $\sqrt[3]{226.981}$ [1]

3 Shivani writes the following numbers in standard form. Correct his mistakes.

a	8000	8×10^4
b	75000	75×10^5

[1]
[1]

c Each side of a dice is 1.6 cm. Write the volume

i in index form [1]

ii as an ordinary number. [1]

d Hal says that $2^3 \times 2^{-3} = 2^{-9}$. Is he correct? Give your reasons. [2]

4 Zak is filling his watering can. The sketch graph shows the height of water in the can at a particular time.

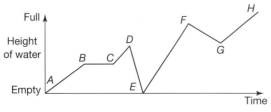

Give reasons for all your answers.

a At what point does Zak's dog knock over the can and spill all water? [2]

b In which section does Zak stop filling the can and water some plants? [2]

4 **c** In which section of the graph is the can filling fastest? [2]

d In which section does Zak stop filling to answer his mobile? [2]

5 **a** Show that the point $(7, -2)$ lies on the line $3y + 4x = 22$. [2]

b Match each equation of a line to its gradient and y-intercept. [3]

i $y = -9 - 5x$ **ii** $y = 2 - 6x$

iii $3x - 2y = 6$ **iv** $4x + 7y = -21$

A $m = -5, c = -9$ **B** $m = 1\frac{1}{2}, c = -3$
C $m = -6, c = 2$ **D** $m = -\frac{4}{7}, c = -3$

c Write down the equations of the straight lines parallel to $y + 4x + 11 = 0$ with these y-intercepts.

i $(0, 3)$ [1]

ii $(0, -7)$ [1]

iii $(0, -245)$ [1]

6 Sarah says that these nets all form cubes. Is she correct? Give reasons for your answers. [3]

7 A ball has an 8.6 cm diameter. Calculate its

a surface area [2]

b volume. [2]

8 A trophy consists of a cuboid, 15 cm by 10 cm by 6 cm. On top of it is a pyramid with base 15 cm by 10 cm. The pyramid is 20 cm high. Find the trophy's total volume. [5]

9 A sphere has a volume of 4000 m³. Using $\pi = 3.142$. Denise calculates its radius to be 9.85 m to 3 sf. Show how she worked this out. [5]

10 The cross section of a prism is a right-angled triangle with hypotenuse 53 cm and the other two sides 28 cm and 45 cm. The thickness of the block is 15 cm. Find

a its volume [4]

b its surface area. [6]

11 A garage checked 104 vehicles for signs of tyre wear.

Depth of tyre tread (mm)	Frequency
$1.0 \leqslant t < 1.6$	19
$1.6 \leqslant t < 2.1$	13
$2.1 \leqslant t < 2.6$	27
$2.6 \leqslant t < 3.1$	25
$3.1 \leqslant t < 3.6$	17
$3.6 \leqslant t < 4.0$	3

a Write down

   i the modal class [1]

   ii the median interval. [2]

b The legal limit for tread on tyres is 1.6 mm. How many tyres failed? [1]

c Calculate an estimate for the mean. [5]

12 The data shows the distances thrown compared with the lengths of the arms of some discus throwers.

Arm length (cm)	58	61	63	67	69	70	72	75	76	77
Distance thrown (m)	52	55	57	58	62	64	67	70	72	74

a Draw a scatter diagram of the data with a line of best fit. [4]

b Comment on the type of correlation. [2]

c Use your line to estimate

   i the distance thrown corresponding to an arm length of 65 cm. [1]

   ii the arm length corresponding to a distance of 65 m. [1]

13 The table shows UK unemployment rates for the period Oct 2012 to Sept 2014.

Month	Oct 2012	Nov 2012	Dec 2012	Jan 2013	Feb 2013	Mar 2013
Rate (%)	7.8	7.7	7.8	7.8	7.9	7.8
Month	Apr 2013	May 2013	Jun 2013	Jul 2013	Aug 2013	Sep 2013
Rate (%)	7.8	7.8	7.8	7.7	7.7	7.6

Month	Oct 2013	Nov 2013	Dec 2013	Jan 2014	Feb 2014	Mar 2014
Rate (%)	7.6	7.4	7.1	7.2	7.2	6.9
Month	Apr 2014	May 2014	Jun 2014	Jul 2014	Aug 2014	Sep 2014
Rate (%)	6.8	6.6	6.5	6.4	6.2	6.0

13 a On the same graph draw a time series graph for Oct 2013 – Sep 2014 and Oct 2012 – Sep 2013. [6]

b Compare unemployment rates in these periods. [1]

14 a A gym has 1.2×10^2 members, each of whom use, on average, 4.7×10^3 units of electricity per year. How many units does the gym use in a year? [2]

b The mass of a nitrogen atom is 2.326×10^{-23} g. One litre of air contains 4.3602×10^{-1} g of nitrogen. How many nitrogen atoms are there in one litre of air? [2]

c Fiona says that 3×10^{-2} is bigger than -3×10^2. Is she correct? Why? [2]

d Work out n for each of the following equations.

   i $2.6 \times 10^n = 2600$ [1]

   ii $3.76 \times 10^n = 3\,760\,000$ [1]

   iii $20.4 \times 10^n = 20\,400$ [1]

15 A bridge goes over a river. y is the height (m), of the arch of the bridge above the river level and x is the distance (m), from the north bank of the river.

$$y = \frac{3x}{2} - \frac{x^2}{40}$$

a Draw the graph of $y = \frac{3x}{2} - \frac{x^2}{40}$ for values of x in tens from 0 to 60. [6]

b The top of the mast of a yacht is 19 m above water level. How far from the bank does the yacht have to be to be able to pass under the bridge? [2]

c What width of river is available if the top of the funnel of a boat is 23 m above water level? [1]

19 Pythagoras and trigonometry

Introduction

The highest mountain in the world is Mount Everest, located in the Himalayas. Its peak is now measured to be 8848 metres above sea level.

The mountain was first climbed in 1953, by Edmund Hilary and Sherpa Tenzing, almost 100 years after its height was first measured as part of the Great Trigonometrical Survey of India in 1856. The original surveyors, who included George Everest, obtained a height of 8840 metres by measuring the distance and angle of elevation between Mount Everest and a fixed location.

What's the point?

Once a right-angled triangle is seen in a particular problem then a mathematician only needs two pieces of information to be able to calculate all the other lengths and angles.

Objectives

By the end of this chapter you will have learned how to …

● Use the formulae for Pythagoras' theorem: $a^2 + b^2 = c^2$.

● Use the trigonometric ratios and apply them to find angles and lengths in right-angled triangles: $\sin \theta = \dfrac{\text{Opp}}{\text{Hyp}}$ $\cos \theta = \dfrac{\text{Adj}}{\text{Hyp}}$ $\tan \theta = \dfrac{\text{Opp}}{\text{Adj}}$

● Know the exact values of $\sin \theta$ and $\cos \theta$ for $\theta = 0°, 30°, 45°, 60°$ and $90°$.

● Know the exact value of $\tan \theta$ for $\theta = 0°, 30°, 45°$ and $60°$.

● Write column vectors and draw vector diagrams.

● Add, subtract and find multiples of vectors.

Check in

1 Work out each of these.

    a 7^2 **b** $4^2 + 6^2$ **c** $3^2 + 5^2$

    d $8^2 - 4^2$ **e** $7^2 - 2^2$ **f** $\sqrt{17^2 - 15^2}$

2 Rearrange these equations to make x the subject.

    a $y = \dfrac{x}{6}$ **b** $y = \dfrac{x}{5}$ **c** $y = \dfrac{x}{10}$

    d $y = \dfrac{2}{x}$ **e** $y = \dfrac{5}{x}$ **f** $y = \dfrac{8}{x}$

Chapter investigation

An engineering company is building a ski lift.
The height of the lift is exactly 45 m.
An engineer suggests using a cable of length 200 m.

The maximum angle of incline is 12°.
Does the engineer's lift meet the criteria?
Design a lift that meets the criteria using the smallest possible length of cable.

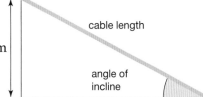

19.1 Pythagoras' theorem

In a right-angled triangle the **hypotenuse** is the longest side. It is opposite the right angle.

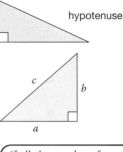

- **Pythagoras' theorem** states
  For any right-angled triangle, $c^2 = a^2 + b^2$ where c is the hypotenuse.

EXAMPLE

Find the length of the hypotenuse.

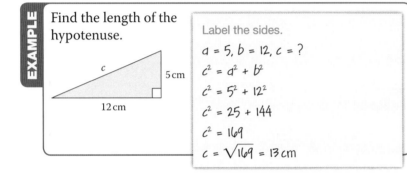

Label the sides.

$a = 5, b = 12, c = ?$
$c^2 = a^2 + b^2$
$c^2 = 5^2 + 12^2$
$c^2 = 25 + 144$
$c^2 = 169$
$c = \sqrt{169} = 13\,cm$

If all three sides of a right-angled triangle are whole numbers, then the numbers make a Pythagorean triple.

You can use Pythagoras' theorem to find any side given two other sides.

- To find the hypotenuse use $c^2 = a^2 + b^2$
- To find a shorter side use $a^2 = c^2 - b^2$ or $b^2 = c^2 - a^2$

Remember that c is the longest side.

EXAMPLE

Calculate the unknown length in these right-angled triangles.

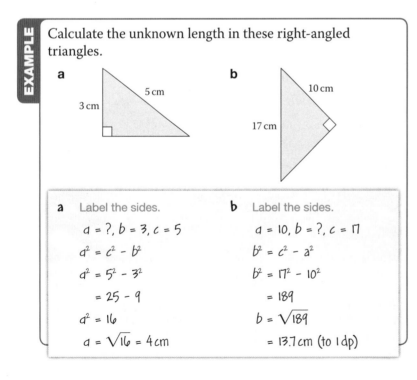

a 5 cm, 3 cm

b 10 cm, 17 cm

a Label the sides.

$a = ?, b = 3, c = 5$
$a^2 = c^2 - b^2$
$a^2 = 5^2 - 3^2$
$\quad = 25 - 9$
$a^2 = 16$
$a = \sqrt{16} = 4\,cm$

b Label the sides.

$a = 10, b = ?, c = 17$
$b^2 = c^2 - a^2$
$b^2 = 17^2 - 10^2$
$\quad = 189$
$b = \sqrt{189}$
$\quad = 13.7\,cm$ (to 1 dp)

Round the answer when it is not exact. One decimal place is sensible here.

Geometry Pythagoras and trigonometry

Exercise 19.1S

1 Calculate the area of these squares. State the units of your answers.

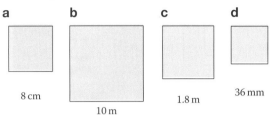

a   b   c   d

8 cm   10 m   1.8 m   36 mm

2 Calculate the length of a side of these squares. State the units of your answers.

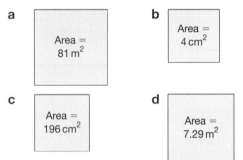

a   Area = 81 m²

b   Area = 4 cm²

c   Area = 196 cm²

d   Area = 7.29 m²

3 Calculate the unknown area for these right-angled triangles.

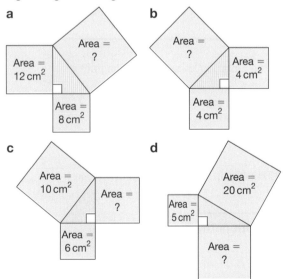

a   Area = ?   Area = 12 cm²   Area = 8 cm²

b   Area = ?   Area = 4 cm²   Area = 4 cm²

c   Area = 10 cm²   Area = ?   Area = 6 cm²

d   Area = 20 cm²   Area = 5 cm²   Area = ?

4 Find the length of the hypotenuse in each of these right-angled triangles.

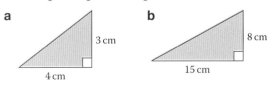

a   3 cm   4 cm

b   8 cm   15 cm

4

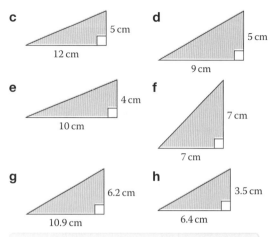

c   5 cm   12 cm

d   5 cm   9 cm

e   4 cm   10 cm

f   7 cm   7 cm

g   6.2 cm   10.9 cm

h   3.5 cm   6.4 cm

> Give answers in this exercise to 1 dp where appropriate.

5 In some of the triangles in question **4**, all three sides have integer (whole number) values. Such sets of three numbers are called Pythagorean triples.
List the Pythagorean triples from question **4**.

6 Find the length of the missing side in each of these right-angled triangles.

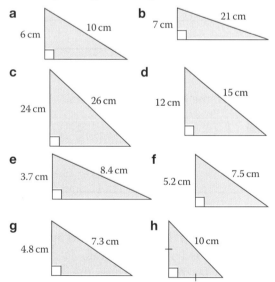

a   6 cm   10 cm

b   7 cm   21 cm

c   24 cm   26 cm

d   12 cm   15 cm

e   3.7 cm   8.4 cm

f   5.2 cm   7.5 cm

g   4.8 cm   7.3 cm

h   10 cm

7 Some of the triangles in question **6** are Pythagorean triples.

   a Write down the Pythagorean triples in question **6**.

   b Compare these with your answers to question **5**.

   c Comment on anything you notice.

 Q 1053, 1112   SEARCH

19.1 Pythagoras' theorem

- **Pythagoras' theorem** states

  For any right-angled triangle, $c^2 = a^2 + b^2$ where c is the hypotenuse.

- To find the length of a shorter side use $a^2 = c^2 - b^2$ or $b^2 = c^2 - a^2$

To solve a problem using Pythagoras' theorem

① Sketch a diagram. Label the right angle and the sides a, b and c.

② Substitute the values into the formula.

③ Round your answer to a suitable degree of accuracy and include any units.

PITAGORA

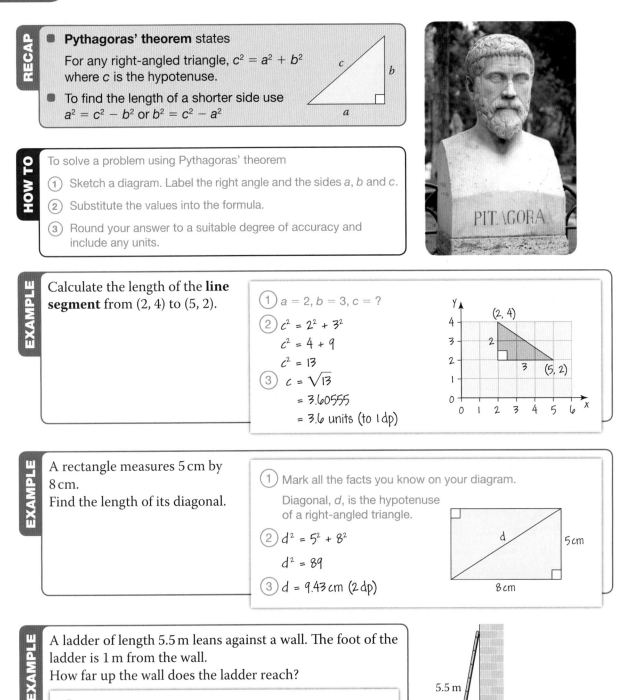

EXAMPLE

Calculate the length of the **line segment** from (2, 4) to (5, 2).

① $a = 2, b = 3, c = ?$

② $c^2 = 2^2 + 3^2$

$c^2 = 4 + 9$

$c^2 = 13$

③ $c = \sqrt{13}$

$= 3.60555$

$= 3.6$ units (to 1 dp)

EXAMPLE

A rectangle measures 5 cm by 8 cm.

Find the length of its diagonal.

① Mark all the facts you know on your diagram.

Diagonal, d, is the hypotenuse of a right-angled triangle.

② $d^2 = 5^2 + 8^2$

$d^2 = 89$

③ $d = 9.43$ cm (2 dp)

EXAMPLE

A ladder of length 5.5 m leans against a wall. The foot of the ladder is 1 m from the wall.

How far up the wall does the ladder reach?

5.5 m

1 m

① The ladder makes a right-angled triangle with the wall.

$a = 1, b = ?, c = 5.5$

② Use the formula $b^2 = c^2 - a^2$

$b^2 = 5.5^2 - 1^2$

$b^2 = 29.25$

③ Round to 1 dp.

$b = 5.4$ m

Exercise 19.1A

1 Calculate the distance between these points. Give your answers to a suitable degree of accuracy.

    a $(1,2)$ and $(4,6)$ **b** $(2,2)$ and $(6,5)$

    c $(1,2)$ and $(2,5)$ **d** $(0,5)$ and $(4,1)$

    e $(3,6)$ and $(6,0)$ **f** $(-3,7)$ and $(3,2)$

2 Find the length of the diagonal of each rectangle.

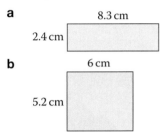

    a 8.3 cm, 2.4 cm

    b 6 cm, 5.2 cm

3 A rectangle has one side 4 cm and diagonal 10.4 cm. Find the length of the other side.

4 Find the length of the diagonal of a square with side length 8 cm.

5 Find the length of the side of a square with diagonal length 8 cm.

6 An isosceles triangle has base 10 cm.

The other two sides of the triangle are each 13 cm.

    a Find the height of the triangle.

    b Find the area of the triangle.

7 The top of a 4-metre ladder leans against the top of a wall. The wall is 3.8 metres high.

How far from the wall is the bottom of the ladder?

3.8 m   4 m   ?

8 The diagram shows a path across a rectangular field.

How much further is it from A to C along the sides of the field than along the path?

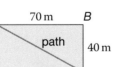

A   70 m   B
path
40 m
D   C

9 PQR and PRS are right-angled triangles. Find the length PS.

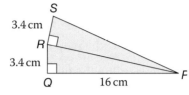

S
3.4 cm
R
3.4 cm
Q   16 cm   F

10 ABC and ACD are right-angled triangles. Find the length AB.

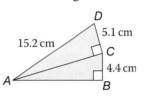

D
5.1 cm
15.2 cm
C
4.4 cm
A   B

11 A square and an isosceles triangle are joined together.
Find the total perimeter of the new shape.

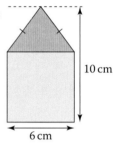

10 cm
6 cm

12 Prove that a triangle with sides 20 cm, 21 cm and 29 cm is right-angled.

13 Jeremy draws this triangle.

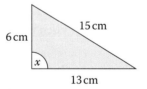

15 cm
6 cm
x
13 cm

    a Explain why angle x cannot be $90°$.

    b Is angle x acute or obtuse?
        Use a sketch to explain your answer.

14 A Pythagorean triple is three whole numbers (a,b,c) that satisfy $a^2 + b^2 = c^2$.

$(3,4,5)$ and $(6,8,10)$ are Pythagorean triples.

$(3,4,5)$ is a primitive Pythagorean triple.

$(6,8,10)$ is not a primitive Pythagorean triple, as it is a multiple of $(3,4,5)$.

There are seven primitive Pythagorean triples with $c < 50$. Can you find them all?

 1053, 1112   SEARCH

19.2 Trigonometry 1

In a right-angled triangle the longest side is the **hypotenuse**.
The side next to the labelled angle is called the **adjacent**.
The side opposite the labelled angle is called the **opposite**.

For a right-angled triangle with angle θ, the ratio $\dfrac{\textbf{opposite}\text{ side}}{\textbf{adjacent}\text{ side}}$
is constant. This ratio is called the **tangent** ratio or **tan**.

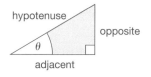

- $\tan \theta = \dfrac{\text{opposite side}}{\text{adjacent side}}$

EXAMPLE

Calculate the value of tan θ in these triangles.

a

b

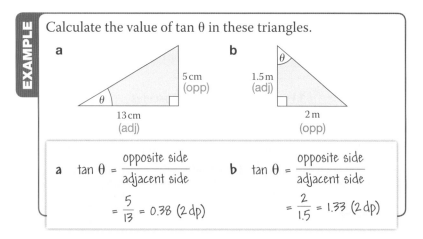

a $\quad \tan \theta = \dfrac{\text{opposite side}}{\text{adjacent side}}$

$\qquad = \dfrac{5}{13} = 0.38 \ (2\,dp)$

b $\quad \tan \theta = \dfrac{\text{opposite side}}{\text{adjacent side}}$

$\qquad = \dfrac{2}{1.5} = 1.33 \ (2\,dp)$

▲ The ancient Egyptians used an early form of trigonometry for building pyramids.

- You can use the tan button on your calculator to work out the value of tan for any angle.

EXAMPLE

Work out the value of tan for each angle. Give your answers to 2 dp.

a $\quad \tan 21°$

b $\quad \tan 53°$

a $\quad \tan 21° = 0.383... = 0.38 \ (2\,dp)$

b $\quad \tan 53° = 1.327... = 1.33 \ (2\,dp)$

Check that you can use the buttons on your calculator to work out the values of tan.

This means that you have these triangles.

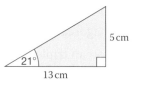

- You can use the inverse function $\tan^{-1}$ to find the angle if you know the opposite and adjacent sides.

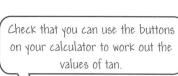

$\tan^{-1}\left(\dfrac{5}{13}\right) = 21.037... = 21° \ (2\,sf)$ $\qquad \tan^{-1}\left(\dfrac{2}{1.5}\right) = 53.130... = 53° \ (2\,sf)$

Geometry Pythagoras and trigonometry

Exercise 19.2S

1 Draw a copy of each triangle.
Label the sides of each triangle with

hypotenuse	adjacent	opposite

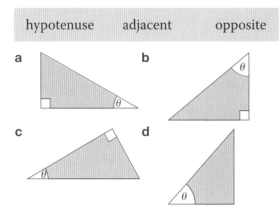

a   **b**   **c**   **d**

2 Use the tan button on your calculator to work out the value of tan for each angle. Give your answers to 2 dp.

a tan 28°   **b** tan 32°   **c** tan 23°

d tan 67°   **e** tan 37°   **f** tan 56°

3 a Work out the value of tan θ in each triangle. Give your answers to 2 dp.

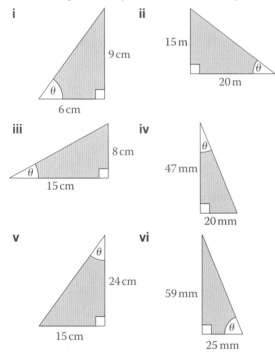

i   **ii**   **iii**   **iv**   **v**   **vi**

b Compare your answers to the values of tan in question **2**. Write down the value of the unknown angle in each triangle.

c Use the tan⁻¹ button on your calculator to check your answers to part **b**.

4 Use the tan⁻¹ button on your calculator to work out the size of the unknown angle in each triangle.

Give your answer to 1 decimal place.

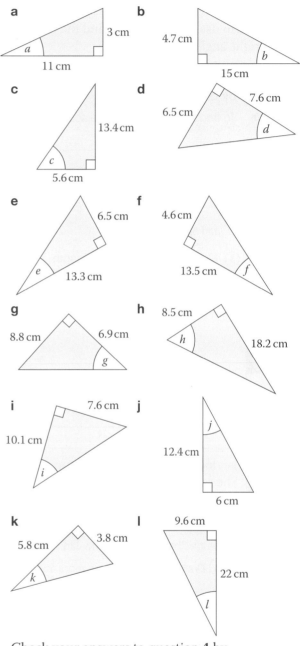

a   **b**   **c**   **d**   **e**   **f**   **g**   **h**   **i**   **j**   **k**   **l**

5 Check your answers to question **4** by constructing an accurate drawing of each triangle.

Could you check your answers using a accurate scale drawing instead?

1133, 1145, 1943   SEARCH

19.2 Trigonometry 1

RECAP

- In a right-angled triangle $\tan \theta = \dfrac{\text{opposite side}}{\text{adjacent side}}$

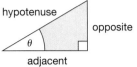

You can use the tan ratio to find the length of an unknown side in a right-angled triangle.

- To find the length of the opposite side, multiply the adjacent by $\tan \theta$.
- To find the length of the adjacent, divide the opposite by $\tan \theta$.

Some values of tan have exact answers.

- $\tan 30 = \dfrac{1}{\sqrt{3}}$
- $\tan 45 = 1$
- $\tan 60 = \sqrt{3}$

You need to learn these for your exam.

HOW TO

1. Draw a copy of the triangle. Label the sides: adjacent, hypotenuse and opposite.

2. Substitute the values that you know into the formula

   $\tan \theta = \dfrac{\text{opposite side}}{\text{adjacent side}}$

3. Find the missing angle or length.

EXAMPLE

Find the missing sides.

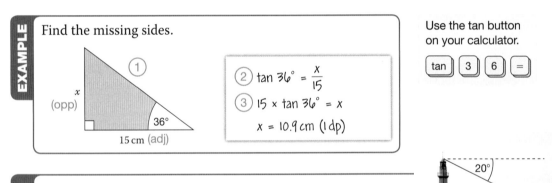

①

② $\tan 36° = \dfrac{x}{15}$

③ $15 \times \tan 36° = x$

$x = 10.9 \, \text{cm} \, (1 \, \text{dp})$

Use the tan button on your calculator.

[tan] [3] [6] [=]

EXAMPLE

Jan looks at a boat from the top of a 40-metre lighthouse. She says the boat is over 100 metres from the lighthouse. Is Jan correct?

State any assumptions you make.

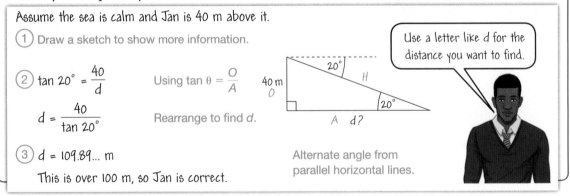

Assume the sea is calm and Jan is 40 m above it.

① Draw a sketch to show more information.

② $\tan 20° = \dfrac{40}{d}$ Using $\tan \theta = \dfrac{O}{A}$

$d = \dfrac{40}{\tan 20°}$ Rearrange to find d.

③ $d = 109.89... \, \text{m}$

This is over 100 m, so Jan is correct.

Alternate angle from parallel horizontal lines.

Use a letter like d for the distance you want to find.

Exercise 19.2A

1 Find the missing side in each triangle.
Give your answers to 3 significant figures.

a

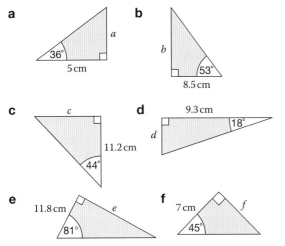

a
36°
5 cm

b

b
53°
8.5 cm

c

c
11.2 cm
44°

d

9.3 cm
18°
d

e

11.8 cm
e
81°

f

7 cm
f
45°

2 Find the missing side in each triangle.
Give your answers to 3 significant figures.

a

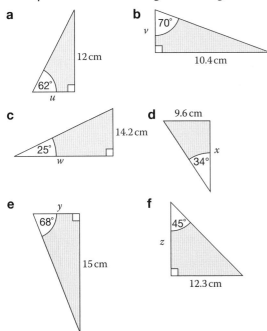

12 cm
62°
u

b

70°
v
10.4 cm

c

25°
14.2 cm
w

d

9.6 cm
34°
x

e

y
68°
15 cm

f

45°
z
12.3 cm

3 A boat is observed from the top of a 50-metre lighthouse. How far is the boat from the cliff?

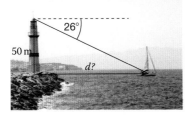

26°
50 m
d?

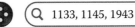

4 Carl looks at Blackpool Tower from the beach. He estimates these measurements.

h?
40°
200 m

a Use Carl's estimates to find the height of Blackpool Tower.

b Why might the answer to **a** be inaccurate?

5 Beth is 1.6 m tall. She looks at the top of a statue that is 8 metres from her. Beth says the statue is over 10 metres tall including the base. Is she correct? Explain your answer.

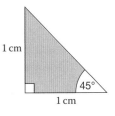

50°
1.6 m
8 m

6 Here is an isosceles right-angled triangle.

a Explain why the angle in the triangle is 45°.

b Use the triangle to explain why tan 45 = 1.

1 cm
45°
1 cm

7 Here is an equilateral triangle with sides 2 cm.
The triangle is folded in half.

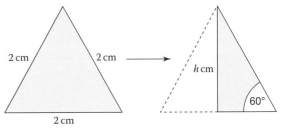

2 cm
2 cm
2 cm

h cm
60°

a Find *h*. Give your answer as an exact square root.

b Use the triangle to explain why tan 60 = $\sqrt{3}$

c Use the triangle to explain why tan 30 = $\frac{1}{\sqrt{3}}$

*8 The diagram shows a flag on top of a building.
How tall is the flag?

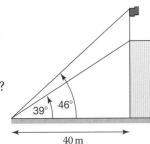

39° 46°
40 m

1133, 1145, 1943   SEARCH

19.3 Trigonometry 2

The **tangent** ratio is: $\tan \theta = \dfrac{\text{opposite side}}{\text{adjacent side}}$

There are two other ratios you can use in **right-angled triangles**.

The **sine** ratio (**sin**) connects the opposite side and the hypotenuse.

The **cosine** ratio (**cos**) connects the adjacent side and the hypotenuse.

> The hypotenuse is the longest side, opposite the right angle.

● **Sine** ratio

$\sin \theta = \dfrac{\text{opposite side}}{\text{hypotenuse}}$

● **Cosine** ratio

$\cos \theta = \dfrac{\text{adjacent side}}{\text{hypotenuse}}$

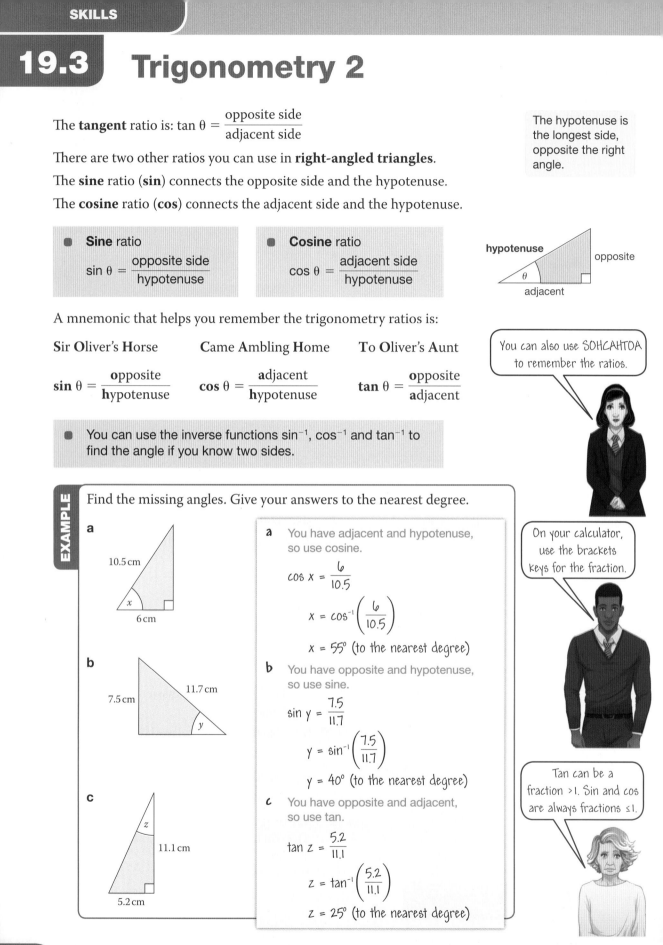

A mnemonic that helps you remember the trigonometry ratios is:

Sir Oliver's Horse	Came Ambling Home	To Oliver's Aunt
$\sin \theta = \dfrac{\text{opposite}}{\text{hypotenuse}}$	$\cos \theta = \dfrac{\text{adjacent}}{\text{hypotenuse}}$	$\tan \theta = \dfrac{\text{opposite}}{\text{adjacent}}$

> You can also use SOHCAHTOA to remember the ratios.

● You can use the inverse functions $\sin^{-1}$, $\cos^{-1}$ and $\tan^{-1}$ to find the angle if you know two sides.

EXAMPLE

Find the missing angles. Give your answers to the nearest degree.

a
10.5 cm
x
6 cm

a You have adjacent and hypotenuse, so use cosine.

$\cos x = \dfrac{6}{10.5}$

$x = \cos^{-1}\left(\dfrac{6}{10.5}\right)$

$x = 55°$ (to the nearest degree)

> On your calculator, use the brackets keys for the fraction.

b
7.5 cm
11.7 cm
y

b You have opposite and hypotenuse, so use sine.

$\sin y = \dfrac{7.5}{11.7}$

$y = \sin^{-1}\left(\dfrac{7.5}{11.7}\right)$

$y = 40°$ (to the nearest degree)

c
z
11.1 cm
5.2 cm

c You have opposite and adjacent, so use tan.

$\tan z = \dfrac{5.2}{11.1}$

$z = \tan^{-1}\left(\dfrac{5.2}{11.1}\right)$

$z = 25°$ (to the nearest degree)

> Tan can be a fraction >1. Sin and cos are always fractions ≤1.

Geometry Pythagoras and trigonometry

Exercise 19.3S

1 Use the cos button on your calculator to work out the value of cos for each angle. Give your answers to 2 dp.

  a cos 20°   **b** cos 35°

  c cos 50°   **d** cos 72°

  e cos 100°   **f** cos 140°

2 Use the $\cos^{-1}$ button on your calculator to work out each calculation.

Give your answer to the nearest whole number.

  a $\cos^{-1}(0.94)$ **b** $\cos^{-1}(0.82)$

  c $\cos^{-1}(0.64)$ **d** $\cos^{-1}(0.31)$

  e $\cos^{-1}(-0.17)$ **f** $\cos^{-1}(-0.77)$

Compare your answers with the angles in question **1**.

3 Use the sin button on your calculator to work out the value of sin for each angle. Give your answers to 2 dp.

  a sin 20°   **b** sin 35°

  c sin 50°   **d** sin 72°

  e sin 100°   **f** sin 140°

4 Use the $\sin^{-1}$ button on your calculator to work out each calculation.

Give your answer to the nearest whole number.

  a $\sin^{-1}(0.34)$ **b** $\sin^{-1}(0.57)$

  c $\sin^{-1}(0.77)$ **d** $\sin^{-1}(0.95)$

  e $\sin^{-1}(0.98)$ **f** $\sin^{-1}(0.64)$

Compare your answers with the angles in question **3**.

5 You can write the values of sin and cos for 90°, 60°, 45°, 30° and 0° exactly.

Match each calculation with one of the exact answers on the cards.

| 0 | $\frac{1}{2}$ | 1 | $\frac{\sqrt{3}}{2}$ | $\frac{1}{\sqrt{2}}$ |

  a cos 90°   **b** sin 90°

  c cos 60°   **d** sin 60°

  e cos 45°   **f** sin 45°

  g cos 30°   **h** sin 30°

  i cos 0°   **j** sin 0°

6 Label the hypotenuse and the adjacent side in each triangle. Use cos to find the missing angle in each triangle.

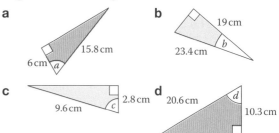

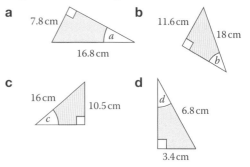

7 Label the hypotenuse and the opposite side in each triangle. Use sin to find the missing angle in each triangle.

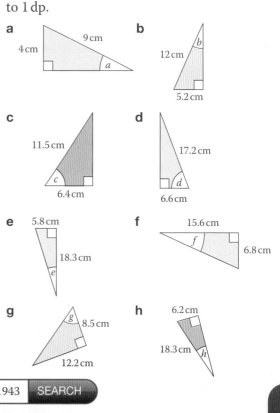

8 Label the adjacent side, opposite side and hypotenuse in each triangle.
Find the missing angle. Give your answer to 1 dp.

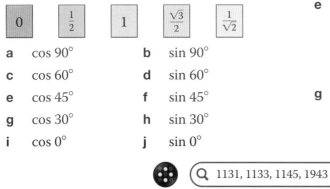

19.3 Trigonometry 2

- The **tan** ratio connects the opposite side and the adjacent side.
- The **sin** ratio connects the opposite side and the hypotenuse.
- The **cos** ratio connects the adjacent side and the hypotenuse.

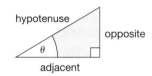

HOW TO

① Label the sides adjacent, hypotenuse and opposite.

Decide which ratio to use.

② Substitute the values that you know into one of the formulae

$$\sin \theta = \frac{\text{opposite}}{\text{hypotenuse}} \quad \cos \theta = \frac{\text{adjacent}}{\text{hypotenuse}} \quad \tan \theta = \frac{\text{opposite}}{\text{adjacent}}$$

③ Find the missing angle or length.

Some values of cos and sin have exact answers.

- $\cos 45 = \frac{1}{\sqrt{2}}$ and $\sin 45 = \frac{1}{\sqrt{2}}$
- $\cos 30 = \frac{\sqrt{3}}{2}$ and $\sin 30 = \frac{1}{2}$
- $\cos 60 = \frac{1}{2}$ and $\sin 60 = \frac{\sqrt{3}}{2}$

You need to learn these for your exam.

EXAMPLE

Find the missing sides.

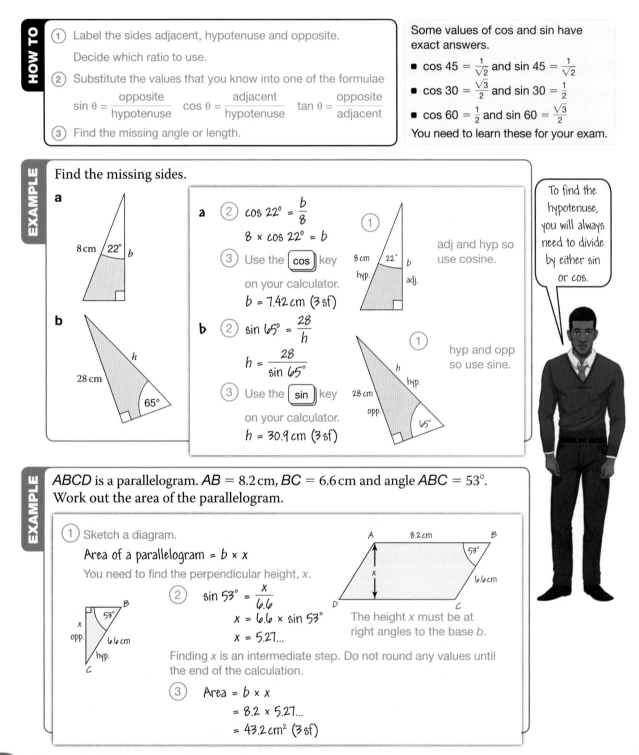

a

8 cm 22° b

b

h

28 cm

65°

a ② $\cos 22° = \dfrac{b}{8}$

$8 \times \cos 22° = b$

③ Use the [cos] key

on your calculator.

$b = 7.42 \text{ cm } (3 \text{ sf})$

b ② $\sin 65° = \dfrac{28}{h}$

$h = \dfrac{28}{\sin 65°}$

③ Use the [sin] key

on your calculator.

$h = 30.9 \text{ cm } (3 \text{ sf})$

① 8 cm 22° b
hyp. adj.

adj and hyp so use cosine.

①
28 cm h hyp.
opp. 65°

hyp and opp so use sine.

To find the hypotenuse, you will always need to divide by either sin or cos.

EXAMPLE

$ABCD$ is a parallelogram. $AB = 8.2 \text{ cm}$, $BC = 6.6 \text{ cm}$ and angle $ABC = 53°$.
Work out the area of the parallelogram.

① Sketch a diagram.

Area of a parallelogram = $b \times x$

You need to find the perpendicular height, x.

② $\sin 53° = \dfrac{x}{6.6}$

$x = 6.6 \times \sin 53°$

$x = 5.27...$

Finding x is an intermediate step. Do not round any values until the end of the calculation.

③ Area = $b \times x$

$= 8.2 \times 5.27...$

$= 43.2 \text{ cm}^2 (3 \text{ sf})$

53° B
x
opp. 6.6 cm
hyp.
C

A 8.2 cm B
53°
x
6.6 cm
D C

The height x must be at right angles to the base b.

Geometry Pythagoras and trigonometry

Exercise 19.3A

1 Find the missing side in each of these right-angled triangles. Give your answers to 3 significant figures.

a

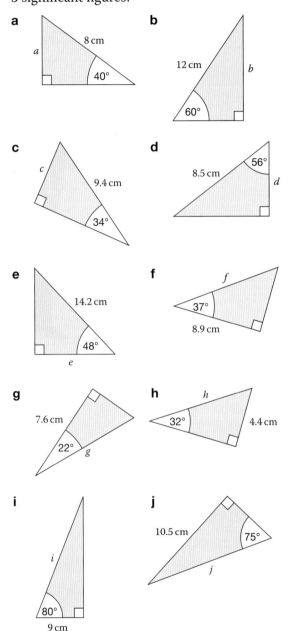

8 cm
a
40°

b

12 cm
b
60°

c

c
9.4 cm
34°

d

56°
8.5 cm
d

e

14.2 cm
48°
e

f

f
37°
8.9 cm

g

7.6 cm
22°
g

h

h
32°
4.4 cm

i

i
80°
9 cm

j

10.5 cm
75°
j

2 A parallelogram has sides of length 5 cm and 9 cm. The smaller angles are both 45°.

Find the area of the parallelogram.

3 A rhombus has side length 8 cm and smaller angle 35°.

Find the area of the rhombus.

4 Here is an isosceles right-angled triangle.

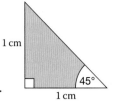

1 cm
1 cm
45°

a Explain why the length of the hypotenuse is $\sqrt{2}$ cm.

b Use the triangle to explain why $\cos 45 = \frac{1}{\sqrt{2}}$ and $\sin 45 = \frac{1}{\sqrt{2}}$.

5 Here is an equilateral triangle with sides 2 cm. The triangle is folded in half.

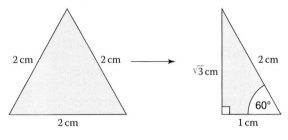

2 cm   2 cm
2 cm

$\sqrt{3}$ cm   2 cm
60°
1 cm

a Use the triangle to explain why $\cos 60 = \frac{1}{2}$ and $\sin 60 = \frac{\sqrt{3}}{2}$

b Use the triangle to explain why $\cos 30 = \frac{\sqrt{3}}{2}$ and $\sin 30 = \frac{1}{2}$

6 A see-saw is 2.8 m long. When one end touches the ground it makes an angle of 35° with it.

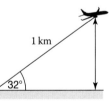

2.8 m
35°

Tina says the other end is less than $1\frac{1}{2}$ metres above the ground. Is she correct?

7 After take-off, a plane climbs at an angle of 32° to the horizontal. Find the height of the plane above the ground when it has travelled 1 kilometre.

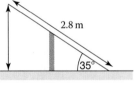

1 km
32°

*8 The angle between the legs of a stepladder must be 74°. A metal bar is needed to keep the legs the correct distance apart. Is a 2 metre bar long enough?

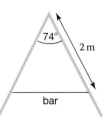

74°
2 m
bar

19.4 Vectors

You describe a translation by a **vector**. You specify the distance moved left or right, and then up or down.

The triangle is translated by the vector $\begin{pmatrix} 4 \\ 3 \end{pmatrix}$.

> The arrowed line shows the direction and distance of the translation.

- A vector has a length and a direction.

You can draw a vector as an arrowed line. Its orientation gives the direction of movement; its length gives the distance.

These lines are all parallel and the same length.

They all represent the same vector **a**.

You can tie a vector to a starting point.

The vector $-\mathbf{a}$ is parallel, the same length, in the opposite direction to **a**.

The line ST represents the vector $\overrightarrow{ST} = \mathbf{s}$.

Note that $\overrightarrow{TS} = -\mathbf{s}$

- You can multiply a vector by a number.

The vector $2\mathbf{p}$ is parallel to the vector **p** and twice the length.

The vector $3\mathbf{p}$ is parallel to the vector **p** and three times the length.

The vector $-\mathbf{p}$ is parallel to the vector **p** and the same length, but in the opposite direction.

- Vectors represented by parallel lines are multiples of each other.

- Vectors can be described using the notation $\overrightarrow{AB}$ or bold type, **a**.
- In handwriting, vectors can be shown with an underline, $\underline{a}$.

> The pairs of letters in the sequence fit together.

You can add and subtract vectors by putting them 'nose to tail'.

The result of the addition or subtraction is called the **resultant** vector.

$$\overrightarrow{AB} + \overrightarrow{BC} + \overrightarrow{CD} + \overrightarrow{DE} + \overrightarrow{EF} = \overrightarrow{AF}$$

EXAMPLE

$\mathbf{m} = \begin{pmatrix} 3 \\ 2 \end{pmatrix}$ $\mathbf{n} = \begin{pmatrix} 2 \\ -1 \end{pmatrix}$

Calculate $\mathbf{m} + \mathbf{n}$ and $\mathbf{m} - \mathbf{n}$.

a $\begin{pmatrix} 3 \\ 2 \end{pmatrix} + \begin{pmatrix} 2 \\ -1 \end{pmatrix} = \begin{pmatrix} 3 + 2 \\ 2 + -1 \end{pmatrix} = \begin{pmatrix} 5 \\ 1 \end{pmatrix}$ 5 right, 1 up

$\begin{pmatrix} 3 \\ 2 \end{pmatrix} - \begin{pmatrix} 2 \\ -1 \end{pmatrix} = \begin{pmatrix} 3 - 2 \\ 2 - -1 \end{pmatrix} = \begin{pmatrix} 1 \\ 3 \end{pmatrix}$ 1 right, 3 up

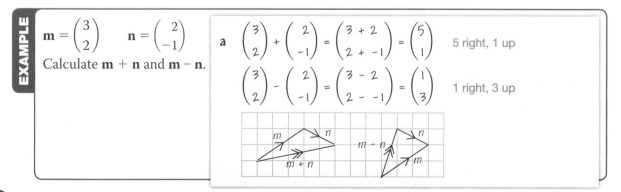

Geometry Pythagoras and trigonometry

Exercise 19.4S

1 Draw these vectors on squared paper.

   a $\begin{pmatrix} 4 \\ 3 \end{pmatrix}$ **b** $\begin{pmatrix} 2 \\ 5 \end{pmatrix}$ **c** $\begin{pmatrix} -1 \\ 4 \end{pmatrix}$

   d $\begin{pmatrix} -3 \\ -3 \end{pmatrix}$ **e** $\begin{pmatrix} 0 \\ 2 \end{pmatrix}$ **f** $\begin{pmatrix} -4 \\ 0 \end{pmatrix}$

2 Use $\overrightarrow{AB}$ notation to identify equal vectors in this diagram.

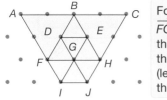

> For example, $\overrightarrow{FG} = \overrightarrow{IJ}$ because they are parallel, in the same direction (left to right) and the same length.

3 ABCDEF is a regular hexagon. X is the centre of the hexagon. $\overrightarrow{XA} = \mathbf{a}$ and $\overrightarrow{AB} = \mathbf{b}$.

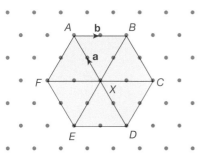

Write all the vectors that are equal to

   a $\mathbf{a}$ **b** $\mathbf{b}$

   c $-\mathbf{a}$ **d** $-\mathbf{b}$

4 JKLMNOPQ is a regular octagon.

   $\overrightarrow{OJ} = \mathbf{j}$ $\overrightarrow{OM} = \mathbf{m}$ $\overrightarrow{OP} = \mathbf{p}$.

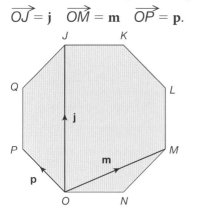

Write all the vectors that are equal to

   a $\mathbf{j}$ **b** $\mathbf{m}$ **c** $\mathbf{p}$

   d $-\mathbf{j}$ **e** $-\mathbf{m}$ **f** $-\mathbf{p}$

5 If $\mathbf{p} = \begin{pmatrix} 3 \\ 4 \end{pmatrix}$ and $\mathbf{q} = \begin{pmatrix} -2 \\ -5 \end{pmatrix}$ draw these vectors on square grid paper.

   a $\mathbf{p} + \mathbf{q}$ **b** $\mathbf{p} + \mathbf{p}$

   c $\mathbf{p} - \mathbf{q}$ **d** $\mathbf{p} + \mathbf{p} + \mathbf{q}$

6 The diagram shows vectors **s** and **t**. On square grid paper draw the vectors that represent

   a $\mathbf{s} + \mathbf{s}$ **b** $\mathbf{s} + \mathbf{t}$

   c $\mathbf{t} + \mathbf{s}$ **d** $\mathbf{s} - \mathbf{t}$

   e $\mathbf{t} - \mathbf{s}$ **f** $\mathbf{t} + \mathbf{t} - \mathbf{s}$

7 The diagram shows vectors **g** and **h**. On isometric paper draw the vectors that represent

   a $\mathbf{g} + \mathbf{g}$ **b** $\mathbf{g} + \mathbf{h}$

   c $\mathbf{h} - \mathbf{g}$ **d** $\mathbf{g} - \mathbf{h} - \mathbf{h}$

8 Write down a vector that is

   a in the same direction as $\begin{pmatrix} -2 \\ 1 \end{pmatrix}$ but 4 times as long

   b half as long as $\begin{pmatrix} -6 \\ 8 \end{pmatrix}$ and in the opposite direction.

9 $\mathbf{a} = \begin{pmatrix} -2 \\ 4 \end{pmatrix}$, $\mathbf{b} = \begin{pmatrix} 3 \\ -1 \end{pmatrix}$ and $\mathbf{c} = \begin{pmatrix} 0 \\ -2 \end{pmatrix}$

   a Calculate

      i $\mathbf{a} + \mathbf{b} + \mathbf{c}$ **ii** $2\mathbf{a} + \mathbf{b}$

      iii $\mathbf{a} - \mathbf{b} - \mathbf{c}$ **iv** $\mathbf{b} - 3\mathbf{c}$

      v $2\mathbf{a} + 3\mathbf{b}$ **vi** $3\mathbf{b} + \frac{1}{2}\mathbf{a} - \frac{1}{2}\mathbf{c}$

   b Use diagrams to check your answers.

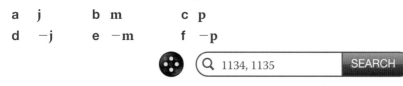

19.4 Vectors

RECAP

- A vector has a fixed length and direction.
- You can add and subtract vectors by putting them 'nose to tail'.
  You can extend addition to more than two vectors.

$$\mathbf{b} + \mathbf{b} + \mathbf{b} + \mathbf{b} = 4\mathbf{b}$$

You can write this vector as $4\mathbf{b}$.

- Vectors represented by parallel lines are **multiples** of each other.

> The vector $4\mathbf{b}$ is **parallel** to the vector $\mathbf{b}$, and four times as long as $\mathbf{b}$.

HOW TO

① Draw a diagram and mark the known vectors.

② Use your knowledge of vector addition and subtraction. Look out for parallel vectors.

③ Answer the question. Make sure you use the correct notation for vectors.

EXAMPLE

$OABC$ is a parallelogram.

$\overrightarrow{OA} = \mathbf{a}$ $\overrightarrow{OC} = \mathbf{c}$

Write the vector that represents the diagonal

a $\overrightarrow{OB}$ **b** $\overrightarrow{CA}$

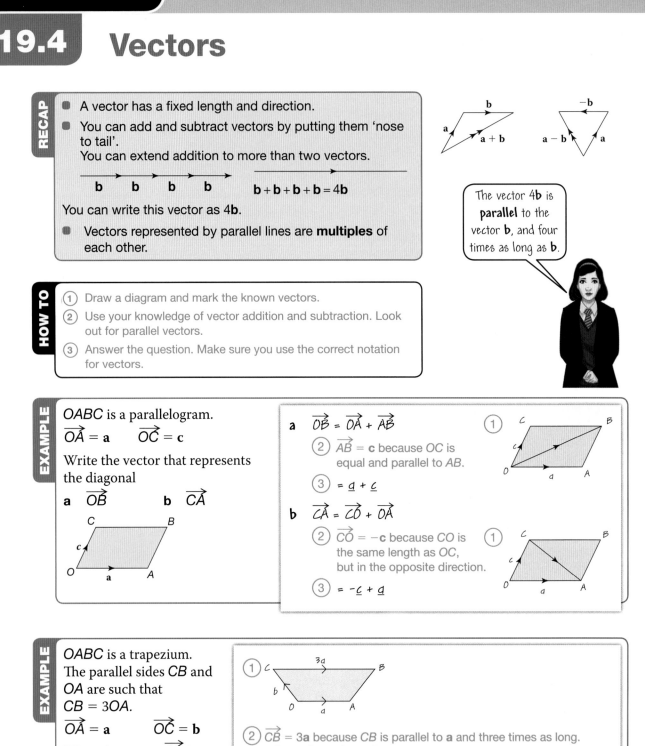

a $\overrightarrow{OB} = \overrightarrow{OA} + \overrightarrow{AB}$ ①

② $\overrightarrow{AB} = \mathbf{c}$ because OC is equal and parallel to AB.

③ $= \underline{a} + \underline{c}$

b $\overrightarrow{CA} = \overrightarrow{CO} + \overrightarrow{OA}$

② $\overrightarrow{CO} = -\mathbf{c}$ because CO is the same length as OC, but in the opposite direction. ①

③ $= -\underline{c} + \underline{a}$

EXAMPLE

$OABC$ is a trapezium. The parallel sides CB and OA are such that $CB = 3OA$.

$\overrightarrow{OA} = \mathbf{a}$ $\overrightarrow{OC} = \mathbf{b}$

Write the vector $\overrightarrow{AB}$ in terms of $\mathbf{a}$ and $\mathbf{b}$.

①

② $\overrightarrow{CB} = 3\mathbf{a}$ because CB is parallel to $\mathbf{a}$ and three times as long.

③ $\overrightarrow{AB} = \overrightarrow{AO} + \overrightarrow{OC} + \overrightarrow{CB}$

$\overrightarrow{AO} = -\mathbf{a}$ because AO is the same length as OA, but in the opposite direction.

$\overrightarrow{AB} = -\underline{a} + \underline{b} + 3\underline{a}$

$\overrightarrow{AB} = 2\underline{a} + \underline{b}$

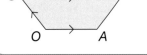

Exercise 19.4A

1 *OPQR* is a rectangle.

$\overrightarrow{OP} = \mathbf{p}$ and $\overrightarrow{OR} = \mathbf{r}$.
Work out the vector, in terms of **p** and **r**, that represents

a $\overrightarrow{PQ}$ **b** $\overrightarrow{OQ}$

c $\overrightarrow{QO}$ **d** $\overrightarrow{RP}$

2 *OJKL* is a rhombus. $\overrightarrow{OJ} = \mathbf{j}$ and $\overrightarrow{OL} = \mathbf{l}$.

Work out the vector, in terms of **j** and **l**, that represents

a $\overrightarrow{JK}$ **b** $\overrightarrow{JL}$

c $\overrightarrow{KO}$ **d** $\overrightarrow{KL}$

3 $\overrightarrow{OP} = \mathbf{p}$ $\overrightarrow{OQ} = \mathbf{q}$

Write, and simplify, the vectors

a $\overrightarrow{OG}$ **b** $\overrightarrow{OL}$ **c** $\overrightarrow{OK}$

d $\overrightarrow{OJ}$ **e** $\overrightarrow{OA}$ **f** $\overrightarrow{OC}$

g $\overrightarrow{OB}$ **h** $\overrightarrow{OF}$ **i** $\overrightarrow{PE}$

j $\overrightarrow{PD}$ **k** $\overrightarrow{EF}$ **l** $\overrightarrow{CA}$

m $\overrightarrow{JK}$ **n** $\overrightarrow{JI}$ **o** $\overrightarrow{JE}$

p $\overrightarrow{FD}$ **q** $\overrightarrow{DF}$ **r** $\overrightarrow{DC}$

s $\overrightarrow{ME}$ **t** $\overrightarrow{HG}$ **u** $\overrightarrow{KD}$

4 The diagram shows vectors **x** and **y**.
On square grid paper draw the vectors

a $2\mathbf{x}$

b $3\mathbf{y}$

c $2\mathbf{x} + 3\mathbf{y}$

d $3\mathbf{x} - \mathbf{y}$

e $\mathbf{y} - 2\mathbf{x}$

f $1\frac{1}{2}\mathbf{x} + 1\frac{1}{2}\mathbf{y}$

g $2(\mathbf{x} + \mathbf{y})$

h $\frac{1}{2}(2\mathbf{x} - 3\mathbf{y})$

i $3\mathbf{x} + 4\mathbf{y}$

5 *OJKL* is a trapezium.
The parallel sides *OJ* and

LK are such that
$LK = \frac{1}{4} OJ$.

$OJ = 6\mathbf{j}$

Write, in terms of **j**, the vectors that represent

a $\overrightarrow{LK}$ **b** $\overrightarrow{KL}$

6 *ABCDEFGH* is a regular octagon.
$\overrightarrow{AB} = \mathbf{a}$ and $\overrightarrow{DE} = \mathbf{d}$.

Work out the vector, in terms of **a** and **d**, that represents

a $\overrightarrow{AH}$ **b** $\overrightarrow{FE}$

c $\overrightarrow{FD}$ **d** $\overrightarrow{HB}$

*7 Show that

a $\mathbf{p} + \mathbf{q} = \mathbf{q} + \mathbf{p}$

b $3(\mathbf{p} + \mathbf{q}) = 3\mathbf{p} + 3\mathbf{q}$

c $-(\mathbf{p} + \mathbf{q}) = -\mathbf{p} - \mathbf{q}$

d $(\mathbf{p} + \mathbf{q}) + \mathbf{r} = \mathbf{p} + (\mathbf{q} + \mathbf{r})$

Summary

Checkout

You should now be able to...

Test it

Questions

✔ Use the formulae for Pythagoras' theorem: $a^2 + b^2 = c^2$,	**1 – 2**
✔ Use the trigonometric ratios and apply them to find angles and lengths in right-angled triangles: $\sin\theta = \dfrac{\text{Opp}}{\text{Hyp}}$ $\cos\theta = \dfrac{\text{Adj}}{\text{Hyp}}$ $\tan\theta = \dfrac{\text{Opp}}{\text{Adj}}$	**3 – 4**
✔ Know the exact values of $\sin\theta$ and $\cos\theta$ for $\theta = 0°, 30°, 45°, 60°$ and $90°$ and $\tan\theta$ for $\theta = 0°, 30°, 45°$ and $60°$.	**5**
✔ Write column vectors and draw vector diagrams.	**6**
✔ Add, subtract and find multiples of vectors.	**7 – 8**

Language   Meaning   Example

Language	Meaning	Example
Hypotenuse	The side opposite the right angle in a right-angled triangle.	hypotenuse
Pythagoras' theorem	For a right-angled triangle, $c^2 = a^2 + b^2$ where c is the hypotenuse.	a c b $a^2 + b^2 = c^2$
Adjacent	The side next to the labelled angle in a right-angled triangle.	hypotenuse, opposite, θ, adjacent
Opposite	The side opposite the labelled angle in a right-angled triangle.	
Sine ratio	The ratio of the length of the opposite side to the hypotenuse in a right-angled triangle.	5 3 θ 4 $\sin\theta = \frac{3}{5} = 0.6$ $\cos\theta = \frac{4}{5} = 0.8$ $\tan\theta = \frac{3}{4} = 0.75$
Cosine ratio	The ratio of the length of the adjacent side to the hypotenuse in a right-angled triangle.	
Tangent ratio	The ratio of the length of the opposite side to the adjacent side in a right-angled triangle.	
Vector	A vector is a quantity with both size and direction.	$\begin{pmatrix} -2 \\ 1 \end{pmatrix}$ 1 −2
Resultant	The vector that is equivalent to adding or subtracting two or more vectors.	b, a, a + b, −b, a − b, a
Multiple	The original vector multiplied by an integer (a whole number).	$3\mathbf{a} = \mathbf{a} + \mathbf{a} + \mathbf{a}$ a a a 3a

Review

1 Calculate the lengths a and b in these right–angled triangles. Give your answers to 1 decimal place.

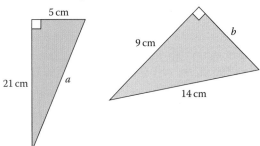

2 a Calculate the lengths x and y.

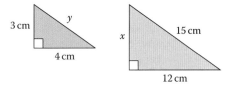

b Are the two triangles similar? Say how you know.

3

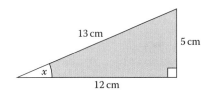

a Write as a fraction the value of
  i $\sin x$ **ii** $\cos x$
  iii $\tan x$.

b Work out the size of the angle x.

4 Calculate the lengths a, b and c to 3 significant figures.

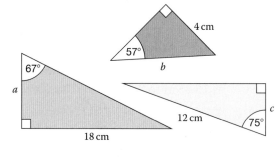

5

Without using a calculator, select the correct value of each ratio.

a $\sin 60°$ **b** $\cos 0°$

6 A ship starts at a port then sails 30 km due south then 50 km due east.

a What is the shortest distance from the final position of the ship back to the port?

b What is the bearing of the ship from the port?

7 Work out these vector sums.

a $\begin{pmatrix} 2 \\ 4 \end{pmatrix} + \begin{pmatrix} 3 \\ 5 \end{pmatrix}$ **b** $\begin{pmatrix} 5 \\ -2 \end{pmatrix} - \begin{pmatrix} -1 \\ 4 \end{pmatrix}$

c $5 \begin{pmatrix} 3 \\ -1 \end{pmatrix}$ **d** $2 \begin{pmatrix} 3 \\ -5 \end{pmatrix} + 5 \begin{pmatrix} -1 \\ 2 \end{pmatrix}$

8 Write down the column vectors to describe vectors **u** and **v**.

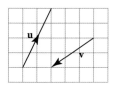

What next?

Score			
	0 – 3		Your knowledge of this topic is still developing. To improve look at MyMaths: 1053, 1112, 1131, 1133, 1134, 1135, 1145
	4 – 7		You are gaining a secure knowledge of this topic. To improve look at InvisiPens: 19Sa – c
	8		You have mastered these skills. Well done you are ready to progress! To develop your exam technique look at InvisiPens: 19Aa – d

Assessment 19

1 a Hannah says that the hypotenuse of this triangle is 181 cm.

Explain Hannah's error and work out the correct hypotenuse. [3]

b Paul says that the missing side in this triangle is 15.45 m to 2 dp.

Explain Paul's error and work out the correct value of the missing side. [4]

2 a Angelina says that 2, 3, 13 is a Pythagorean triple. Explain why she is wrong. [2]

b Darshna says that 20, 99 and 100 is a Pythagorean triple. Explain which number is wrong and find the correct third value in the triple. [2]

3 Laura plots the points, $(-3, -5)$ and $(6, -1)$. Find the distance between the points. [3]

4 There are three rectangles with area $28\,m^2$. The sides of the rectangle are a whole number of metres.

a Find the side lengths of the three possible rectangles. [3]

b Calculate the length of the diagonal of each of these rectangles. Give your answer to 4 sf. [6]

5 Douglas draws five triangles with the following sides:

a 9, 12, 15 [3]   **b** 9, 14, 17 [3]   **c** 1.6, 3.0, 3.4 [3]   **d** 11, 19, 22 [3]   **e** 3.6, 7.7, 8.5 [3]

Which of these are right angled triangles? Explain your answers.

6 Ainslie is competing in a yacht race. He starts the harbour, H, and sails 2.5 km on a bearing of 062° until he reaches a buoy, B.

He then sails for 3.6 km on a bearing of 152° until he reaches the lighthouse, L. He then returns to the Harbour.

The race course is shown on the diagram.

a Prove that the angle x is a right angle. [4]

b Calculate the distance from the lighthouse to the harbour. [3]

c Hence calculate the total length of the race. [1]

d The angle y is 55°. Hence find the bearing from L to H. [2]

7 a Erica is given the three triangles shown.
All lengths are in cm and angles in degrees
correct to 1 decimal place.

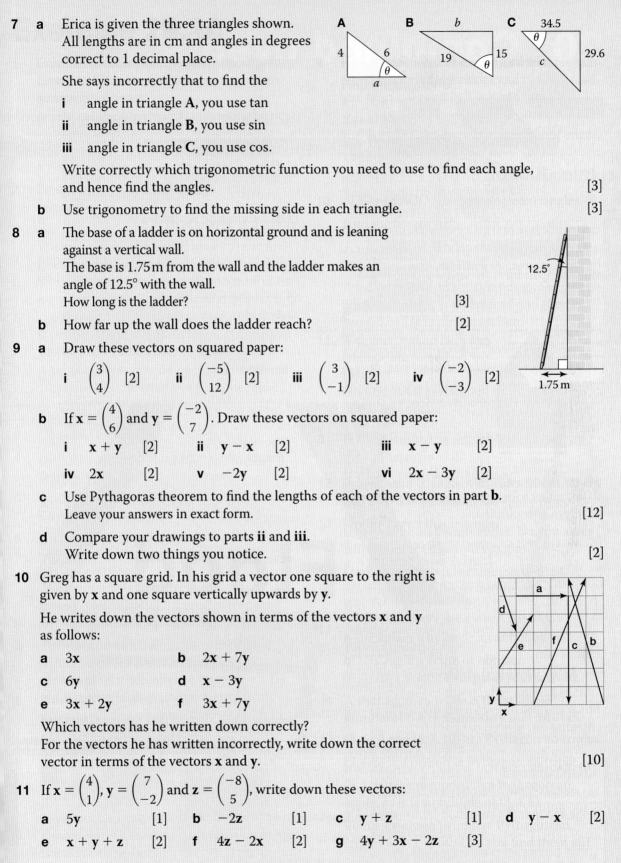

She says incorrectly that to find the

  i angle in triangle **A**, you use tan

  ii angle in triangle **B**, you use sin

  iii angle in triangle **C**, you use cos.

Write correctly which trigonometric function you need to use to find each angle,
and hence find the angles.   [3]

b Use trigonometry to find the missing side in each triangle.   [3]

8 a The base of a ladder is on horizontal ground and is leaning
against a vertical wall.
The base is 1.75 m from the wall and the ladder makes an
angle of 12.5° with the wall.
How long is the ladder?   [3]

b How far up the wall does the ladder reach?   [2]

9 a Draw these vectors on squared paper:

  i $\begin{pmatrix} 3 \\ 4 \end{pmatrix}$ [2]   **ii** $\begin{pmatrix} -5 \\ 12 \end{pmatrix}$ [2]   **iii** $\begin{pmatrix} 3 \\ -1 \end{pmatrix}$ [2]   **iv** $\begin{pmatrix} -2 \\ -3 \end{pmatrix}$ [2]

b If $\mathbf{x} = \begin{pmatrix} 4 \\ 6 \end{pmatrix}$ and $\mathbf{y} = \begin{pmatrix} -2 \\ 7 \end{pmatrix}$. Draw these vectors on squared paper:

  i $\mathbf{x} + \mathbf{y}$ [2]   **ii** $\mathbf{y} - \mathbf{x}$ [2]   **iii** $\mathbf{x} - \mathbf{y}$ [2]

  iv $2\mathbf{x}$ [2]   **v** $-2\mathbf{y}$ [2]   **vi** $2\mathbf{x} - 3\mathbf{y}$ [2]

c Use Pythagoras theorem to find the lengths of each of the vectors in part **b**.
Leave your answers in exact form.   [12]

d Compare your drawings to parts **ii** and **iii**.
Write down two things you notice.   [2]

10 Greg has a square grid. In his grid a vector one square to the right is
given by **x** and one square vertically upwards by **y**.

He writes down the vectors shown in terms of the vectors **x** and **y**
as follows:

  a $3\mathbf{x}$ **b** $2\mathbf{x} + 7\mathbf{y}$

  c $6\mathbf{y}$ **d** $\mathbf{x} - 3\mathbf{y}$

  e $3\mathbf{x} + 2\mathbf{y}$ **f** $3\mathbf{x} + 7\mathbf{y}$

Which vectors has he written down correctly?
For the vectors he has written incorrectly, write down the correct
vector in terms of the vectors **x** and **y**.   [10]

11 If $\mathbf{x} = \begin{pmatrix} 4 \\ 1 \end{pmatrix}$, $\mathbf{y} = \begin{pmatrix} 7 \\ -2 \end{pmatrix}$ and $\mathbf{z} = \begin{pmatrix} -8 \\ 5 \end{pmatrix}$, write down these vectors:

  a $5\mathbf{y}$ [1]   **b** $-2\mathbf{z}$ [1]   **c** $\mathbf{y} + \mathbf{z}$ [1]   **d** $\mathbf{y} - \mathbf{x}$ [2]

  e $\mathbf{x} + \mathbf{y} + \mathbf{z}$ [2]   **f** $4\mathbf{z} - 2\mathbf{x}$ [2]   **g** $4\mathbf{y} + 3\mathbf{x} - 2\mathbf{z}$ [3]

20 Combined events

Introduction

There is an old British myth, dating back hundreds of years, that says if it rains on St Swithin's Day (15th July) it will rain for 40 days afterwards. Surprisingly the myth retains its popularity despite statistics showing that it is untrue. Long-scale weather forecasting is fairly unreliable, particularly in the UK where the summer weather is largely determined by the 'jet stream'. However people will always look for tell-tale signs to help their predictions. In probability language, if a particular event occurs (rain on St Swithin's Day), does this increase the probability of another event occurring (a wet summer)?

What's the point?

Quite often, the occurrence of one event will significantly increase the probability of another event occurring. For example, lung cancer appears unpredictably, but its occurrence is greatly increased if a particular person is a smoker. Understanding the probabilities attached to linked events helps us to evaluate everyday risks.

Objectives

By the end of this chapter you will have learned how to …

● Use Venn diagrams to record outcomes and calculate probabilities of events.
● Construct possibility spaces and use these to calculate probabilities.
● Use tree diagrams to show the frequencies or probabilities of two events.
● Use tree diagrams to calculate the probability of independent and dependent events.

Check in

1 Cancel these fractions to their simplest form.

a $\frac{6}{9}$ **b** $\frac{5}{10}$ **c** $\frac{15}{20}$ **d** $\frac{2}{8}$ **e** $\frac{20}{20}$

2 Calculate

a $\frac{3}{4} + \frac{1}{4}$ **b** $\frac{7}{10} + \frac{3}{10}$ **c** $1 - \frac{7}{10}$ **d** $1 - \frac{3}{5}$

3 Calculate

a $1 - 0.1$ **b** $1 - 0.6$ **c** $1 - 0.15$

4 Choose a number from the rectangle that is

a prime   **b** square

c triangular   **d** a multiple of 4

e a factor of 10.

> 6
> 5   2
>   4
>   10   8
> 1
>   7   9   3

Chapter investigation

Two six-sided dice, labelled 1 to 6 are thrown. The score is given by adding the two numbers that appear uppermost.

Which result is most likely to occur, or are all possible results equally likely?

What if, instead of adding the two numbers, you multiply them together?

20.1 Sets

- A **set** is a collection of numbers or objects.
- The objects in the set are called the **members** or **elements** of the set.

If the set X is 'the factors of 6', then you can write $X = \{1, 2, 3 \text{ and } 6\}$.

$3 \in X$ means that 3 is an element (member) of the set X.

If the set Y is 'the even numbers', then you can write $Y = \{2, 4, 6, 8, ...\}$.

- The universal set, which has the symbol ξ, is the set containing all the elements.
- The empty set, $\varnothing$, is the set with no elements.

You can use a Venn diagram to show the relationship between sets.

- The **intersection** of two sets, A∩B, consists of the elements common to both sets.

- The **union** of two sets, A∪B, consists of the elements which appear in at least one of the sets.

- The **complement** of a set, A', consists of the elements which are not in A.

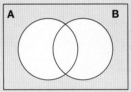

EXAMPLE

$\xi = \{1, 2, ... 11, 12\}$
$A = \{\text{factors of 12}\}$
$B = \{2, 3, 5, 6, 11\}$.
Find **a** A∩B
      b A∪B
      c (A∪B)'

a $A = \{1, 2, 3, 4, 6, 12\}$
    $A \cap B = \{2, 3, 6\}$
b $A \cup B = \{1, 2, 3, 4, 5, 6, 11, 12\}$
c $(A \cup B)' = \{7, 8, 9, 10\}$

- You list the elements of each set   or   show the number of elements in each region

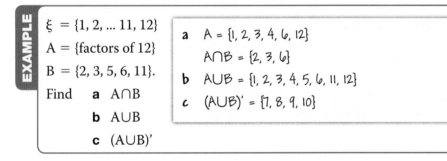

In the example,
$P(A) = \dfrac{6}{12} = \dfrac{1}{2}$

- You can use Venn diagrams to work out probabilities.
- $P(A) = \dfrac{\text{number of elements in set A}}{\text{total number of elements in } \xi}$

Exercise 20.1S

1 List the elements of these sets.

    a P = the first ten square numbers

    b R = countries in North America

    c S = the first ten prime numbers

    d T = factors of 36

2 Using the sets in question **1**, give the sets

    a P∩T   **b** S∩T

    c P∩S   **d** P∪S

3 Give a precise description of each set.

    a {1, 2, 5, 10}

    b {2, 4, 6, 8, 10, 12, ...}

    c {a, e, i, o,u}

    d {HH, HT, TH, TT}

    e {1p, 2p, 5p, 10p, 20p, 50p, £1, £2}

    f {3, 6, 9, 12, 15, 18, 21, 24, 27, 30}

4 List the elements of these sets.

    a A = the first ten positive integers

    b B = single digit odd numbers

    c C = single digit prime numbers

    d D = single digit square numbers

5 Using the sets in question **4**, give the sets

    a B∩C   **b** B∩D

    c B∪D   **d** C∪D

6 Say why B, C and D must be subsets of A for the sets in question **4**.

7 A = {even numbers}

B = {odd numbers}

C = {multiples of 5}

D = {prime numbers}

E = {multiples of 3}

F = {square numbers}

G = {factors of 36}

For these pairs of sets, state if they have any elements in common, if they do, then list them.

7 **a** A, C   **b** A, D   **c** F, G

    d E, F   **e** A, B   **f** C, G

    g D, F   **h** D, E   **i** D, E and F

8 The Venn diagram shows information about the sport that 25 students are playing in PE this term.

The results are shown on the Venn diagram.

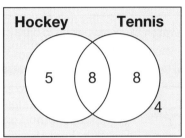

    a How many students are

       i in the intersection of hockey and tennis

      ii playing hockey

      iii not playing tennis

      iv in the union of hockey and tennis.

    b Describe the shaded region in words.

    c What fraction of students are playing either hockey or tennis, but not both?

9 An insurance company surveys 50 customers. The customers are sorted into

P = {pet insurance}

H = {home insurance}

The results are shown on the Venn diagram.

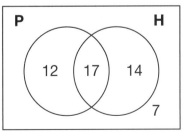

    a A customer is chosen at random. Find

       i P(P)   **ii** P(H')

      iii P(P ∩ H)   **iv** P(P ∪ H)

    b How many customers had home insurance but not pet insurance?

 🔍 1262, 1921, 1922   SEARCH

20.1 Sets

- The **intersection** of two sets, A∩B, consists of the elements common to both sets.
- The **union** of two sets, A∪B, consists of the elements which appear in at least one of the sets.
- The **universal set**, which has the symbol ξ, is the set containing all the elements.
- The **empty set**, ∅, is the set with no elements.
- **Venn diagrams** can be used to represent the relationships between sets.

HOW TO

1. Decide which elements belong to each set.
2. Draw a Venn diagram.
3. Use the Venn diagram to calculate probabilities.

Descriptions often use properties of numbers like primes, multiples etc.

EXAMPLE

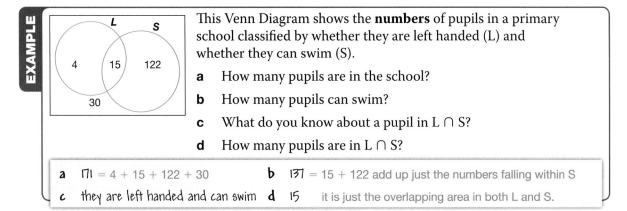

This Venn Diagram shows the **numbers** of pupils in a primary school classified by whether they are left handed (L) and whether they can swim (S).

a How many pupils are in the school?

b How many pupils can swim?

c What do you know about a pupil in L ∩ S?

d How many pupils are in L ∩ S?

a |ξ| = 4 + 15 + 122 + 30   **b** |S| = 15 + 122 add up just the numbers falling within S

c they are left handed and can swim **d** 15   it is just the overlapping area in both L and S.

EXAMPLE

A teacher is organising a school trip to Rome for the students in year 11.

There are 120 students in the school year.

72 students study history. 90 students are visiting Rome.

12 students don't study history and aren't visiting Rome.

Find the probability that a student is studying history and visiting Rome.

Draw a Venn diagram to show the number of students in each region. Let the number of students studying history and visiting Rome be x. The total number of students is 120

$$72 - x + x + 90 - x + 12 = 120$$

$$174 - x = 120$$

$$x = 174 - 120 = 54$$

P (history and Rome) $= \dfrac{54}{120}$

$= 0.45$

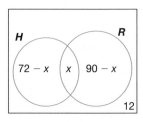

Exercise 20.1A

1   P = {factors of 42} and Q = {factors of 63}

   a   Find the set P ∩ Q

   b   What is the largest element in P ∩ Q?

2   The Venn diagram shows pupils in a primary school class. Girls are in set Q and pupils with dark hair are in set P.

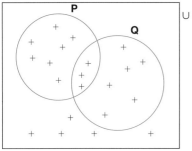

   a   How many pupils are in the class?

   b   How many girls have dark hair?

   c   How many boys have dark hair?

   d   Describe in words a pupil in P' ∩ Q'

3   The Venn Diagram shows the **numbers** of pupils in Year 11 in a school.

   G = {Pupils who are taking Biology GCSE}

   H = {pupils who walk to school}

   L = {pupils who are an only child}

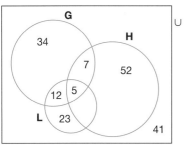

   a   How many pupils are taking GCSE Biology?

   b   How many pupils do not walk to school?

   c   How many pupils are there in G ∩ H ∩ L?

   d   What do you know about a pupil in G ∩ H ∩ L?

4   U = {all triangles}, E = {equilateral triangles}

   I = {isosceles triangles},

   R = {right-angled triangles}

   a   Sketch a member of I ∩ R

   b   Explain why E ∩ R = ∅

5   Elsie sorts a group of objects into the sets A and B.

   She draws a Venn diagram to show her results.

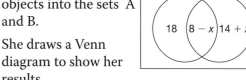

   a   Explain why Elsie must have started with 42 objects.

   b   Use the value $x = 4$ to find

     i   P(A)   **ii**   P(B')

     iii   P(A ∪ B)   **iv**   P(A ∩ B)

   c   If A and B are mutually exclusive, find the value of x.

   d   If $x = -14$, what can you say about the sets A and B?

6   David is organising a family reunion.

   His relatives can take part in two activities, archery or paintball.

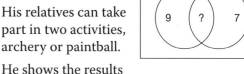

   He shows the results on a Venn diagram.

   Archery costs £22 and paintball costs £18.

   David collects £524 to pay for the activities.

   How many relatives sign up for both activities?

7   At Newtown School there are 27 students in class 11B.

   17 students play tennis, 11 play basketball and 2 play neither game.

   How many students play both tennis and basketball?

     🔍 1262, 1921, 1922   SEARCH

20.2 Possibility spaces

- The list or table of all of the possible outcomes of a trial is called a **possibility space** or **sample space**.

When two ordinary dice are thrown, there are 36 possible outcomes. If the dice are fair then the 36 outcomes are equally likely.

- You can use a sample space to calculate probabilities.

	1	2	3	4	5	6
1	1,1	1,2	1,3	1,4	1,5	1,6
2	2,1	2,2	2,3	2,4	2,5	2,6
3	3,1	3,2	3,2	3,4	3,5	3,6
4	4,1	4,2	4,3	4,4	4,5	4,6
5	5,1	5,2	5,3	5,4	5,5	5,6
6	6,1	6,2	6,3	6,4	6,5	6,6

EXAMPLE

Jasper throws two dice and adds the results.

Lena throws two dice and multiplies the results.

a Draw a possibility space for Jasper's and Lena's experiments.

b Find the probability that Jasper scores 8.

c Find the probability that Lena scores 6.

d Find the probability that Lena scores 5 or less.

a Jasper

+	1	2	3	4	5	6
1	2	3	4	5	6	7
2	3	4	5	6	7	8
3	4	5	6	7	8	9
4	5	6	7	8	9	10
5	6	7	8	9	10	11
6	7	8	9	10	11	12

Lena

×	1	2	3	4	5	6
1	1	2	3	4	5	6
2	2	4	6	8	10	12
3	3	6	9	12	15	18
4	4	8	12	16	20	24
5	5	10	15	20	25	30
6	6	12	18	24	30	36

b The same sum appears on each diagonal.

$$P(8) = \frac{5}{36}$$

c $P(6) = \frac{4}{36} = \frac{1}{9}$

d $P(5 \text{ or less}) = \frac{10}{36} = \frac{5}{18}$

- You can use the possibility space to write the set of all possible outcomes.

EXAMPLE

A fair coin is tossed and a fair die is thrown.

- If a head is seen then the score on the dice is doubled.

- If a tail is seen then the score is just the number on the dice.

a Show the possibility space in a grid.

b Write the set of all possible outcomes.

c Find P(6).

a

	1	2	3	4	5	6
T	1	2	3	4	5	6
H	2	4	6	8	10	12

b The set of all outcomes is {1, 2, 3, 4, 5, 6, 8, 10, 12}

c $P(6) = \frac{2}{12} = \frac{1}{6}$

Probability Combined events

Exercise 20.2S

1 Using Jasper's table shown opposite find the probability that the **sum of the scores** seen on two fair dice is

   a exactly 10   **b** at least 10

   c a square number   **d** less than 5.

   e Write the set of all possible outcomes.

2 Using Lena's table shown opposite find the probability that the **product of the scores** seen on two fair dice is

   a exactly 10   **b** at least 10

   c a square number   **d** less than 5.

   e Write the set of all possible outcomes.

3 Two fair dice are thrown and the difference between the scores showing on the two dice is recorded.

   a Make a table to show the possibility space.

   b Write the set of all possible outcomes.

   c Find the probability that the difference is

     i 0   **ii** 3

     iii 6   **iv** a prime number.

4 A fair coin is tossed three times and the outcome recorded (for example HHT).

   a Write the set of the 8 possible outcomes.

   b In how many of these are exactly two heads seen?

   c In how many do you see three of the same?

5 Two fair spinners are used. On one the possible scores are 1, 2 and 4, on the other the scores are 1, 3 and 5. The sum of the scores on the two spinners is recorded.

   a Make a table to show the possibility space.

   b Write the set of all possible outcomes.

   c Find the probability that the score is

     i 2   **ii** 3   **iii** even.

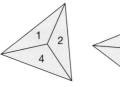

6 The two spinners in question **5** are used again but the score recorded is the product of the two scores.

   a Make a table to show the possibility space.

   b Write the set of all possible outcomes.

   c Find the probability that the score is

     i 2   **ii** 3   **iii** even.

7 For the spinners used in questions **5** and **6**, if you wanted to get an even number, would you be better to use the sum or the product of the scores on the two spinners?

8 A pair of unbiased dice are thrown and the sum and product of the scores are recorded in two lists. The dice are thrown 100 times.

   a Estimate the number of times a sum of exactly 10 will be seen.

   b Estimate the number of times a product of exactly 10 will be seen.

   c Would you expect to see 6 in the list of sums more often, less often or about the same number of times as in the list of products?

   d Would you expect to see 3 in the list of sums more often, less often or about the same number of times as in the list of products?

9 Two fair dice are thrown together. One is an ordinary dice with the numbers 1 to 6, and the other has faces labelled 1, 2, 2, 3, 3, 3.

The possibility space for the sum of scores is shown.

	1	2	3	4	5	6
1	2	3	4	5	6	7
2	3	4	5	6	7	8
2	3	4	5	6	7	8
3	4	5	6	7	8	9
3	4	5	6	7	8	9
3	4	5	6	7	8	9

   a Find the probability that the score is

     i 6   **ii** 7   **iii** 9   **iv** 3

   b What other scores have the same probability as 6?

 1199, 1263   SEARCH

20.2 Possibility spaces

- When dealing with equally likely outcomes for a single or a combined experiment, a list or table showing the outcomes is helpful.

> Once you've drawn a table, the individual cells give you the outcomes.

HOW TO

(1) If you are constructing a list, work systematically so you can generate them all in sequence.
For tossing three coins, THH, HTH, TTT, HHT, HHH, THT are 6 of the 8 possibilities – but what are the other 2?

(2) When combining two simple experiments, a table allows you to enter the 'score' in the cell while the row and header column still tells you what each outcome was.

EXAMPLE

Tara has 6 cards with the numbered from 1 to 6.
She takes two cards without replacement.
A = {product is even} B = {sum is even}
Show that P(A) = 2P(B).

Could you work out what the probabilities of even sum or product is without finding all the sums and products?

(1) Draw a possibility space.

You cannot take the same card twice so there are 30 outcomes.

(2) Use the sample space to calculate probabilities.

$P(A) = \dfrac{24}{30} = \dfrac{4}{5}$ There are 24 cells with an even product.

	1	2	3	4	5	6
1		2	3	4	5	6
2	2		6	8	10	12
3	3	6		12	15	18
4	6	8	12		20	24
5	5	10	15	20		30
6	6	12	18	24	30	

$P(B) = \dfrac{12}{30} = \dfrac{2}{5}$ There are 12 cells with an even sum.

$P(A) = 2P(B)$

	1	2	3	4	5	6
1		3	4	5	6	7
2	3		5	6	7	8
3	4	5		7	8	9
4	5	6	7		9	10
5	6	7	8	9		11
6	7	8	9	10	11	

EXAMPLE

A fair coin is tossed and a fair die is thrown. If a head is seen then the score on the dice is squared. If a tail is seen then the score is just the number on the dice.

Show the possibility space in a grid. Explain why 1 and 4 are the most likely scores to be recorded.

(1) Draw a possibility space.

	1	2	3	4	5	6
T	1	2	3	4	5	6
H	1	4	9	16	25	36

(2) 1 and 4 are the only square numbers that fall within the range 1-6.

Probability Combined events

Exercise 20.2A

1 Two cards are taken from a set of cards showing the numbers 1 to 6. Find the probability that the **difference** in the value of the two cards is

   a   3

   b   a factor of 6

   c   at least 1.

2 A fair coin is tossed until a head is seen. The number of times it has been tossed is recorded as the score on that trial.

   What is the probability that the score recorded is at least 3?

   Hint: consider the possible outcomes if a fair coin is tossed twice.

3 A fair dice in thrown and the spinner shown is spun. If the spinner lands on yellow you miss a turn (move 0 squares) otherwise you move the number of squares given by the product of the scores on the spinner and the dice.

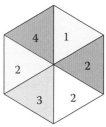

   a   Using labels R1, Y2 etc for the outcomes on the spinner construct a table to show the possible scores.

   b   What is the probability that the score recorded is

      i   4   **ii**   more than 6?

4 Cards numbered 1 to 100 are put in a box and Alessandra is asked to pick one at random. What is the probability that she chooses

   a   a single digit number

   b   a two digit number

   c   a number containing at least one 3?

   Don't write out the whole list, but imagine how many cards satisfy each condition.

5 Two fair spinners are used – one has sections showing the numbers 1, 1, 2, 3, 5 and the other has sections showing 3, 4, 5, 7, 8, 9.

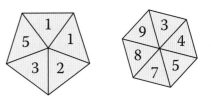

   a   What is the probability that the total score on the two spinners is

      i   6   **ii**   even?

   b   What is the probability that the score on one spinner is at least twice the score on the other spinner?

*6 Jack and Jill each roll a fair dice. Whoever gets the larger score wins the game.

   a   If a draw is allowed, what is the probability that Jill wins the game?

   b   A draw is not allowed and if the two dice show the same they roll again until one wins.
      What is the probability it will be Jill?

*7 A blue and a red dice are both fair and are thrown together. The following events are defined on the scores seen

   A – the dice show the same score

   B – the total score is at least 10

   C – the total score is odd

   D – the high score is a 4

   E – the score on one dice is a proper factor of the score on the other (a proper factor is a factor which is not 1 or the number itself)

   a   find these probabilities

      i $P(A \cap B)$

      ii $P(E)$

      iii $P(C \cap D)$

   b   Find at least two pairs of mutually exclusive events (there are 4 pairs).

    Q 1199, 1263   SEARCH

20.3 Tree diagrams

● You can use a **frequency tree** to show the outcomes of two events.

A factory employs 300 workers. 100 workers are skilled and the rest are unskilled. 25 skilled workers work part time. 120 unskilled workers work full time.

a Draw a frequency tree to show this information.

b How many part-time workers are there altogether?

c If a worker is chosen at random what is the probability they are full-time?

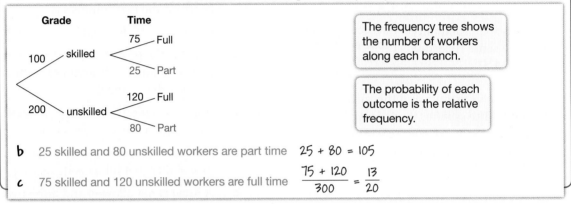

The frequency tree shows the number of workers along each branch.

The probability of each outcome is the relative frequency.

b 25 skilled and 80 unskilled workers are part time $25 + 80 = 105$

c 75 skilled and 120 unskilled workers are full time $\dfrac{75 + 120}{300} = \dfrac{13}{20}$

You can use a tree **diagram** to show the probabilities of two events.
● Write the outcomes at the end of each branch
● Write the probability on each branch
● The probabilities on each set of branches should add to 1

When you give a probability as a fraction, try to reduce it to its simplest form.

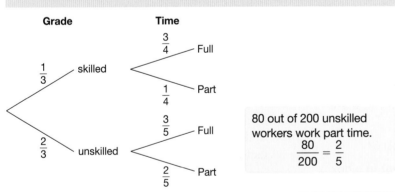

80 out of 200 unskilled workers work part time.
$\dfrac{80}{200} = \dfrac{2}{5}$

To find probabilities when an event can happen in different ways
● Multiply the probabilities along the branches.
● Add the probabilities for the different ways of getting the chosen event.

$P(\text{full time}) = P(\text{skilled and full time}) + P(\text{unskilled and full time})$

$$= \left(\frac{1}{3} \times \frac{3}{4} \right) + \left(\frac{2}{3} \times \frac{3}{5} \right)$$

$$= \frac{3}{12} + \frac{6}{15} = \frac{13}{20}$$

Probability   Combined events

Exercise 20.3S

1 A company wants to classify its employees by whether they are male or female, and by whether or not they are part of the company's pension scheme.

    a Draw a frequency tree that could be used.

    The company has 360 employees, of whom 120 are male. 80% of the male employees and 70% of the female employees are in the pension scheme.

    b Show the numbers of employees on the branches of your diagram.

2 Lydia travels to work by car two days a week and by train on the other three. She is late for work 10% of the time when she travels by car, and late 20% of the time when she travels by train. She works 150 days during the first 8 months of a year.

    a Draw a possibility tree that could be used.

    b Show the numbers of days Lydia travels to work by car and train, and on which days she is late and on time.

    c Estimate how many days she is late for work during this period.

    d What is the probability that Lydia is late for work on a day chosen at random during this period?

3 A university collects some data about student debt among their first year undergraduates. There are 3250 students of whom approximately 2000 live at home. In their sample, the university found that a quarter of those living at home, and three in five of those living away from home, felt that debt was a problem.

    a Draw a frequency tree that could be used.

    b If the whole year is similar to the sample data the university collected, put in the numbers on the branches.

    c If a first year undergraduate is selected at random what is the probability that they are concerned about debt?

4 A red and a blue dice are thrown together. A is the event 'the red dice shows an even number'. B is the event 'the blue dice shows a multiple of 3'.

    a Describe the event $A \cap B$ in words.

    b Calculate $P(A \cap B)$ directly.

    c Calculate $P(B) \times P(A)$

    d Are events A and B independent of one another?

5 The MOT test examines whether cars are roadworthy. 10% of cars fail because there is something wrong with their brakes. 40% of the cars with faulty brakes also have faulty lights, while 20% of cars whose brakes are satisfactory have faulty lights.
In March an MOT test centre deals with approximately 3000 cars.

    a Draw a frequency tree showing the numbers of cars failing with brakes and lights in March.

    b How many cars fail on at least one of brakes and lights?

    c Tommy says that this means the rest of the 3000 cars passed the MOT test. Why is Tommy wrong?

6 Athletes are regularly tested for performance drugs. If an athlete is taking the drug, the test will give a positive result 19 times out of 20. However one in fifty tests on athletes who are not taking the drug is also positive. It is thought that around 20% of athletes in a particular event are taking the drug. The authorities carry out tests on 500 athletes during the season.

    a Draw a frequency tree showing the numbers of athletes expected to be taking or not taking the drug, and the results of the test.

    b Calculate the total number of positive tests expected to be seen.

    c How many of the positive tests came from athletes not taking the drug?

 Q 1208, 1935, 1954, 1966 SEARCH

20.3 Tree diagrams

- You can use tree **diagrams** to show the probabilities of two events.
  - Write the outcomes at the end of each branch
  - Write the probability on each branch
  - The probabilities on each set of branches should add to 1
- To find probabilities when an event can happen in different ways
  - Multiply the probabilities along the branches.
  - Add the probabilities for the different ways of getting the chosen event

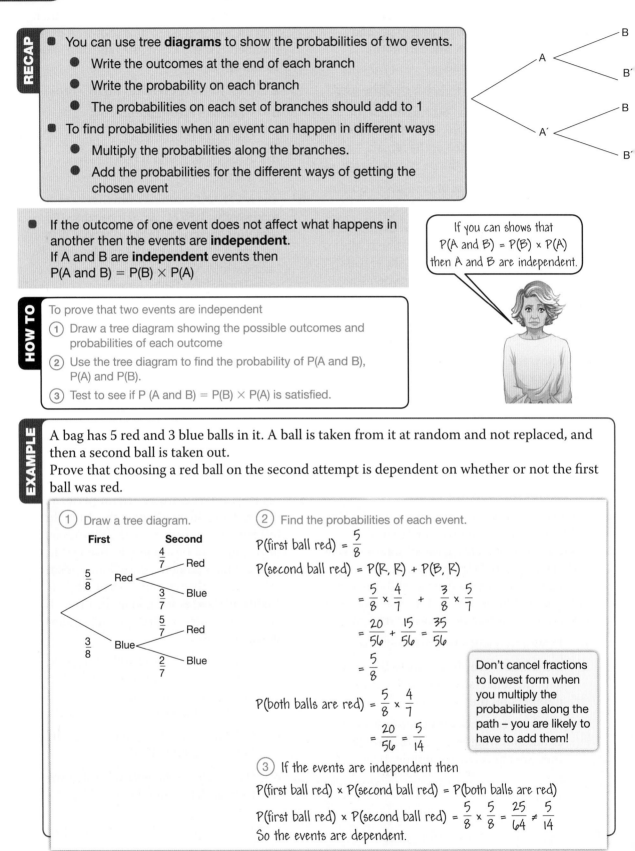

- If the outcome of one event does not affect what happens in another then the events are **independent**.
  If A and B are **independent** events then
  P(A and B) = P(B) × P(A)

> If you can shows that
> P(A and B) = P(B) × P(A)
> then A and B are independent.

HOW TO

To prove that two events are independent
1. Draw a tree diagram showing the possible outcomes and probabilities of each outcome
2. Use the tree diagram to find the probability of P(A and B), P(A) and P(B).
3. Test to see if P (A and B) = P(B) × P(A) is satisfied.

EXAMPLE

A bag has 5 red and 3 blue balls in it. A ball is taken from it at random and not replaced, and then a second ball is taken out.
Prove that choosing a red ball on the second attempt is dependent on whether or not the first ball was red.

1. Draw a tree diagram.

First **Second**

$\frac{5}{8}$ Red
  $\frac{4}{7}$ Red
  $\frac{3}{7}$ Blue

$\frac{3}{8}$ Blue
  $\frac{5}{7}$ Red
  $\frac{2}{7}$ Blue

2. Find the probabilities of each event.

P(first ball red) = $\frac{5}{8}$

P(second ball red) = P(R, R) + P(B, R)

$= \frac{5}{8} \times \frac{4}{7} + \frac{3}{8} \times \frac{5}{7}$

$= \frac{20}{56} + \frac{15}{56} = \frac{35}{56}$

$= \frac{5}{8}$

P(both balls are red) = $\frac{5}{8} \times \frac{4}{7}$

$= \frac{20}{56} = \frac{5}{14}$

> Don't cancel fractions to lowest form when you multiply the probabilities along the path – you are likely to have to add them!

3. If the events are independent then

P(first ball red) × P(second ball red) = P(both balls are red)

P(first ball red) × P(second ball red) = $\frac{5}{8} \times \frac{5}{8} = \frac{25}{64} \neq \frac{5}{14}$

So the events are dependent.

Probability Combined events

Exercise 20.3A

1 In a league, teams are awarded 3 points for a win, one for a draw and none for a loss. Amelie thinks that

- her team has a probability of 0.6 of winning any match, and a probability of 0.3 for a draw
- the result of any game is independent of other results.

a Find the probability that her team has at least three points after two games.

b How have you used Amelie's assumption that the results of the game are independent in your answer to part **a**?

c Do you think that Amelie's assumption that the results of the game are independent is reasonable? Give a reason for your answer.

2 A Year 13 pupil is taking their driving test. Records show that people taking the test at that age have a 70% chance of passing on the first attempt and 80% on any further attempts needed.

a Show this information on a tree diagram showing up to three attempts.

b Find the probability that

i the pupil passes at the second attempt

ii the pupil has still not passed after three attempts.

3 Denzel is going to the airport to catch a flight. He needs to travel on a bus and then catch a train to the airport.
He catches a bus which has a probability of 0.8 of making a connection with a train which always gets to the airport on time. The next train has a probability of 0.7 of getting him to the airport on time.

a Show this information on a tree diagram.

b Find the probability that Denzel gets to the airport in time for his flight.

4 A bag has 4 red and 4 blue balls in it. A ball is taken from it at random and not replaced and then a second ball is taken out.

X = the second ball is red

Y = the two balls are the same colour

Z = both balls are red

a Ellie says that $Z = (X \cap Y)$. Is she correct?

b Construct a tree diagram and show that X and Y are independent events.

5 A bag has 10 white and 5 black balls in it. A ball is taken from it at random, the colour noted and the ball is replaced and then a second ball is taken at random.

A = the two balls are different colours

B = at least one ball is black.

Construct a tree diagram and decide if A and B are independent events.

*6 A spinner has a probability of $\frac{1}{6}$ of landing on blue.
Green is three times as likely to occur as blue.
Red, black and yellow are equally likely to occur.

a Calculate the probability that the spinner lands on red.

Sara is playing a game where she spins the spinner and if it lands on red, she takes double the score seen when she throws a fair dice.
Sara needs a 4 to finish.

b Draw a probability tree showing the outcomes of the spinner and the dice.

c What is the probability Sara finishes on her go?

d Is the use of the spinner a help or a hindrance to Sara getting a 4 to finish, or does it not matter? Give a reason.

e How would Sara's experiment change if she needed

i a 5 to finish?

ii an 8 to finish?

 Q 1208, 1935, 1954, 1966 SEARCH

Summary

Checkout

You should now be able to...

Test it

Questions

✔ Use Venn diagrams to record outcomes and calculate probabilities of events.	1
✔ Construct possibility spaces and use these to calculate probabilities.	2
✔ Use tree diagrams to show the frequencies or probabilities of two events.	3
✔ Use tree diagrams to calculate the probability of independent and dependent events.	4 – 5

Language   Meaning   Example

Language	Meaning	Example
Set	A collection of numbers or objects	{a, e, i, o, u} is the set of all vowels
Member **Element**	A member or element of a set is one of the objects contained in that set.	a, e, i, o and u are all members of the set of all vowels.
Universal set, ξ	Once defined this is the set containing all the elements.	ξ = {positive numbers less than ten}
Empty set, ∅	The empty set has no members.	The set {all odd numbers divisible by 2} is empty.
Intersection, ∩	The intersection of 2 or more sets is the single set containing only members that are common to all.	A = {even numbers less than ten} B = {multiples of 3 less than nine}
Union, ∪	The union of 2 or more sets is the single set containing all of the members of the original sets.	A ∩ B = {6} A ∪ B = {2, 3, 4, 6, 8, 9, 10}
Complement	The complement of a set is all members which are in that set but are in the universal set. The complement of A is A'.	ξ = {positive numbers less than ten} A = {even numbers less than ten} A' = {1, 3, 5, 7, 9}
Venn diagram	Shows the relationship between sets.	See lesson 20.1
Possibility space **Sample space**	A list or table that shows all the possible outcomes of one or two events.	See lesson 20.2
Frequency tree	A tree diagram which shows the outcomes of two events.	See lesson 20.3
Tree diagram	Shows the probabilities of two or more events. To find the probability of an event happening multiply the probabilities along the branches.	See lesson 20.3
Independent	The outcome of one event does not affect what happens to the other. P(A and B) = P(A) × P(B)	There are 2 red sweets and 3 green sweets in a bag. I take one sweet, put it back and then take another. Taking one sweet does not affect taking the other so they are **independent** events.
Dependent	The outcome of one event affects what happens to the other. P(A and B) ≠ P(A) × P(B)	I take one sweet and then another. Taking the first sweet affects taking the second as there is now one less sweet. These are **dependent** events.

Review

1 **a** Copy and complete the Venn diagram by meeting all the integers up to and including 12.

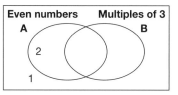

   b How many numbers are in the intersection of A and B?

   c A number is chosen at random. Find P(A and B).

2 David has a choice of red, blue and green plates and cups. He randomly chooses one plate and one cup.

   a Draw a table to show all the possible combinations of colours.

There are the same number of each colour of plate and the same number of each colour of cup.

   b What is the probability that David chooses

     i a red plate and a green cup

     ii a blue cup

     iii a plate and a cup of the same colour?

3 Evie must pass through two sets of traffic lights on her way to work each day.
The probability the first set is red when she approaches is 0.2 and the probability the second set is red is 0.3.
She works 150 days in the first 8 months of the year.

3 **a** Copy and complete the tree diagram to show all the possible outcomes.

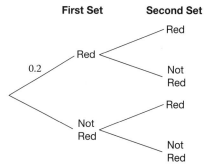

   b Show the number of days she does or does not meet a red light on her way to work.

   c Calculate the probability that

     i both sets of lights are on red

     ii neither light is on red.

4 The probability that Robbie jumps more than 4 m in a competition is 0.6.
The probability that Rachel jumps more than 4 m in the competition is 0.1.
The probabilities are independent.
Calculate the probability that

   a Robbie and Rachel both jump more than 4 m

   b Only one of them jumps more than 4 m.

5 A bag contains 3 red and 2 black counters. A counter is removed at random and *not* put back in. A second counter is then removed. Calculate the probability that

   a both counters are red

   b both counters are black

   c the counters are different colours.

What next?

Score			
	0 – 2		Your knowledge of this topic is still developing. To improve look at MyMaths: 1199, 1208, 1262, 1263, 1334, 1921, 1922, 1935
	3 – 4		You are gaining a secure knowledge of this topic. To improve look at InvisiPens: 20Sa – g
	5		You have mastered these skills. Well done you are ready to progress! To develop your exam technique look at InvisiPens: 20Aa – d

Assessment 20

1 a List the elements of each set.   [4]

    i N = {the first 10 prime numbers}

    ii M = {the first 10 multiples of 3}

    iii P = {the individual letters in the word *parallelepiped*}

    iv F = {the elements of the title of the book *Fahrenheit 451*}

b List the elements in the sets

    i N∩M   **ii** N∪M   **iii** N∩F   **iv** P∩F

    v P∪F   **vi** N∩M∩F   **vii** M∩P   **viii** N∪M∪P∪F   [8]

2 A residential area contains 34 cats. A survey found that 18 cats ate just Kattibix and 13 cats ate just Mice Pudding. Two ate both.

  a Draw a Venn diagram to show this information.   [4]

  b How many cats did not eat either Kattibix or Mice Pudding?   [1]

  c How many cats ate either Kattibix or Mice Pudding, but not both?   [2]

3 The table shows the groupings of 80 people in a local Gymnastics club.

	Children under 12		Teenagers 13 to 19		Adults 20 to 30	
	Male	Female	Male	Female	Male	Female
Number of people	4	9	17	24	15	11

Teenagers (T)   Male (M)

  a Use the information in the table to complete the Venn diagram.   [4]

  b Calculate these probabilities.

    i P(T)   **ii** P(M')   **iii** P(T∪M)   **iv** P(T∩M) [4]

4 Dominos is game that consists of a set of tiles divided into two squares, each with a number of dots on it.
Alyssa has a fair coin and these five domino tiles.
Alyssa throws the coin and chooses a domino tile.
If the coin shows heads, then she adds the dots on the two squares to give a score.
If the coin shows tails, then she multiples the dots on the two squares to give a score.

  a Complete the sample space diagram to show the possible scores.   [4]

		Domino tile				

Coin	Heads	8				
	Tails	16				

  b What is the probability that Alyssa score is 7?   [1]

  c What is the probability that Alyssa's score is even?   [1]

  d What is the probability that Alyssa's score is 10 or more?   [1]

  e What is the probability that Alyssa's score is a factor of 3?   [1]

5 Victoria has two unbiased spinners. Each spinner is divided into eight equal sectors containing the numbers 1 to 8. The spinner are spun and their scores are added together.

a Draw a sample space diagram for these spinners. [3]

b Use your diagram to find

    i P(3) [1]   **ii** P(8) [1]   **iii** P(12) [1]

c What is the most likely score? [1]

d How has the assumption that the spinners are unbiased affected your answers to parts **b** and **c**? [1]

6 ξ = {Whole numbers from 1 to 20}, T = {factors of 18}, F = {factors of n} for some integer n.
P(F) = 0.2   P(T∩F) = 0.1 and P(T∪F) = 0.45

a Draw a Venn diagram to show the number of elements in each region. [4]

b What is the value of n? [3]

7 32 editors in a publishing company went for a break in the staff room. The staff room has tea and coffee and a box of chocolate and plain biscuits.

17 editors had a cup of tea.
12 of the editors who had tea chose a chocolate biscuit.
22 editors ate chocolate biscuits altogether.

Complete the frequency tree. [4]

8 The probability that my postman delivers my mail between 10 and 10:30 in the morning is 0.15. The probability it comes before 10 am is 0.1.

a Complete the tree diagram by writing the probabilities on each branch. [4]

b Calculate the probability that the postman delivers my post before 10:30 on both days. [2]

c Calculate the probability that the postman delivers my post before 10:30 on at least one day. [3]

9 Laura eats boiled, poached or scrambled eggs for breakfast.
The probability that she chooses scrambled eggs is 0.65
The probability she chooses poached eggs is 0.25

a What is the probability Laura chooses boiled eggs? [1]

b Draw a tree diagram to show Laura's various choices on two consecutive days. Write the probabilities along each branch. [5]

c Calculate the probability that

    i Laura chooses boiled eggs on the first day and poached eggs on the second [2]

    ii Laura chooses scrambled eggs on both days [2]

    iii Laura chooses poached eggs on the second day only [3]

    iv Laura chooses boiled eggs on at least one day. [4]

Life skills 4: The launch party

Now that the business is set up, the restaurant is ready to start receiving customers. Abigail, Raheem, Mike and Juliet plan a grand opening. They expect more people to come than they can fit in the restaurant, so they plan to hire a marquee for the car park. They also continue to plan the future growth of their business.

Task 1 – Number of guests

The friends send emails to 128 people about the opening night.

They ask each person contacted to forward the email to 5 other people in exchange for entry into a prize draw for a free meal for two.

Based on email invite only, and the assumptions in the box on the right, how many people would you expect to attend the opening night?

Assumptions

- Half of the 128 people forward the email to 5 others
- $\frac{1}{4}$ of these forward the email to 5 more people
- No one receives the email more than once
- 10% of the people who receive the email go to the opening night

Task 2 – Marquees

They consider various marquees to hire.

One marquee is in the shape of a cuboid, the other is cylindrical.

The marquee company charges a flat fee of £500 for hire of all marquees, plus £1 per cubic metre for marquees larger than the standard of size of 300 m³.

a   Which marquee is cheaper to hire?

Mike wants to work out how much marquee they get for their money.

b   What is the total surface area of each marquee?

c   Using Mike's ratio, which marquee is better value for money?

d   Suggest a better calculation for Mike to work out how much marquee he gets for every pound spent.

After some thought, they go with the cuboid marquee.

e   Draw a plan, front elevation and side elevation of the chosen marquee.

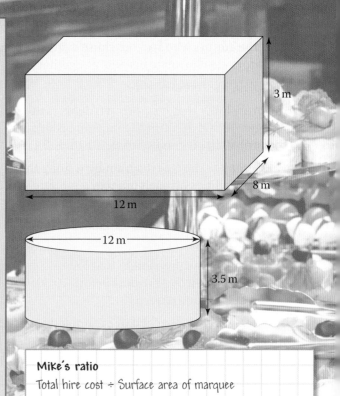

Mike's ratio
Total hire cost ÷ Surface area of marquee

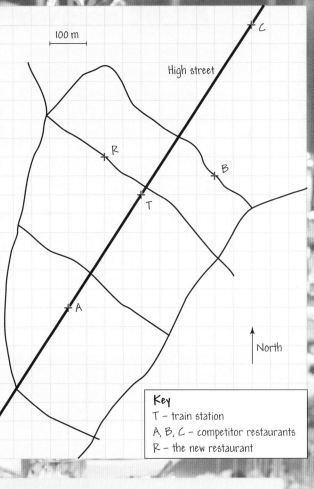

100 m

High street

C

R

B

T

A

North

Key
T – train station
A, B, C – competitor restaurants
R – the new restaurant

Task 3 – Marketing slogan

The friends make the following claim in their marketing material for the opening night.

'Closest restaurant to the train station!'

Their restaurant is shown on the map by '*R*'.

a By drawing a suitable triangle, use trigonometry to find the acute angle between the High Street and North.

b Use Pythagoras' theorem to find the distances *RT*, *BT*, *CT* and *AT*, and so determine if their marketing claim is true.

c If someone walks at 4 km/hour, how long will it take them to walk directly from the station to restaurant *C*?

Task 4 – Forecasting

Main courses will be available at three different prices.

£15 (expensive)

£13 (medium)

£11 (cheap)

Assume that all customers will have a main course.

Use the market research and the prices to estimate the expected amount that a customer spends on a main course.

Task 5 – Future growth

The friends have used the outcome of the opening night, and some additional market research, to make some projections about the growth of the business. They create a table as part of a report to the bank who gave them the business loan.

Based on their projections, copy and complete the following table for their report.

Number of customers in 4th month	
Total number of customers in the first six months	

Results of market research

- The probability that a customer has an expensive main course is 0.2

- The probability that a customer has a medium main course is 0.5

- The probability that a customer has a cheap main course is 0.3

Projections

Number of customers in first month = 700
Growth of 5% per month

21 Sequences

Introduction

Musical scales are typically written using eight notes: the C Major scale uses C D E F G A B C. The interval between the first and last C is called an octave.

The pitch of a musical note, measured in Hertz (Hz), corresponds to the number of vibrations per seconds.

The frequencies of the corresponding notes in each octave follow a geometric sequence. If the C in one octave is 130.8 Hz then the C in the next octave is $2 \times 130.8 = 261.6$ Hz, the next C is $2 \times 261.6 = 523.2$ Hz, etc.

What's the point?

Understanding the relationship between terms in a sequence lets you find any term in the sequence and begin to understand its properties.

Objectives

By the end of this chapter you will have learned how to …

- Find terms of a linear sequence using a term-to-term or position-to-term rule.
- Recognise special types of sequence and find terms using either a term-to-term or position-to-term rule.
- Find terms of a quadratic sequence using a term-to-term or position-to-term rule.

Check in

1 Work out the difference between each of these pairs of numbers.

   a 5 and 8   **b** 3 and 9   **c** 6 and 11   **d** -4 and $+2$

2 Find the difference between each of these pairs of numbers.

   a 10 and 7   **b** 9 and 5   **c** 12 and 7   **d** 3 and -1

3 Write down the first six multiples of each of these numbers.

   a 4   **b** 3   **c** 5   **d** 6

4 Substitute $n = 10$ into each of these expressions.

   a $4n$ **b** $1 + 3n$ **c** $3n - 4$ **d** n^2

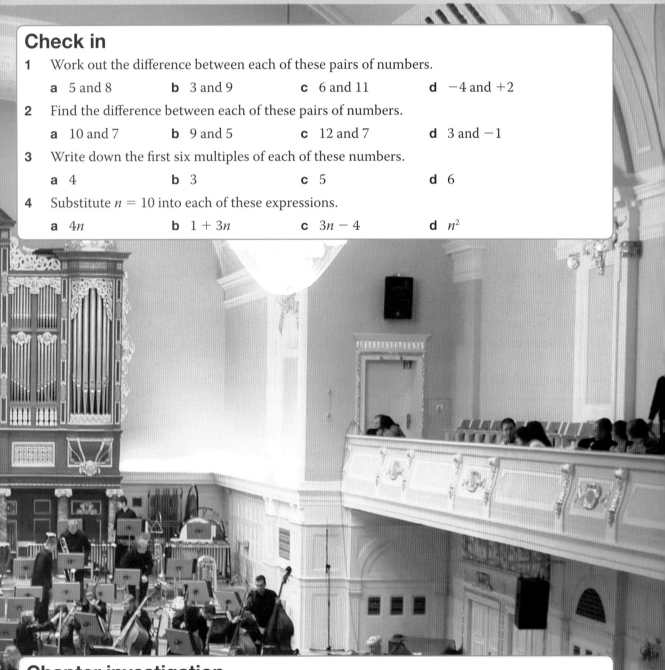

Chapter investigation

Abi, Bo and Cara are making patterns with numbers.

Abi makes a sequence by adding a fixed number onto a starting number.

1, 4, 7, 10, 13, 16, 19, 22, 25, 28, ...   Start with 1, add 3 each time

Bo makes a second sequence by adding Abi's sequence onto another starting number.

1, 2, 6, 13, 23, 36, 52, 71, 93, 118, ...   Start with 1, then add 1, then add 4, then add 7 then add 10, etc.

Cara takes the first two numbers in Bo's sequence and makes a third term by adding these together, then a fourth term by adding the second and third terms, etc.

1, 2, 3, 5, 8, 13, 21, 34, 55, 89, 144, ... $1 + 2 + 3, 2 + 3 = 5, 3 + 5 = 8$, etc.

How do sequences like these behave?

Sequence rules

> The third term in the sequence 2, 6, 10, 14, 18 ... is 10.

- The numbers in a **sequence** follow a pattern.
  Each number in a sequence is called a **term**.

You can describe a **sequence** by giving the first **term** and the term-to-term **rule**.

The rule tells you how to work out each term from the one before.

EXAMPLE

Write the first five terms in the sequence with first term 4 and term-to-term rule 'add 3'.

Start with 4; add 3 each time.

4, 7, 10, 13, 16

- In an **increasing** sequence, the terms are getting larger. For example 5, 9, 13, 17, 21, ...
- In a **decreasing** sequence, the terms are getting smaller. For example 12, 10, 8, 6, 4, ...

You can work out the term-to-term rule for a sequence and use it to find more terms.

EXAMPLE

For each sequence, work out the two missing terms.
Describe the sequence in words.

a   19, 9, − 1, − 11, ___

b   2, ___, 14, ___, 26, 32

a   19, 9, − 1, − 11, ___, ___
     The first term is 19 and the terms decrease by 10 each time.
     The next two terms are −21 and −31.

b   From 26 to 32 is an increase of 6.
     The sequence is 2, 8, 14, 20, 26, 32.
     The first term is 2 and the terms increase by 6 each time.

A **linear** sequence increases or decreases in equal-sized steps.
The size of the 'step' is called the **common difference**.

For example: 7,   3,   −1,   −5,   −9   Common difference = −4

$$\underbrace{7,\quad 3,\quad -1,\quad -5,\quad -9}_{-4\quad -4\quad -4\quad -4}$$

In some sequences, the 'steps' from one term to another are not equal.

EXAMPLE

Describe how this sequence is increasing: 3, 4, 7, 12, 19, ...

Work out the next two terms in the sequence.

3,   4,   7,   12,   19, ...
   +1   +3   +5   +7

> Arithmetic sequence is another name for a linear sequence.

The first difference is 1 and the difference increases by 2 each time.
To find the next term, +9:
19 + 9 = 28
To find the one after, +11:
28 + 11 = 39
The next two terms are 28 and 39.

Algebra   Sequences

Exercise 21.1S

1 Find the first four terms of these sequences using the term-to-term rules.

	First term	Rule
a	2	Add 6
b	3	Add 5
c	20	Subtract 2
d	−1	Subtract 4
e	10	Multiply by 5
f	0.5	Multiply by 2
g	64	Divide by 2
h	2	Divide by 0.5

2 Find the next two terms for each sequence.

    a 3, 5, 7, 9, ...   **b** 2, 5, 8, 11, ...

    c 1, 5, 9, 13, ...   **d** 4, 6, 8, 10, ...

3 Find the next two terms for each sequence.

    a 26, 21, 16, 11, ...   **b** 32, 28, 24, 20, ...

    c 22, 19, 16, 13, ...   **d** 86, 76, 66, 56, ...

4 Find the next two terms for each sequence.

    a 6, 9, 12, 15, ...   **b** 15, 13, 11, 9, ...

    c 2, 9, 16, 23, ...   **d** 50, 42, 34, 26, ...

5 Write the next two terms of the sequence 15, 17, 19, 21, ...

Explain why 72 is **not** a term in this sequence.

6 For each sequence

    i work out the differences between consecutive terms

    ii follow the pattern to work out the next two terms.

    a 2, 3, 5, 8, 12, ...   **b** 21, 20, 18, 15, ...

    c 4, 5, 8, 13, 20, ... **d** 20, 15, 11, 8, ...

7 Follow the patterns in these sequences to work out the next two terms.

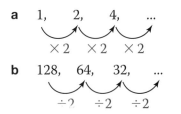

    a 1, 2, 4, ...

        × 2 × 2 × 2

    b 128, 64, 32, ...

        ÷ 2 ÷ 2 ÷ 2

    c 5, 10, 20, ...   **d** 81, 27, 9, ...

8 Write the first five terms of each of these sequences.

1st term	Rule
6	Increase by 4 each time
26	Decrease by 5 each time
10	Decrease by 3 each time
−6	Decrease by 2 each time
−23	Decrease by 7 each time

9 Write the first five terms in each of these sequences.

    a Even numbers

    b Odd numbers larger than 16

    c Multiples of 4

    d Multiples of 6 greater than 20

    e Two more than the 5 times table

    f Square numbers

    g One more than square numbers

    h Powers of 2

10 Write the first five terms of these sequences

    a 2nd term 7, increases by 3 each time

    b 2nd term 19, decreases by 6 each time

    c 3rd term 12, increases by 4 each time

    d 3rd term 14, decreases by 8 each time

    e 5th term is 8, increases by 6 each time

11 For each sequence

    i work out the two missing terms

    ii describe the sequence.

    a 3, 10, 17, 24, ____, ____

    b −7, 1, 9, 17, ____, ____

    c −19, −16, −13, −10, ____, ____

    d 7, ____, 15, 19, 23, ____

    e 3, ____, ____, 27, 35, 43

    f 16, 11, 6, ____, ____, −9

    g ____, −8, −5, −2, ____, 4

    h ____, 6, ____, −8, −15

Q 1173, 1945   SEARCH

21.1 Sequence rules

- The numbers in a sequence follow a pattern.
- Each number in a sequence is called a term.
- You can see how a sequence grows by looking at the differences between consecutive terms.

HOW TO

① Use the sequence of numbers or patterns to find the term-to-term rule.

② Use the term-to-term rule to continue the sequence (the more terms you have, the easier it is to spot patterns).

③ Find a rule that links the numbers or patterns to their position in the sequence. You could use a table to set out your workings.

EXAMPLE

Here is a sequence of patterns made from stars.

a   Draw the next two patterns in the sequence.

b   Complete the table for the sequence.

Pattern number	1	2	3	4	5
Number of stars					

c   Work out the number of stars in the 10th pattern.

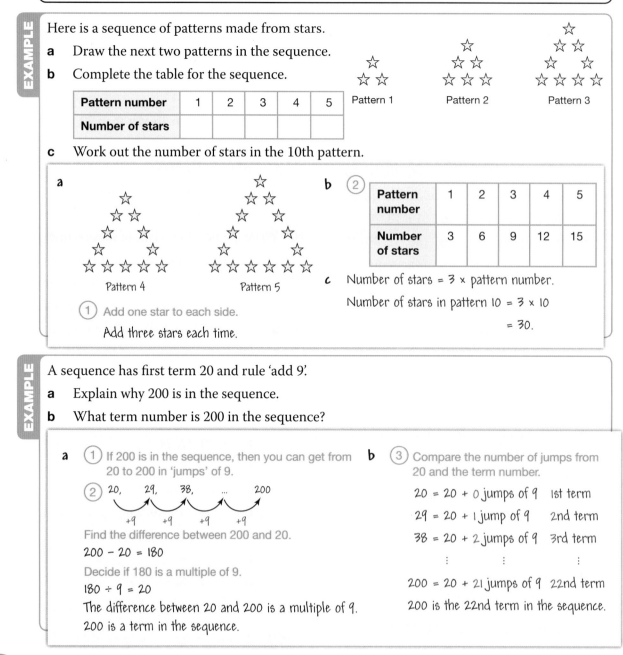

a   Pattern 4   Pattern 5

① Add one star to each side.

Add three stars each time.

b ②

Pattern number	1	2	3	4	5
Number of stars	3	6	9	12	15

c   Number of stars = 3 × pattern number.

Number of stars in pattern 10 = 3 × 10

= 30.

EXAMPLE

A sequence has first term 20 and rule 'add 9'.

a   Explain why 200 is in the sequence.

b   What term number is 200 in the sequence?

a ① If 200 is in the sequence, then you can get from 20 to 200 in 'jumps' of 9.

② 20,   29,   38,   ...   200

+9   +9   +9   +9

Find the difference between 200 and 20.

200 − 20 = 180

Decide if 180 is a multiple of 9.

180 ÷ 9 = 20

The difference between 20 and 200 is a multiple of 9.

200 is a term in the sequence.

b ③ Compare the number of jumps from 20 and the term number.

20 = 20 + 0 jumps of 9   1st term

29 = 20 + 1 jump of 9   2nd term

38 = 20 + 2 jumps of 9   3rd term

⋮   ⋮   ⋮

200 = 20 + 21 jumps of 9   22nd term

200 is the 22nd term in the sequence.

Algebra   Sequences

Exercise 21.1A

1 Draw the next pattern in each sequence.

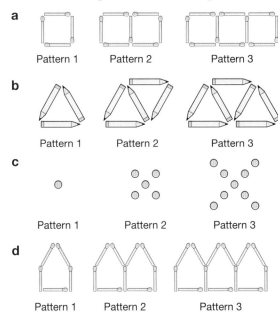

a

Pattern 1   Pattern 2   Pattern 3

b

Pattern 1   Pattern 2   Pattern 3

c

Pattern 1   Pattern 2   Pattern 3

d

Pattern 1   Pattern 2   Pattern 3

2 a Copy and complete this table for pattern **a** in question **1**.

Pattern number	1	2	3	4	5
Number of matches					

b Copy and complete this table for pattern **b** in question **1**.

Pattern number	1	2	3	4	5
Number of pencils					

c Copy and complete this table for pattern **c** in question **1**.

Pattern number	1	2	3	4	5
Number of dots					

d Copy and complete this table for pattern **d** in question **1**.

Pattern number	1	2	3	4	5
Number of matches					

3 How many matches, pencils or dots will there be in pattern number 10 for each sequence in question **1**?

4 Here are some patterns of dots.

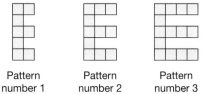

Pattern 1   Pattern 2   Pattern 3

a Draw pattern number 4.

b Explain how you would work out the number of dots in pattern number 10.

5 Here is a sequence of patterns made from squares.

Pattern number 1   Pattern number 2   Pattern number 3

a Draw the next pattern in the sequence.

b How many squares will there be in pattern number 10?

6 A sequence of numbers is generated by this rule.

> Start with 5, multiply by 2 and add 3 to generate the next term.

a Find the first five terms.

b The term 1021 is in the sequence. Calculate the term that is immediately *before* 1021.

7 Is 1000 a term in the sequence with

a first term 1005 and rule 'subtract 2'

b first term 8000 and rule 'subtract 100'

c first term 28 and rule 'add 7'

d first term 115 and rule 'add 15'?

8 A sequence has first term 10 and rule 'add 3'. What term number is 124 in the sequence?

9 A sequence has first term 300 and rule 'subtract 10'. What term number is 150 in the sequence?

*10 A sequence has 31st term 159. The rule for the sequence is 'add 5'. Find the first term in the sequence.

11 Research and produce a presentation about the classic 'Rice and Chessboard' Problem.

Q 1173, 1945   SEARCH

21.2 Finding the *n*th term

- A **position-to-term** rule links a term with its position in the sequence.

The **general term** or ***n*th term** of the 4 times table is T(*n*) = *n* × 4 or 4*n*.

You can generate a sequence from the general term.

EXAMPLE

Find the first three terms and the 10th term of the sequence with general term 3*n* + 2.

Ist term 3 × 1 + 2 → 5

2nd 3 × 2 + 2 → 8

3rd 3 × 3 + 2 → 11

10th 3 × 10 + 2 → 32

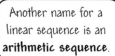

You could write the rule as T(n) = 3n + 2.

Here are some linear sequences you have met before.

Name of sequence	General term	Sequence	Common difference
Multiple of 2	2*n*	2, 4, 6, 8, 10, ...	+2
Multiple of 3	3*n*	3, 6, 9, 12, 15, ...	+3
Multiple of 4	4*n*	4, 8, 12, 16, 20, ...	+4
Multiple of 5	5*n*	5, 10, 15, 20, 25, ...	+5
Multiple of −3	−3*n*	−3, −6, −9, −12, −15, ...	−3

Another name for a linear sequence is an **arithmetic sequence**.

- For a **linear** sequence, the **common difference** tells you the multiple of *n* in the general form.

EXAMPLE

Find the general term of the sequence 5, 8, 11, 14, 17, ...

The common difference is +3.

The *n*th term contains the term 3*n*.

Compare the sequence to the multiples of 3:

3*n*	3	6	9	12	15
Term	5	8	11	14	17

Each term is 2 more than a multiple of 3.

The general term is 3*n* + 2.

Check:

n = 1 → 3 + 2 = 5

n = 2 → 6 + 2 = 8

n = 3 → 9 + 2 = 11

...

- To find the general term of a linear sequence

  - work out the common difference

  - write the common difference as the coefficient of *n*

  - compare the terms in the sequence to the multiples of *n*.

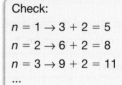

Linear sequences have *n*th term in the form T(*n*) = *an* + *b*.

Exercise 21.2S

1 Find the first three terms and the 10th term of these sequences.

   a $5n + 1$ **b** $3n + 8$

   c $8n - 4$ **d** $6n - 8$

   e $24 + 2n$ **f** $15 + 5n$

   g $7n - 20$ **h** $4n - 6$

   i $3n + 8$ **j** $6n - 15$

2 Write the first five terms for the sequence with nth term

   a $10 - 2n$ **b** $2 - n$

   c $18 - 4n$ **d** $20 - 7n$

   e $16 - 3n$ **f** $6 - 5n$

   g $-2n - 5$ **h** $30 - 5n$

   i $8 - 3n$ **j** $14 - 10n$

3 Sort these sequences into a copy of this table.

Linear Sequence	Non–linear sequence

   a $1, 6, 11, 16, 21, ...$ **b** $1, 3, 6, 10, 15, ...$

   c $1, 2, 4, 8, 16, ...$ **d** $3, 6, 9, 12, 15, ...$

   e $\frac{1}{2}, 1, 1\frac{1}{2}, 2, 2\frac{1}{2}, ...$ **f** $10, 8, 6, 4, 2, ...$

   g $3, -3, 3, -3, 3, ...$ **h** $1, 4, 9, 16, ...$

   i $-2, -5, -8, -11, ...$

4 **a** Find the common difference for the series $5, 9, 13, 17, 21, ...$

   b Copy and complete this statement:

      The nth term contains the term $\square n$.

   c Copy and complete this table to show the sequence and the multiples of n.

Sequence				
$\square n$				

   d Compare the terms in the sequence to the multiples of n and write the general term for the sequence.

5 Follow the steps in question **4** to find the general terms for these sequences.

   a $11, 17, 23, 29, 35, ...$

   b $1, 10, 19, 28, 37, ...$

   c $15, 22, 29, 36, 43, ...$

   d $-10, -6, -2, 2, 6, ...$

   e $20, 17, 14, 11, 8, ...$

   f $15, 11, 7, 3, -1, ...$

   g $16, 8, 0, -8, -16, ...$

   h $31, 23, 15, 7, -1, ...$

6 Find the nth term for each of these arithmetic sequences.

   a $7, 11, 15, 19, 23, ...$

   b $-6, -2, 2, 6, 10, ...$

   c $32, 23, 14, 5, -4, ...$

   d $15, 9, 3, -3, -9, ...$

7 Find the nth term of these sequences.

   a $1, -4, -9, -14, -19, ...$

   b $2, 3.5, 5, 6.5, 8, ...$

   c $8, 6.5, 5, 3.5, 2, ...$

   d $1.4, 2, 2.6, 3.2, 3.8, ...$

   e $2, 2\frac{1}{2}, 3, 3\frac{1}{2}, 4, ...$

   f $3, 2\frac{1}{2}, 2, 1\frac{1}{2}, 1, ...$

8 The terms of a sequence can be generated using this rule: $T(n) = 4n + 7$. Calculate the

   a 10th term

   b 15th term

   c 100th term

9 **a** Predict the 10th term for the sequence $2, 6, 10, 14, 18, ...$

   b Find the nth term for the sequence $2, 6, 10, 14, 18, ...$

   c Use your answer to part **b** to evaluate the accuracy of your prediction.

10 Is 75 a term in the sequence described by the nth term $5n - 3$?

   Show your workings.

21.2 Finding the *n*th term

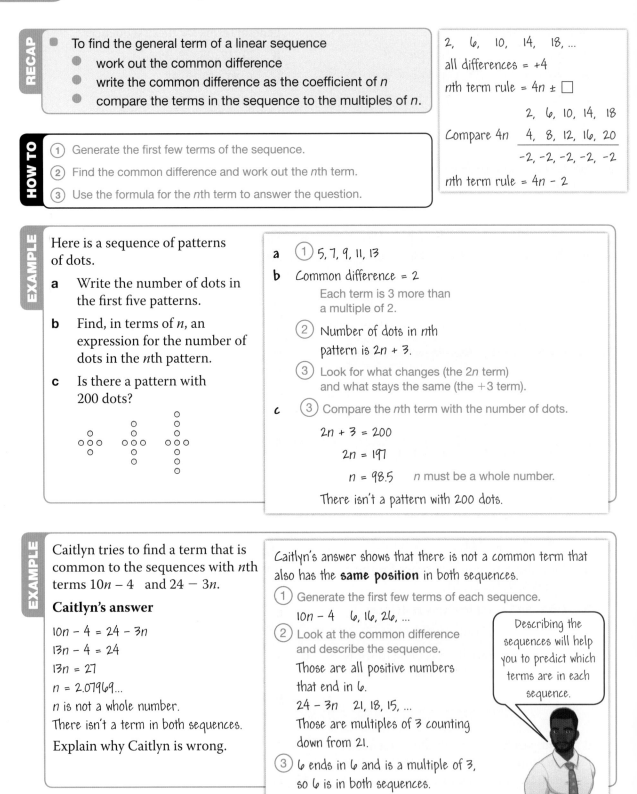

RECAP

- To find the general term of a linear sequence
  - work out the common difference
  - write the common difference as the coefficient of *n*
  - compare the terms in the sequence to the multiples of *n*.

2, 6, 10, 14, 18, ...

all differences = +4

*n*th term rule = 4*n* ± ☐

$$\begin{array}{c} 2, \ 6, \ 10, \ 14, \ 18 \\ \text{Compare } 4n \quad \underline{4, \ 8, \ 12, \ 16, \ 20} \\ -2, -2, -2, -2, -2 \end{array}$$

*n*th term rule = 4*n* − 2

HOW TO

① Generate the first few terms of the sequence.

② Find the common difference and work out the *n*th term.

③ Use the formula for the *n*th term to answer the question.

EXAMPLE

Here is a sequence of patterns of dots.

a Write the number of dots in the first five patterns.

b Find, in terms of *n*, an expression for the number of dots in the *n*th pattern.

c Is there a pattern with 200 dots?

a ① 5, 7, 9, 11, 13

b Common difference = 2
Each term is 3 more than a multiple of 2.

② Number of dots in *n*th pattern is 2*n* + 3.

③ Look for what changes (the 2*n* term) and what stays the same (the +3 term).

c ③ Compare the *n*th term with the number of dots.

$$2n + 3 = 200$$
$$2n = 197$$
$$n = 98.5 \quad \text{*n* must be a whole number.}$$

There isn't a pattern with 200 dots.

EXAMPLE

Caitlyn tries to find a term that is common to the sequences with *n*th terms 10*n* − 4 and 24 − 3*n*.

Caitlyn's answer

10*n* − 4 = 24 − 3*n*
13*n* − 4 = 24
13*n* = 27
n = 2.07969...
n is not a whole number.
There isn't a term in both sequences.

Explain why Caitlyn is wrong.

Caitlyn's answer shows that there is not a common term that also has the **same position** in both sequences.

① Generate the first few terms of each sequence.

10*n* − 4   6, 16, 26, ...

② Look at the common difference and describe the sequence.

Those are all positive numbers that end in 6.

24 − 3*n*   21, 18, 15, ...

Those are multiples of 3 counting down from 21.

③ 6 ends in 6 and is a multiple of 3, so 6 is in both sequences.

> Describing the sequences will help you to predict which terms are in each sequence.

Algebra Sequences

Exercise 21.2A

1 Here is a sequence of patterns of squares.

a Write the number of squares in the next two patterns.

Explain how you worked them out.

b Find, in terms of n, an expression for the number of dots in the nth pattern.

c Find the number of squares in the 50th pattern.

2 Repeat question **1** for these sequences of patterns.

a
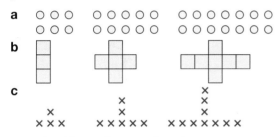

b

c

3 Jas is building a house of cards.

Stage 1   Stage 2   Stage 3

There are 52 cards in a pack. If Jas continues his pattern, can he use them all? Explain your answer.

4 Dave is building a fence from vertical and horizontal posts.
The fence grows like this:

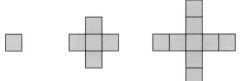

1 metre   2 metres   3 metres

Dave's garden is 26 metres long. How many posts will he need?

5 How many squares are there in the 15th diagram?

6 Draw a set of repeating patterns to represent the sequence described the nth term $4n + 3$.

7 For each pattern, there is a formula. Explain why each formula works.

a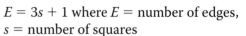

$E = 3s + 1$ where E = number of edges, s = number of squares

b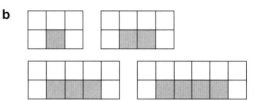

$W = B + 4$ where W = number of white tiles, B = number of coloured tiles

c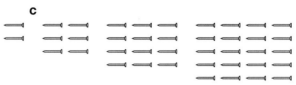

$M = L(L + 1)$ where M = number of nails and L = width of rectangle

8 Is 75 a term in the sequence described by the nth term $5n - 3$?
Explain your reasoning.

9 Sandra is generating a sequence using the rule $T(n) = 4n - 2$. She thinks that every term will be an even number. Do you agree with Sandra? Give your reasons.

10 Find the term that is common to the sequences with nth terms $60 - 12n$ and $7n + 1$.

*11 A sequence has first three terms 140, 133, 126 ...

A second sequence has first three terms $-59, -54, -49$ ...

a Find the nth term of each sequence and prove that the 17th term is the same in both sequences.

b There are 6 terms that are common to both sequences.

Can you find them all without writing out both sequences?

> **Hint:** Think of the numbers in the second sequence as 'one more than a multiple of 5'.

21.3 Special sequences

- Square, cube and triangular numbers are associated with geometric patterns.

Square numbers

1   4   9   16   25 ...

Cube numbers

1   8   27   64   125 ...

Triangular numbers

1   3   6   10   15 ...

EXAMPLE

Write the fifth

a   square number   **b**   cube number   **c**   triangular number.

a $5 \times 5 = 25$ **b** $5 \times 5 \times 5 = 125$ **c**

1   3   6   10   15

+2   +3   +4   +5

- **Arithmetic** (linear) sequences have a constant difference between terms.
- **Geometric** sequences have a constant ratio between terms.

EXAMPLE

Are these sequences arithmetic or geometric?

a   5, 8, 11, 14, ...

b   1, −2, 4, −8, ...

a $8 - 5 = 11 - 8 = 14 - 11 = 3$
    Arithmetic   Constant difference, +3

b $-2 \div 1 = 4 \div -2 = -8 \div 4 = -2$
    Geometric   Constant ratio, −2

- In a **Fibonacci**-type sequence each term is the sum of the two previous terms.

EXAMPLE

a   Is this a Fibonacci-type sequence?
    2, 1, 3, 4, 7, 11

b   What is the next term?

a   Check that each term, beyond the second, is the sum of the previous two terms.

$3 = 2 + 1$ ✓ $4 = 1 + 3$ ✓

$7 = 3 + 4$ ✓ $11 = 4 + 7$ ✓

b $7 + 11 = 18$

- In a **quadratic** sequence the differences between terms form an arithmetic sequence; the second differences are constant.

EXAMPLE

Is this a quadratic sequence?
2, 6, 12, 20, 30, 42, 56

2   6   12   20   30   42   56

   4   6   8   10   12   14   Arithmetic sequence

    2   2   2   2   2   Constant difference +2

Yes

Exercise 21.3S

1 Write the sixth term of the

    a triangular number sequence

    b square number sequence

    c cube number sequence.

2 Draw the first five terms of the triangular number sequence.

3 Describe these sequences using one of the words in the coloured panel.

> arithmetic   geometric
>
> quadratic   Fibonacci-type

    a 2, 5, 8, 11, ...   **b** 7, 11, 15, 19, ...

    c 2, 3, 5, 8, 13, ...   **d** 2, 5, 10, 17, ...

    e 2, 6, 18, 54,...   **f** 18, 15, 12, 9, ...

    g 1, 4, 5, 9, ...   **h** 1, 2, 4, 8, ...

    i 3, 7, 13, 21, ...   **j** 0.5, 2, 3.5, 5, ...

    k $\frac{1}{4}, \frac{1}{2}, 1, 2, ...$ **l** $3, 1, \frac{1}{3}, \frac{1}{9}, ...$

4 Find the next three terms of the following sequences using the properties of the sequence.

    a Arithmetic   2, 4, ☐, ☐, ☐

    b Geometric   2, 4, ☐, ☐, ☐

    c Fibonacci   2, 4, ☐, ☐, ☐

    d Quadratic   2, 4, ☐, ☐, ☐

5 Find the missing term in each of these sequences.

    a 5, 10, ☐, 20, 25, ...

    b 5, 10, ☐, 40, 80, ...

    c 5, 10, ☐, 25, 40, ...

    d 5, 10, ☐, 26, 37, ...

6 Alice thinks that the sequence 1, 4, 9, 16, 25, ... is a quadratic sequence.
Bob thinks that the sequence 1, 4, 9, 16, 25, ... is a square sequence.
Who is correct?
Explain your reasoning.

7 Drew thinks that the square number sequence can be generated using this rule.
    $T(n) = n^2$
Do you agree with Drew? Give your reasons.

8 Emily thinks that the triangular number sequence can be generated using this rule.
    $T(n) = n^3$
Do you agree with Emily?
Give your reasons.

9 Tyler thinks that the triangular number sequence can be generated using this rule.
    $T(n) = 12n(n+1)$
Do you agree with Tyler?
Give your reasons.

10 Generate the first four terms of a geometric sequence using the following facts.

	First term	Multiplier
a	3	2
b	10	5
c	3	0.5
d	2	−3
e	$\frac{1}{2}$	$\frac{1}{2}$
f	−3	−2
g	4	$\sqrt{3}$

11 The third term of a Fibonacci-type sequence is 12.
The first term is a and the second term is b.
Find three possible values for a and the corresponding value for b.

12 a Find the total of the first five terms of the geometric sequence $4, 2, 1, \frac{1}{2}, \frac{1}{4}, ...$

    b Find the total of the first ten terms of the geometric sequence $4, 2, 1, \frac{1}{2}, \frac{1}{4}, ...$

    c Comment on your results.

▲ The Fibonacci sequence can also be seen in the way tree branches form or split.

   🔍 1053, 1054, 1920, 1946 SEARCH

21.3 Special sequences

● Square numbers	1, 4, 9, 16, ...	general form 'number × number'
● Cube numbers	1, 8, 27, 64, ...	general form 'number × number × number'
● Triangular numbers	1, 3, 6, 10, 15, ...	differences 2, 3, 4, 5, ...
● Arithmetic sequence		constant differences
● Geometric sequences		constant ratio
● Quadratic sequence		differences form an arithmetic sequence
● Fibonacci-type sequence		each term is the sum of the previous two terms

① Know the square, triangular and cube number sequences.

② Generate an arithmetic sequence by adding the same constant to terms.

③ To generate a quadratic sequence, first create a linear sequence then add the linear sequence to successive terms in the quadratic sequence.

④ Generate geometric sequences by multiplying terms by the same constant.

⑤ Generate Fibonacci-type sequences by adding the previous two terms to create the next term.

Create two sequences with the following properties.

a Arithmetic sequence with starting term 5   **b** Geometric sequence with starting term 3

c Fibonacci-type sequence with starting term 4   **d** Quadratic sequence with starting term 4

a $5, 8, 11, 14, 17, ...$ ② Constant difference, $+ 3$

    $5, 2, -1, -4, -7, ...$ Constant difference, $- 3$

b $3, 6, 12, 24, 48, ...$ ④ Constant ratio, $\times 2$

    $3, 30, 300, 3000, 30000, ...$ Constant ratio, $\times 10$

c $4, 5, 9, 14, 23, ...$ ⑤ Pick a second term, 5, then $9 = 4 + 5$, $14 = 5 + 9$, $23 = 9 + 14$

    $4, 8, 12, 20, 32, ...$ Pick a second term, 8, then $12 = 4 + 8$, $20 = 8 + 12$, $32 = 12 + 20$

d $4, 7, 12, 19, 28, ...$ ③ Pick a linear sequence,   3,   5,   7,   9

                      then $7 = 4 + 3$, $12 = 7 + 5$, $19 = 12 + 7$, $28 = 19 + 9$

    $4, 10, 18, 28, 40, ...$ Pick a linear sequence,   6,   8,   10,   12

                      then $10 = 4 + 6$, $18 = 10 + 8$, $28 = 18 + 10$, $40 = 28 + 12$

Hannah is designing square frames for different sized pictures.

How many squares will she need to make a frame for a picture with dimensions 2×15?

Pattern, n	1, 2, 3, ...	Identify the pattern.
Number of squares, $T(n)$	12, 14, 16, ...	② Constant difference, 2 – linear

$T(n) = 2n + 10$

2×15 frame has $n = 14$ $T(14) = 2 \times 14 + 10 = 28 + 10 = 38$

Hannah needs 38 squares.

Algebra   Sequences

Exercise 21.3A

1 Hannah would like to create arithmetic sequences using the rule 'add 4'.

Write four possible sequences.

2 George would like to create geometric sequences using the rule 'multiply by 2'.

Write four possible sequences.

3 Dan has found a number that is in both the square number and cube number sequences.

    a What is the number?

    b Write its position in

        i the square number sequence

        ii the cube number sequence.

4 Is this statement true or false?

> The sum of two terms of the triangular number sequence equals one term of the square number sequence.

Show your reasoning.

5 Jenny thinks that the triangular number sequence can be created by starting with the number 1 and then adding on 2, adding on 3 and so on.
Do you agree with Jenny?
Explain your reasoning.

6 The first term of a sequence is 4.

    a Create five terms of a sequence which is

        i arithmetic   **ii** geometric

        iii Fibonacci-type   **iv** quadratic.

    b Explain how you created each sequence.

7   **a** Research and produce a presentation about how Leonardo Fibonacci, also known as Leonardo of Pisa, discovered the Fibonacci sequence 1, 1, 2, 3, 5, …

    b How is the Fibonacci sequence linked to

        i Pascal's Triangle

        ii the golden ratio?

8 Research other special sequences, such as pentagonal, hexagonal and tetrahedral numbers.

9 Carl Gauss was one of the world's most famous mathematicians. One day at school, he was asked to find the sum of $1+2+3+ … + 100$. Within seconds, he gave the correct total of 5050. Explore how he could calculate the sum so quickly.

10 Research and produce a presentation about the 'Handshakes' problem.

11 Hannah and Sam are given these options for money as a gift to celebrate their birthdays.
Option 1 £500
Option 2 £100 for the month containing the birthday, £200 the next month, £300 the next month until the end of the year
Option 3 During the month of their birthday 1p on Day 1, 2p on Day 2, 4p on Day 3, 8p on Day 4, …
Hannah's birthday is 20th February. Sam's birthday is 5th September.

    a Which option should Hannah choose? Explain your reasoning.

    b Which option should Sam choose? Explain your reasoning.

12 Using the cards below, match the sequence with the correct description, rule and visual representation. Complete any missing cards.

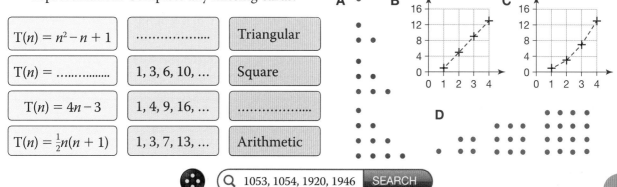

$T(n) = n^2 - n + 1$	....	Triangular
$T(n) = $	1, 3, 6, 10, …	Square
$T(n) = 4n - 3$	1, 4, 9, 16, …	....
$T(n) = \frac{1}{2}n(n + 1)$	1, 3, 7, 13, …	Arithmetic

Q 1053, 1054, 1920, 1946   SEARCH

Summary

Checkout

You should now be able to...

✓ Generate terms of a sequence from both a term-to-term and a position-to-term rule.	1 – 3
✓ Write a formula for the *n*th term of a linear sequence.	4 – 5
✓ Recognise special sequences and use them to solve problems.	6 – 8

Language	Meaning	Example
Sequence	A set of numbers or other objects arranged in order that follow a rule.	Sequence: square numbers 1 4 9 16 ...
Term	One of the separate items in a sequence.	
Position	A number that counts where a term appears in a sequence.	$T(1) = 1$ First term, position 1 $T(2) = 4$ Second term, position 2
Term-to-term rule	A rule that links a term in a sequence with the previous term.	Sequence: 3, 5, 7, 9, 11, 13 …. Term-to-term rule, 'add 2'
Position-to-term rule	A rule that links a term in a sequence with its position in the sequence.	Position-to-term rule, $T(n) = 2n + 1$ $T(n)$ is the term in position n.
Linear/ Arithmetic	A sequence that has a constant difference between the terms. If plotted on a graph, a linear relationship gives a straight line.	4, 9, 14, 19, 24, ... +5 +5 +5 +5 Common difference = 5 $T(n) = 5n - 1$ is the nth term
Common difference	The difference between each term and the previous term in a linear sequence	$T(1) = 5 \times 1 - 1 = 4$ is the 1st term $T(2) = 5 \times 2 - 1 = 9$ is the 2nd term
General term ***n*th term**	A general expression that can be used to find all the terms of the sequence.	$T(10) = 5 \times 10 - 1 = 49$ is the 10th term
Cube numbers	The sequence formed by multiplying the position number by itself three times.	1, 8, 27, 64, 125, ... $1 = 1 \times 1 \times 1 = 1^3$ $8 = 2 \times 2 \times 2 = 2^3$
Triangular numbers	Form a triangle. Each successive term is formed by adding on another layer to the triangle.	1, 3, 6, 10, 15, ...
Geometric sequence	A sequence that has a constant ratio between terms.	1, 3, 9, 27, 81, ... The constant ratio is 3.

Fibonacci-type sequence	Each term is a sum of previous terms.	1, 1, 2, 3, 5, 8, 13, ...
		   $2 = 1 + 1, 3 = 1 + 2, 5 = 2 + 3$ etc.
		 1, 1, 2, 4, 7, 13, 24, ...
		   $4 = 1 + 1 + 2, 7 = 1 + 2 + 4,$
		   $13 = 2 + 4 + 7$ etc.

Quadratic sequence	A sequence in which the differences between terms form an arithmetic sequence.	4, 9, 16, 25, ... +5, +7, +9, ...

Review

1 **a** What are the next three terms of these sequences?

    i   5, 9, 13, 17, ...

    ii   44, 34, 24, 14, ...

    iii   0.3, 0.9, 1.5, 2.1, ...

b Write the term–to–term rule for each of the sequences in part **a**.

2 Calculate the 9th term for the sequences with these position–to–term rules.

    a $T(n) = 3n + 2$

    b $T(n) = 8n - 11$

    c $T(n) = 12 - n$

    d $T(n) = 4 + \frac{1}{2}n$

3 The nth term of a sequence is given by $2 + n^2$.

Calculate

    a   the 5th term   **b**   the 12th term

    c   the 100th term.

4 Write a rule for the nth term of these sequences.

    a   3, 6, 9, 12...   **b**   4, 9, 14, 19, ...

    c   8, 14, 20, 26, ...   **d**   20, 18, 16, 14, ...

5 Matchsticks are arranged into squares as shown.

    a   Draw the fourth pattern in the sequence.

5 **b** Copy and complete the table using the pattern.

Number of Squares	Number of Matchsticks
1	4
2	
3	
4	
5	
6	

    c   Write down a formula that links the number of squares, s, with the number of matchsticks, m.

6 This sequence is formed by doubling the current term to get the next term.
3, 6, 12, ...

    a   Write down the next three terms of the sequence.

    b   Name the type of sequence.

7 **a** Write down the next two terms of these sequences.

    i   1, 8, 27, 64, ...

    ii   1, 3, 6, 10, ...

    b   Name the type of sequence.

8 Describe these sequences using one of the words in the coloured panel.

    Arithmetic   Quadratic

    Fibonnaci-type   Geometric

    a $\frac{3}{4}$, $1\frac{1}{2}$, 3, 6, 12, ...

    b   0, 2, 2, 4, 6, ...

    c   4, 5.5, 7, 8.5, 10, ...

    d   5, 6, 9, 14, 21, ...

What next?

Score			
	0 – 3		Your knowledge of this topic is still developing. To improve look at MyMaths: 1053, 1054, 1165, 1173, 1920
	4 – 7		You are gaining a secure knowledge of this topic. To improve look at InvisiPens: 21SA – h
	8		You have mastered these skills. Well done you are ready to progress! To develop your exam technique look at InvisiPens: 21Aa – e

Assessment 21

1   Jill is given the following sequences, all with terms missing.
For each sequence find and write down

   i   the next two terms

   ii   any missing terms

   iii   if the sequence is ascending or descending

   iv   the differences between consecutive terms

   v   the rule explaining how the sequence works.

   a   ☐,   20,   10,   0,   −10   [5]

   b   44,   66,   88,   110,   132,   ☐,   166   [5]

   c   −15,   −12,   ☐,   −6,   ☐,   0   [5]

   d   45,   34,   23,   ☐,   1   [5]

   e   24,   29,   ☐,   ☐,   44   [5]

   f   1.1,   1.5,   ☐,   2.3,   2.7   [5]

   g   12,   ☐,   36,   48,   ☐,   72   [5]

   h   1.12,   1.11,   1.1,   ☐,   1.08   [5]

   i   −0.90,   −0.95,   ☐,   −1.05,   ☐   [5]

   j   999 993,   999 996,   ☐,   ☐,   1 000 005   [5]

   k   3,   2.25,   1.5,   ☐,   0   [5]

   l   −10,   −17,   ☐,   −31   [5]

   m   6,   ☐,   6.25,   6.375,   6.5   [5]

2   Chris has the terms and descriptions of some sequences.
He can't remember which description matches which group of terms.
Match each group of terms to the correct description.   [7]

   a   The first five even numbers less than 5.   **i**   −3   −1   1   3   5

   b   The first five multiples of 7.   **ii**   5   10   15   20   25   30

   c   The first five factors of 210.   **iii**   11   13   17   19   23   29

   d   The first five triangular numbers.   **iv**   7   14   21   28   35

   e   Odd numbers from −3 to 5.   **v**   4   2   0   −2   −4

   f   Prime numbers between 10 and 30.   **vi**   1   3   6   10   15

   g $15 <$ square numbers $\leqslant 81$.   **vii**   1   2   3   5   7

   h   Multiples of 5 between 3 and 33.   **viii**   16   25   36   49   64   81

3   Nathan is given this sequence.

   1   11   21   31   41   ☐   ☐

He says that the common difference of this sequence is +11.

   a   Work out the correct common difference.   [1]

   b   He also says, 'the nth term of this sequence is…'
Complete his sentence. Show your working.   [2]

4   Jack and Dawn are looking at the start of this
sequence of patterns.

a   Jack says the next pattern will have 7 dots.
Dawn says that it will have 9 dots.
Who is correct? Draw the next pattern.   [1]

b   Describe the rule that tells you the number of dots in each pattern.
Show your working and give reasons for your answer.   [2]

c   Use your formula to find the number of dots in pattern number 10.   [1]

5   a   Complete this sequence with the missing numbers.

1   4   9   ☐   25   36   ☐   64   100   121   ☐   169   ☐   225   [2]

b   What is the name of this sequence. Give reasons for your answer.   [1]

6   For this set of patterns how many red and
blue squares are there in the

a   10th   [1]   b   50th   [1]

c   100th   [1]   d nth pattern?   [1]

7   Here is a sequence of patterns.

a   Find the nth term for the

i   perimeter   [2]   ii   area.   [2]

b   A pattern in the sequence has an area of 48 cm². What is its perimeter?   [1]

c   A pattern in the sequence has a perimeter of 48 cm. What is its area?   [1]

8   The first four terms of a sequence are   2   5   8   11

a   Which one of the expressions below gives the nth term of the sequence?
Give reasons for your choice.   [1]

$n + 2$ $n + 5$ $n + 10$ $n + 13$ $2n$ $5n$ $2n + 5$ $2n + 12$ $3n - 1$

The first four terms of another sequence are   27   21   15   9

b   Write down the next two terms of this sequence.
Give reasons for your answer.   [2]

c   Find the nth term of this sequence.   [2]

d   The 50th term of this sequence is -267.
Write down the 52nd term of this sequence.   [2]

9   Kerry is given some nth terms, $T(n)$, for some sequences. Which of her statements are correct
and which are not? Give reasons for your answers and rewrite any incorrect statements.

a $T(n) = 2n + 7$ the 10th term is 27.   [1]

b $T(n) = 6n - 5$ the first 3 terms are $-5, 1, 7$.   [1]

c $T(n) = 13 - 3n$ the 100th term is 287.   [1]

d $T(n) = n^2 - 10$ the 10th term is 100.   [1]

e $T(n) = 15 - 3n^2$ the 100th term is $-29\,985$.   [1]

10   The diagram shows a sequence of patterns.
Helen says that the ratio of red squares to
blue squares stays the same, as the shapes get bigger.
Is she correct? Give reasons for your answer.   [2]

22 Units and proportionality

Introduction

The half-life of a radioactive isotope is the time taken for half its radioactive atoms to decay. The number of radioactive isotopes remaining after each half-life forms a geometric sequence. Comparing the proportion of remaining radioactive isotopes to the geometric sequence, allows you to estimate the age of an artefact – even something that is millions of years old.

Scientists can estimate the age of a dinosaur fossil by analysing the proportion of radioactive uranium atoms in the surrounding layers of volcanic rock. The oldest dinosaur fossils are thought to be more than 240 million years old.

What's the point?

Understanding proportion and modelling growth and decay can help you understand the past and make predictions about the future.

Objectives

By the end of this chapter you will have learned how to ...

- Calculate with standard and compound units.
- Compare lengths, areas and volumes of similar shapes.
- Solve direct and inverse proportion problems.
- Describe direct and inverse proportion relationships using equations.
- Interpret the gradient of a straight line graph as a rate of change.
- Set up, solve and interpret growth and decay problems.

Albertosaurus

Check in

1 Neil buys 2 pizzas at a cost of £7.00.
What is the cost of 8 pizzas?

2 John weighs 50 kg. Kevin weighs 75 kg.
How many times heavier than John is Kevin?

3 Krishna earns £8.50 per hour.
How much does he get paid for 8 hours work?

Chapter investigation

Jamie and Jeannie both earn £20 000 per year, in different companies.

Jeannie's pay increases by 8% per year.

Jamie's pay increases by £2000 per year.

Who earns the most after one year?

At what point, if ever, does the other person's salary overtake?

Centrosaurus

22.1 Compound units

Compound measures describe one quantity in relation to another.

These are examples of compound measures.

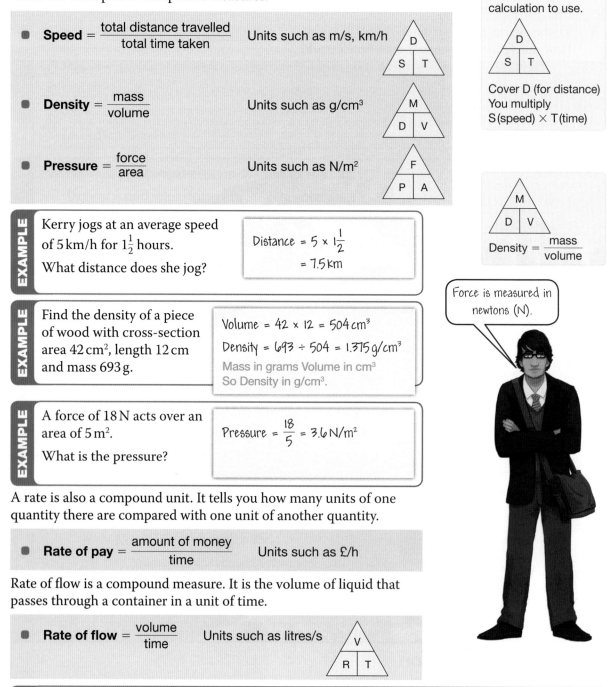

Use the triangle to work out which calculation to use.

Cover D (for distance)
You multiply
S (speed) × T (time)

● **Speed** = $\dfrac{\text{total distance travelled}}{\text{total time taken}}$ Units such as m/s, km/h

● **Density** = $\dfrac{\text{mass}}{\text{volume}}$ Units such as g/cm³

Density = $\dfrac{\text{mass}}{\text{volume}}$

● **Pressure** = $\dfrac{\text{force}}{\text{area}}$ Units such as N/m²

EXAMPLE

Kerry jogs at an average speed of 5 km/h for $1\frac{1}{2}$ hours.

What distance does she jog?

Distance = $5 \times 1\frac{1}{2}$
= 7.5 km

Force is measured in newtons (N).

EXAMPLE

Find the density of a piece of wood with cross-section area 42 cm², length 12 cm and mass 693 g.

Volume = 42 × 12 = 504 cm³
Density = 693 ÷ 504 = 1.375 g/cm³
Mass in grams Volume in cm³
So Density in g/cm³.

EXAMPLE

A force of 18 N acts over an area of 5 m².

What is the pressure?

Pressure = $\dfrac{18}{5}$ = 3.6 N/m²

A rate is also a compound unit. It tells you how many units of one quantity there are compared with one unit of another quantity.

● **Rate of pay** = $\dfrac{\text{amount of money}}{\text{time}}$ Units such as £/h

Rate of flow is a compound measure. It is the volume of liquid that passes through a container in a unit of time.

● **Rate of flow** = $\dfrac{\text{volume}}{\text{time}}$ Units such as litres/s

EXAMPLE

Water empties from a tank at a rate of 1.5 litres per second.

It takes 10 minutes to empty the tank.

How much water was in the tank?

Use the triangle to work out which formula to use.

Volume = rate × time

Convert the time to seconds.

10 minutes = 10 × 60 s = 600 s

Volume in tank = 1.5 × 600 = 900 litres

Exercise 22.1S

1 The winners' times in some of the races at a sports day are

 a 100 metres in 13 seconds

 b 200 metres in 28 seconds

 c 400 metres in 58.4 seconds

 d 1500 metres in 4 minutes 52 seconds.

 Calculate the speed of each winner in m/s, correct to 1 dp.

2 Work out the distance travelled in

 a 2 hours at 80 km/h

 b 7 hours at 23 mph

 c 6 seconds at 9 m/s

 d 1 day at 12 mph.

3 Work out the time it takes to travel

 a 180 kilometres at 60 km/h

 b 280 miles at 70 mph

 c 8 kilometres at 24 km/h

 d 15 miles at 60 mph.

4 A cube with volume 640 cm³ has a mass of 912 g. Find the density of the cube in g/cm³.

5 An emulsion paint has a density of 1.95 kg/litre. Find

 a the mass of 4.85 litres of the paint.

 b the number of litres of the paint that would have a mass of 12 kg.

6 The table shows the densities of different metals.

Metal	Density
Zinc	7130 kg/m³
Cast iron	6800 kg/m³
Gold	19 320 kg/m³
Tin	7280 kg/m³
Nickel	8900 kg/m³
Brass	8500 kg/m³

 Use the information in the table to find

 a the mass of 0.8 m³ of zinc

 b the mass of 0.5 m³ of cast iron

 c the mass of 3.2 m³ of gold

 d the volume of 910 g of tin

 e the volume of 220 g of nickel

 f the volume of a brass statue that has mass 17 kg.

7 The table shows the pressure, force and area of different materials.

 Complete the table. Include the correct units in your answers.

	Pressure	Force	Area
a		12.9 N	10 m²
b		482.5 N	25 cm²
c	2560 N/m²	1200 N	
d	512 N/mm²		14.5 mm²
e	17.8 N/cm²	225 N	
f	24.6 N/m²		2.8 m²

8 If 4 metres of fabric costs £8.40, find the price of the fabric in pounds per metre.

9 Jane is paid £478 a week. Each week she works 40 hours. What is her hourly rate of pay?

10 Find the rate of flow for pipes A and B in litres/s.

 a Pipe A: 20 litres of water in 8 seconds.

 b Pipe B: 48 litres of water in 30 seconds.

11 Water empties from a tank at a rate of 2 litres/s. It takes 10 minutes to empty the tank.

 How much water was in the tank?

12 An electric fire uses 18 units of electricity over a period of 7.5 hours.

 a What is the hourly rate of consumption of electricity in units per hour?

 b How many units of electricity are used in 24 hours?

13 A car has fuel efficiency of 8 litres per 100 km.

 a How far can the car travel on 40 litres of petrol?

 b How many litres of petrol would be needed for a journey of 250 miles?

 c What is the car's rate of fuel consumption in km per litre?

14 Rose received 75 US dollars in exchange for £50.

 a Calculate the rate of exchange in US dollars per £.

 b How many US dollars would she get for £125?

 c If Rose received $120 US dollars, how many pounds did she exchange?

Q 1061, 1121, 1246, 1970, 1971 SEARCH

22.1 Compound units

Compound units describe one quantity in relation to another.

- The density of a material is its mass divided by its volume.
- Speed is the distance travelled divided by the time taken.
- Pressure is the force divided by the area.
- A formula triangle is a useful way to remember the relationships between the different parts.
- A rate is also a compound unit.

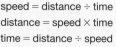

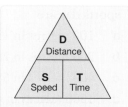

speed = distance ÷ time
distance = speed × time
time = distance ÷ speed

HOW TO

① Draw a formula triangle and write the correct formula.

② Convert units or work out quantities to apply the formula.

③ Work out the answer, making sure the units are correct.

EXAMPLE

A train leaves Norwich at 13:40 and arrives in Cambridge at 15:00. If the distance is 90 km find the average speed of the train.

① Draw a formula triangle and write the formula for speed.

② Work out the time in hours.

③ Work out the answer using the correct units.

$Speed = \dfrac{distance}{time}$

$Time = 1\ hour\ 20\ min = 1\frac{1}{3}\ h = 1.333...\ h$

$Speed = \dfrac{90}{1.333} = 67.5$

The average speed of the train is 67.5 km/h.

EXAMPLE

Sand was falling from the back of a lorry at a rate of 0.4 kg/s. It took 20 minutes for all the sand to fall from the lorry.

How much sand was the lorry carrying?

① Draw a formula triangle and write the formula for mass.

② Convert the time to seconds.

③ Work out the answer using the correct units.

The rate of flow is in kg/s, which is mass divided by time.

Mass = rate × time

20 minutes = 20 × 60 s = 1200 s

Mass = 0.4 × 1200 = 480

The lorry was carrying 480 kg of sand.

EXAMPLE

A metal cuboid has a length of 7 cm, a width of 5 cm and a height of 4 cm. It has a mass of 1.470 kg. Find its density in g/cm³.

① Draw a formula triangle and write the formula for density.

② Work out the volume of the cuboid and convert mass to grams.

③ Work out the answer using the correct units.

$Density = \dfrac{mass}{volume}$

Volume of cuboid = length × width × height
= 7 × 5 × 4 cm³ = 140 cm³

Mass = 1.470 kg = 1470 g

$Density = \dfrac{1470}{140} = 10.5\ g/cm^3$

Ratio and proportion Units and proportionality

Exercise 22.1A

1 A train leaves Euston at 8:57 am and arrives at Preston at 11:37 am.
If the distance is 238 miles find the average speed of the train.

2 Sand falls from the back of a lorry at a rate of 0.2 kg/s.
It took 25 minutes for all the sand to fall from the lorry.

How much sand was the lorry carrying?

3 A metal cuboid has a length of 9 cm, a width of 5 cm and a height of 4 cm.
It has a mass of 1.53 kg.

Find its density in g/cm^3.

4 A car travels 24 miles in 45 minutes.
Find the average speed of the car in miles per hour (mph).

5 Copy and complete the table to show speeds, distances and times for five different journeys.

Speed (kmph)	Distance (km)	Time
105		5 hours
48	106	
	84	2 hours 15 minutes
86		2 hours 30 minutes
	65	1 hours 45 minutes

6 A cube of side 2 cm weighs 40 grams.

  a Find the density of the material from which the cube is made, giving your answer in g/cm^3.

    Volume of cube = length³.

  b A cube of side length 2.6 cm is made from the same material.
Find the mass of this cube, in grams.

7 A solid block has a length and width of 22.50 mm, and a height of 3.15 mm. It has a mass of 9.50 g.

  a Find the density of the metal from which the block is made, giving your answer in g/cm^3.

  b How many blocks can be made from 1 kg of the material?

8 In this question, give your answers in kg/m^3.

  a The volume of 31.5 g of silver is 3 cm^3.
Work out the density of silver.

  b The volume of 18 g of titanium is 4 cm^3.
Work out the density of titanium.

  c A sheet of aluminium foil has volume 0.4 cm^3 and mass 1.08 g. Work out the density of aluminium foil.

9 Grace earns £340 per week for 40 hours work.
If Grace works overtime, she is paid 1.5 times her standard hourly rate.

  a How much is Grace paid for 7 hours of overtime work?

  b Grace earned £531.25 last week. How many hours of overtime did she work?

10 The toll charged for a car travelling on a motorway was £33.60 for a journey of 420 km.
Cars with trailers are charged double.

How much would it cost for a car with a trailer to travel 264 km?

11 A yacht race has three legs of 8 km, 6 km and 10 km.
The average speed for the winning yacht was 6.2 km/h.
The second yacht finished 8 minutes after the winner.
How long did it take the second place yacht to finish the race?

12 a Julia is wearing high-heeled shoes. Each heel has an area of 1 cm^2. Julia weighs 550 N.

How much pressure does Julia's heel exert when she has one heel on the ground?

  b An elephant's foot is 45 cm across and is approximately circular.
An elephant walks with two feet on the ground at a time.
An elephant weighs 55 000 N.

How much pressure does an elephant's foot exert when the elephant has two feet on the ground?

1061, 1121, 1246, 1970, 1971   SEARCH

22.2 Direct proportion

A ratio tells you how many times bigger one number is compared with another number.
You can calculate the ratio by dividing one number by the other.

● Two variables are in **direct proportion** if the ratio between
then stays the same as the actual values vary.

EXAMPLE

A pipe 2.5 metres long weighs 35 kilograms.
How much would 5.5 metres of the same pipe weigh?

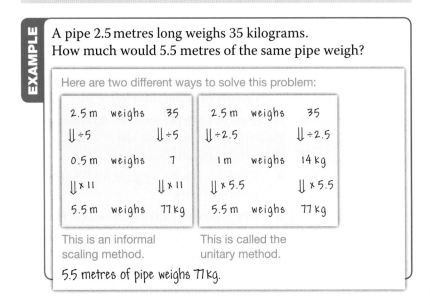

Here are two different ways to solve this problem:

2.5 m	weighs	35
$\Downarrow \div 5$		$\Downarrow \div 5$
0.5 m	weighs	7
$\Downarrow \times 11$		$\Downarrow \times 11$
5.5 m	weighs	77 kg

2.5 m	weighs	35
$\Downarrow \div 2.5$		$\Downarrow \div 2.5$
1 m	weighs	14 kg
$\Downarrow \times 5.5$		$\Downarrow \times 5.5$
5.5 m	weighs	77 kg

This is an informal
scaling method.

This is called the
unitary method.

5.5 metres of pipe weighs 77 kg.

● When two quantities are in direct proportion, then their
graph is a straight line that passes through the origin.

● The line has the equation $y = kx$.

● The gradient of the line, k, is the rate of change between the
two quantities.

Two points is enough to plot
a straight line. The third point
checks the line is accurate.

The graph has £s on the
horizontal axis and euros on the
vertical axis. You could use either
axis for either currency.

Davina buys some euros for a trip to France.
The **exchange rate** is £1 = €1.20.

She draws a conversion graph to help her
convert prices.

First she works out some simple **conversions**
to plot on the graph by writing a formula:
number of euros = 1.20 × number of pounds

She wants to include prices up to £40.
Using the formula,
1.20 × 20 = 24, and 1.20 × 40 = 48

£20 = €24, and £40 = €48.
The euros scale needs to go up
to at least €50.

She chooses the scale so her
graph fits her paper.

£1 = €1.20	£0 = €0
	£10 = €12
	£20 = €24

This point represents
£10 = €12

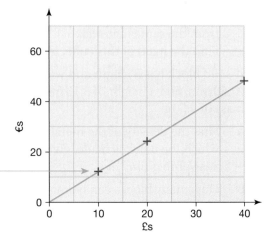

Exercise 22.2S

1 Ribbon costs £2.75 per metre. Find the cost of these lengths of ribbon.

   a 3 m   **b** 4.5 m

   c 6.85 m   **d** 27.55 m

2 A shop sells shelving at £3.45 per metre. Find the cost of these lengths of shelving.

   a 5 m   **b** 3.45 m

   c 2.25 m   **d** 4.85 m

3 A 2 m length of pipe weighs 8 kg. How much does a 3 m length of the same pipe weigh?

4 Four buckets of water weigh 60 kg. How much would 5 buckets of water weigh?

5 400 g of powder paint costs £2.40.

   a Find the cost of 100 g of the paint.

   b Use your answer to part **a** to find the cost of 300 g of the paint.

6 300 g of sherbet drops cost £1.20.

   a How much do 100 g of sherbet drops cost?

   b How much do 700 g of sherbet drops cost?

7 A pack of 250 tea bags contains 130 g of tea and costs £7.50. Calculate

   a the cost of one tea bag.

   b the weight of tea in one bag.

8 A shop sells five different types of luxury tea. Calculate the cost of 100 g of each brand, given that

   a 200 g of brand A costs £3.75

   b 500 g of brand B costs £7.40

   c 300 g of brand C costs £5.20

   d 250 g of brand D costs £5.10

   e 350 g of brand E costs £6.50.

9 Alan and Barry buy sand from a builders' merchant.

Alan buys 35 kg of sand for £4.55.

Barry buys 28 kg of the same sand.

How much does Barry pay?
Show your working.

10 The conversion rate for millimetres to centimetres is 1 cm = 10 mm

   a Work out two simple conversions you could plot for a millimetres to centimetres conversion graph.

   b The graph needs to convert distances up to 10 cm. What is the highest value needed on the mm scale?

11 The conversion rate for pounds (lb) to kilograms (kg) is 1 kg = 2.2 lbs

   a Copy and complete these conversions.

     0 kg = _____ lbs

     10 kg = _____ lbs

     5 kg = _____ lbs

   b Use your conversions from part **a** to draw a conversion graph from pounds to kilograms on a grid from 0 kg to 10 kg.

   c Use your graph to convert

     i 10 lbs to kg   **ii** 5 lbs to kg

     iii 3 kg to lbs   **iv** 2.5 kg to lbs.

12 a This graph is a conversion graph for miles to kilometres and vice versa.

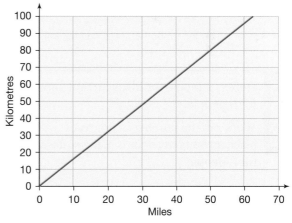

Use the graph to convert

     i 20 miles into kilometres

     ii 60 kilometres into miles.

   b If Dan ran 30 miles and Charlie ran 50 kilometres, who ran further?

   c By finding the gradient of the line, give a formula to connect the number of miles (x) with the number of kilometres (y).

 1036, 1059, 1948   SEARCH

22.2 Direct proportion

RECAP

- Numbers or quantities are in **direct proportion** when the ratio of each pair of corresponding values is the same.
- When two quantities are in direct proportion, then their graph is a straight line that passes through the origin.
- The line has the equation $y = kx$.
- The gradient of the line, k, is the rate of change between the two quantities.

HOW TO

(1) Use the values in the question or the gradient of the graph to find the rate of conversion for the quantities.

(2) Divide and/or multiply to keep the ratio between the quantities the same.

EXAMPLE

Two bottles of olive oil are on sale in a supermarket.

The 500 ml bottle is priced at £2.99.

The 750 ml bottle is priced at £4.29.

Which is better value?

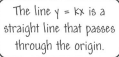

The line $y = kx$ is a straight line that passes through the origin.

(1) Find the cost per ml, k, for each bottle.

500 ml bottle $299 = k \times 500$, so $k = 0.598$

750 ml bottle $429 = k \times 750$, so $k = 0.572$

(2) Compare the cost per ml.

The 750 ml bottle costs less per ml, so it is better value.

EXAMPLE

Here is a recipe for blackcurrant squash for 5 people.

Work out the number of grams of blackcurrants needed to make squash for 8 people.

Blackcurrant squash (for 5 people)

400 g of blackcurrants
1200 ml of water
100 g sugar
250 ml blackcurrant juice

The ratio of people to blackcurrants stays the same.

(1) Set out the problem.

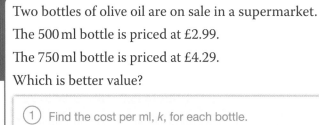

Number of people   Grams of blackcurrants

(2) Divide by 5 to find the value for 1 person.
Then multiply by 8 to find the value for 8 people.

(3) Write the answer.

So for 8 people you need 640 g blackcurrants.

Exercise 22.2A

1 A shop sells drawing pins in two different sized packs.

> Pack A contains 120 drawing pins and costs £1.45.
>
> Pack B contains 200 of the same drawing pins, and costs £2.30.

Calculate the cost of one drawing pin from each pack, and say which pack is better value.

2 A store sells packs of paper in two sizes.

> **Regular**
> 150 sheets
> Cost £1.05

> **Super**
> 500 sheets
> Cost £3.85

Which of these two packs gives better value for money?

You must show all of your working.

3 A recipe for cake uses 400 g of sugar for 5 people.

What mass of sugar is needed for

  a 8 people

  b 12 people

  c 14 people

  d 30 people?

4 A recipe for three bean chilli uses 840 g of beans for 7 people.

What mass of beans is needed for

  a 6 people

  b 17 people

  c 24 people

  d 100 people?

5 Although builders often mix materials according to their volumes, these materials are usually bought according to their mass. The conversions they must make are based upon mass being directly proportional to the volume.

1 cubic metre of sand has mass 1.7 tonnes.

1 cubic metre of cement has mass 1.4 tonnes.

Mortar is made by mixing sand and cement in the ratio $3:1$.

  a What is the total mass of mortar made using 1 cubic metre of cement?

  b Find and simplify the ratio *by mass* of cement to sand in a standard mortar mix.

6 Vince works for 4 hours. He gets paid £24.

Vince works for 8 hours. He gets paid £50.

Is Vince's pay directly proportional to the time he works? Give a reason for your answer.

7 Barry can fit 12 radiators in one day. The number of radiators that Barry can fit is directly proportional to the number of days he works.

Does this formula calculate the number of radiators that Barry can fit?

number of radiators = days worked + 12

Give a reason for your answer.

8 The table shows standard paper sizes.

Paper size	Width (mm)	Height (mm)
A0	841	1189
A1	594	841
A2	420	594
A3	297	420
A4	210	297
A5	148	210

  a Find the ratio of width:height for each paper size. Is the height directly proportional to the width?

  b Plot a graph with width on the x-axis and height on the y-axis.
    Does the graph show that the height is directly proportional to the width?

   1036, 1059, 1948   SEARCH

22.3 Inverse proportion

- When variables are in **inverse proportion**, one of the variables increases as the other one decreases, and vice-versa.

For example, the number of bricklayers building a wall and the time taken are in inverse proportion.

5 bricklayers × 4 hours = 20   5 people take 4 hours.

10 bricklayers × 2 hours = 20   10 people take 2 hours.

When one quantity doubles, the other halves.

1 bricklayer   × 20 hours = 20   1 person takes 20 hours.

When one quantity is divided by 10, the other is multiplied by 10.

> The amount of labour needed is often called 'man-hours'.

- If two variables are in inverse proportion, the **product** of their values will stay the same.

- '*y* is inversely proportional to *x*' can be written as '*y* is proportional to $\frac{1}{x}$' or $y = \frac{k}{x}$, where *k* is the constant of proportionality.
- When two quantities are in inverse proportion, then their graph is a reciprocal curve.
- The product of each *x*- and *y*-coordinate on the curve is the same.

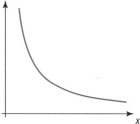

EXAMPLE

x is inversely proportional to *y* with $y = \dfrac{20}{x}$.

a   Complete the table.

x	1	2	4	5	10	20
y						

b   Plot each pair of values for *x* and *y*. Join the points with a smooth curve.

> The product of each *x*- and *y*-coordinate on the curve is 20.
>
> You could use the equation to work out the number of bricklayers needed to build the wall in the previous example.

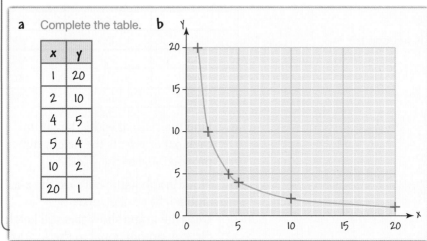

a   Complete the table.

x	y
1	20
2	10
4	5
5	4
10	2
20	1

b

Exercise 22.3S

1 You are told that y is **directly** proportional to x.
Explain what will happen to the value of y when the value of x is

  a doubled   **b** halved

  c multiplied by 6   **d** divided by 10

  e multiplied by a factor of 0.7.

2 You are told that w is **inversely** proportional to z.
Explain what will happen to the value of w when the value of z is

  a doubled   **b** halved

  c multiplied by 6   **d** divided by 10

  e multiplied by a factor of 0.7.

3 It takes 12 hours for 5 builders to build a wall.

  a How many 'man-hours' does it take to build the wall?

  b How many builders are needed to build the same wall in 6 hours?

  c How long will it take 15 builders to build the same wall?

4 A ship has enough food to supply 600 passengers for 3 weeks.

  a How long would the food last for one passenger?

  b How long would the food last for 300 passengers?

  c How long would the food last for 1200 passengers?

5 It takes 20 hours for 5 people to decorate 100 cakes.

  a How many people are needed to decorate 100 cakes in 10 hours?

  b How many hours will it take 4 people to decorate 100 cakes?

  c How long will it take 5 people to decorate 200 cakes?

> Take care! Hours and people are inversely proportional. People and cakes are directly proportional.

6 Sort each of the formulae into one of the categories in the table.

$$y = 4x \qquad y = \frac{20}{x} \qquad y = \frac{x+1}{10}$$

$$y = \frac{13.5}{x} \qquad y = 5x + 1 \qquad y = \frac{x}{10}$$

Inverse proportion	Direct proportion	Neither

> Hint: There are two of each type.

7 x is inversely proportional to y with $y = \dfrac{16}{x}$.

  a Complete the table

x	$\frac{1}{4}$	1	2	4	8	16
y						

  b Plot each pair of values for x and y. Join the points with a smooth curve.

  c Pick any point on your line. Is the product of the x and y-coordinate 16?

*8 h is inversely proportional to b, with $h = \dfrac{60}{b}$.

  a Find b when $h = 5$.

  b Explain why you can use this equation to check your answers to question **3**.

  c Can you write an equation $p = \dfrac{?}{w}$ that you can use to check your answers to question **4**?

22.3 Inverse proportion

- When variables are in **inverse proportion**, one of the variables increases as the other one decreases, and vice-versa.
- 'y is inversely proportional to x' can be written as $y = \frac{k}{x}$, where k is the **constant** of proportionality.

The graph of $y = \frac{k}{x}$ is a reciprocal graph.

EXAMPLE

The time taken for a journey is
- directly proportional to the distance travelled
- inversely proportional to the average speed.

A pilot flies at a moderate speed and completes a journey of 1200 km in 3 hours.

Can the pilot complete a 3600 km journey in less than 5 hours, if he doubles the average speed of the plane?

The distance has tripled. Time and distance are directly proportional.

The time will also triple.   3 hours × 3 = 9 hours

The speed has doubled. Time and speed are inversely proportional.

The time will halve.   9 hours ÷ 2 = 4.5 hours

Yes, the pilot can complete the journey in less than 5 hours.

HOW TO

Show that two quantities are inversely proportional
1. Use the fact that if two variables are in inverse proportion, the product of their values will stay the same.
2. Plot of graph of the variables. The graph should be a reciprocal curve.

EXAMPLE

Stefan carries out an experiment to confirm Ohm's law. This states that for a constant voltage, the current in a circuit is inversely proportional to the resistance. Stefan varies the resistance and records the current. His results are shown in the table.

Stefan thinks that one of the readings is wrong. Which reading could be wrong?

Resistance (ohms)	Current (amps)
5	16
10	8.1
15	4.2
20	4.0
25	3.2
30	2.6

1. If two quantities are inversely proportional, then the product of the values will stay the same.

Resistance (ohms)	Current (amps)	Resistance × current
5	16	80
10	8.1	81
15	4.2	63
20	4.0	80
25	3.2	80
30	2.6	78

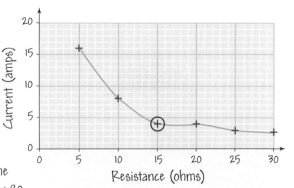

The reading at 15 ohms could be wrong as all the products of the other readings are approximately 80.

2. The graph should be a reciprocal curve – the point (15, 4.2) does not fit the pattern.

Exercise 22.3A

1 Rebecca travels at 40 km/h, the journey takes her 3 hours.

How long will it take Carly to make the same journey if she travels at 60 km/h?

2 Masood assumes that when cooking in a microwave, the cooking time is inversely proportional to the power level.

Level	Power (watts)
Full	600
Roast	400
Simmer	200
Defrost	100
Warm	50

a A pizza takes 8 minutes on 'Roast'. How long will it take on 'Full'?

b Lasagna takes 12 minutes on 'Warm'. How long will it take on 'Simmer'?

3 The safeload of a wooden beam is

● directly proportional to the height

● inversely proportional to the distance between its supports.

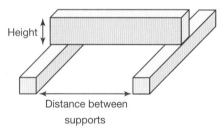

Height

Distance between supports

a The safeload of a particular beam is 500 kg.

i What will the safeload be if the height of the beam is halved?

ii What will the safeload be if the distance between the supports is doubled?

b The safeload of a beam is 800 kg. An engineer plans to double the height of the beam and halve the distance between the supports. Will the modifications be enough to support 3500 kg?

4 z is inversely proportional to y.

y	0.5	1.5				
z			12		6		3.2

Use the numbers on the cards to complete the table. Which card is left over?

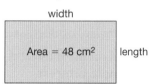

2   3   4   4.8   5

7.5   8   10   16   48

5 A rectangle has area 48 cm².

width

Area = 48 cm²   length

a Explain why the length of the rectangle is inversely proportional to the width.

b Complete the table to show some of the possible dimensions and perimeter for the rectangle.

Length (cm)	Width (cm)	Perimeter (cm)
1	48	98
2	24	52
3	16	38
4		
6		
8		
12		
16		
24		
48		

Jennifer says

> The perimeter of the shape decreases as the length increases.
> The perimeter must be inversely proportional to the length.

Explain why Jennifer is wrong.

22.4 Growth and decay

12.3

An increase of 20% means that the new amount is 120% of the old amount. You can find this by multiplying by 1.2

A decrease of 20% means the new amount is 80% of the old amount. You can find this by multiplying by 0.8

1.2 and 0.8 are called **multipliers**.

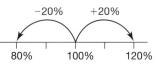

−20%   +20%

80%   100%   120%

▲ To increase something by 20%, multiply by 1.2
To reduce something by 20%, multiply by 0.8

- To increase something by r%, multiply by $\dfrac{100 + r}{100} = 1 + \dfrac{r}{100}$

- To decrease something by r%, multiply it by $\dfrac{100 - r}{100} = 1 - \dfrac{r}{100}$

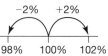

−2%   +2%

98%   100%   102%

▲ To increase something by 2%, multiply by 1.02
To reduce something by 2%, multiply by 0.98

Sometimes percentage increases or decreases are repeated.

When interest is calculated on the interest it is called **compound interest** otherwise it is **simple interest**.

18.2

EXAMPLE

100 red squirrels are delivered to a forest as part of a re-introduction program.
This population increases by 45% each year.

This is called **exponential growth**.

a Draw a table to show how the population increases in the next 6 years.

b Draw a graph to illustrate this growth.

The population at the end of each year is 145% of the population at the beginning of the year.

a Multiplier = 1.45   Multiplier = $\dfrac{110}{100}$

Years	Population
0	100
1	100 × 1.45 = 145
2	145 × 1.45 = 210
3	210 × 1.45 = 305
4	305 × 1.45 = 442
5	442 × 1.45 = 641
6	641 × 1.45 = 929

Try using the 'Ans' key on your calculator to repeat calculations.

b

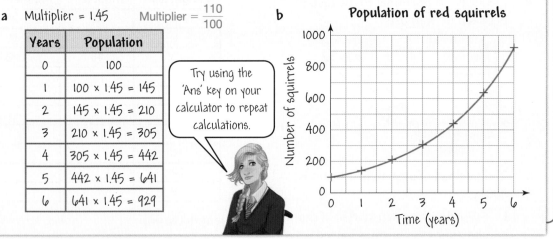

Population of red squirrels

(Graph: Number of squirrels vs Time (years))

EXAMPLE

The value of a car is £15 000. The value decreases by 12% each year.

a Find a formula for the value of the car after n years.

This is called **exponential decay**.

b Use your formula to find the value of the car after 10 years.

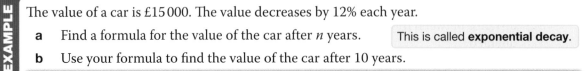

a 100% − 12% = 88%, the value at the end of each year is 88% of its value at the start of that year.
Multiplier = 0.88
Value after n years = £15 000 × 0.88^n Multiply by 0.88 for each year.

b Value after 10 years = £15 000 × 0.88^{10}

You can check this by multiplying by 0.88 ten times.

= £4178 (nearest £)

Ratio and proportion   Units and proportionality

Exercise 22.4S

1 Find the multiplier that gives the amount after each change. Write your answers as decimals.

    a Increase of 30%   **b** Decrease of 30%

    c Increase of 3%   **d** Decrease of 3%

    e Increase of 25%   **f** Decrease of 25%

    g Increase of 2.5%   **h** Decrease of 2.5%

2 A bacteria population doubles every 20 mins.

    a Copy and extend this table to 180 mins.

Minutes	Population
0	1
20	2
40	4

    b Draw a graph to illustrate this growth.

    c Estimate the number of bacteria after

       i 150 minutes   **ii** 170 minutes.

    d What has happened to the population between 150 minutes and 170 minutes?

3 The half-life of a radioactive substance is the time it takes for the amount to go down by a half. The half-life of caesium-137 is 30 years.

    a Copy and extend this table to 6 half-lives.

No of half-lives	Time (years)	Amount (grams)
0	0	1000
1	30	500
2	60	

    b Draw a graph of amount against time.

    c Use your graph to estimate the time taken for the amount to go down from 400 g to 200 g.

4 **a** A shop decreases its prices by 20% on each day of a sale. Copy and complete this table for Monday to Saturday.

Day	Price of boots
Monday	£50
Tuesday	

    b By what percentage is the Monday price reduced on Saturday?

5 The number of trout in a lake is 800. The number decreases by 15% each year.

    a Copy and extend this table to 8 years.

Year	Number of trout
0	800

    b Draw a graph to illustrate your values.

    c **i** Show that the number of trout after n years will be 800×0.85^n

      ii Use this formula to check the values in your table.

6 The rate at which a bacteria colony increases depends on the conditions. The table gives information about two colonies.

Colony	Population now	Increase per hour
A	200	50%
B	400	35%

    a Show that the population of Colony A after n hours is 200×1.5^n

    b Find an expression for the population of Colony B after n hours.

    c When does the population of Colony A become bigger than that of Colony B?

7 Sadie invests £2000 in a savings account. The bank adds 4% **compound** interest at the end of each year. Sadie does not add or take any money from the account for 10 years.

    a Copy and extend this table to show how Sadie's investment grows.

End of year	Amount in the account (£)
1	$2000 \times 1.04 = 2080$
2	

    b Work out the percentage interest that Sadie's investment earns in 10 years.

  *c £P is invested with compound interest r% added at the end of each year.

    i Show that the total amount at the end of n years is $A = P\left(1 + \dfrac{r}{100}\right)^n$

    ii Use this formula to check the last amount in your table in part **a**.

Q 1070, 1238   SEARCH

22.4 Growth and decay

- To increase something by $r\%$, multiply by $1 + \dfrac{r}{100}$
- To reduce something by $r\%$, multiply it by $1 - \dfrac{r}{100}$
- When a principal amount £P is invested with compound interest $r\%$ added at the end of each year, the total amount at the end of n years is

$$A = P\left(1 + \frac{r}{100}\right)^n$$

-5% $+5\%$

95%   100%   105%

▲ To increase something by 5%, multiply by 1.05
To decrease something by 5%, multiply by 0.95

Money earns **simple** interest when the interest is *taken out* of the account at the end of each time period.

Money earns **compound** interest when the interest is *left in* the account.

HOW TO

To solve a repeated percentage change problem
① Find the multiplier.
② Work out the value you need using the formula
③ Answer any related questions.

EXAMPLE

Harry invests £4000 in an account. Compound interest of 3% is added at the end of each year.
  a i   How much is in the account at the end of 5 years?
     ii How much compound interest has been earned?
  b How much longer will it take for Harry to have more than £5000 in the account?
  c What assumptions have you made?

Banks round money to the nearest pence.

① Multiplier = $1 + \dfrac{3}{100}$ = 1.03

a i   Amount after 5 years = 4000 × 1.03⁵ = £4637.10   ②
   ii Interest = £4637.10 − £4000 = £637.10   ③ Take away the original amount.

b Continuing to multiply by 1.03 gives
   4776.21, then 4919.50, then 5067.08   ②
   It takes 3 more years.

c This assumes Harry does not add or take out any money from the account.   ③

EXAMPLE

Town planners use the formula 240×0.9^n to estimate the population of a village n years from now.
  a What is the population of the village now?
  b Explain what the 0.9 tells you about the planners' assumptions.
  c Sketch a graph that shows how the planners expect the population to change.

a Population now = $240 \times 0.9^0 = 240 \times 1$
   Population now = 240

Any number to the power 0 is 1.

b 0.9 = 90% is the multiplier.   ①
   This means the population at the end of each year will be 90% of what it was at the beginning of the year.
   The planners assume the population   ②
   will decrease by 10% each year.

c Population

240

0   Time (years)   ③

Exercise 22.4A

1 For each account in the table

   i find the amount in the account after the given number of years

   ii find the compound interest earned.

Account	Original amount	Compound interest rate	Number of years
a	£1000	5% per year	3
b	£250	4% per year	6
c	£840	2.5% per year	10
d	£45 000	3.25% per year	12

2 Lily works out the compound interest earned by £500 invested for 4 years at a rate of 6%. Here is Lily's working.

> Interest in 1 year = 6% of £500
>   = 0.06 × £500 = £30
> Interest for 4 years = 4 × £30 = £120

   a Why is Lily's working incorrect?

   b Work out the correct answer.

3 The value of a new car is £16 000. Each year the car loses 15% of its value at the start of the year.

   a Work out the value of the car when it is 4 years old.

   b After how many complete years will the car's value drop below £4000?

4 The population of a town is 52 000. Assume that the population increases by 1.5% each year.

   a Work out the population 6 years from now.

   b When will the population reach 60 000?

   c Why might the answers to parts **a** and **b** be incorrect?

5 A company uses the formula 350×0.96^n to estimate the number of workers needed in their factory n years from now.

   a How many workers are there now?

   b Explain what the 0.96 tells you about the company's assumptions.

   c Sketch a graph that shows how the company expects the number of workers to change.

6 A road planner uses the formula 2400×1.08^n to estimate the number of vehicles per day that will travel on a new road n months after it opens.

   a Describe two assumptions the planner has made.

   b Sketch a graph to show what the planner expects to happen.

   c Give reasons why the planner's assumptions may not be appropriate.

7 Mark has £50 000 to invest. He finds the interest rates for two accounts.

Account A	Account B
4.5% per year interest	Year 1 5.0% interest
	Year 2 4.5% interest
	Year 3 4.0% interest

   a Calculate which account would give more compound interest on £50 000 invested for 3 years.

   b Give reasons why Mark may not decide to use this account.

8 Match a graph to each statement.

   a Sales went down by 40 each month.

   b Sales went up by 40% each month.

   c Sales went down by 40% each month.

   d Sales went up by 40 each month.

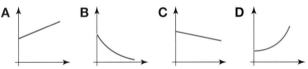

*9 A building society offers two accounts. Karen says that these accounts would give the same interest on an investment. Is Karen correct? Explain your answer.

> **Easy Saver**
> 4% interest added at the end of each year
>
> **Half-yearly saver**
> 2% interest added at the end of every 6 months

Summary

Checkout

You should now be able to...

✔ Calculate with standard and compound units.	1 – 4
✔ Compare lengths, areas and volumes of similar shapes.	5
✔ Describe and solve direct and inverse proportion relationships using equations.	6 – 7
✔ Interpret the gradient of a straight line graph as a rate of change	8
✔ Interpret graphs that illustrate direct and inverse proportion.	9
✔ Set up, solve and interpret growth and decay problems.	10

Language	Meaning	Example
Rate	One quantity measured per unit of time or per unit of another quantity.	Total pay for 3 hours work = £16.50 Rate of pay $= \dfrac{16.50}{3}$ $= £5.50$ per hour
Proportion	Proportion is the size of something compared with the size of something else.	If there are 6 eggs in a carton Total number of eggs = 6 × number of cartons The number of cartons is proportional to the number of eggs.
Proportional	Two quantities are proportional if one is always the same multiple of the other.	
Direct proportion	A set of quantities is in direct proportion with another set of quantities if the ratio of each pair of corresponding values is the same.	The number, n, of eggs is in direct proportion to the number, c, of cartons.
Constant of proportionality	The multiplier, k, between the quantities.	
Inverse proportion	Two quantities are in inverse proportion if one increases as the other decreases.	For a rectangle of width w and length l and area $8\,\text{m}^2$, w Area 8 m^2 l w is inversely proportional to l.
Varies	If x varies with y then x changes when y changes.	$y = kx$ direct proportion $y = \dfrac{k}{x}$ inverse proportion

Review

1 Convert

   a 140 mm to m

   b 1000 s to minutes and seconds

   c 3 litres to cm^3

   d 5 m/s to km/h.

2 A force of 10 N acts over an area of 4 m^2. What is the pressure?

3 A box of cereal costs £1.80 for 750 g. What is the cost per 100 g?

4 Convert 1.8 m^2 into cm^2.

5 A cube of side length 4 cm is enlarged by scale factor 3.

   a What is the volume of

      i the small cube

      ii the enlarged cube?

   b What is the surface area of

      i the small cube

      ii the enlarged cube?

6 Two variables, x and y are directly proportional to each other. When $x = 3$, $y = 12$.

   a Write a formula to link x and y.

   b What is the value of

      i y when $x = 8$

      ii x when $y = 44$?

7 X is inversely proportional to Y. When $X = 5$, $Y = \frac{2}{5}$.

   a Write a formula to link X and Y

   b What is the value of

      i Y when $X = 3$ **ii** X when $Y = \frac{2}{9}$?

8 The graph shows the price paid for a given number of magazines.

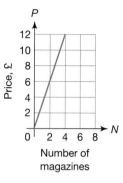

Number of magazines

   a Write an equation linking price, P, and the number of magazines, N.

   b What is the gradient of the graph?

   c What does the gradient represent?

9

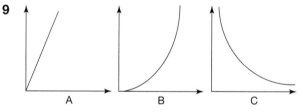

   Which of these graphs shows the relationship in

   a question 6   **b** question 7?

10 Isabella invests £2000 in a bank account that pays 2% interest per year.

   a How much is in the account after

      i one year   **ii** three years?

   b Write a formula to find the value of the account, £V, after, t years.

What next?

Score			
	0 – 4		Your knowledge of this topic is still developing. To improve look at MyMaths: 1036, 1048, 1059, 1061, 1070, 1121, 1238, 1246
	5 – 8		You are gaining a secure knowledge of this topic. To improve look at InvisiPens: 22Sa – e
	9 – 10		You have mastered these skills. Well done you are ready to progress! To develop your exam technique look at InvisiPens: 22Aa – f

Assessment 22

1 The distance from Carlisle through Lancaster to Preston is 92 miles.
Carlisle to Lancaster is 72 miles.

   a Keanu drives from Lancaster to Preston at an average speed of 40 mph.
How long does Keanu's journey take? Give your answer in hours and minutes. [2]

   b Demi meets Keanu in Preston and they drive back to Lancaster together.
They leave Preston at 18:15 and arrive in Lancaster at 18:55.
Calculate their average speed. [2]

   c Keanu and Demi leave Lancaster at the same time.
Keanu drives back to Carlisle and Demi travels back to Preston by train.
They both get to their destinations at the same time.
Demi's journey takes 1 hour and 20 minutes.
Calculate the difference in their average speeds. [3]

2 Light travels at 186 000 miles per second. The Sun is 93 000 000 miles from Earth.
Calculate how long it takes for light from the Sun to reach the Earth.
Give your answer in minutes and seconds. [3]

3 **a** 1 inch is equivalent to 2.54 cm. How many inches are there in 1 m? [2]

   b A pack of 8 batteries cost £6.50. How much do 12 batteries cost? [2]

   c Nigel can cook 8 omelettes in 15 minutes. How long would it take
him to cook 28 omelettes? [2]

4 The Olympic sprinter Usain Bolt can run 100 m in 9.8 seconds.
A wombat can maintain a speed of 40 km/h for 150 m.
Who would win a 100 m race, Usain Bolt or a wombat? [4]

5 **a** A statue has a mass of 3850 grams and volume of 529 cm³. What is its density? [2]

   b A silver bar has a mass of 250 g and a density of 10.5 g/cm³. What is its volume? [2]

   c A litre of milk has a density of 1.03 g/cm³. What is the mass of the milk? [2]

6 John's dog Muttley eats three tins of dog food in two days.

   a How many tins does Muttley eat in 30 days? Give your answer to the nearest tin. [2]

   b John says that 100 tins will feed Muttley for 67 days. Is John correct? [2]

   c John buys 30 tins of dog food to feed Muttley and Muttley's friend Fido.
Fido eats twice as much dog food as Muttley each day.
Does John have enough food to feed Muttley and Fido for two weeks? [4]

7 Adam Titchswamp has a cylindrical water container in his garden.
The container was full of rainwater, but Adam left the tap running and the container emptied.
It took $17\frac{1}{2}$ minutes for the container to empty.
The rainwater flowed out at an average rate of 150 ml/second.

   a How many litres of rainwater does Adam's container hold when full? [3]

   b The base of the cylinder is 23 cm. How high is it? [3]

8   A cuboid has volume 9600 cm³. The density of the cuboid is 7.2 g/cm³.

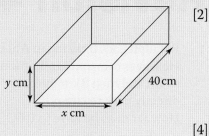

   a   Find the mass of the cuboid in kg.   [2]

   The pressure that the base of the cuboid exerts on the floor is 0.7 N/cm².
   The force of the cuboid on the floor can be calculated using the formula: Force = 9.8 m, where m is the mass in kg.
   The cuboid has depth 40 cm, length x cm and height y cm.

   b   Find the values of x and y to the nearest cm.   [4]

9   **a**   Complete this table for tan x. Give your answers to 4 dp.

x	0°	1°	2°	3°	4°	5°	6°	7°	8°	9°	10°
tan x	0	0.0175									

   [3]

   b   Plot a graph of $y = \tan x$ for values of x from 0° to 10°.   [3]

   c   Clara says that for small angles, tan x° is approximately directly proportional to x. Explain why Clara is correct.   [1]

10   Nina believes that monthly rent for a one bedroom flat is inversely proportional to the distance in km to the city centre. Nina collects this data from eight available flats.

	Flat 1	Flat 2	Flat 3	Flat 4	Flat 5	Flat 6	Flat 7	Flat 8
Distance to city centre (km)	2.1	2.8	3.7	4.1	4.7	5.6	6.2	6.5
Monthly rent (£)	900	625	475	450	425	325	300	275

   a   Draw a graph to show Nina's data.   [4]

   b   Which flat does not support Nina's conclusion that the rent and distance are inversely proportional?   [1]

   c   Nina has a maximum budget of £525 per month to spend on rent. How close to the city centre can Nina afford to live?   [1]

11   Miranda invests £2500 in an account. Compound interest of 3.2% is added at the end of each year.

   a   How much money is in Miranda's account at the end of 5 years?   [3]

   b   After 6 years, Miranda withdraws £500 from the savings account. She then leaves the remaining money in her account for a further 4 years. How much money is in the savings account at the end of the 10 years?   [4]

12   The value, V, of Jenna's car can be calculated using the formula $V = 27\,000 \times 0.95^t$, where t is the number of years since Jenna first bought the car.

   a   Complete this sentence: Jenna originally bought the car for £_____ and the value of the car decreases by ___% each year.   [2]

   b   How much is the car worth after 10 years?   [2]

   c   After 10 years the value of the car decreases by 10% each year. How much will Jenna's car be worth after 15 years?   [3]

Revision 4

1 Ranulph goes trekking in Antarctica. He leaves camp and treks 6 km due south and then 3 km due west. He treks directly back to camp. How far does he have to trek? [3]

2 a Pumba walks 7.6 km through the jungle on a bearing of 200°. How far south is he from his starting point? [3]

b The sun casts a shadow on Dan 230 cm long. The shadow makes an angle with the ground of 35°. How tall is Dan, to the nearest cm? [2]

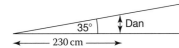

c A plane at P is coming into land and has 10 miles to go. It is currently 1.5 miles above the ground. Calculate its angle of descent, θ. [2]

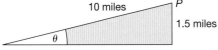

3 Rachel has an isosceles triangle, ABC, in which $AB = AC = 15$ cm, and $BC = 8$ cm. The foot of the perpendicular from BC to A is point M.

a She says that the length of $BM = 4$ cm. Explain why she is correct. [1]

b Find the distance AM. [3]

4 D is the midpoint of AB in the triangle ABC.

a Calculate the length, x. [7]

b Find the value of θ. [6]

5 M and N are the midpoints of PQ and PR. Helen says that $\overrightarrow{RQ} = -\overrightarrow{NM}$. Give the correct relationship, and express the vectors in terms of **p** and **q**. [3]

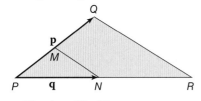

6 Louis has two vectors $a\mathbf{p} + 35\mathbf{q}$ and $24\mathbf{p} + b\mathbf{q}$ that are parallel. Find an equation connecting a and b. [2]

7 $\mathbf{a} = \begin{pmatrix} 3 \\ -6 \end{pmatrix}$, $\mathbf{b} = \begin{pmatrix} -3 \\ 9 \end{pmatrix}$ and $\mathbf{c} = \begin{pmatrix} -3 \\ -3 \end{pmatrix}$.

Write down the vectors

a $5\mathbf{a}$ [1]   **b** $-2\mathbf{b}$ [1]

c $3\mathbf{c}$ [1]   **d** $\mathbf{a} - \mathbf{b}$ [2]

e $\mathbf{b} - \mathbf{a}$ [2]   **f** $\mathbf{a} + \mathbf{b} - \mathbf{c}$ [2]

g $4\mathbf{a} + 2\mathbf{c}$ [2]   **h** $\mathbf{c} - 12\mathbf{a}$ [2]

i $\dfrac{\mathbf{a}}{2} + \dfrac{\mathbf{b}}{3}$ [2]   **j** $2\mathbf{a} + \mathbf{b} + \mathbf{c}$. [2]

8 Ally plays tennis, netball or squash every morning. The probability of choosing each sport is:

Tennis 0.28   Netball 0.35   Squash P

a Show that $P = 0.37$ [2]

b Draw a tree diagram showing her possible choices over two consecutive days. [3]

c Find the probability Ally

i plays tennis on day 1 and squash on day 2 [1]

ii plays netball at least once [1]

iii doesn't play tennis on either day. [1]

9 Ben writes a sequence that begins 2 6 12 20 .... and says that $T(n) = n^2 + n$.

a Is he correct? Show your working. [2]

b Work out the 50th and 100th terms. [2]

A second sequence is formed of the differences of each pair of terms in the first sequence.

c Work out the nth term of this new sequence. [4]

d Work out the 50th and 100th terms. [2]

10 James is cooking pancakes for 9 people. Pancakes for 6 people requires 4 eggs. How many eggs does James need? [2]

11 The supermarkets 'Liddi' and 'Addle' sell teacakes. 'Liddi' sells a pack of 5 for £1.35 and 'Addle' sells a pack of 4 for £1.10. Which supermarket offers the best value? [2]

12

Kilometres	Metres	Centimetres
Millimetres	Tonnes	Kilograms
Grams	Litres	Millilitres

Write down which of the above units you would use to measure the

a length of a shoe [1]

b mass of a suitcase [1]

c volume of a raindrop [1]

d thickness of a sheet of glass [1]

e capacity of a swimming pool [1]

f distance from Earth to Mars [1]

g mass of a mouse [1]

h length of a rugby pitch [1]

i mass of the London Eye. [1]

13 The diagram shows a pattern of black and white triangles.

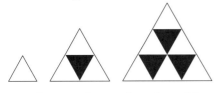

a Copy and complete the table. [5]

Rows (*r*)		1	2	3	4	5
White triangles		1	3			
Black triangles		0	1			
Total number of triangles		1	4			

b The formula for the number of white triangles is $n = \dfrac{r(r + 1)}{2}$.

Find the number of white triangles in a pattern with

i 50 rows [2] ii 100 rows. [1]

c Write down the formula for the total number of triangles in row *r*. [1]

d Use the formulae from parts **b** and **c** to derive a formula for the number of black triangles in any row. Give your answer in its simplest form. [3]

e Test your formula by working out the number of black triangles in row 2. [1]

13 f Use your formula to find the number of black triangles in a pattern with 60 rows. [1]

14 a Ellie says that there are 13 ml in 1.3 l. Correct her answer. [1]

b Lewis says that there are 7.6 kg in 76 g. Correct his answer. [1]

15 Norwich to Harwich is 72 miles.

a Richard drives this journey in 108 minutes. What is his average speed? Give your answer in mph. [2]

b Jeremy drove this journey at an average speed of 50 mph. How long, to the nearest minute, did his journey take? [2]

c Jeremy drove the return journey at an average speed 4 mph faster than his outward journey. How many minutes did he save on the return journey? [3]

16 a An object has a mass of 1.26 kg and volume 180 cm³. What is its density? [2]

b A cylindrical metal rod with radius 3.5 cm and length 12.5 cm has a density of 11.4 g/cm³. What is the mass of the bar? [3]

c A silver bar has a mass of 30 g and density 10.5 g/cm³. What is its volume? [2]

17 A team of 4 scaffolders can build 6 m of scaffolding in 4 hours.

a How many metres of scaffolding could the team build in 7 hours? [2]

b How long would it take a team of 8 scaffolders to build 6 m of scaffolding? [1]

c How long would it take a team of 12 scaffolders to build 18 m of scaffolding? [3]

d Gina says that the size of the team and the size of the scaffolding are directly proportional to each other. Is Gina correct? Give a reason for your answer. [2]

e Gina says that the size of the team and the time taken to build 10 m of scaffolding are directly proportional to each other. Is Gina correct? Give a reason for your answer. [2]

Formulae

Make sure that you know these formulae.

Circles

Circumference of a circle $= 2\pi r = \pi d$

Area of a circle $= \pi r^2$

Pythagoras' theorem

In any right-angled triangle

$$a^2 + b^2 = c^2$$

Trigonometry formulae

In any right-angled triangle $\quad \sin A = \dfrac{a}{c} \quad \cos A = \dfrac{b}{c} \quad \tan A = \dfrac{a}{b}$

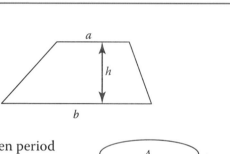

Make sure that you know and can derive these formulae.

Perimeter, area, surface area and volume formulae

Area of a trapezium $= \dfrac{1}{2}(a+b)h$

Volume of a prism $=$ area of cross section $\times$ length

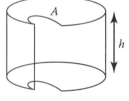

Compound interest

If P is the principal amount, r is the interest rate over a given period and n is number of times that the interest is compounded then

$$\text{Total accrued} = P\left(1 + \frac{r}{100}\right)^n$$

Probability

If P(A) is the probability of outcome A and P(B) the probability of outcome B then P(A or B) = P(A) + P(B) − P(A and B)

These formulae will be provided for students within the relevant examination questions.

Perimeter, area, surface area and volume formulae

Curved surface area of a cone $= \pi rl$

Surface area of a sphere $= 4\pi r^2$

Volume of a sphere $= \dfrac{4}{3}\pi r^3$

Volume of a cone $= \dfrac{1}{3}\pi r^2 h$

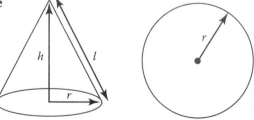

Key phrases and terms

Use our handy grid to understand what you need to do with a maths question.

Circle	Draw a circle around the correct answer from a list.
Comment on	Give a judgement on a result. This could highlight any assumptions or limitations or say whether a result is sensible.
Complete	Fill in any missing information in a table or diagram.
Construct	Draw a shape accurately.
… using ruler and compasses	Compasses must be used to create angles. Do not erase construction lines.
Criticise	Make a negative judgement. This could highlight any assumptions or limitations or say why a result is not sensible.
Describe	Usually describing a graph, one mark per descriptive sentence. When describing transformations give all relevant information.
Diagram not accurately drawn	Calculate any angles or sides, do *not* measure them on the diagram.
Draw	Accurately plot a straight line, bar chart or transformation.
Draw and label	Draw… and mark on values.
Estimate	Simplify a calculation, by rounding, to find an approximate value
Expand	Multiply out brackets.
Factorise fully	Put in brackets with the highest common factor outside the brackets.
Give a reason	Write an explanation of your argument, including the information you use and your reasons or rules used.
Give the exact value	Do not use rounding or approximations in your calculations. You should give your answers as fractions, surds and multiples of π.
Give your answer in terms of π.	Do not use a numerical approximation to π, instead treat it as a letter (algebraic variable).
Give your answer in its simplest form/ simplify	Collect all like terms, any fractions or ratios should be in lowest terms.
Give your answer to an appropriate degree of accuracy	For example, if the numbers in the question are given to 2 decimal places, give your answer to 2 decimal places.
Measure	Use a ruler or protractor to accurately measure lengths or angles.
Rearrange	Change the subject of a formula.
Shade	Use hatching to indicate an area on a diagram.
Show	Present information by drawing the required diagram.
Show that	Obtain a required result showing each stage of your working.
Show working	Usually asked for to support a decision.
Sketch	Represent using a diagram or graph. This should show the general shape and any important features, such as the position of intercepts, using labels as necessary. It does not require an accurate drawing.
Solve	Find an answer using algebra or arithmetic. This often means find the value of x in an equation.

State	Write one sentence answering the question.
Tick a box	Choose the correct answer from a list.
Work out	Find the answer showing your working.
Write down	Write down your answer, written working out is not usually required.
Use an approximation	Estimate using rounding.
Use a line of best fit	Draw a line of best fit and use it.
Use	Use supplied information or the preceding result(s) to answer the question.
You must show your working.	Marks will be lost for not writing down how you found the answer to the question.

Chapter 1

Check in 1

1 304
2 a 70   **b** -12
3 $-8, -5, -3, -1, 2, 4$

1.1S

1 a 87   **b** 143   **c** 406   **d** 460
   e 2053   **f** 8503   **g** 8530   **h** 34,640
   i 30,464   **j** 417.3   **k** 537.403
2 a four hundred and fifty six
   b thirteen thousand. two hundred
   c one hundred and fifteen thousand and twenty
   d four hundred and sixty thousand, three hundred and forty
   e four million, three hundred and twenty-five thousand, four hundred
   f fifty-five million, six hundred and seventy thousand, three hundred and forty-five
   g forty-five point eight
   h three hundred and sixty-seven point zero three
   i four thousand, five hundred and three point three four
   j two thousand seven hundred point zero two
3 a 9, 34, 56, 89, 112, 139, 178
   b 1784, 1990, 2372, 2386, 3022, 3233
   c 4005, 4555, 40500, 40545, 44054, 45045
   d 24454, 42024, 204044, 240440, 242404, 245004
4 a 5.007, 5.099, 5.103, 5.12, 5.2
   b 0.5, 0.509, 0.525, 0.545, 0.55
   c 7.058, 7.302, 7.35, 7.387, 7.403
   d 0.4, 0.42, 2.4, 4.2, 42
   e 26.9, 26.97, 27.06, 27.1, 27.6
   f 13.19, 13.3, 13.43, 14.03, 14.15
5 a $=$ **b** $\neq$ **c** $\neq$ **d** $=$
6 a $<$ **b** $>$ **c** $<$ **d** $<$
   e $>$ **f** $<$ **g** $<$ **h** $>$
   i $>$ **j** $<$
7 a $-13, -12, -6, 0, 15, 17$ **b** $-8, -7, -6, -5, -3, 0$
   c $-9, -1, 2, 3, 6, 8$ **d** $-4.5, -3, -2.5, -1, 0, 5.5$
   e $-6, -5.8, -5.7, -5.4, -5.1, -5$
8 a 25.5   **b** 1.85   **c** 50   **d** 4.95
   e 1.375   **f** 0.705   **g** -2.55 **h** -35
   i -2.2 **j** -8.525
9 a 120   **b** 400   **c** 32   **d** 46
   e 300   **f** 46   **g** 2.3   **h** 65.9
   i 34000   **j** 356   **k** 2.36   **l** 34.5
   m 0.0124   **n** 8.14

1.1A

1 a 3.8   **b** 4.25   **c** 540   **d** 4.8
2 a 6.5 cm   **b** 4.75 kg   **c** 1154°C   **d** 3.24 tonnes
3 a Gold: Reese; Silver: Sokolova; Bronze: DeLoach
   b Yes, bronze
4 £3.60
5 a $21   **b** 14 packs
6 a 97531   **b** 31579   **c** 59731
7 Jessica can place the $\leqslant$, $=$ and $\geqslant$ between pair 2, as opposed to $<$ for pair 1.
8 a 1, 2, 3, 12, 13, 21, 23, 31, 32, 123, 132, 213, 231, 312, 321
   b 4, 7, 8, 47, 48, 74, 78, 84, 87, 478, 487, 748, 784, 847, 874
   c 24

1.2S

1 a 50   **b** 90   **c** 480   **d** 790
   e 2640   **f** 6190
2 a 300   **b** 500   **c** 900   **d** 2700
   e 5700   **f** 16500
3 a 3000   **b** 3000   **c** 5000   **d** 37000

   e 63000   **f** 262000
4 a i 3000   **ii** 3500   **iii** 3470
   b i 81000   **ii** 814000   **iii** 81380
   c i 1000   **ii** 1200   **iii** 1240
   d i 0   **ii** 300   **iii** 280
   e i 14000   **ii** 14000   **iii** 14000
   f i 10000   **ii** 10000   **iii** 10000
5 a 5   **b** 4   **c** 12   **d** 25
   e 16   **f** 436
6 a 4   **b** 9   **c** 19   **d** 69
   e 110   **f** 7
7 a 0.3   **b** 2.9   **c** 3.8   **d** 12.5
   e 0.3   **f** 2.9   **g** 3.8   **h** 14.5
   i 3.7   **j** 28.8   **k** 468.6   **l** 369.3
8 a i 3.447   **ii** 3.45   **iii** 3.4
   b i 8.948   **ii** 8.95   **iii** 8.9
   c i 0.128   **ii** 0.13   **iii** 0.1
   d i 28.387   **ii** 28.39   **iii** 28.4
   e i 17.999   **ii** 18.00   **iii** 18.0
   f i 10.000   **ii** 10.00   **iii** 10.0
   g i 0.004   **ii** 0.00   **iii** 0.0
9 a 3000   **b** 3000   **c** 15000   **d** 60000
   e 90000   **f** 90000   **g** 3000   **h** 70000
10 a 480   **b** 1200   **c** 490   **d** 14000
   e 530   **f** 15000
11 a 0.36   **b** 0.42   **c** 0.057   **d** 0.0047
   e 1.4   **f** 0.0000042
12 a i 8.37   **ii** 8.4   **iii** 8
   b i 18.8   **ii** 19   **iii** 20
   c i 35.8   **ii** 36   **iii** 40
   d i 279   **ii** 280   **iii** 300
   e i 1.39   **ii** 1.4   **iii** 1
   f i 3890   **ii** 3900   **iii** 4000
   g i 0.00837   **ii** 0.0084   **iii** 0.008
   h i 2400   **ii** 2400   **iii** 2000
   i i 8.99   **ii** 9.0   **iii** 9
   j i 14.0   **ii** 14   **iii** 10
   k i 1400   **ii** 1400   **iii** 1000
   l i 140000   **ii** 140000   **iii** 100000
13 a 9.73   **b** 0.36   **c** 147.5   **d** 29
   e 0.53   **f** 4.20   **g** 1245.400   **h** 0.004
   i 270   **j** 460.0

1.2A

1   Mumbai   16370000
   New York   8090000
   Beijing   7440000
   London   7170000
   Paris   2140000
2 a Con $-$ 26000   Lab $-$ 26000
     UKIP $-$ 6000   Lib $-$ 3000
   b Conservatives and Labour would have the same amount of votes
   c Yes, both still round to 26000
3 a 17.80 m²   **b** 18 m²
   c **b**, more accurate estimate of area
   d b, allows for spare
4 a £21.87 with 4p left   **b** 7 trips
   c 4 coaches   **d** 35 crates
5 4.48 and 4.53
6 minimum 11.55 cm   maximum 11.65 cm
7 a 27000 (to 3 sf)   **b** $27100 - 27000 = 100$

1.3S

1 a 270   **b** 80   **c** 330   **d** 40
   e 3800   **f** 70   **g** 430   **h** 390
2 a 8   **b** -2 **c** -13 **d** -14
   e 12   **f** 3   **g** -5 **h** -6
   i -9 **j** -23 **k** -23 **l** -23

3 **a** 355 **b** 560 **c** 950 **d** 808
 e 567 **f** 889 **g** 14.3 **h** 20.7
4 **a** 15 **b** 15 **c** 131 **d** 124
 e 161 **f** 202
5 **a** 131 **b** 141 **c** 202 **d** 248
6 **a** 48.6 **b** 48.3 **c** 28.6 **d** 45.2
7 **a** 6.2 **b** 7.4 **c** 18.4 **d** 25.8
 e 22.2 **f** 22.6
8 **a** 8.12 **b** 16.52 **c** 23.31 **d** 104.19
9 **a** 13.08 **b** 13.13 **c** 18.84 **d** 18.88
10 **a** 26.27 **b** 56.92 **c** 3.52 **d** 36.72
 e 27.98 **f** 18.88
11 **a** 36.8 **b** 78.3 **c** 27.1 **d** 42.1
12 **a** 6.2 **b** 7.1 **c** 8.5 **d** 27.9
13 **a** 10.72 **b** 19.72 **c** 18.11 **d** 141.99
14 **a** 8.05 **b** 4.73 **c** 38.74 **d** 18.68
15 **a** 34.06 **b** 44.71 **c** 3.51 **d** 24.59
 e 17.98 **f** 28.89
16 **a** 22.08 **b** 47.34 **c** 83.17

1.3A

1 **a** 248 km **b** 97 **c** 1.102 kg
2 **a**

```
                25
          12.7     12.3
       5.8     6.9     5.4
    3.7     2.1     4.8     0.6
```

 b

```
              -5
          -2      -3
      -7      5      -8
```

 c

```
                 -24
             -20      -4
         -12      -8      4
      -5      -7      -1      5
```

3 **a**
```
   2 1.7
   9 1.5
 + 9 4.9
 ─────────
   2 0 8.1
```
 b
```
   3 6.5 8
 - 2 7.7 9
 ─────────
     8.7 9
```

4 **a** £5.49 **b** £172.38
5 **a** 100.97 Ten not carried
 b 245.1 Ten not carried
 c 161.28 One not removed
 d 87.82 One not removed
6 Various answers possible, e.g. $621 - 534 = 87$
7 **a** 1026 and 342 **b** 724 and 1448
 c 432 & 342 and 1026 & 522
8 0.2 g

1.4S

1 **a** -25 **b** 32 **c** -72 **d** -20
 e 30 **f** 49 **g** 16 **h** -20
 i -18 **j** 26 **k** -42 **l** -48
2 **a** -2 **b** -5 **c** 5 **d** 4
 e -22 **f** -1 **g** 40 **h** 4
 i 5 **j** -17 **k** -3 **l** 27
3 **a** 98 **b** 152 **c** 273 **d** 323
4 See answers to questions **1–3**.
5 **a** 24.91 **b** 4.284 **c** 105.84 **d** 130.8985
 e 42.9442 **f** 369.5328
6 **a** 3.87 **b** 0.775 **c** 0.916 **d** 7.53
 e 18.13 **f** 3.45 **g** 4.15 **h** 7.74
 i 4.08 **j** 2.35
7 **a** 5.26 **b** 28.88 **c** 1384.29 **d** 175.56
 e 28.65 **f** 111.51
8 See answers to questions **5–7**.
9 **a** 28.81 **b** 28.81 **c** 4.3 **d** 0.067
 e 10

10 **a** 26 **b** 37 **c** 52 **d** 10
 e 33 **f** 5 **g** 180 **h** 9
11 **a** 28 **b** 72 **c** 5 **d** 16
 e 2 **f** 75
12 **a** 1 **b** 2 **c** 2 **d** 14
 e 40 **f** 7
13 **a** 14 **b** 10 **c** 2 **d** 12
 e 91 **f** 112 **g** 70 **h** 37
 i 1 **j** 3
14 **a** 170 **b** 0.58 **c** 1.78

1.4A

1 **a** £1.37
2 $\frac{36}{(2+3)} = 7.2$ but $\frac{36}{2} + \frac{36}{3} = 30$
3 **a** $5 \times (2 + 1) = 15$ **b** $5 \times (3 - 1) \times 4 = 40$
 c $20 + (8 \div 2) - 7 = 17$ **d** $2 + (3^2 \times (4 + 3)) = 65$
 e $2 \times (6^2 \div 3) + 9 = 33$ **f** $((4 \times 5) + 5) \times 6 = 150$
4 **a** 437×87 **b** 851×59
 c $429 \times 39 = 3861 + 12870 = 16731$
5 **a** $2414 \div 17 = 142$ **b** $277.59 \div 19$
6 **a** max $= 72$ min $= -54$
 b $(3 \times 4) + (4 \times -3) = 0$
 c 3 correct 0 incorrect, 6 correct 4 incorrect,
 9 correct 8 incorrect
7 Answer will be the same as original number.

Review 1

1 **a** 6700 **b** 85.2 **c** 240 **d** 5
2 **a** 45 **b** 6.21 **c** 0.079 **d** 0.006
3 **a** $905 < 961$ **b** $14.7 < 14.9$
 c $0.7 > 0.09$ **d** $0.214 < 0.22$
4 **a** 53 099, 53 909, 503 099, 503 909, 530 909
 b 4.09, 4.289, 4.29, 4.3, 4.32 **c** $-14, -8, -4, 0, 9$
5 **a** 850 **b** 25 **c** 0.8 **d** 62.94
6 **a** 400 **b** 5100 **c** 45.7 **d** 0.08
 e 0.090 **f** 1
7 **a** -1 **b** -5 **c** 6 **d** 2
 e 0 **f** -3
8 **a** 1358 **b** 38.4 **c** 914.07 **d** 7.401
9 **a** 513 **b** 268 **c** 219.2 **d** 3.67
10 **a** -21 **b** 32 **c** -5 **d** 13
11 **a** 240 **b** 400 **c** 40 **d** 8
12 **a** 2961 **b** 14976 **c** 28 **d** 26
13 **a** 17.92 **b** 11.637 **c** 2.45 **d** 5.65
14 **a** 33 **b** 17 **c** 55 **d** 45
 e 18 **f** 14
15 **a** 36 **b** 2

Assessment 1

1 **a** $-33 < 8$; $-33, 8, 19, 44, 303, 576$
 b $-576 < -19$; $-576, -19, 8, 33, 44, 303$
2 Yes, 0.42, 3, 4.236, 51.6, 4200, 216 000.
3 **a** $-2, -2, -4$ **b** $-16, -25, -39$
4 **a** 24 **b** 7, 12, 5; 6, 8, 10; 11, 4, 9
5 **a** No, LB $= 271.75$ cm < 271 cm.
 b No, Yao Defen UB $= 233.345$ cm > 233.341 cm.
6 **a** Yes **b** No, Dave 40, Jane 50 (1 sf).
 c Dave 45, Jane 53
7 **a** Abena 13 000 (2 sf), Edward 8100 (2 sf)
 b 15 000 km (14 522.324)
 c The new approximation would be $3 \times (10 000 - 8000) = 6000$
8 **a** C **b** B **c** B **d** A
9 A, C, F, E, H, B, D, G
10 2.125 km
11 **a** 25 sweets **b** 25 000 (nearest 1000)
12 **a** $(3 + 4) \times 5 + 2 = 35$ **b** $60 \div (5 + 7) + 5 = 10$

Chapter 2

Check in 2

1 a 9 **b** 4 **c** 16
2 15
3 a 7 **b** −1 **c** 2 **d** −3
4 a i 1, 2, 3, 6, 9, 18 **ii** 1, 2, 3, 4, 6, 12
  iii 1, 2, 3, 4, 6, 8, 12, 24
  b 1, 2, 3, 6 **c** 6

2.1S

1 a £2.40 **b** £0.3n
2 20x pence
3 $n + 2$ = a number add 2
  $n - 2$ = a number subtract 2
  $2n$ = a number multiplied by 2
  $\frac{n}{2}$ = a number divided by 2
  $2 - n$ = 2 take away a number
4 $4x + 2$ = x multiplied by 4 add 2
  $2x + 4$ = x multiplied by 2 add 4
  $2(x + 4)$ = x add 4 multiplied by 2
  $4(x + 2)$ = x add 2 multiplied by 4
5 No I disagree with Liam as $7a = 7 \times 3 = 21$ not 73
6 a 6 **b** 4 **c** 12 **d** 1
  e 9 **f** 5
7 a 7 **b** 3 **c** 9 **d** −3
  e 4 **f** −3 **g** 6 **h** −10
8 a 15 **b** 9 **c** 13 **d** 10
  e 7 **f** 5 **g** 36 **h** 72
9 a 9 **b** 18 **c** 25 **d** 50
  e 16 **f** 32 **g** 36 **h** 72
  i 360
10 a −1 **b** −1 **c** 11 **d** 8
  e 0 **f** 12 **g** 6 **h** −9
11 a $\frac{12}{5}$ **b** $\frac{40}{10} = 4$ **c** $\frac{12}{3} = 4$ **d** $\frac{13}{7}$
  e $\frac{30}{2} = 15$ **f** $\frac{89}{8}$ **g** $\frac{20}{10} = 2$ **h** $\frac{16}{2} = 8$
12 a $\frac{4}{2} = 2$ **b** $\frac{12}{3} = 4$ **c** $\frac{-3}{3} = -1$ **d** $\frac{6}{3} = 2$
  e $\frac{-8}{2} = -4$ **f** $\frac{8}{3}$ **g** $\frac{6}{-3} = -2$ **h** $\frac{-12}{2} = -6$
  i $\frac{6}{-3} = -2$
13 a Expression **b** Equation **c** Formula **d** Expression
  e Formula **f** Equation
14 a 2p, 4q **b** 3x, y **c** 2a, −5b, 3c

2.1A

1 a n oranges at $5p = 5n$ **b** $7 \times m = 7m$
  c $n + n = 2n$ **d** $4 \times m = 4m$
  e 3 chews at n pence each = $3 \times n$
  f 6 stamps at n pence each = $6n$
  g m toys each at £3.00 each = $3m$
2 $F = 2 + 4n$ $F = 6 + 3n$
  $= 2 + (4 \times 7)$ $= 6 + (3 \times 7)$
  $=$ £30 $=$ £27
  Carla's Cabs is the cheapest.
3 a No, it increases by £20 **b** $30 + (20 \times 10) = £230$
  c $260 = 30 + 20n \rightarrow n = 11.5 \rightarrow$ max = 11 days
4 H = 28 O = 16 T = −10 L = 33
  A = 6 C = 1 E = 46
  CHOCOLATE
5 $2x - 3$
6 $4x + 8$
7 $6c + 10d$
8 a £0.5f + £0.3g **b** £0.8j + £0.4k
  c £0.5x + £0.6y + £0.3z **d** £0.6p + £0.8q + £0.4r
9 Paul is right as you first square the number, then multiply by two.
10 a Audrey is incorrect, Cerys is correct

b A = £5 B = £15 C = £20
  Total = £40

2.2S

1 a 3m **b** 7n **c** 6y **d** 3z
2 a 13a **b** 17n **c** 15t **d** 19x
  e 4r **f** 4f **g** 7g **h** 8r
  i 23n **j** 17c
3 a 4m **b** 3p **c** rs **d** 12q
  e 5g **f** 4c **g** 8d **h** jk
  i n^2 **j** 15e **k** t^2 **l** 6mn
4 a 15n **b** 12m **c** 17p **d** 18q
  e x **f** 4w **g** 7a **h** 5b
  i j **j** 2k
5 a $\frac{d}{4}$ **b** $\frac{x}{3}$ **c** $\frac{y}{7}$ **d** $\frac{t}{9}$
  e $\frac{2a}{3}$ **f** $\frac{3n}{4}$ **g** $\frac{5p}{7}$ **h** $\frac{2v}{4} = \frac{v}{2}$
6 a $2a + 2b$ **b** $4c + d$ **c** $7e + 5f$ **d** $7g + 8h$
  e $7i + 6j$ **f** $4u + 7v$
7 a $a + 3b$ **b** $4x + 6y$ **c** $7m + 2n$ **d** $9s + t$
  e $2p + 5q$ **f** $5c + d$
8 a $a + 2b$ **b** $c + d$ **c** $5u + v$ **d** $x + 6y$
  e $9m + 5n$ **f** $6p + 2q$
9 a $3e - f$ **b** $g - 2h$ **c** $5j - k$ **d** $r - 3s$
  e $3t - 2u$ **f** $12v - 7w$
10 a $5p + 6q$ **b** $9x + 7y$ **c** $2m + 8n$
  d $x + 5y$ **e** $8r - 6s$ **f** $2g - 4f$
  g $2a + 6b + 5c$ **h** $5u - 2v + 3w$
  i $3x - 4y + 5z$ **j** $6r + 5s + 2t$
11 a 9r **b** $4m^2$ **c** 8tv **d** 10mn
  e $6xy^2$ **f** $2x^2 + x$ **g** $9w - 8$ **i** $z^3 + 3z + 1$

2.2A

1 a S = $x + y$ G = $2x + y$ K = $3x + 2y$ S + G = K
  They have the same number of peanuts as Kofi.
  b S + G + K = $6x + 4y$
2 $3x + 5y - x + 2y = 7y + 2x = 3y + 7x + 4y - 5x$
  $2x - 4y + 3x + 2y = 5x - 2y = 2x - 4y + 2y + 3x$
  $2x + 4y - x = 4x + 4y - 3x = 3x + 6y - 2x - 2y$
  $2y + 3x - x + 3y = 2x + 5y = 7y - 3x + 5x - 2y$
3 Abdul
4 a $7a = 5a + 3a - a$
  b $3x + 2y + x - 3y - 2x - 3y = 2x - 4y$
  c $3a \times 6 + 2a + 10b \div 2 - b = 20a + 4b$
5 a i $16 + 8p = 8(2 + p)$ **ii** 32p
  b i 64p
    ii Check students' rectangles have perimeter $32 + 8p$.
6 3x by 2y
7 $7y \times 3x - 4y \times x = 17xy$
8 a $2(5y + 2x) + 2(4y + x) = 18y + 6x = 3z$
  $z = 6y + 2x$
  b Trapezium. $x > 0$ as 3x is a length so $10y + 8x > 10y + 7x$.

2.3S

1 a y^4 **b** m^6
2 a $3t^2$ **b** $4pq^2$ **c** $6v^2w^3$ **d** $2r^4s$
  e $6m^2n$ **f** $8y^3z^2$
3 a $6m^2$ **b** $12p^3$ **c** $6xy^2$ **d** $10s^2r^2$
4 No because you add the indices: $a^5 \times a^2 = a^7$
5 a n^5 **b** s^7 **c** p^4 **d** t^4
6 a x^7 **b** x^8 **c** x^9 **d** x^7
7 a r^2 **b** r **c** r^5 **d** r^3
8 a m^4 **b** x **c** t^2 **d** y^3
9 a x **b** m^2 **c** s^3 **d** v^3
  e q^3 **f** t^4 **g** p **h** y^2
10 No I do not as you add the indices but multiply the numbers: $= 8y^7$
11 a $3x^7$ **b** $5y^7$ **c** $12b^8$ **d** $10p^{11}$
  e $30h^{11}$ **f** $12s^3t^4$
12 No, correct answer is $4p^8$

13 a $2y^4$ **b** $2a^6$ **c** $5k^4$ **d** $3p^5$
e $5x^6$ **f** $0.5x^8y^4$
14 a a^6 **b** y^{12} **c** k^{15} **d** p^{56}
e a^{21} **f** a^{21}
15 a $4a^6$ **b** $729y^{12}$ **c** $25k^6$ **d** $216p^{21}$
e $128a^{21}$ **f** $256a^{16}$
16 a $8a^{12}$ **b** m^7 **c** $4y^6$ **d** $9y^4$
17 a g^3 **b** h^{-6} **c** b^{-12} **d** j^{-6}
e t^{10} **f** n^{-2}
18 a $0.25p^{-16}$ **b** $60r^{-1}$ **c** $27h^{-9}$ **d** $6b^8$

2.3A

1 $2n^3 = 2 \times n^3$ $n^2 = \frac{n^4}{n^2}$ $5 \times n = 5n$ $2n^2 = 2 \times n \times n$
2 $8x^3y^6\,\text{cm}^3$
3 a $24x^2y^3 + 8x^2y^3 = 32x^2y^3$ **b** $20a^2b^4 - a^2b^4 = 19a^2b^2$
4 $10p^2q \times 2 + 2 \times 2p^2q = 24p^2q$
5 $4ab^3$
6 a $(pq)^3 \times p^2q = p^5q^4$ **b** $(xy + xy)^2 \times xy = 4x^3y^3$
7 x^{40}
8 False, x^3y^6 not equal to x^2y^6
9 A: $12a^7b^5 \to$ Multiplied b's indices instead of adding.
 H: $5p^4q^3 \to$ Divided p's indices instead of subtracting.
 E: $9x^2y^3 \to$ Only add numbers not indices.
 S: $16m^4n^6 \to$ squared n's indices instead of multiplying them.
10 $x^{-2} \div x^{-6} = (x^2)^2 = x^4$
 $x^6 \div x^5 = x^{-2} \times x^3 = x$
 $(x^{-4} \times x^4)^2 = x^3 \times x^{-3} = 1$
11 $0.5 \times 15w^3 \times 4w^6 = 30w^9$
12
$$x$$
$$x^3 \qquad x^{-2}$$
$$x^2 \qquad x \qquad x^{-3}$$
13 $5x^3 \quad 6x^{-2} \quad 2x^{-1}$
 $3x^{-4} \qquad 10x^4$
 $4x \quad 5x^2 \quad 3x^{-3}$

2.4S

1 a $3m + 6$ **b** $4p + 24$ **c** $2x + 8$ **d** $5q + 5$
e $12 + 2n$ **f** $6 + 3t$ **g** $12 + 4s$ **h** $8 + 2v$
2 a $6q + 3$ **b** $8m + 4$ **c** $12x + 9$ **d** $6k + 2$
e $10 + 10n$ **f** $12 + 6p$ **g** $4 + 12y$ **h** $10 + 8z$
3 a $5p + 9$ **b** $7m + 8$ **c** $2x + 4$ **d** $5k + 10$
e $9t + 10$ **f** $4r + 7$
4 a 2 **b** $2, 4$ **c** $2, 5, 10$ **d** $2, 3, 6$
e 3 **f** 2 **g** 2 **h** $2, 4, 8$
5 a 3 **b** 2 **c** 2 **d** 4
6 a y **b** s **c** m **d** $2y$
7 a $2(x + 5)$ **b** $3(y + 5)$ **c** $4(2p - 1)$ **d** $3(2 + m)$
e $5(n + 1)$ **f** $6(2 - t)$ **g** $2(7 + 2k)$ **h** $3(3z - 1)$
i $3(4m + 5)$ **j** $2(14 - 3y)$ **k** $8(5z - 3)$ **l** $6(3 - 5b)$
8 a $3r + 6 + 2r - 2 = 5r + 4$ **b** $4s + 4 - 2s - 4 = 2s$
c $6j + 9 - 2j - 4 = 4j + 5$
d $12t - 6 + 3t - 3 = 15t - 9 = 3(5t - 3)$
9 a $2n + 6 + 3n + 6 = 5n + 12$
b $4p + 4 + 6 + 2p = 6p + 10 = 2(3p + 5)$
c $8x + 4 + 2x + 6 = 10x + 10 = 10(x + 1)$
d $6n + 4 + 12n + 3 = 18n + 7$
10 a $4x^2 + x$ **b** $m^3 + 2m$ **c** $2t^3 + 8t$ **d** $3p^3 + 3p$
11 a $4m^2 - 12m$ **b** $2p^2 - 12p$ **c** $-3x - 6$ **d** $-10m + 20$
12 a $w(w + 1)$ **b** $z(1 - z)$ **c** $y(4 + y)$ **d** $m(2m - 3)$
e $p(4p + 5)$ **f** $k(7 - 2k)$ **g** $n(3n - 2)$ **h** $r(5 + 3r)$
13 a $4(y - 3)$ **b** $x(2x + 3)$ **c** $(\sqrt{3}y + 1)(\sqrt{3}y - 1)$
d $5(3 + t^2)$ **e** $3m(1 + 3m)$ **f** $2r(r - 1)$
g $v(4v^2 + 1)$ **h** $3w(w + 1)$
14 $4(x + 3) = 4x + 12$
 $3(x - 4) = 3x - 12$
 $x(4 + 3x) = 4x + 3x^2$
 $x(4x - 3) = 4x^2 - 3x$

2.4A

1 $L = n$ $M = n + 5$ $N = 3(n + 5)$
2 a $J = n$ $S = n + 4$ $M = 3(n + 4)$ $F = 4n$
 $U = 4n - 2$ $G = 2(4n - 2)$
b $8n - 4 - 3n - 12 = 5n - 16$
3 a Kate
b Both Debbie and Bryan have removed the wrong factor.
4 a $3(x + 4) = 3x + 12$ **b** $2(6x + 5) = 12x + 10$
c $2(x + 4) + 3 = 2x + 11$ **d** $4(2x - 1) = 8x - 4$
e $6(x - 4) + 5(2x + 1) = 16x - 19$
f $2(x + 1) + 6(3x - 1) = 20x - 4 = 4(5x - 1)$
g $3(4x + 3) + 2(x - 1) = 7(2x + 1)$
h $5(5x + 6) - 7(3x + 2) = 4(x + 4)$
5 a $3(2x - 1)$ **b** $6x - 3$ **c** $6x - 3 = 15$, so $6x - 18 = 0$
6 a $8(3x + 2)$ **b** $(12x + 8) + (12x + 8)$
7 $0.5(2 \times 2y) + (y \times 2y) = 2y(y + 1)$
8 a $8 + 2(2x - 6) = 4x - 4 = 4(x - 1)$ **b** $20(b + 2)$
9 $4(3x + 2) - 2(2x - 1) = 8x + 10 = 2(4x + 5)$
10 $(2x, 5y, 2 + 5x)$ $(5x, 2y, 2 + 5x)$ $(10x, y, 2 + 5x)$
11 a $(4x, 4x, y - 4)$ $(16x, x, y - 4)$ $(16, x^2, y - 4)$
b As otherwise you would have a negative volume of the cuboid, which is impossible.

Review 2

1 a $13y$ **b** $7xy$ **c** $3x$ **d** y^3
e $\frac{1}{2}x$ **f** $5xy$
2 a £1.35 **b** $15y$
3 a 35 **b** 15 **c** -4 **d** 49
4 a -28 **b** -12 **c** 32 **d** 18
e 1 **f** 34
5 a 6 **b** -4 **c** -2 **d** 2
e -2 **f** 6
6 a $3x - 2 = 7$ **b** $S = \frac{D}{T}$
c $5x - y$
7 a $5f, 6g, -2h$ **b** $5p, -6q, q^2$
8 a $5ab$ **b** $4c$ **c** d^3 **d** $10fg$
e $8e - 15$ **f** $a^3 + a^2 + 3$
9 a $2r$ **b** $5a + 6b$ **c** $5d - 7e$ **d** $3x^2 + x$
10 b and c
11 a c^6 **b** d^5 **c** r^{12}
d t^6 **e** $6u^8$ **f** $3v$
12 a $2a + 2$ **b** $32b - 16c$
c $-15d + 20$ **d** $h^2 + 2h$
13 a $7(2b + 1)$ **b** $4(2 - c)$ **c** $3x(x + 2)$ **d** $4b(a - 3)$

Assessment 2

1 They are both right, $c + c + c + c$ means 'add c to itself four times which is the same as $4c$.
2 a $V - 4$ years old. **b** $\frac{1}{2}V$ or $\frac{V}{2}$ years old.
c $\frac{1}{2}V + 5$ years old. **d** $V + \frac{1}{2}V = \frac{3V}{2}$ years old.
3 a The DVD costs £8.
b The Blu-ray is £7 more expensive than the DVD.
c The book is half the cost of the DVD.
d The DVD and the Blu-ray together cost £23.
e The total cost of all three items is £27.
4 a $2l + 2w$ **b** $2(l + 5) + 2(w - 5)$
c Yes. $2(l + 5) + 2(w - 5) = 2l + 10 + 2w - 10 = 2l + 2w$.
5 a $157\,\text{cm}^2$ **b** $3.58\,\text{m}$ **c** $3.27\,\text{in}$
6 Fiona is correct. $9x$ and $4x$ are like terms and 2 is not.
7 a $C = 62S + 93L$
b $C = 4 \times 62 + 2 \times 93 = 248 + 186 = 434$ pence = £4.34
c £5.89 **d** £20.77
e 558p or £5.58
8 $10z^3$
9 a $2y^2 + 80y$
b No, $V = 20y^2 = 200\,\text{cm}^3$, $y^2 = 10$ so $y = \sqrt{10}$.

Answers

10 a $5(W - 2)$ **b** $4(2W + 1)$
11 No, $(10 - w) + (3w - 5) = w$, $w = -5$, w cannot be negative as it is a length.
12 $2p(16p - 11)$
13 a No, $DC = a + 13$.   **b** $2(2a + 13)$
  c Area of a rectangle = height × width.
    Area of $AEFD = a(a + 4)$, area of $EBCF = 9a$.

Chapter 3

Check in 3

1 a 57 mm   **b** 5.7 cm
2 a around 40° **b** around 130–140°
3 a 180°   **b** 270°   **c** 90°   **d** 180°

3.1S

1 a Acute   **b** Obtuse   **c** Reflex   **d** Right angle
2 a Right angle **b** Acute   **c** Obtuse   **d** Reflex
  e Reflex   **f** Acute   **g** Obtuse   **h** Reflex
  i Reflex   **j** Acute   **k** Obtuse   **l** Reflex
3 a see students' diagrams   **b** see students' diagrams
4 a 180°   **b** 360°
5 a 40°, angles around a point sum to 360°.
  b 240°, angles around a point sum to 360°.
  c 130°, angles on a line sum to 180°.
  d 50°, angles on a line sum to 180°.
6 a 110°, corresponding angles.   **b** 47°, alternate angles.
  c 115°, alternate angles.   **d** 63°, corresponding angles.
7 a 110°, angles around a point sum to 360°.
  b 135°, angles around a point sum to 360°.
  c 45°, angles on a line sum to 180°.
  d 45°, angles on a line sum to 180°.
8 $i = 14°$, $h = 14°$, $g = 166°$, $j = 37°$, $l = 143°$, $k = 143°$; angles on a line sum to 180°, vertically opposite angles are equal.
9 a $h = 50°$, $i = 50°$; angles on a straight line sum to 180°, corresponding angles are equal.
  b $j = 63°$, $k = 63°$; alternate angles are equal, vertically opposite angles are equal.

3.1A

1 Angles on a straight line add up to 180°, these angles add up to 170°.
2 a $a = 17°$, $b = 17°$, $c = 163°$; angles on a straight line are equal, alternate angles are equal, corresponding angles are equal.
  b $d = 125°$, $e = 105°$; alternate angles are equal, vertically opposite angles are equal.
  c $f = 134°$, $g = 134°$, $h = 24°$; vertically opposite angles are equal, corresponding angles are equal, angles on a straight line sum to 180° and angles in a triangle sum to 180°.
  d $i = 140°$, $j = 40°$, $k = 140°$, $l = 40°$; vertically opposite angles are equal, interior angles sum to 180° (used three times).
3 a 275°   **b** 68°
4 a 284°   **b** 263°   **c** 117°
5 158° to the right
6 a Missing angle in a triangle = $180° - (a + b)$, exterior angle $= 180° - [180° - (a + b)] = (a + b)$
  b $c + (a + b) = 180°$, angles on a straight line sum to 180°.
7 Let the interior angles be w, x, y, and z (clockwise). $w = 180° - x = y$, interior angles in parallel lines sum to 180°.
8 $a = 80°$, $b = 110°$, $c = 70°$, $d = 30°$, $e = 150°$

3.2S

1 a Right angled triangle   **b** Equilateral triangle
  c Kite   **d** Rhombus
2 a 50°   **b** 70°   **c** 80°   **d** 60°
  e 70°   **f** 125°
3 a a   **b** None

4 a Equilateral **b** Scalene   **c** Isosceles   **d** Isosceles
5 a 90°, right-angled   **b** 70°, isosceles
  c 60°, equilateral   **d** 80°, scalene
  e 90°, right-angled   **f** 130°, scalene
6 a $c = 80°$, $b = 100°$ **b** $e = 55°$, $d = 125°$
  c $e = 105°$ **d** $f = 110°$, $g = 70°$, $h = 110°$
7 a 90°, rectangle   **b** 115°, kite
  c 106°, parallelogram   **d** 108°, rhombus
  e 67°, isosceles trapezium   **f** 130°, arrowhead

3.2A

1 50°, 60°, 70°
2 a $m = 52°$, $n = 38°$ **b** $o = 59°$, $p = 61°$
  c $q = 115°$, $r = 35°$ **d** $s = 69°$, $t = 39°$
3 a e.g. (1, 1)   **b** e.g. (−1, 3)
  c (−1, 2) or (3, 2) or (−1, −6) or (3, −6) or (1, 0) or (1, −4)
  d e.g. (2, 1)   **e** (1, 1.5) or (1, −5.5)
  f e.g. (3, 0) or (−1, 0)
4 a 70°   **b** 50°   **c** 100°
5 a $a = 36°$ alternate angles are equal, $b = 63°$ alternate angles are equal, $c = 81°$ angles on a straight line sum to 180°.
  b $a = 61°$ corresponding angles are equal, $b = 49°$ corresponding angles are equal, $c = 70°$ angles in a triangle sum to 180°.
  c $a = 113°$ alternate angles are equal, $b = 67°$ angles on a straight line sum to 180°, $c = 113°$ corresponding angles are equal, $d = 67°$ angles on a straight line sum to 180°, $e = 113°$ interior angles sum to 180°.
6 a 75°; angles on a straight line sum to 180°, angles in a triangle sum to 180°.
  b 34°; angles on a straight line sum to 180°, vertically opposite angles are equal, angles in a triangle sum to 180°, vertically opposite angles are equal.
7 a True, two pairs of parallel sides, four right angles.
  b False, some kites have two different side lengths.
  c False, some rhombuses have angles not equal to 90°.
8 Using the 30° alternate angles and the sum of angles in a triangle, the two unmarked angles in the pink triangle are both 30°, so it is an isosceles triangle.

3.3S

1 D and H – rectangle   L and M – arrowhead
  J and K – kite   A and G – rhombus
  C and F – square   I and B – parallelogram
  E and N – isosceles trapezium
2 C, D
3 B
4 a $a = 40°$, $b = 50°$ **b** $c = 20°$, $d = 40°$
  c all angles 60°
  d $h = 75°$, $i = 75°$, $j = 30°$, $k = 75°$, $l = 75°$
5 a – scale factor = 2   **b** – scale factor = 5
  e – scale factor = 4
6 a scale factor = 2, length = 8 cm
  b scale factor = 3, length = 12 cm

3.3A

1 Three pairs of congruent triangles.
2 Both triangles have angles 90°, 38° and 52°. Triangles satisfy ASA.
3 Satisfies SSS.
4 a Does not satisfy SSS.   **b** Satisfies SSS.
5 a See students' diagrams.
  b First is a line of symmetry, second is not.
6 a $a = 6$ cm   **b** $b = 12$ cm
7 a 9.6 cm   **b** 6.7 cm
8 Mini and medium, small and extra large.
9 Circles, equilateral triangles, squares, regular hexagons . The angles are fixed and the ratio between side lengths is fixed in each case.

3.4S

1 **a** 3   **b** 4   **c** 5   **d** 6
   e 7   **f** 8
2 **a** 3   **b** 4   **c** 5   **d** 6
   e 7   **f** 8
3 **a** 60°   **b** 60° and 120°   **c** 360°
4 Exterior angle: 120°, 90°, 72°, 60°, 51.4°, 45°, 40°, 36°
   Interior angle: 60°, 90°, 108°, 120°, 128.6°, 135°, 140°, 144°
5 **a** 156°   **b** $180 - 360 \div 15 = 156$
6 **a** 160°   9 144°
7 $6 \times 180° = 1080°$ 10 106°
8 $x = 141°$ 11 **a** 40°   **b** 70°

3.4A

1 **a** 18°   **b** 360°   **c** 20   **d** 20
2 **a** 24°   **b** 156°
3 **a** 146°   **b** 115°
4 **a** 45°   **b** 135°   **c** octagon
5 **a** $x = 125°, y = 50°$ **b** $x = 110°, y = 65°$
6 $x = 135°$, smallest angle $= 85°$
7 **a** 12   **b** 8
8 Interior angle $= 120°$, which is a factor of 360°.
9 No, interior angle $= 108°$ which is not a factor of 360°.
10 **a** 150°   **b** 12-sided regular polygon, or any shapes interior angles totalling 150°.
11 $x = 120°$
12 $\angle ABC = 120°, \angle BCA = (180 - 120) \div 2 = 30°,$
   $\angle ACD = 120 - 30 = 90°$

Review 3

1 $a = 70°$ (ASL), $b = 120°$ (AP), $c = 65°$ (VO), $d = 115°$ (ASL)
2 $a = 55°$ (CA), $b = 100°$ (AA)
3 35°
4 DF
5 $2 \times 180° = 360°$
6 $a = 50°$ $b = 130°$
7 **a** Parallelogram   **b** Trapezium   **c** Kite
8 Isosceles
9 Yes, $\angle ZXY = \angle WUV, XY = UV, \angle XYZ = \angle UVW$, (ASA).
10 10 cm
11 540°

Assessment 3

1 **a** $\frac{2}{3}$ **b** $\frac{3}{4}$ **c** 3
2 Check students' reasoning.
3 60° clockwise.
4 **a** 130°   **b** 255°   **c** 215°
5 Rafa was south of Sunita.
6 No, $64° + 59° + 56° = 179°$ (ASL).
7 No $a = 34°$ (AST), Yes $b = 75°$ (AST), Yes $c = 46°$ (AST),
   No $d = 74°$ (AST), Yes $e = 121°$ (VO), No $f = 59°$ (ASL).
8 **a** False, 2 obtuse angles $> 180°$ which is the sum of the three angles in a triangle.
   b True.
   c False, One right-angle $= 90°$, obtuse angle $> 90°$, right-angle + obtuse-angle $> 180°$.
   d True.
9 $\hat{X} = 48°, \hat{Y} = 65°, \hat{Z} = 67°$
10 15 m
11 **a** 93°   **b** 170°   **c** $w = 73°$ **d** 174°
12 **a** No, UV is not parallel to WY and VW is not parallel to UY.
   b **i** 60°   **ii** 120°
   c 90°   **d** No, equilateral triangle.

Answers

Chapter 4

Check in 4

1 **a** 6, 17, 19, 26, 29, 30, 37, 42   **b** 106, 115, 118, 121, 130, 135
   c 144, 145, 154, 155, 156, 165, 166
2 **a** 121   **b** 144   **c** 252   **d** 413
   e 68   **f** 23   **g** 49   **h** 82
   i 189   **j** 266
3 **a** 90°   **b** 130°
4 **a** 120   **b** 45   **c** 60   **d** 72
   e 6   **f** 20   **g** 30   **h** 18
   i 10

4.1S

1 **a** 10, 7, 4, 2   **b** 23
2 **a** 10, 8, 3, 9   **b** 30   **c** Tomato
3 **a** 21, 11, 3, 4, 1   **b** 40   **c** 73
4 **a** Biased, as each name is not equally likely to be chosen.
   b Biased, as each name is not equally likely to be chosen.
   c Random, as each card is equally likely to be picked.
   d Biased, as each height is not equally likely to be chosen.
   e Biased, as each person is not equally likely to find the star.
   f Biased, as only students 1 to 6 will have an equal chance of being selected.
5 Difficult to make sure that the whole population is included – census, disadvantage.
   Less data to analyse – sample, advantage.
   Unbiased – census, advantage.
   Not everyone is represented – sample, disadvantage.

4.1A

1 **a** Not true. The raw data may differ as different members of the population may be chosen for the sample.
   b Not true. A sample can be any size.
   c True. A sample contains less data so is faster to collect.
   d Not true. A larger sample is a better representation of the population.
2 A sample. Taking a census of all the batteries would be impossible. A sample is faster and less expensive to carry out.
3 The people included would be those visiting the cinema so not typical of the population.
4 People in an athletics club will probably play sport more often than those who aren't.
5 **a** It is not representative of the people who use the school tuck shop.
   b It is not representative of the whole school.
   c Pick names out of a hat, or take every 20th person on a list of all the people in the school.
6 **a** His friends are not representative of the whole population, they might particularly like (or dislike) travelling to see bands.
   b People listening to MP3 players may be more interested in music than is typical.
7 All girls/all friends so may have same taste in music/small sample.
8 Cars passing at similar time/small sample.
9 **a** People at bus stops are more likely to take the bus to work.
   b Biased against people who are ex-directory, don't have a landline or aren't in when phoned.
10 It is a leading question. People at a netball club are likely to find netball most exciting, so they are not representative.
11 Not representative of the whole population, students taking media studies may have similar viewing habits.
12 Billie's sample is very small. Her friend's parents may not be representative of all adults. She may record the results incorrectly the next day.

4.2S

1 **a** 17, 11, 4, 8 **b** 17   **c** 40
2 **a** 10, 12, 9, 9 **b** 9

3 **a** 78   **b** 20   **c** 43   **d** 55
    e 98

4 **a**

0	5 7 4
10	2 5 0 8 4 9
20	5 7 2 0 8 5
30	5 4 0 2 6

b

0	4 5 7
10	0 2 4 5 8 9
20	0 2 5 5 7 8
30	0 2 4 5 6

5

43.0	0 2 3 8 8 9
44.0	0 1 3 4 5 7
45.0	0 0 1 2 6
46.0	0 1 3 5 5 9 9

Key: | 44.0 | 7 | means 44.7 seconds

6

40	0 1 2 4 5 6 6 7 7 8 8 9
50	0 3 3 4 5 5 6 6 7
60	0 1 1 3 4 5 5
70	0 0

Key: 50 | 3 means 53

7

200	4 6 7 8 9 9
210	0 1 4 6 6 7
220	5
230	0 0 0 2 3 6 7 8
240	0 1 2 4

Key: 200 | 3 means 203

1 More brown eyed boys (5) than blue-eyed girls (3).
2 8 girls study Spanish.
3 Emily is wrong. There are 17 girls and 19 boys taking part.
4 **a** 28   **b** 67
5 Agree – out of the 60 red cars, 55 were speeding.
6 **a** Stem-and-leaf diagram displays all the raw data. The table will have too many categories.
    b Advantage – grouping the data makes it easier to spot trends. Disadvantage – the raw data is not recorded – i.e. a loss of information.
    c Disadvantage – the data set is large so it is difficult to spot trends. Advantage – all the data is recorded.

1 Library: 3 computers, school: $4\frac{1}{2}$ computers, work: $8\frac{1}{2}$ computers
2 Tea: 4 cups, coffee: 3 cups, hot chocolate: $2\frac{1}{2}$ cups, soup: $1\frac{1}{2}$ cups, other: 1 cup
3 **a** Italian   **b** 7   **c** $7 - 2 = 5$ **d** 23
4

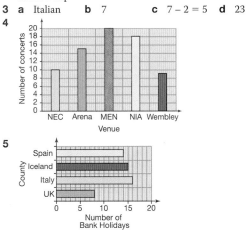

5

6

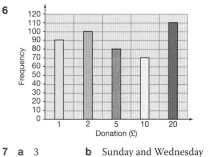

7 **a** 3   **b** Sunday and Wednesday   **c** 14

1 70 people
2 **a** 1990
    b Population in village A has increased steadily. Population in village B increased until the 1980s then decreased.
3 The lengths of the leaves are in proportion to the lengths of bars in a grouped bar chart.
4 Check students' bar charts. Check that total frequencies add up to 28, silver medals has frequency of 7, bronze frequency is greater than gold.
5 **a** Time consuming to draw lots of symbols for large data sets.
    b The vertical scale does not start at 0.
6 World Single Trip $\times$ £20, which is cheaper than World Annual insurance.

1 Tuna: 1 sector, cheese and tomato: 2 sectors, chicken: 3 sectors, corned beef: 2 sectors
2 **a** 12   **b** 30°   **c** Boys 210°, Girls 150°
    d Pie chart with angles from part c
3 **a** 6°
    b Sunny 90°, Cloudy 108°, Rainy 84°, Snowy 18°, Windy 60°
    c Pie chart with angles given in part b
4 **a** 360 minutes
    b Bat the Rat: 30°, Hook a Duck: 25°, Smash a Plate: 35°, Roll a Coin: 80°, Tombola: 70°, Break 1: 60°, Break 2: 60°
5 **a** **i** $\frac{1}{2}$ **ii** $\frac{1}{4}$ **iii** $\frac{1}{4}$
    b **i** 50   **ii** 25   **iii** 25
6 **a** 1 loaf = 30°
    b **i** 5   **ii** 3   **iii** 4
7 **a** 20°
    b **i** 7   **ii** 10   **iii** 1

1

Region	Number of tourists	Angle
London	225000	225°
Southern	60000	60°
South East	35000	35°
South West	40000	40°
	360000	

2 Too many categories.
3 Frequencies are similar, so difficult to compare on a pie chart.
4 **a** No, the angle for German is the same as the angle for Spanish.
    b No, Lydia doesn't know how many students in the schools are represented in the pie charts.
5 **a** Monday 60°, Tuesday 100°, Wednesday 60°, Thursday 40°, Friday 80°, Saturday 20°.
    b Monday 80°, Tuesday 40°, Wednesday 0°, Thursday 40°, Friday 80°, Saturday 120°.
    c Pie chart – the angles represent the proportion.
    d Bar chart – the heights of the bars show the frequencies.

1 **a** 8 **b** 9 **c** 2 **d** 6
 e 2 **f** 24 **g** 18 **h** 104
 i 15 **j** 5
2 **a** **i** 3, 3, 4, 5, 7, 8, 16 **ii** 9, 9, 10, 11, 12
 iii 24, 34, 35, 37, 38 **iv** 95, 97, 97, 98, 99, 101, 103
 v 0, 0, 1, 1, 2, 2, 2, 3, 3
 b **i** 5 **ii** 10 **iii** 35 **iv** 98 **v** 2
3 **a** Mode = 1, Range = 10 **b** Mode = 8, Range = 3
 c Mode = 11, Range = 4 **d** Mode = 25, Range = 15
 e Mode = 8, Range = 3 **f** Mode = 5, Range = 3
4 **a** 3, 3, 3, 3, 4, 4, 5, 5
 b Range = 2, Mode = 3, Median = 3.5, Mean = 3.75
5 **a** 1, 1, 1, 1, 2, 2, 2, 2, 2, 3, 3, 4, 4, 4, 4, 5, 5, 5, 5, 5, 5, 5
 b 8
 c Range = 4, Mode = 5, Median = 4, Mean = 3.28
6 **a** **i** 7 **ii** 6 **iii** 5.82 **iv** 6
 b **i** 75 **ii** 63 **iii** 60.1 **iv** 63
 c **i** 8 **ii** 96 **iii** 95.6 **iv** 96
 d **i** 71 **ii** 22, 37 **iii** 40.4 **iv** 37
 e **i** 26 **ii** 88, 89 **iii** 84.2 **iv** 87
 f **i** 72 **ii** 27 **iii** 46.9 **iv** 34
 g **i** 8 **ii** 105 **iii** 105.2 **iv** 105
7 **a** 1, 6, 8, 2, 8, 5, 6, 9, 3, 5, 7, 4, 4, 5, 5
 b **i** 8 **ii** 5 **iii** 5.2 **iv** 5
 c Range stays the same.

1 **a** 2, 4
 b Number 47's data is more spread out than Number 45's.
2 **a** 1, 3, 4, 1, 1 **b** 1.8, 2, 2
 c On average, houses in Ullswater Drive have more cars than those in Ambleside Close.
3 23 or 42
4 −3.1, −2.6, 3.5, 4.1, 4.1
5 **a** 3, 4, 0, 3, 3, 3, 3, 0
 b Mean = 5.85, Mode = 4, Median = 6, Range = 7
 c Mean = 6.05, Mode = 4, Median = 6, Range = 8
6 41 years old **7** 14 raisins **8** 78.6%

1

Number	Frequency
1	2
2	4
3	8
4	5

2 **a** 30 **b** 13 **c** 2 **d** 27
3 **a** **i** 16 **ii** 9
 b **i** $2\frac{1}{2}$ footballs **ii** $3\frac{3}{4}$ footballs
4 Check students' bar-line chart.
5 **a** 40 **b** 10
6 Pie chart with angles 36°, 108°, 216°.
7 **a** 6 **b** 6.5 **c** 7 **d** 6
8 **a** 44 **b** The sample is biased and too small.
9 **a** 35 **b** 34 **c** 11 **d** 12
 e The 2nd Zoo has a higher median number of birds per aviary and its range is smaller.

1 **a**

Milk	Frequency
0	6
1	14
2	13
3	7
4	7
5	2
6	1

b

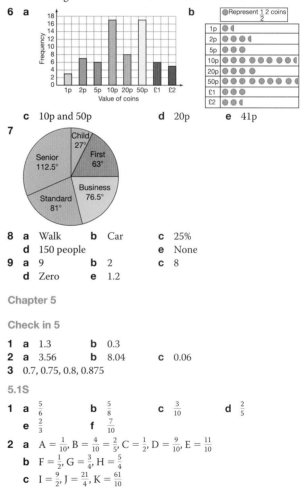

2 **a** **i** Yes **ii** Look at the cars in the car park.
 iii Take measurements in the middle of day when all teachers are present.
 b **i** Yes **ii** Measure the students in class.
 iii Have measurements of each student taken multiple times by separate people.
 c **i** No
 ii Gather data from the Met Office, academic institutions etc.
 iii Use data collected from the same locations using the same method each year.
 d **i** No
 ii Offer subjects a variety of lengths to estimate.
 iii Sample subjects randomly from the available population.
 e **i** No
 ii Send electronic questionnaires to students via email.
 iii Randomly sample students from the whole school.
3 **a** 35 **b** 16 **c** 3
 d Some students access information in more than one way.

4

	Glasses	No glasses
Left-handed	1	4
Right-handed	14	16

5 **a** **i** 103 miles **ii** 99.5 miles **iii** 88 miles
 b You cannot tell because you do not know how many journeys he makes in one day.
 c 107 miles
 d Mean increases, because the total has increased but not the number of journeys. Median increases, because 98 replaces 90 as one of the two middle numbers. Mode, no mode as each number occurs once. Range, stays the same because the highest and lowest numbers are unaffected.
6 **a**

b

	Represent 1.2 coins
1p	
2p	
5p	
10p	
20p	
50p	
£1	
£2	

 c 10p and 50p **d** 20p **e** 41p
7

Pie chart: Child 27°, First 63°, Business 76.5°, Standard 81°, Senior 112.5°

8 **a** Walk **b** Car **c** 25%
 d 150 people **e** None
9 **a** 9 **b** 2 **c** 8
 d Zero **e** 1.2

Chapter 5

Check in 5

1 **a** 1.3 **b** 0.3
2 **a** 3.56 **b** 8.04 **c** 0.06
3 0.7, 0.75, 0.8, 0.875

1 **a** $\frac{5}{6}$ **b** $\frac{5}{8}$ **c** $\frac{3}{10}$ **d** $\frac{2}{5}$
 e $\frac{2}{3}$ **f** $\frac{7}{10}$
2 **a** $A = \frac{1}{10}, B = \frac{4}{10} = \frac{2}{5}, C = \frac{1}{2}, D = \frac{9}{10}, E = \frac{11}{10}$
 b $F = \frac{1}{2}, G = \frac{3}{4}, H = \frac{5}{4}$
 c $I = \frac{9}{2}, J = \frac{21}{4}, K = \frac{61}{10}$

3 a i $\frac{8}{12}$ ii $\frac{2}{3}$ **b** i $\frac{14}{16}$ ii $\frac{7}{8}$
  c i $\frac{12}{20}$ ii $\frac{3}{5}$ **d** i $\frac{11}{15}$ ii $\frac{11}{15}$

4 a $\frac{1}{3}$ **b** $\frac{3}{4}$ **c** $\frac{3}{5}$ **d** $\frac{4}{9}$
  e $\frac{5}{8}$ **f** $\frac{1}{3}$ **g** $\frac{4}{9}$ **h** $\frac{23}{93}$

5 a $\frac{3}{2}$ **b** $\frac{11}{3}$ **c** $\frac{35}{8}$ **d** $\frac{20}{9}$
  e $\frac{41}{7}$ **f** $\frac{39}{5}$ **g** $\frac{96}{11}$ **h** $\frac{88}{7}$

6 a $1\frac{1}{4}$ **b** $1\frac{3}{5}$ **c** $1\frac{4}{7}$ **d** $2\frac{1}{4}$
  e $2\frac{1}{5}$ **f** $2\frac{6}{7}$ **g** $4\frac{3}{5}$ **h** $3\frac{1}{9}$

7 a 8 **b** 27 **c** 56 **d** 56
  e 2 **f** 90 **g** 85 **h** 7

8 a 2.07, 2.09, 2.12, 2.13, 2.2 **b** 0.3, 0.309, 0.325, 0.345, 0.35
  c 1.058, 1.32, 1.35, 1.387, 1.4 **d** 5.288, 5.29, 5.3, 5.306, 5.308

9 a 4.2 **b** 4.38 **c** 6.8

10 b i Marked at 0.8 ii Marked at 0.6
   iii Marked at 1.3

11 a $\frac{1}{10}$ **b** $\frac{3}{5}$ **c** $\frac{3}{4}$ **d** $\frac{5}{4}$
   e $\frac{3}{2}$ **f** $\frac{6}{5}$

12 a 0.3 **b** 0.8 **c** 0.25 **d** 1.1
   e 1.4 **f** 1.6

5.1A

1 a $\frac{15}{28}$ **b** $\frac{13}{28}$
2 a $\frac{8}{15}$ **b** $\frac{7}{15}$
3 a $\frac{91}{300}$ **b** $\frac{1}{5}$
4 a $\frac{8}{11}$ **b** $\frac{3}{11}$
5 a $\frac{11}{37}$ **b** $\frac{15}{37}$ **c** $\frac{9}{37}$
6 $1\frac{1}{2} = \frac{9}{6} = 1.5 = \frac{30}{20}$, $2.20 = 2\frac{1}{5} = \frac{11}{5} = \frac{22}{10}$, $0.625 = \frac{5}{8} = \frac{20}{32} = \frac{10}{16}$
7 $\frac{2}{5}$ is larger; illustrated by students' diagrams
8 a $\frac{2}{5}$ **b** $\frac{2}{3}$ **c** $\frac{4}{7}$ **d** $\frac{5}{6}$
  e $\frac{4}{7}$ **f** $\frac{10}{7}$
9 a $\frac{3}{15}, \frac{1}{3}, \frac{2}{5}$ **b** $\frac{1}{2}, \frac{15}{28}, \frac{4}{7}$ **c** $\frac{4}{7}, \frac{5}{8}, \frac{9}{14}$
10 a 3, 1, 2 **b** 2, 1, 3
11 a 6 sectors shaded.
   b $\frac{3}{4}$ of 9 is not a whole number of squares.
12 $\frac{1}{2}a + 0.4b - 0.1a + \frac{1}{4}b = 0.4a + 0.65b$
   $1.8a + 0.4b - 1\frac{2}{5}a + \frac{1}{5}b = 0.4a + 0.6b$
   $\frac{1}{10}a + \frac{7}{4}b + \frac{2}{5}a - 1.1b = 0.5a + 0.65b$

5.2S

1 a 20 **b** 5 **c** $\frac{5}{8}$ **d** $\frac{12}{13}$
2 a $\frac{5}{2}$ **b** 2 **c** $\frac{8}{3}$ **d** $\frac{13}{7}$
  e 2 **f** $\frac{4}{3}$
3 a 4 **b** $\frac{15}{4}$ **c** $\frac{4}{5}$ **d** $\frac{14}{3}$
  e $\frac{9}{4}$ **f** $\frac{112}{5}$ **g** $\frac{40}{3}$ **h** $\frac{77}{9}$
4 a $\frac{2}{5}$ **b** 40% **c** 7 squared shaded.
5 a £30 **b** 20 **c** 136p **d** 4
  e 37p **f** £7 **g** 6kg **h** £15.50
  i 2 **j** 4.2m
6 a £4 **b** £34 **c** £6 **d** 42km
  e \$106 **f** £75 **g** £6.40 **h** 1.4m
7 a €12 **b** £28 **c** $\frac{75}{2}$ m **d** $\frac{256}{7}$ km
  e £375 **f** $\frac{175}{3}$ mm **g** 1375m **h** $\frac{240}{13}$ g
8 a 264kg **b** \$4500 **c** 4.44kg
  d 952 cups **e** 21.67 tonnes **f** 96°
  g 139.35° **h** 52 minutes **i** £260.67
9 a £1.98 **b** 4380km **c** £2.40 **d** €60
  e 4.2m **f** 5.2cm **g** 36m **h** £2.10
  i 75kg **j** 1680km
10 a £2.04 **b** 13.92km **c** £3.04 **d** €108.80

e 11.05m **f** 33.58cm **g** 125.8m **h** £1.53
i £21.25 **j** £1.14

5.2A

1 a 0.8kg **b** 3.5kg **c** 7.2kg **d** 14.4kg
2 25% of £520 = £130   20% of £640 = £128
  30% of £450 = £135
3 a £104 **b** 35 weeks
4 a 34 models **b** 56 litres
5 40.83 m³   **10** £199.17
6 £425   **11** Merton (£71.25 < £72.25)
7 £58.75   **12 a** 48 **b** 40 **c** 2
8 649 boys   **13** 120
9 £49090.60

5.3S

1 a 5, 8 **b** 9, 50, 100 **c** 8, 15, 20, 80
  d 16, 15, 50, 120
2 a $\frac{3}{6}$ and $\frac{2}{6}$ **b** $\frac{3}{15}$ and $\frac{5}{15}$ **c** $\frac{5}{10}$ and $\frac{2}{10}$
  d $\frac{8}{12}$ and $\frac{3}{12}$ **e** $\frac{9}{30}$ and $\frac{10}{30}$ **f** $\frac{16}{20}$ and $\frac{5}{20}$
3 a $\frac{2}{3}$ **b** $\frac{3}{5}$ **c** 1 **d** $\frac{3}{4}$
  e $\frac{4}{7}$ **f** $\frac{8}{15}$
4 a $1\frac{1}{4}$ **b** $1\frac{4}{5}$ **c** $1\frac{5}{8}$ **d** $4\frac{1}{4}$
5 a $\frac{7}{4}$ **b** $\frac{23}{16}$ **c** $\frac{14}{9}$ **d** $\frac{18}{7}$
6 a $\frac{7}{3}$ **b** $\frac{25}{7}$
7 a $\frac{5}{6}$ **b** $\frac{17}{20}$ **c** $\frac{4}{15}$ **d** $\frac{18}{35}$
  e $\frac{23}{24}$ **f** $\frac{38}{45}$
8 a $\frac{3}{5}$ **b** 2 **c** $\frac{10}{3}$ **d** $\frac{15}{7}$
  e 2 **f** $\frac{13}{3}$
9 a 2 **b** 4 **c** $\frac{10}{3}$ **d** $\frac{7}{12}$
  e $\frac{3}{5}$ **f** $\frac{187}{8}$
10 a 8 **b** 10 **c** 14 **d** 20
   e 48 **f** 220
11 a $\frac{31}{12}$ **b** $\frac{66}{35}$ **c** $\frac{31}{15}$ **d** $\frac{15}{8}$
   e $\frac{19}{12}$ **f** $\frac{43}{63}$
12 a $\frac{12}{5}$ **b** 14 **c** 48 **d** $\frac{45}{7}$
   e 2 **f** $\frac{15}{7}$
13 a $\frac{3}{10}$ **b** $\frac{9}{20}$ **c** $\frac{15}{28}$ **d** $\frac{12}{35}$
   e $\frac{2}{3}$ **f** $\frac{7}{24}$ **g** $\frac{2}{3}$ **h** $\frac{9}{4}$
   i $\frac{9}{24}$ **j** $\frac{1}{2}$ **k** $\frac{7}{3}$ **l** $\frac{77}{40}$
14 a 10 **b** $\frac{2}{5}$ **c** $\frac{16}{15}$ **d** $\frac{6}{7}$
   e $\frac{27}{28}$ **f** $\frac{9}{2}$ **g** $\frac{1}{7}$ **h** $\frac{4}{35}$
   i $\frac{4}{55}$ **j** $\frac{8}{5}$ **k** $\frac{7}{6}$ **l** $\frac{27}{25}$
   m 2 **n** $\frac{28}{8}$ **o** $\frac{28}{15}$

5.3A

1 a false **b** true **c** true **d** true
  e true **f** false **g** true **h** true
2 $5\frac{11}{12}$ miles
3 $1\frac{13}{16}$ lb
4 $1\frac{79}{80}$ kg
5 $\frac{2}{15}$
6 a $\frac{14}{5}$ kg **b** $\frac{15}{7}$ m **c** $\frac{28}{5}$ litres
7 a 3 **b** 4
8 a $\frac{821}{28}$ feet **b** $\frac{2794}{45}$ feet **c** $\frac{59}{3}$ m
9 $\frac{2}{3}, \frac{3}{10}, \frac{1}{2}, \frac{3}{20}, \frac{2}{3}, \times$
10 $\frac{17}{35}$
11 Many examples, such as $\frac{1}{3}, \frac{1}{10}, \frac{1}{15}$

5.4S

1 a 0.67 **b** 0.78 **c** 0.99 **d** 0.7

e	0.39	**f**	0.88	**g**	1.5	**h**	1.25	
i	0.999	**j**	1.1	**k**	0.75	**l**	0.376	

2 a 32% **b** 22% **c** 85% **d** 3%
  e 54% **f** 63% **g** 38% **h** 37.5%
  i 33.3% **j** 125% **k** 0.15% **l** 99.5%

3 a $\frac{4}{5}$ **b** $\frac{7}{25}$ **c** $\frac{13}{40}$ **d** $\frac{1}{20}$
  e $\frac{3}{25}$ **f** $\frac{3}{8}$

4 a 0.3 **b** 0.28 **c** 0.58 **d** 0.6
  e 2.14 **f** 2.11

5 a $\frac{1}{4}$ **b** $\frac{2}{5}$ **c** $\frac{13}{20}$ **d** $\frac{3}{20}$ **e** $\frac{29}{20}$

6 a $\frac{3}{10}$ **b** $\frac{3}{5}$ **c** $\frac{16}{25}$ **d** $\frac{9}{20}$
  e $\frac{3}{8}$ **f** $\frac{27}{25}$ **g** $\frac{259}{80}$ **h** $\frac{49}{16}$
  i $\frac{17}{4}$

7 a 0.3 **b** 0.44 **c** 1.04 **d** 0.62
  e 0.45 **f** 0.52 **g** 0.28 **h** 3.35
  i 3.56

8 a 0.44 **b** 0.67 **c** 1.35 **d** 0.73
  e 1.14 **f** 1.4 **g** 2.17 **h** 0.85

9 a $\frac{2}{5}$ **b** $\frac{9}{10}$ **c** $\frac{7}{20}$ **d** $\frac{13}{20}$
  e $\frac{1}{100}$ **f** $\frac{181}{50}$ **g** $\frac{61}{400}$ **h** $\frac{17}{800}$

10 a 54% **b** 40% **c** 85% **d** 52%
  e 66.67% **f** 24% **g** 120% **h** 44%

11 a 0.37 **b** 0.07 **c** 1.89 **d** 0.45

12 a 72% **b** 20% **c** 125% **d** 3%

13 a 68.6% **b** 64% **c** 89.5% **d** 191.7%
  e 26.3%

14 a $0.\dot{6}$ **b** $0.2\dot{7}$ **c** $0.\dot{2}$ **d** $0.\dot{4}2857\dot{1}$

5.4A

1 a 55% **b** 75% **c** 60% **d** 20%

2 a 47%, $\frac{12}{25}$, 0.49 **b** 78%, $\frac{4}{5}$, 0.81
  c $\frac{7}{12}$, $\frac{5}{8}$, 66% **d** 29%, 0.3, $\frac{5}{16}$, $\frac{7}{22}$

3 a German **b** Sarah's class has fewer students who do not like eating meat than the rest of the school.

4 a $\frac{3}{5}$, 61%, 0.63 **b** $\frac{17}{25}$, 69%, $\frac{7}{10}$, 0.71
  c 34%, $\frac{7}{20}$, 0.36, $\frac{3}{8}$, $\frac{2}{5}$ **d** $\frac{2}{5}$, 42%, $\frac{3}{7}$
  e 0.14, 15%, $\frac{3}{19}$, $\frac{1}{5}$ **f** 81%, $\frac{8}{9}$, 0.9, 0.93, $\frac{19}{20}$

5 a i $\frac{3}{5}$ **ii** 60%
  b i $\frac{13}{25}$ **ii** 52%
  c i $\frac{3}{20}$ **ii** 15%
  d i $\frac{7}{10}$ **ii** 70%

6 a 0.1428571429…, 0.2857142857…, 0.4285714286…, 0.5714285714…, 0.7142857143…, 0.8571428571…
  b numbers repeat after every 7th number
  c 0.07692307692…, 0.1538461538…, 0.2307692308…, 0.3076923077…, 0.3846153846…, 0.4615384615…, 0.5384615385…, 0.6153846154…, 0.6923076923…, 0.7692307692…, 0.8461538462…, 0.9230769231… numbers also repeat after every 7th number

7 a 0.33, 33.3%, $33\frac{1}{3}$%, 33
  b $0.\dot{4}$, 44.5%, 0.45, 0.454
  c 22.3%, 0.232, 23.22%, 0.233, $0.2\dot{3}$
  d $0.6\dot{5}$, 0.66, 66.6%, 0.6666, $\frac{2}{3}$
  e 14%, 14.$\dot{1}$%, 0.142, $\frac{1}{7}$, $\frac{51}{350}$
  f $\frac{5}{6}$, $\frac{6}{7}$, 86%, 0.866, 0.86

Review 5

1 a $\frac{7}{5}$ **b** $\frac{25}{7}$

2 a $2\frac{1}{4}$ **b** $1\frac{5}{6}$

3 a $\frac{3}{10}$ **b** $\frac{1}{4}$ **c** $\frac{22}{25}$ **d** $\frac{1}{20}$

4 a 0.5 **b** 0.7 **c** 0.02 **d** 0.625

5 a $\frac{2}{7}$ **b** $\frac{8}{3}$

6 a 9 **b** 12 **c** 8 **d** 36

---

4 a 16 **b** 48 **c** 12 **d** 99

8 a $\frac{7}{11}$ **b** $\frac{7}{10}$ **c** $1\frac{1}{24}$ **d** $1\frac{3}{20}$
  e $2\frac{11}{12}$ **f** $4\frac{1}{2}$

9 a $2\frac{1}{2}$ **b** $\frac{2}{35}$ **c** $\frac{3}{10}$ **d** $1\frac{7}{18}$

10 a $\frac{2}{25}$ **b** $\frac{2}{3}$ **c** 20 **d** $2\frac{4}{5}$
  e $1\frac{1}{4}$ **f** $12\frac{1}{4}$

11 $\frac{3}{5}$, 0.6, 60%; $\frac{1}{100}$, 0.01, 1%; $\frac{13}{20}$, 0.65, 65%; $\frac{6}{5}$, 1.2, 120%

Assessment 5

1 0, $\frac{1}{6}$, $\frac{2}{5}$, 0.5, $\frac{2}{3}$, $\frac{8}{10}$, 1

2 a No, $\frac{4}{5} = \frac{12}{15}$.
  b Yes, in a fraction the numerator is divided by the denominator, any number goes into itself exactly once.

3 a $\frac{1}{4}$
  b Ben has treated the 6s as factors. Only factors can be cancelled, 16 is not 1×6 and 64 is not 4×6.
  c Students' answers, for example, $\frac{15}{65} \neq \frac{1}{6}, \frac{15}{65} = \frac{3}{13}$.

4 a 5580, 558.0, 55.80, 5.580, 0.5580, 0.05580
  b i Yes **ii** No, 9.55 **iii** No, 2.205 **iv** No, 6.995

5 a 40% **b** 15% **c** $\frac{9}{20}$
  d Bananas 45%, Apples 40%, Pears 15%

6 First pair, one quarter = 25% > 20%.

7 a £120 **b** £37.80 **c** £31.50 **d** £14.30
  e £0.36 **f** £380

8 60

9 a Redyonder **b** Air **c** Chat-Chat
  d Incorrect, 8 is 32% of 25 and 40% of 20.

10 a 36.4% (3 sf) **b** 14%

11 a 0.12 **b** 4th **c** 3rd

12 $1\frac{1}{2}$ hectares

13 8.5 yards

14 a 145.8 m² **b** 142 m²

15 $\frac{10}{33}$

16 Yes, % increase = 60%.

17 a $0.\dot{3}$ **b** $0.\dot{5}$ **c** $0.\dot{8}5714\dot{2}$ **d** $0.6\dot{3}$

18 33.3%, $33\frac{1}{3}$%, 0.334, 0.34, $\frac{5}{14}$, $\frac{3}{8}$

Lifeskills 1

1 a

Key   2|4 means 24 years old.

Women		Men
9 8	1	
8 3 2 2 0	2	0 1 3 4 6
8 5 2	3	1 7 9
7 4	4	0 2 7
8 2	5	1 5
1	6	2 6

Overall, the men interviewed were older than the women.
  b Yes. Women £28.60, Men £30.

2 a £29.30 (nearest penny) **b** £222222
  c $P = R - G - S - C$ **d** £52222
  e $S = R - G - G - P$

3 38, 94, 36, 32

Task 4

a $\frac{7}{40}$ **b**

Pie chart: Juliet 63°, Abigail 144°, Raheem 90°, Mike 63°

c Abigail £20000, Raheem £21500, Mike £8750, Juliet £8750

5 a 0.06
  b i £13365.05   **ii** £12885.48
  c i £11223.29   **ii** £11641.01

Chapter 6

Check in 6

1 a 5   **b** 9   **c** 4   **d** 10
2 a $3x + 3$ **b** $2x - 2$ **c** $8x + 12$ **d** $12x - 6$
3 a $4(x + 2)$ **b** $2(3x + 1)$ **c** $3(y - 3)$

6.1S

1 a 60p   **b** 126p
  c cost of ribbon = length of ribbon × price per metre
2 Pay = number of hours worked × hourly rate
3 Cost of the repair = cost of parts + number of hours worked × hourly rate for labour
4 a $c = l + p$, c = repair cost, l = labour cost, p = parts cost
  b $c = p \times l$, c = cost of electric cable, p = price per metre, l = length in metres
  c $m = a \div 12$, m = monthly cost, a = annual cost
  d $c = p \times n$, c = cost of apples, p = price per kg, n = number of kg
  e $m = k \times 1000$, m = distance in metres, k = distance in kilometres
  f $m = c \div 100$, m = distance in metres, c = distance in centimetres
5 a $P = 3x$ **b** 12
6 a $P = 4y$ **b** 28
7 a 20   **b** 15   **c** 50   **d** 12.5
8 a 6   **b** 28   **c** 13.5   **d** 5
  e 2.5
9 a 18   **b** 10   **c** 6   **d** 34
  e 16   **f** 36   **g** 64   **h** 38
10 a 50 miles per hour   **b** 95 km per hour
  c 50 metres per second   **d** 120 km per hour

6.1A

1 a 12p   **b** 48p   **c** £4.80
  d Total value of row = £4.80 × length of row in metres
2 $2k^2\,\text{m}^2$
3 a $C = 20t + 35$ **b** £95
4 a $C = 0.6m + 2$ **b i** £5   **ii** £11
5 a $C = 35t + 75$ **b** €320   **c** 9 days
6 a $N = 2b + 2$ **b** 12   **c** 18
  d $2b + 2 = 2(b + 1)$
7 $s = 6$, switched a and t, $0 \times 3 \neq 3$, $2 \times 3^2 \neq 6^2$

6.2S

1 a 20   **b** 16   **c** 16   **d** 7
  e 60   **f** 6   **g** 8   **h** 31
2 a ×3   **b** ×7   **c** ×6   **d** −5
  e +5   **f** +13   **g** ÷3   **h** ÷4
3 a ÷2   **b** −4   **c** +3   **d** ×6
  e +7   **f** ÷5   **g** ×2   **h** −11
4 a $a = H - 2$ **b** $a = H + 2$
  c $a = \frac{H}{2}$ **d** $a = 2H$
5 a $x = y - 3$ **b** $x = y - c$ **c** $x = y + 4$ **d** $x = y + c$
  e $x = \frac{y}{5}$ **f** $x = \frac{y}{c}$ **g** $x = 6y$ **h** $x = cy$
6 a $x = \frac{y - c}{m}$ **b** $t = \frac{v - u}{a}$ **c** $x = 2(y - d)$
  d $y = \frac{4 - x}{3}$
7 a $t = \frac{s}{3} + 2$ **b** $t = \frac{2x - 9}{5}$ **c** $t = \frac{3x}{5} - 6$
8 a $y = \frac{1 - 2x}{3}$ **b** $y = \frac{2x}{3} + 2$ **c** $y = \frac{3x - z}{5}$
9 a $b = H - a^2$ **b** $b = A + a^2$ **c** $b = \frac{Q - a^2}{2}$ **d** $b = \frac{F + a^2}{2}$
  e $b = M - p^2$ **f** $b = \frac{L}{p^2}$ **g** $b = \frac{T}{a^2}$ **h** $b = Wa^2$
10 No, $x = 10 - y$
11 a $x = 5 - y$ **b** $x = 20 - y$ **c** $x = m - y$ **d** $x = 2b - y$
  e $x = s^2 - y$ **f** $x = \sqrt{p} - y$

12 $T = \frac{D}{S}$

6.2A

1 a 13   **b** 20
2 a $-6 + 5$ **b** $\div 2 + 5$ **c** $\times 3 + 5$
3 a $n = \frac{P - 1}{4}$
  b i 9   **ii** 13   **iii** 57
4 a 200   **b** 150   **c** 570
5 a $C = 15t + 45$ **b** £180
6 a $C = 1.6m + 2$ **b i** £10   **ii** £26
7 a $C = 35t + 75$ **b** €320   **c** 12 days
8 $\frac{ab}{b} \equiv b$
9 $\frac{8(D + k)}{ab} = c$, $8(D + k) = abc$, $D + k = \frac{1}{8}abc$, $D = \frac{1}{8}abc - k$

6.3S

1 a $7x + 28$ **b** $3x - 6$ **c** $6 + 2x$ **d** $10 - 5x$
2 a $4(2m + 1)$ **b** $3(4n - 3)$ **c** $5(3p + 11)$ **d** $q(q + 2)$
  e $4(4r - 7)$ **f** $2q(2p - 5)$
3 a Identity   **b** Function   **c** Formula   **d** Equation
  e Expression   **f** Formula   **g** Formula   **h** Function
4 a Formula   **b** Identity   **c** Equation   **d** Identity
  e Equation   **f** Formula   **g** Formula   **h** Identity
  i Equation
5 a $4a + 8 + 2a + 2$ **b** $3x + 6 + 4x - 4$
  c $5y - 10 + 3y - 9$ **d** $y^2 + 3y + 2y + 6$
  e $x^2 - 4x + x^2 + 2x$
6 a $a = 7, b = 9$ **b** $a = 7, b = 6$
  c Many answers possible, e.g. $a = -2, b = 3$
  d $a = 1, b = 2$ **e** $a = 3, b = 10$
7 a False, $a^2 + 7a + 10$ **b** False, $x^2 + 7x + 12$
  c True   **d** True   **e** False, $y^2 - y - 6$
  f False, $p^2 - 5p + 6$
8 a $x^2 + 2xy + y^2$ **b** $x^2 - 2xy + y^2$
9 a $x^2 + 3x - 3x + 9$ **b** $y^2 - 4y + 4y - 16$
  c $25 + 5a - 5a - a^2$ **d** $b^2 - ab + ab - b^2$
10 $a = 3, b = 11$
11 $P = 2a + 2b = 2 \times a + 2 \times b = 2(a + b)$

6.3A

1 a i $5(a + 2) = 5a + 10$ **ii** $5(a + 2) = 80$
  b i $7(b + 5) = 7b + 35$ **ii** $7(b + 5) = 105$
  c i $12(c + 10) = 12c + 120$ **ii** $12(c + 10) = 240$
  d i $2(d + 1) = 2d + 2$ **ii** $2(d + 1) = 24$
  e i $4(e + 3) = 4e + 12$ **ii** $4(e + 3) = 44$
  f i $3(1 + f) = 3 + 3f$ **ii** $3(f + 1) = 24$
2 a 12   **b** 9   **c** 6, 72   **d** 4, 4
  e 11, 3   **f** 5, 7
3 a 6, 2   **b** 4, 8   **c** 2, 8   **d** 4, 4
  e 2, 1   **f** 8, 3, 67
4 a $7 - 7 = 0$ **b** $3^2 = 9$ **c** $2 \times 3 = 6$ **d** $3 \times 4 = 12$
5 b $5n + 10 = 5(n + 2)$ has a factor of 5 for all n
6 a $2n + 2m = 2(n + m)$ has a factor of 2 for all n, m
  b $(2n)^2 = 4n^2$ has a factor of 4 for all n
  c $2n + 2m + 1 = 2(n + m) + 1$
  d $n(n + 1) - n = n^2 + n - n = n^2$
7 a $1^2 = 1$ **b** $0 \leqslant x \leqslant 1$
8 a Sometimes true   **b** True   **c** Sometimes true
  d Sometimes true   **e** False   **f** False
  g Sometimes true   **h** Sometimes true

6.4S

1 a $x^2 + 5x + 6$ **b** $p^2 + 11p + 30$
  c $w^2 + 5w + 4$ **d** $c^2 + 10c + 25$
  e $x^2 + 2x - 8$ **f** $y^2 + 5y - 14$
  g $t^2 + 4t - 12$ **h** $x^2 - 7x + 10$
  i $y^2 - 14y + 40$ **j** $w^2 - 3w + 2$
  k $p^2 - 10p + 25$ **l** $q^2 - 24q + 144$
2 a $6x^2 + 17x + 7$ **b** $10p^2 + 19p + 6$
  c $6y^2 + 11y + 4$ **d** $4y^2 + 24y + 36$

e $10t^2 + 12t - 16$
f $15w^2 + 42w - 9$
g $6x^2 - 6y^2$
h $9m^2 - 24m + 16$
i $6p^2 - pq - 40q^2$
j $4m^2 - 12nm + 9n^2$
3 a $x^2 - 1$ **b** $25x^2 - 1$ **c** $4x^2 - 9$ **d** $x^2 - y^2$
4 a $2x^2 - 2x - 24$
b $12p^2 + 23p + 10$
c $6m^2 - 32m + 42$
d $10y^2 + 17y - 63$
e $9t^2 - 12t + 4$
f $2x^2 + 4x - 15$
g $-1 + 10b - 25b^2$
5 a $(x + 2)(x + 4)$
b $(x + 3)(x + 7)$
c $(x + 4)(x + 7)$
d $(x + 3)(x + 8)$
e $(x - 2)(x - 6)$
f $(x - 3)(x - 6)$
g $(x - 4)(x - 9)$
h $(x + 4)(x - 3)$
i $(x - 7)(x + 5)$
j $(x + 9)(x - 3)$
k $(x + 2)(x - 16)$
l $(x + 20)(x - 2)$
6 a $(x + 12)(x - 6)$
b $(x - 12)(x + 2)$
c $(x - 5)(x - 15)$
d $(x + 16)(x - 4)$
e $(x + 8)(x - 8)$
f $(x - 4)(x - 25)$
7 a $(x + 2)(x - 2)$
b $(2x + 1)(2x - 1)$
c $(4x + 3)(4x - 3)$
d $(a + b)(a - b)$
e $(10x + 5)(10x - 5)$
f $(p^2 + q^2)(p^2 - q^2)$
8 a $(x + 2)(x + 19)$
b $x(5x + y + 5)$
c $(x + 11)^2$
d $(x + 9)(x - 2)$
e $(p + 3)(p + 11)$
f $x(2x + 3y)$

6.4A

1 a $(x + 6)(x - 3)$
b $(2m - 3)^2$
c $(2x + 3)(3x - 1)$
d $\frac{(3x - 4)(5x + 2)}{2}$
2 a $x^2 + 9x - 22$
b $(x + 11)(x - 2)$
c Perimeter would equal zero
3 $(x + 3)^2$
4 a $(x + 1)(x + 5) = x^2 + 6x + 5$
b $x^2 + 5x + 6 = (x + 2)(x + 3)$
5 a $(a + b)^2 = a^2 + 2ab + b^2$ **b** 16
c Students' answers
6 $(2.3 + 1.7)^2 = 4^2 = 16$
7 a 3, 18 **b** 2, 2 **c** 4, 2, 14
8 $2(x^2 + 17)$
9 $4x^2 + 8x, 2x, 2x + 4, x, 2, x + 2$

Review 6

1 a $3\,\text{g/cm}^3$ **b** $12.5\,\text{g/cm}^3$
2 a 16 **b** 22.5
3 a i $2x$ **ii** $x - y$
b £15
4 a $A = b - 3$ **b** $A = \frac{d}{2}$ **c** $A = \frac{F + c}{5}$
d $A = 2J - h$ **e** $A = \sqrt{L + 2K}$ **f** $A = \frac{2}{b}$
5 a $2y + 3 = 7$ **b** $3b \times 4b = 12b^2$
c $4z + 2$ **d** $F = ma$
e $y = 3x + 4$
6 $\equiv 6x + 2 - 4 \equiv 6x - 2$
7 a $x^2 + 8x + 15$ **b** $x^2 - 8x + 12$
c $x^2 - 3x - 28$ **d** $6x^2 + 13x - 5$
8 a $x(x + 5)$ **b** $3x(4x - 1)$
9 a $(x + 4)(x + 1)$ **b** $(x - 6)(x - 1)$
c $(x - 4)(x + 2)$ **d** $(x + 5)(x - 2)$
10 a $(x + 6)(x - 6)$ **b** $(2x + 5)(2x - 5)$

Assessment 6

1 a $m = \frac{t}{5}$ or $t = 5m$
b i 3 miles **ii** 0.5 miles **iii** $12\frac{1}{2}$ s
2 a $J = 0.45R + 0.55S$ **b** $1.8\,\text{kg}$ **c** $3.3\,\text{kg}$
d Raspberries $4.05\,\text{kg}$, Sugar $4.95\,\text{kg}$ **e** $5\,\text{kg}$
3 a i 13 hrs 30 min **ii** 7 hrs 30 min
b i 12 **ii** 8
c 1 hrs 30 min. No, this is too little sleep.
4 a i Yes **ii** No, ± 11.

5 a i No, $p^2 - 11p + 28$. **ii** No, $26v^2 - 38v - 47$.
b i Yes **ii** No, $(v + 10)(v - 10)$.
6 a $W = 7D$ **b** $C = 50 + 250D$
7 a i $P = 15m$ **ii** $P = 12m$ **iii** $P = 14n$
b i £1.72 **ii** £5.47
c £8.30
9 a F $5 \times 4 = 20$ **b** F 5 has two factors, 1 and 5.
c T Let the two numbers be $2x$ and $2y$. $2x + 2y = 2(x + y)$ so the sum is even.
d F $0^2 = 0$ **e** F $2 \times 7 = 14$
f T Let the numbers be $2x$, $2x + 2$ and $2x + 4$. $2x + 2x + 2 + 2x + 4 = 6x + 6 = 6(x + 1)$, is divisible by 6.
g T p^2 has 3 factors, 1, p and p^2.
10 a $y = 4x + 1$ **b** $y = \frac{x}{3} - 2$
11 a 58.8 m (1 dp) **b** 44.7 m **c** 101 km/h

Revision 1

1 $42 \div 3\frac{1}{2} = 12$ mph, $15 \div 10 = 1\frac{1}{2}$ hours,
$(42 + 15) \div \left(3\frac{1}{2} + 1\frac{1}{2}\right) = 11.4$ mph

2 5.625 kg

3 If Jenni bought a weekly season ticket she would save
£17.50 − £15 = £2.50

4 a COOL value of 9 = 9, COOL value of 28 = 1
b 1, 4, 7 and 9 **c** 1 and 9.
d 13 and 14. **e** 20
5 a 95 **b** 355
c $5x + 30$ **d** 13
e T_{20} is NOT possible. The T shape will not fit onto the grid because 20 is in the left hand column so you can't take a number to the left of it.
f $x = 78$. $T_{78} = 77 + 78 + 79 + 88 + 98 = 420$
6 a $y = 180 - 130 = 50°$ (Angles on a straight line).
$z = 180 - 2 \times 50 = 80°$ (Isosceles triangle)
b $y = 32°$, $x = 148°$
c An equilateral triangle. $\angle \textbf{PRQ} = y = 180 - 120 = 60$,
$x = 180 - 2 \times 60 = 60$.
7 a 55
b Mean = 4.85 (3 sf), median = 5
c 8
8 a $10y + 8$ **b** $84y + 40$ **c** $3x + 2 = 35, x = 11$
9 a Correct. HBD is an isosceles right angled triangle.
b Incorrect. HBDF is a square.
c Incorrect. HBCG is an isosceles trapezium.
d i One of: OPF, OBQ, ODQ
ii One of: HOF, HOB, BOD; FOD; BHF; BDF; HBD; HFD; ACO; CEO; EGO; GAO.
iii Any triangle with letter vertices.
iv One of: *ABC, HBD, GBE, FEG, FHD, FAC, AOC, GOE.*
e i 135° **ii** 90° **iii** 67.5°
iv 45° **v** 45° **vi** 135°
10 a £1.50 **b** £2.50 **c** £3.50
11 a 108 **b** 18% **c** 261
12 a $\frac{1}{5}$ **b** $\frac{4}{5}$
c 16 ml acid and 4 ml water.
d A: Acid 64 ml, Water 16 ml. B: Acid 16 ml, Water 84 ml.
e i 4 : 1 **ii** 4 : 21
13 a 2.27 (3sf) **b** $1 = \sqrt{\frac{A}{4\pi}}$; $A = 1^2 \times 4\pi = 12.6$ (3sf)
c $A = 4\pi r^2$ **d** $2980\,\text{cm}^2$ (3sf)
14

Chapter 7

Check in 7

1 a $9\frac{3}{5}$ b $22\frac{1}{2}$ c $10\frac{1}{2}$
2 a 7.8   b 15.5   c 9
3 a 45 mm   b 4.5 cm

7.1S

1 a 1800 mm   b 4.5 cm   c 3.5 m
   d 2 km   e 3.5 km   f 4.5 m
2 a 4.7 cm, 6.6 cm, 3.4 cm, 8.9 cm, 2.1 cm, 6.9 cm, 2.5 cm, 2.7 cm, 4.6 cm, 4.1 cm
   b 47 mm, 66 mm, 34 mm, 89 mm, 21 mm, 69 mm, 25 mm, 27 mm, 46 mm, 41 mm
3 a Line AB exactly 9 cm.   b midpoint = 4.5 cm
4 a Acute   b 30°   c 30°
5 a Acute   b 50°   c 45°
6 a Obtuse   b 110°   c 120°
7 a Reflex   b 330°   c 330°
8 a acute   b obtuse   c right angle
   d acute   e obtuse   f acute
   g obtuse   h acute   i obtuse
   j acute   k reflex   l reflex
   m reflex   n reflex   o reflex
9 a 50°   b 319°
10 a Bearing at 70° from north   b Bearing at 155° from north
   c Bearing at 340° from north   d Bearing at 260° from north

7.1A

1 a 056°   b 170°   c 238°   d 275°
   e 349°
2 Kim is wrong.
3 a St. Mawes positioned at the intersection of the bearings.
   b 2.2 km   c 13.8 km
4 a Check students' drawings. $SY = 4$ cm, horizontal. $SH = 5$ cm. Angle $HSY = 48°$.
   b 170°
5 a i Check students' drawings   ii 328°
     iii 8.6 km
   b i Check students' drawings   ii approximately 356°
     iii 7.3 km

7.2S

1 a 15 cm²   b 27 m²   c 200 cm²   d 81 cm²
2 a 16 cm²   b 100 m²   c 9 cm²   d 64 m²
3 a 40 cm²   b 16 m²   c 18 cm²   d 36 mm²
4 a 6 square units   b 12 square units
5 a 6 square units   b 6 square units
6 a 80 cm²   b 800 m²   c 120 mm²   d 384 cm²
7 a 50 cm²   b 375 mm²   c 28 m²   d 160 cm²
8 a 150 cm²   b 150 cm²
9 a 6 cm   b 14 cm   c 8 mm   d 8 cm

7.2A

1 a perimeter = 42 cm   b perimeter = 40 cm
    area = 74 cm²   area = 72 cm²
   c perimeter = 33 cm   d perimeter = 36 cm
    area = 32 cm²   area = 44 cm²
2 area of shape = 121 cm²
   area remaining = 479 cm²
3 205.5 cm²
4 a Check students' arrow.   b, c 19 cm²
5 54 cm
6 a 20 cm
   b Many possible solutions. Check the perimeter of students' shapes is 28 cm.
7 a Many answers possible check base × height = 36
   b Many answers possible check base × height = 48
   c Many answers possible check height × sum of parallel sides = 40
8 No. Smallest diameter possible is a square with sides 12 m and the perimeter of this would be 48 m, which would cost £480.
9 11.5 cm²

7.3S

1 a Mirror lines vertical and horizontal
   b Mirror line vertical only
2 a Reflected in top right and bottom left quadrants
   b Reflected in bottom right and top left quadrants
3 a 180°   b 90° anti-clockwise
   c 90° clockwise   d 270° clockwise
4 a   b   c   d

5 C, G, I
6 a $\begin{bmatrix} 3 \\ 1 \end{bmatrix}$ b $\begin{bmatrix} 5 \\ 3 \end{bmatrix}$ c $\begin{bmatrix} -3 \\ 1 \end{bmatrix}$ d $\begin{bmatrix} 5 \\ -3 \end{bmatrix}$ e $\begin{bmatrix} 2 \\ -4 \end{bmatrix}$ f $\begin{bmatrix} -4 \\ 2 \end{bmatrix}$
7 a Translation made 1 right 2 up   b Translation made 1 right
   c Translation made 1 down   d Translation made 1 down
   e Translation made 2 right   f Translation made 1 right 1 down

7.3A

1 a $x = 2$ b $y = 3$
2 a Rotate 90° clockwise around (0, 0)
   b Rotate 180° around (0, 0)
   c Rotate 90° anti-clockwise around (0, 0)
3 Rotate 135° clockwise
4 a Reflected into the bottom right quadrant with vertices $(2, -1)$ $(4, -1)$ $(4, -4)$
   b Rotated into the top left quadrant with vertices $(-2, 1)$ $(-4, 1)$ $(-4, 4)$
   c reflection in $x = 0$
5 a Vertices at $(1, -2)$, $(1, 0)$, $(2, 1)$ and $(4, 1)$
   b Vertices at $(0, 0)$, $(0, 2)$, $(1, 3)$ and $(3, 3)$
   c translate $\begin{bmatrix} -4 \\ 1 \end{bmatrix}$
6 No, the shapes are not congruent so they are not related by any rotation.
7 i rotation 180° around (3, 3)   ii reflection in $(x = 3)$
   iii translation $\begin{bmatrix} 4 \\ 0 \end{bmatrix}$

7.4S

1 a A = no   C = no   E = no
    B = yes, × 2   D = yes, × 3
   b B, D
2 a   c

   b   d

3 a Ensure axes extended to 16 and drawn to scale.

b Vertices at (4, 4), (4, 10) and (8, 4)

c Vertices at (6, 6), (6, 15) and (12, 6)

4 a Ensure axes extended and drawn to scale.

b Vertices at $(1, -1)$, $(3, 3)$, $(1, 5)$ and $(-1, 3)$

c Vertices at $(1, -5)$, $(5, 3)$, $(1, 7)$ and $(-3, 3)$

5 a Ensure axes extended and drawn to scale. Check points are plotted correctly.

b Vertices of R: $(3, 3)$, $(4, 4)$, $(3, 4)$ and $(3\frac{2}{3}, 5)$

c Vertices of S: $(1, 2)$, $(0.5, 3.5)$, $(-0.5, 2)$ and $(-0.5, -0.5)$

6 b Vertices of U: $(1, 2)$, $(5, 2)$ and $(2, 3)$

c Vertices of V: $(2, 3)$, $(6, 3)$ and $(3, 4)$

7.4A

1 a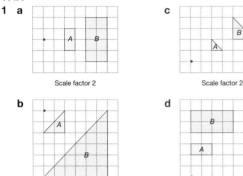

Scale factor 2

c

Scale factor 2

b

Scale factor 3

d

Scale factor 2

2 a scale factor 3, centre of enlargement (0, 0)

b scale factor $\frac{1}{3}$, centre of enlargement (0, 0)

3 a scale factor 2, centre of enlargement $(-3, 0)$

b scale factor $\frac{1}{2}$, centre of enlargement $(-3, 0)$

4 a i scale factor $\frac{1}{2}$, centre of enlargement (0, 0)

ii scale factor 2, centre of enlargement (0, 0)

iii scale factor 4, centre of enlargement (0, 0)

b perimeter of C = 4 × perimeter of A

5 b Vertices of Y: $(-2, 2)$, $(0, 6)$, $(6, 4)$

c Vertices of Z: $(-3, 3)$, $(0, 9)$, $(9, 6)$

d scale factor 3, centre of enlargement (0, 0)

e scale factor $\frac{1}{3}$, centre of enlargement (0, 0)

f Same centre, scale factors which are reciprocals of one another.

6 a The value of each coordinate is tripled.

b Enlargement of scale factor 3 with centre (0, 0).

Review 7

1 a $55°$ **b** 5.1 cm

2 Check $AC = 5.7$ cm, $\angle BAC = 73°$, $\angle BCA = 42°$.

3 a $28\,m^2$ **b** $6\,cm^2$ **c** $20\,mm^2$ **d** $25\,cm^2$

e $55\,m^2$

4 a, b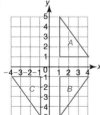

c Reflection in the y-axis.

5 a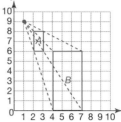

b Rotation 90° CW (or 270° ACW), about (0, 0).

6 a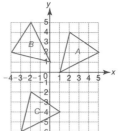

b Enlargement scale factor $\frac{1}{3}$, centre of enlargement (1, 9).

Assessment 7

1 a Karl   **b** Neither, multiply by 1 000 000.

2 a Check students' drawing.

b i 6.5 cm ± 3 mm   **ii** $\hat{A} = 60° ± 3°$ **iii** 135° ± 3°

c Kite   **d** Rectangle

3 a Yes, $90° < a < 180°$.   **b** No, $b = 90°$, right-angle.

c Yes, $c < 90°$.   **d** Yes, $d > 180°$.

e Yes, $90° < e < 180°$.   **f** No, $f < 90°$, acute.

g No, $g = 90°$, right-angle.   **h** No, $h < 90°$, acute.

i Yes, $90° < i < 180°$.   **j** No, $j < 90°$, acute.

k No, $k > 180°$, reflex.

4 a 195°   **b** 110°   **c** 015°   **d** 290°

e 245°

5 a Yes   **b** No, $2.625\,cm^2$.

c Yes, $294\,cm^2$.   **d** No, $21.12\,cm^2$.

e No, $181.5\,cm^2$.   **f** No, $4.64\,cm^2$.

g No, 22 cm.   **h** No, 22 cm.

i No, 12 cm.   **j** No, 20 cm.

6 a 49

b No, $2 × 2 = 4$ and $3 × 3 = 9$ are not factors of 49.

c No, each side is 7 m so must include two 2 × 2 slabs and a 3 × 3 slab. There are three ways of arranging these slabs around the edges, all of which have gaps or overlaps.

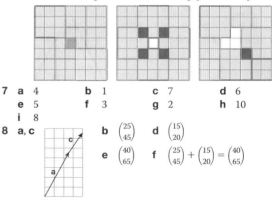

7 a 4   **b** 1   **c** 7   **d** 6

e 5   **f** 3   **g** 2   **h** 10

i 8

8 a, c   **b** $\binom{25}{45}$ **d** $\binom{15}{20}$

e $\binom{40}{65}$ **f** $\binom{25}{45} + \binom{15}{20} = \binom{40}{65}$

9 a Rotation 180° about the point (0, 0).

b, c

d Reflection in the x-axis.

Chapter 8

Check in 8

1 a $\frac{2}{3}$ **b** $\frac{4}{5}$ **c** $\frac{1}{4}$ **d** $\frac{3}{5}$

   e 1

2 a 0.2, 0.25, 0.3   **b** 0.7, 0.75, 0.8

   c 0.8, 0.85, 1

3 a 1   **b** 1   **c** $\frac{1}{10}$ **d** $\frac{1}{5}$

   e $\frac{1}{4}$

4 a 0.8   **b** 0.3   **c** 0.1

8.1S

1 a i Even chance, Sebastian won half the matches last year.

     ii Unlikely, Sebastian only won half the matches last year.

   b Certain. Every number on the dice is less than 7.

   c i Likely, the average temperature in Edinburgh in Feb is $4\,°C < 10\,°C$.

     ii Impossible. There was no February the 29th in 1913!

   d Unlikely, 4 even numbered cards $< \frac{1}{2}$

2 a i $\frac{1}{5}$ **ii** 0.2   **iii** 20%

   b i $\frac{7}{40}$ **ii** 0.175   **iii** 17.5%

   c i $\frac{13}{40}$ **ii** 0.325   **iii** 32.5%

   d i $\frac{3}{10}$ **ii** 0.3   **iii** 30%

3 a Bag A, P(Red) $= \frac{1}{3} >$ Bag B, P(Red) $= \frac{1}{5}$.

   b Bag B, Bag A only has one red ball.

4 a 50 times

   b i $\frac{9}{50} = 0.18 = 18\%$ **ii** $\frac{14}{50} = \frac{7}{25} = 0.28 = 28\%$

     iii $\frac{27}{50} = 0.54 = 54\%$

   c Students' answers, for example, Red 2, Green 3, Blue 5.

   d Complete more trials.

5 a Check students' results $\approx$ 0.5, 0.25, 0.13, 0.06, 0.03, 0.02, 0.01, 0, ... (2 dp)

   b P(Head on 1st toss) is the highest.

6 a Check students' results $\approx$ 0.67, 0.14, 0.12, 0.1, 0.08, 0.07, 0.06, 0.05, 0.04, 0.03, 0.03, 0.02, 0.02, 0.02, 0.01, 0.01, 0.01, 0.01, 0.01, 0.01 (2 dp)

   b P(6 on 1st toss) is the highest.

8.1A

1 No, the relative frequency will only approaches the theoretical probability with a high number of trials.

2 a His sample size is too small to give a reliable result.

   b No, relative frequency is an estimate.

   c i $\dfrac{\text{Total number of red balls}}{\text{Total number of trials}} = \dfrac{4 + 16 + 95}{5 + 20 + 100} = \dfrac{115}{125} = 0.92$

     ii Probably, this is only an estimate.

3 a Frequency 10, Relative frequency 0.4, the relative frequencies sum to 1.

   b A and C

   c Any spinner that has red, white and blue sections.

4 a **B** and **C**, **A** does not have any 4s.

   b Correct net with at least one 1, 2, 3 and 4.

   c C, it has more 1s and 3s.

5 No, the relative frequencies are very similar, B 0.34, W 0.32, G 0.34

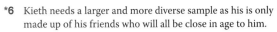

*6 Kieth needs a larger and more diverse sample as his is only made up of his friends who will all be close in age to him.

8.2S

1 a 5   **b** ≈ 2

2 a 15   **b** ≈ 5

3 a ≈ 7 **b** ≈ 13

4 a 25   **b** ≈ 13

5 ≈ 4 people

6 ≈ 4 people

7 ≈ 19 people

8 Score 1 = 25, Score 2 = 50, Score 3 $\approx$ 63, Score 4 $\approx$ 38, Score 5 = 25, Score 6 = 50

9 a 44   **b** 33

   c 44   **d** 67

10 9 days

11 14 shares

8.2A

1 a 12 times   **b** 24 times

2 a 5 points

   b Students' answers, for example, less because it is difficult for counters to land near the sides of the box.

3 60

4 2 games

5 15

6 Purple 15, Blue 10, Red 15, Brown 5, Green 15

*7 a 10 times   **b** More often

*8 a 6 times   **b** £6

8.3S

1 a $\frac{1}{6}$ **b** $\frac{1}{3}$ **c** $\frac{1}{2}$ **d** $\frac{1}{3}$

2 a Yes

   b No, P(Point down) $<$ P(Point up).

   c No, planned births don't happen at the weekend.

   d No, P(Head on 1st toss) $= \frac{1}{2}$ so the other outcomes cannot be equally likely.

   e No, some letters are more common than others.

3 6

4 a Highest 9, Lowest 3   **b** 0.34 (2 dp)

5 a $\frac{1}{6}$ **b** $\frac{4}{9}$ **c** 1   **d** 0.2 (est)

   e 1

6 a $\frac{5}{10} = 0.5,\ \frac{9}{20} = 0.45,\ \frac{14}{30} = 0.4\dot{6},\ \frac{17}{40} = 0.425,\ \frac{19}{50} = 0.38,$
$\frac{23}{60} = 0.38\dot{3},\ \frac{29}{70} = 0.4\dot{1}4285\dot{7},\ \frac{32}{80} = 0.4,\ \frac{35}{90} = 0.3\dot{8},\ \frac{38}{100} = 0.38,$
$\frac{43}{110} = 0.3\dot{9}\dot{0},\ \frac{47}{120} = 0.391\dot{6},\ \frac{49}{130} = 0.3\dot{7}6923\dot{0},\ \frac{53}{140} = 0.378\dot{5}7142\dot{8}$

   b 0.38 (2 dp)

7 0.24 (2 dp)

8.3A

1 a $\frac{7}{15}$ **b** $\frac{3}{10}$

2 Yes, relative frequency of 5 $= 0.29 > \frac{1}{6}$.

3 a $\frac{1}{2}$ **b** $\frac{1}{3}$

4 $\frac{3}{8}$

5 0.79 (2dp)

6 a $\frac{3}{5}$ **b** $\frac{13}{32}$

7 $\frac{1}{4}$

8 If the spinner is fair, the results indicate 7.6 black sectors and 7.4 white sectors which is impossible.

8.4S

1 a No   **b** Yes   **c** No   **d** No

2 a No   **b** Yes   **c** Yes   **d** Yes

3 a 0   **b** $\frac{2}{3}$

4 a $\frac{1}{6}$ **b** $\frac{1}{2}$

5 a $\frac{1}{3}$ **b** $\frac{1}{2}$

6 a 0.4   **b** 0.8   **c** 0.2

7 a $\frac{1}{7}$ **b** $\frac{3}{7}$ **c** $\frac{11}{21}$

8 a 0.2   **b** 0.6   **c** 0.5

8.4A

1 a

	1	2	3	4	5	6
1	0	1	2	3	4	5
2	1	0	1	2	3	4
3	2	1	0	1	2	3
4	3	2	1	0	1	2
5	4	3	2	1	0	1
6	5	4	3	2	1	0

b i $\frac{5}{18}$ ii $\frac{1}{6}$ iii $\frac{5}{9}$

2 No, $\frac{1}{4} + \frac{1}{3} + \frac{1}{2} = \frac{3}{12} + \frac{4}{12} + \frac{6}{12} = \frac{13}{12} > 1$

3 a Yes, $\frac{1}{6} + \frac{1}{3} + \frac{1}{8} + \frac{3}{8} = \frac{4}{24} + \frac{8}{24} + \frac{3}{24} + \frac{9}{24} = \frac{24}{24} = 1$

   b There are no other colours because the probabilities sum to 1.

4 a $\frac{1}{2}$ **b** $\frac{3}{4}$ **c** $\frac{5}{12}$

   The probabilities sum to more than 1 because the events are not mutually exclusive and exhaustive.

5 a C and D **b** C and D

   c 2 is both prime and even.

6 a True, cannot be M and F simultaneously.

   b True, can only be M or F.

   c False, can be both M and D.

   d False, can be both F and R.

7 a

	1	2	3	4	5	6
1	0	−1	−2	−3	−4	−5
2	1	0	−1	−2	−3	−4
3	2	1	0	−1	−2	−3
4	3	2	1	0	−1	−2
5	4	3	2	1	0	−1
6	5	4	3	2	1	0

b i $\frac{1}{9}$ ii $\frac{1}{36}$ iii $\frac{35}{36}$

8 a 0.1 **b** 0.3 **c** 0.5

Review 8

1 a 0.4 **b** 120

2 a 0.2 **b** 24

3 a 0.1 **b** $\frac{1}{6}$ **c** $\approx \frac{1}{6}$

4 a $\frac{3}{10}$ **b** 0 **c** 1

5 0.25

6 $\frac{8}{11}$

7 a $\frac{2}{7}$ **b** $\frac{4}{7}$

Assessment 8

1 No, the probability of this event is greater than zero.

2 P(H) = 0.5 because each throw is an independent event.

3 $\frac{51}{100}$

4 a 200 **b** 2 **c** 4

5 a If $x = 0.2$ then the total probability = 1.02, $x = 0.19$

   b 52 **c** 128 **d** 118

6 a i 0.18 ii 0.17

   b Ben's, because he sampled more packets.

7 a $\frac{1}{5}$ **b** $\frac{1}{2}$

8 9 packets

9 a Yes, P(Late) = 0.51 > 0.49 **b** 70

10 a $\frac{3}{25}$ **b** $\frac{8}{25}$ **c** $\frac{7}{25}$

   d $\frac{11}{25}$ **e** P(Blue 4) = 0

11 a $\frac{2}{5}$ **b** $\frac{1}{5}$ **c** $\frac{1}{2}$ **d** $\frac{3}{5}$

12 a i No, 102 is a multiple of 3. ii No, 5 is prime.

   iii Yes, no white numbers are multiples of 7

   b i No, 81 is a multiple of 3. ii No, 29 is prime.

   iii No, 84 is a multiple of 7.

13 No, sunny and snowing are not mutually exclusive events.

14 a P(blue) = $\frac{72}{360} = \frac{1}{5}$

   b You have assumed that all angles are equally likely to be chosen.

Chapter 9

Check in 9

1 a 400 **b** 14 000 **c** 31 **d** 1340

   e 6300 **f** 40 **g** 60 **h** 4.3

   i 0.64 **j** 0.31

2 −7°C

3 15 miles

9.1S

1

	(i)	(ii)	(iii)
a	8.37	8.4	8
b	18.8	19	20
c	35.8	36	40
d	279	280	300
e	1.39	1.4	1
f	3890	3900	4000
g	0.00837	0.0084	0.008
h	2400	2400	2000
i	8.99	9.0	9
j	14.0	14	10
k	1400	1400	1000
l	140000	140000	100000

2 a 240 **b** 2030 **c** 5000 **d** 8000

3 a 8 **b** 500 **c** 0.008 **d** 10

   e 4 **f** 40

4 a 50 **b** 1000 **c** 60 **d** 60

   e 200 **f** 72 **g** 5 **h** 50

5 a $400 \div 20$ **b** 40×40 **c** $1000 \div 100$

   d $4000 + 10\,000$ **e** $100 + (2000 \div 50)$

6 a 16 **b** 400 **c** 5 **d** 30

   e 40 **f** 20

7 a $\frac{50 \times 20}{10}$ **b** $\frac{20 \times 5}{50}$ **c** $\frac{70 + 50}{4 \times 3}$

8 a 2 **b** 2 **c** 50 **d** 100

   e 1 **f** 9

9 a 900 **b** 60 **c** 12 100

10 a 1.6 **b** 160 **c** $\frac{28}{15}$

11 a 1.4 **b** 2.8 **c** 3.2 **d** 3.9

   e 4.5 **f** 5.1 **g** 5.7 **h** 6.7

   i 8.4

12 a 7.1 **b** 7.1 **c** $8\frac{1}{3}$

9.1A

1

	a	b	c
(i)	$260 + 360 = 620$	exact	620
(ii)	$60 \div 30 = 2$	underestimate	2.37
(iii)	$60 \times 200 = 12000$	underestimate	13279
(iv)	$100 - 65 = 35$	overestimate	31.9

2 a Zero **b** More than 100 **c** Negative

3 30 slabs

4 a Accept 40000−52000 grains

   b Yes, accept 850 g − 1200 g as estimate for weight.

5 a 75 photos **b** 4 photos

6 a 4 jars **b** 16.99 ml

7 a £2000 **b** £500 **c** £3500

8 a 5 hours **b** 35 litres **c** £45

   d less

9.2S

1 a 21, 20.1 **b** 19, 18.6 **c** 13, 12.5 **d** 9, 9.8

   e 10, 11.1 **f** 3, 3.1

2 a 71.1 **b** 29.624 **c** 2.07885304659
d 186.408 **e** 0.1508856039 **f** 19.05
3 a 5.8 **b** 1.3 **c** 1.7 **d** 2.2
4 a 5.6 **b** 73 **c** 0.085
d −35 **e** 110 **f** 38
5 a 1.83 **b** 2.03 **c** 50.96 **d** 94.75
e 1.06 **f** 10.02
6 a 729.11 **b** 58.19 **c** 10983.46
7 Compare estimates.
8 a 464.5923967 **b** 0.4536084142
9 a 178.4123835 **b** 0.1967089505
c 3.210178253 **d** 3.350190476
e 1.157007415 **f** 0.1356045007
10 a 7.04 **b** 6.96 **c** 8.11

9.2A

1 a $2.4 \times (4.2 + 3.7) = 19.2$ **b** $6.8 \times (3.75 − 2.64) = 7.548$
c $(3.7 + 2.9) \div 1.2 = 5.5$ **d** $(2.3 + 3.4^2) \times 2.7 = 37.422$
e $5.3 + 3.9 \times (3.2 + 1.6) = 24.02$
f $3.2 + 6.4 \times (4.3 + 2.5) = 46.72$
2 a $3.4 \times (2.3 + 1.6) = 13.26$ **b** $3.5 \times (2.3 − 1.04) = 4.41$
c $(2.6 + 6.5) \div 1.3 = 7$ **d** $(1.4^2 − 1.2) \times 2.3 = 1.748$
e $(2.4^2 \div 1.8) \times (3.2 + 1.6) = 15.36$
f $(3.2 + 5.3) \times (2.4 − 1.2) = 10.2$
3 a i $15.23\,\text{m}^2$ **ii** £103.41
b £66.67
4 a 30 hours 21 mins **b** 1821 mins
5 a £21.13 **b** £16.37
c Yes − in both cases it would save money to use the new offer.
6 a £187 **b** £28.20
7 a i 1087.2 feet per second **ii** 1122.4 feet per second
b i 741.3 mph **ii** 765.3 mph

9.3S

1 a centimeters (cm) **b** milliliters (ml)
c kilograms (kg) **d** centimeters (cm)
e kilograms (kg) **f** kilometers (km)
g milliliters (ml) **h** liters (l)
i tonne (t) **j** gram (g)
2 a 200 cm **b** 4 m **c** 4.5 m **d** 4 km
e 5 mm **f** 4500 g **g** 6 kg **h** 6.5 kg
i 2.5 tonnes **j** 3000 ml
3 2.5 mph
4 37.5 miles
5 2.5 hours (2 hr 30 mins)
6 a $5\,\text{g/cm}^3$ **b** 87.88 g
7 a 9.4575 kg **b** 6.15 l
8 a 5.75 − 5.85 m **b** 16.45 − 16.55 l
c 0.85 − 0.95 kg **d** 6.25 − 6.35 N
e 10.05 − 10.15 s **f** 104.65 − 104.75 cm
g 15.95 − 16.05 km **h** 9.25 − 9.35 m/s
9 a 6.65 − 6.75 m **b** 7.735 − 7.745 l
c 0.8125 − 0.8135 kg **d** 5.5 − 6.5 N
e 0.0005 − 0.0015 s **f** 2.535 − 2.545 cm
g 1.1615 − 1.1625 km **h** 14.5 − 15.5 m/s
10 a Max: 174 kg, min: 162 kg **b** Max: 28.4 kg, min: 27.6 kg
11 a 32.5 − 37.5 mm **b** 37.5 − 42.5 mm
c 107.5 − 112.5 mm
d 42.5 − 47.5 mm

9.3A

1 a 80 km/hr **b** 62.5 km/hr **c** 15 mins **d** 120 m
e 20 m/s **f** 0.0025 m/s **g** 125 sec **h** 20 km
2 a 5400 m/h **b** 90 m/min **c** 1.5 m/s
3 a 141 kg **b** $8.4\,\text{m}^2$ **c** £13.63
d 0.8 mins = 45 seconds
4 Max: $526.75\,\text{cm}^2$, min: $481.75\,\text{cm}^2$

5 Max: 65 g, min: 63 g
6 14 boxes
7 a Steel **b** 4470 kg
8 a 3408 ml **b** 1592 ml **c** 4 bottles
9 a £4.36 **b** £0.66 **c** £2.31 **d** £2.40
e £1.40

Review 9

1 a i 8750 **ii** 8800 **iii** 9000
b i 15.0 **ii** 15 **iii** 20
c i 0.0682 **ii** 0.068 **iii** 0.07
d i 0.509 **ii** 0.51 **iii** 0.5
2 a 200 **b** 103 **c** **d** 4
3 a 8, 8.114056225 **b** 80, 81.312
c 1, 1.052060738 **d** 300, 255.9035917
4 a cm **b** cm^3 **c** litres **d** kg
e m^2
5 a 3.5 m **b** 0.145 m **c** 200 m **d** 9.32 m
6 1 h 52 min
7 a £72.50 **b** 9.5 hours/ 9 hours 30 minutes
8 25 s
10 $89.5 \leqslant x < 90.5$
11 Nearest kg

Assessment 9

1 a 32.3 (3 sf) **b** 5700 (nearest 100)
c 0.2 (1 sf) **d** 310 (nearest 10)
e 290 (2 sf) **f** 256 000 (nearest 1000)
2 a B (37.7) **b** B (113.12)
c C (69.52) **d** A (5.076923̇)
e C (8.681318̇)
3 $6 \times 60 \times 25 = 9000$, Mia used 25 as an estimate for the number of hours in a day.
4 a 6 and 7 **b** $6^2 = 36$ and $7^2 = 49$
c 7, 46 is closer to 49.
5 a $2 \times 2^2 = 8$, 2 is a better estimate.
b 0.808629471
6 40
7 a $94.18\,\text{m}^2$ **b** $11.7725\,\text{m}^3$
c i $300\,\text{kg/m}^3$ **ii** 3.53 tonnes
8 a $4.6 + (5.3 \times 2.6) = 18.38$
b $14.9 − (6.8 \div 2.5) = 12.18$
c $(3.4 \times 1.6) + (5.9 − 2.8) = 8.54$
d $2.6 + (7.56 \div 1.8) − 0.72 = 6.08$
e $(12.3 − (5.2 \times 1.6) + 3.4) \times 2 = 14.76$
9 a $\approx 100 − 1000$ litres **b** ≈ 20 kilograms
c ≈ 2 grams **d** ≈ 150 metres
e $\approx 50 − 300$ tonnes **f** ≈ 30 millilitres
g ≈ 15 millimetres **h** ≈ 40 kilometres
i ≈ 15 centimetres **j** ≈ 300 millilitres
10 a £25.30 **b** 25 m 30 cm
c 25 kg 300 g **d** 25 cl 3 ml
e 25 hrs 18 min
11 a 66 km/h **b** 150 km/h (3 sf)
c 481.25 km **d** 3410 miles
e 17 min 52 s
12 a Yes, Density = $0.81\,\text{g/cm}^3 < 1$.
b $2.28\,\text{cm}^3$ (3 sf) **c** 5400 tonnes
13 a LB 221455, UB 221465 **b** LB 85 cm, UB 95 cm
c LB 452.5 g, UB 457.5 g
d LB 3 min 28.75 s, UB 3 min 28.85 s
e LB 238 bags, UB 242 bags
f LB 27.5 tonnes, UB 28.5 tonnes
g LB 585.5 mm, UB 586.5 mm
14 a $390.1625\,\text{cm}^3$, $261.4375\,\text{cm}^3$
b $0.293\,\text{g/cm}^3$ (3 sf), $0.194\,\text{g/cm}^3$ (3 sf)

Chapter 10

Check in 10

1 a 3, 4   **b** 5, 7

2 a x **b** m **c** $3n$ **d** $2p$

3 a 6   **b** 11   **c** 9   **d** 6

4 a 7   **b** -1 **c** 12   **d** $2\frac{1}{4}$

10.1S

1 a 4   **b** 5   **c** 5   **d** 8
   e 6   **f** 6   **g** 15   **h** 12

2 a 3   **b** 11   **c** 5   **d** 15
   e -9 **f** 6   **g** 11   **h** 15

3 a 3   **b** 7   **c** 3   **d** 9
   e 7   **f** 4   **g** 2   **h** 2.5

4 a 25   **b** 36   **c** 8   **d** 21
   e 45   **f** 24   **g** 50   **h** 27

5 a 4   **b** 15   **c** 36   **d** 3
   e 7   **f** -4 **g** 0   **h** 10

6 a 6   **b** -3 **c** -5 **d** -40
   e 2   **f** $\frac{4}{3}$ **g** 36   **h** -5
   i -2.5 **j** 14

7 a 5   **b** 2   **c** 6   **d** 10

8 a 10   **b** 5   **c** -2 **d** 7
   e 7   **f** 0   **g** 5   **h** 6

9 a 0.5   **b** 1   **c** 0.5   **d** 2.5
   e -1 **f** 0.5

10 a 10   **b** 9   **c** 2   **d** 15
   e 70   **f** 30   **g** 60   **h** 5
   i 1.5   **j** -2

11 a 5   **b** 5   **c** 8   **d** 6
   e 5.5   **f** 50   **g** 18   **h** 40
   i 3   **j** 2

12 a 7   **b** -7 **c** 2   **d** 5
   e -11 **f** 6   **g** 5   **h** 8

10.1A

1 a 23   **b** 40

2 $6p + 12 = 16$

3 a 5   **b** 12

4 a 5 cm   **b** 6 cm
   c 8 cm. Shape C is a regular hexagon since all sides are equal lengths.

5 Spare solutions are 4, 6, -1, 0.5

6 Sarah: 3, Josh: 6, Millie: 2

7 £12.50

8 Side Lengths: 14 cm, 1 cm

9 51°

10.2S

1 a $2a + 6$ **b** $4b - 20$ **c** $6c + 3$ **d** $-5d + 50$
   e $16e - 8$ **f** $20 - 10f$

2 a 2   **b** 8   **c** 4   **d** 5
   e 2   **f** 0

3 a 4   **b** 6   **c** 3   **d** -6

4 a -6 **b** -1 **c** 2   **d** -2

5 a 1.5   **b** 10   **c** 0.75   **d** -0.5

6 a 5   **b** 3   **c** 3   **d** 7
   e 8   **f** 8

7 a -3 **b** -4 **c** -2 **d** -1

8 a 5   **b** -4 **c** 2.5   **d** -3

9 a 7.5   **b** -3 **c** 0.25   **d** -1.5

10 a 4   **b** -3 **c** 2   **d** -5

11 a 3   **b** -2 **c** 5   **d** -4

12 a 6   **b** -6 **c** 5.5   **d** 2

13 a 3.5   **b** 0.5   **c** -1.5 **d** -1

14 a 9   **b** -8 **c** -18 **d** 20

15 a 9   **b** 6   **c** -10 **d** 12

16 a 6   **b** 6   **c** 15   **d** -8

17 a 1   **b** 11   **c** -17 **d** 6

18 a 12   **b** 6   **c** 7   **d** 5

10.2A

1 6

2 15

3 10

4 1

5 8

6 Solutions: 8, 2.5, 1.5, 0.4, 1.25, -0.25, -1.5, $-\frac{5}{9}$, no solution

7 5

8 2

9 6 buttons

10 a 10   **b** 6   **c** -4

11 9 cm

12 36 cm

13 a 8 cm.
   b Since the sides are length 13 cm the shape is actually a rectangle, so perimeter is $13 + 13 + 5 + 5 = 36$ cm

10.3S

1 a 6, 2   **b** 3, 8   **c** 9, 4   **d** 11, 5
   e $-3, -4$ **f** $-6, -4$ **g** 7, -4 **h** $-5, 3$

2 a $(x + 6)(x + 2)$ **b** $(x + 8)(x + 3)$
   c $(x + 4)(x + 9)$ **d** $(x + 5)(x + 11)$
   e $(x - 3)(x - 4)$ **f** $(x - 6)(x - 4)$
   g $(x + 7)(x - 4)$ **h** $(x - 5)(x + 3)$

3 a $-3, -4$ **b** $-6, -2$ **c** $-5, -5$ **d** $-7, 2$
   e 5, -1 **f** 3, 2   **g** 3, 7   **h** 8, -5

4 a $-1, -17$ **b** $-13, -2$ **c** $-5, 3$ **d** $-8, 2$
   e $-6, 3$ **f** $-11, 2$

5 a 3, -2 **b** 4, -3 **c** 5, -3 **d** 8, -2
   e 15, -2 **f** 7, -4

6 a 0, 8   **b** 0, -4 **c** 0, 6   **d** 0, -5
   e 0, 9   **f** 0, 12   **g** 0, -4 **h** 0, 6

7 a 0, 3   **b** 0, -8 **c** 0, 4.5   **d** 0, 3
   e 0, 5   **f** 0, 7   **g** 0, 12   **h** 0, 2
   i 0, 6   **j** 0, 3   **k** 0, 7

8 a $-3, -4$ **b** $-6, -2$ **c** $-5, -5$ **d** $-5, 3$
   e $-7, 2$ **f** 5, -1 **g** 6, -1 **h** 6, 6
   i 6, 2   **j** 2, 3   **k** $-7, -3$

9 a $-4, -5$ **b** $-12, -1$ **c** $-10, 2$ **d** $-9, 3$
   e 6, 4   **f** 7, 5   **g** $-1, -5$ **h** 3, 3

10 a 5, -4 **b** $-12, 1$ **c** $-4, 2$ **d** $-7, 3$
   e 9, 5   **f** 0, -20

11 a 4, -4 **b** 8, -8 **c** 5, -5 **d** $\frac{2}{3}, -\frac{2}{3}$
   e 0.5, -0.5 **f** 13, -13 **g** 2.5, -2.5 **h** 2, -2

10.3A

1 a Multiply and rearrange to form a quadratic equal to zero, then factorise and solve.
   b $-5, 3$

2 5, 1

3 a -10 and -6 or 10 and 6   **b** -6 and 2 or -2 and 6

4 a $-8, 5$ **b** -8 and -10, or 5 and 16

5 a -10 and 5   **b** $t(t + 5) = 50$, therefore $t^2 + 5t - 50 = 0$
   c $t = -10$ is invalid as a length cannot be negative

6 10

7 15, 25, 20 or 24, 16, 20

8 $x = 12°$, rhombus.

9 3 cm

10 a $x^2 + x^2 + 2x + 1 = 25$, $x^2 + x - 12 = 0$
   b Perimeter = 12 cm

11 4

12 a 7   **b** Once, $(7, -3)$

10.4S

1 a $x = 4, y = 3$ **b** $x = 2, y = -3$
   c $y = 4, x = -1$ **d** $x = 3, y = 0.5$
   e $n = 2, m = 5$ **f** $x = -2, y = 1$

2 a $x = 7, y = -1$ **b** $x = 3, y = 0.5$

c $a = 3, b = -1$ **d** $y = 5, x = 2$

e $x = 3.25, y = -\frac{2}{7}$ **f** $p = 2, q = 7$

3 **a** $x = 5, y = -2$ **b** $x = 2, y = 3$

   c $b = 3, a = -2$ **d** $v = 4, w = 2$

   e $p = 2, q = 2$ **f** $y = 2, x = 5$

4 **a** $b = \frac{4}{3}, a = \frac{29}{3}$ **b** $w = 4, v = 7$

5 **a** $x = 1, y = 3$ **b** $x = 5, y = 2$

   c $q = 7, p = 2$ **d** $b = -1, a = 5$

6 **a** $x = 6, y = 1$ **b** $x = 7, y = -2$

   c $q = 1, p = 6$ **d** $b = 1, a = 7$

7 **a** **i** $x = 2.5, y = 2.5$ **ii** $x = 1, y = 4$

     iii $x = 1.2, y = 0.8$ **iv** $x = 0.2, y = 1.8$

   b Because the lines do not intersect.

8 **a** $x = 3, y = 7$ **b** $x = 1, y = 1$

9 $(-0.3, 1.5)$

10 **a** $x = \frac{10}{7}, y = \frac{6}{7}$ **b** $x = \frac{4}{7}, y = \frac{19}{7}$

   c $x = -0.5, y = -2.5$ **d** $x = 2.5, y = 0.5$

10.4A

1 a, b

2 **a** $x = 12, y = -16$ **b** $x = 7, y = -1$

   c $a = 3, b = -2$ **d** $v = 3, w = 2$

   e $q = 3, p = 0$ **f** $x = 5, y = 2$

3 Blue and red $x = -1, y = 14$

   Blue and orange $x = 5, y = 2$

   Blue and green $x = 5, y = 2$

   Red and orange $x = -67, y = -52$

   Red and green $x = -2.2, y = 12.8$

   Orange and green $x = 5, y = 2$

4 **a** £0.17   **b** 6.4 cm

5 **a** $x = 14, y = 9$ **b** $x = 4, y = -2$

6 $s = 29, m = 17$

7 73 children

8 **a** $x = 2, y = 0$ **b** Isla 3, James 1

9 **a** The lines are parallel so they do not intersect, therefore no solution.

   b If either of the equations are not linear there could be more solutions.

10 The lines may look parallel, but in reality are not since the gradient isn't the same. They would eventually intersect and so would have a solution.

   $x = 36, y = 82$

10.5S

1 **a**   **b**   **c**

   d   **e**   **f**

   g   **h**

2 **a** (ii)   **b** (v)   **c** (vi)   **d** (i)

   e (iv)   **f** (iii)

3 **a** $-10 \leq x < 5$ **b** $5 < y < 12$

   c $-2 < z \leq 6$ **d** $-4 \leq t \leq 2$

4 **a**   **b**

   c   **d**

5 **a**

   b $-4, -3, -2, -1, 0, 1, 2, 3, 4$

6 **a** **i** $2x > 10$ **ii** $4x > 20$

b **i** $3y \leq 18$ **ii** $5y \leq 30$

   c $5x \geq -20$ **d** $6m < -18$

7 **a** $x \leq 2$ **b** $x < 5$ **c** $x > -2$ **d** $x \geq -4$

8 **a** If $3x + 2 > 11$ then $3x > 9$ and $x > 3$

   b If $7x - 4 > 31$ then $7x > 35$ and $x > 5$

   c If $2x + 9 \leq 11$ then $2x \leq 2$ and $x \leq 1$

   d If $5x - 3 \geq 12$ then $5x \geq 15$ and $x \geq 3$

9 **a** $x \leq 5$ **b** $x \geq 6$ **c** $x \geq 2$ **d** $x < 4$

   e $x \geq -1$ **f** $x \leq 3$ **g** $x > 3$ **h** $x \leq 2$

10 **a** $x \geq 3$ **b** $x > 2$ **c** $x \geq 0$ **d** $x > -1$

   e $x \geq -2$ **f** $x \leq 4$ **g** $x \geq -2.5$ **h** $x < -7.5$

11 1, 2, 3, 4

12 **a** **i**   **ii** $-2, -1, 0, 1, 2$

   b **i**   **ii** $-1, 0, 1, 2, 3, 4$

   c **i**   **ii** 2, 3, 4, 5

   d 2

10.5A

1 a, b, c, e

2 Max − incorrect − sign needs to be changed

   Natalie − incorrect − bringing 'x' over inequality

   Owen − incorrect − divide by negative without changing inequality

   Pritesh − correct

3 **a** $x < 3$ **b** $x \leq -4$ **c** $x > -8$ **d** $x \geq -5$

   e $x < 4$ **f** $x \geq -3$

4 **a** $30 < x < 75$ **b** $45 < y < 81$

5 $6 < x$

6 **a,b** $x > 5$ so smallest value of x is 6

7 $-\frac{1}{3} < x \leq 2$

8 **a** $-1 < x < 4$ **b** $-5 < x < 18$

9 y cannot be less than or equal to 6, and greater than 6, at the same time

10 **a** $x + 2y = 180$

     $x = 180 - 2y$

     $x < 30$ therefore $180 - 2y < 30$

   b $y > 75$

   c Because $2 \times 90 = 180$, but the total angle in a triangle is 180, therefore there couldn't be a third angle.

Review 10

1 **a** 48   **b** 5   **c** 7   **d** 16.5

   e 100   **f** -2

2 **a** 6   **b** 2   **c** -7

3 **a** $2x + 24 = 42$ **b** Vicky $= 9$, mum $= 33$

4 **a** $-3, -5$ **b** 1, 5   **c** $-3, 2$ **d** $-8, 8$

   e 0, 12

5 $-2, 10$

6 **a** $x = 3, y = 5$ **b** $a = 1, b = -2$

   c $x = 9, y = 3$ **d** $v = 2, w = -3$

7 **a** $2c + w = 3.75, 3c + 2w = 6$

   b $c = 1.5$ kg, $w = 0.75$ kg

8 $x = 1, y = 3$

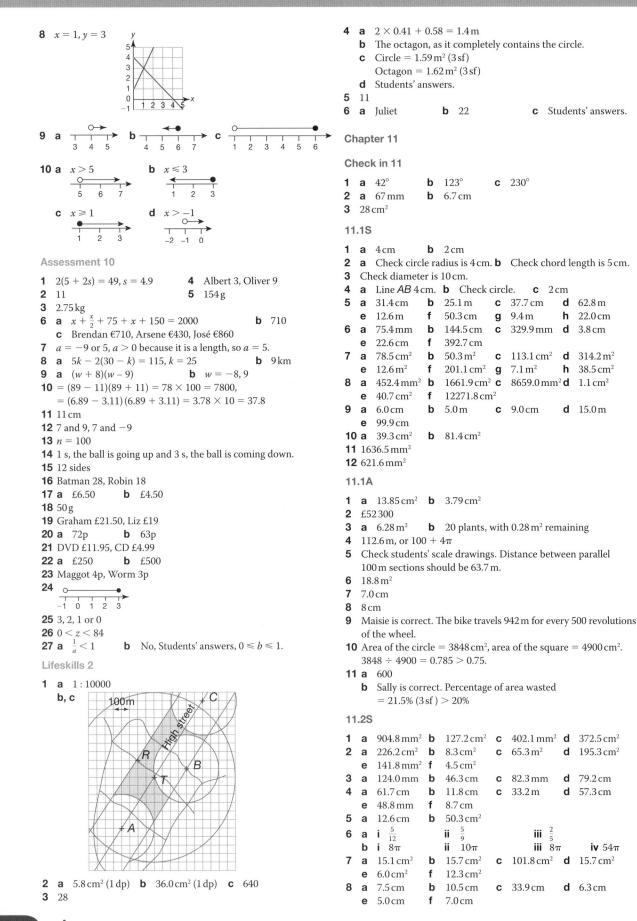

9 a **b** **c**

10 a $x > 5$ **b** $x \leqslant 3$

   c $x \geqslant 1$ **d** $x > -1$

Assessment 10

1 $2(5 + 2s) = 49, s = 4.9$ **4** Albert 3, Oliver 9
2 11   **5** 154 g
3 2.75 kg
6 a $x + \frac{x}{2} + 75 + x + 150 = 2000$ **b** 710
   c Brendan €710, Arsene €430, José €860
7 $a = -9$ or 5, $a > 0$ because it is a length, so $a = 5$.
8 a $5k - 2(30 - k) = 115, k = 25$ **b** 9 km
9 a $(w + 8)(w - 9)$ **b** $w = -8, 9$
10 $= (89 - 11)(89 + 11) = 78 \times 100 = 7800$,
    $= (6.89 - 3.11)(6.89 + 3.11) = 3.78 \times 10 = 37.8$
11 11 cm
12 7 and 9, 7 and -9
13 $n = 100$
14 1 s, the ball is going up and 3 s, the ball is coming down.
15 12 sides
16 Batman 28, Robin 18
17 a £6.50   **b** £4.50
18 50 g
19 Graham £21.50, Liz £19
20 a 72p   **b** 63p
21 DVD £11.95, CD £4.99
22 a £250   **b** £500
23 Maggot 4p, Worm 3p
24

25 3, 2, 1 or 0
26 $0 < z < 84$
27 a $\frac{1}{a} < 1$ **b** No, Students' answers, $0 \leqslant b \leqslant 1$.

Lifeskills 2

1 a 1 : 10000
   b, c

2 a 5.8 cm² (1 dp) **b** 36.0 cm² (1 dp) **c** 640
3 28

4 a $2 \times 0.41 + 0.58 = 1.4$ m
   b The octagon, as it completely contains the circle.
   c Circle $= 1.59$ m² (3 sf)
      Octagon $= 1.62$ m² (3 sf)
   d Students' answers.
5 11
6 a Juliet   **b** 22   **c** Students' answers.

Chapter 11

Check in 11

1 a 42°   **b** 123°   **c** 230°
2 a 67 mm   **b** 6.7 cm
3 28 cm²

11.1S

1 a 4 cm   **b** 2 cm
2 a Check circle radius is 4 cm. **b** Check chord length is 5 cm.
3 Check diameter is 10 cm.
4 a Line AB 4 cm. **b** Check circle.   **c** 2 cm
5 a 31.4 cm   **b** 25.1 m   **c** 37.7 cm   **d** 62.8 m
   e 12.6 m   **f** 50.3 cm   **g** 9.4 m   **h** 22.0 cm
6 a 75.4 mm   **b** 144.5 cm   **c** 329.9 mm   **d** 3.8 cm
   e 22.6 cm   **f** 392.7 cm
7 a 78.5 cm²   **b** 50.3 m²   **c** 113.1 cm²   **d** 314.2 m²
   e 12.6 m²   **f** 201.1 cm²   **g** 7.1 m²   **h** 38.5 cm²
8 a 452.4 mm² **b** 1661.9 cm² **c** 8659.0 mm² **d** 1.1 cm²
   e 40.7 cm²   **f** 12271.8 cm²
9 a 6.0 cm   **b** 5.0 m   **c** 9.0 cm   **d** 15.0 m
   e 99.9 cm
10 a 39.3 cm²   **b** 81.4 cm²
11 1636.5 mm²
12 621.6 mm²

11.1A

1 a 13.85 cm² **b** 3.79 cm²
2 £52 300
3 a 6.28 m²   **b** 20 plants, with 0.28 m² remaining
4 112.6 m, or $100 + 4\pi$
5 Check students' scale drawings. Distance between parallel
   100 m sections should be 63.7 m.
6 18.8 m²
7 7.0 cm
8 8 cm
9 Maisie is correct. The bike travels 942 m for every 500 revolutions
   of the wheel.
10 Area of the circle $= 3848$ cm², area of the square $= 4900$ cm².
   $3848 \div 4900 = 0.785 > 0.75$.
11 a 600
   b Sally is correct. Percentage of area wasted
      $= 21.5\%$ (3 sf) $> 20\%$

11.2S

1 a 904.8 mm² **b** 127.2 cm² **c** 402.1 mm² **d** 372.5 cm²
2 a 226.2 cm² **b** 8.3 cm²   **c** 65.3 m²   **d** 195.3 cm²
   e 141.8 mm² **f** 4.5 cm²
3 a 124.0 mm   **b** 46.3 cm   **c** 82.3 mm   **d** 79.2 cm
4 a 61.7 cm   **b** 11.8 cm   **c** 33.2 m   **d** 57.3 cm
   e 48.8 mm   **f** 8.7 cm
5 a 12.6 cm   **b** 50.3 cm²
6 a i $\frac{5}{12}$ **ii** $\frac{5}{9}$ **iii** $\frac{2}{5}$
   b i 8π **ii** 10π **iii** 8π **iv** 54π
7 a 15.1 cm²   **b** 15.7 cm²   **c** 101.8 cm² **d** 15.7 cm²
   e 6.0 cm²   **f** 12.3 cm²
8 a 7.5 cm   **b** 10.5 cm   **c** 33.9 cm   **d** 6.3 cm
   e 5.0 cm   **f** 7.0 cm

9

	Angle	Radius	Arc Length	Area
a	40°	18 mm	12.6 mm	113.1 mm²
b	135°	1.6 m	3.8 m	3.0 m²
c	252°	3.5 cm	15.4 cm	26.9 cm²
d	312°	0.46 km	2.50 km	0.58 km²

10 a 45.4 mm **b** 14.8 cm **c** 29.5 cm

11.2A

1 123.6 cm²
2 23.6 cm²
3 a 11.06 m²
  b 12.9 bricks (assuming inner radius of brick is 30 cm long)
4 Yes 311.8 > 300.
5 10.9 cm²

6 143°
7 229.2°
8 50 cm²
9 $(24\pi + 96)$ inches
10 a Perimeter $= 20\pi$ cm
  b Check students' area and factorising. Area of leaf $= 200(\pi - 2)$ cm²

11.3S1

1 a 8.2 cm   **b** 5.3 cm   **c** 11.6 cm
2 a 5.6 cm, 5.6 cm **b** 7.7 cm, 5.1 cm **c** 3.4 cm, 5.7 cm
3 a 60°   **b** 75°   **c** 90°
4 a 5.7 cm, 55° **b** 5.9 cm, 44° **c** 6.0 cm, 67°
5 a Check angles are 30°, 61° and 89°.
  b Check third side is 2 − 2.1 cm.
  c Check angles are 37°, 48° and 95°.
  d Check third side is 7.7 cm
  e Check angles are 32°, 39° and 109°.
  f Check 90° and two sides 2.8 cm.
6 a Check angles are 69°, 69° and 42°.
  b Check angles are 89°, 46° and 46°.
7 a Check angles are 60°.
  b, c Check constructions.
8 a Check angles are 67°, 23° and 90°.
  b Right-angled triangle.
9 a Check angles are 53°, 37° and 90°.
  b, c Rectangle with sides 4 cm and 3 cm.

11.3S2

1 a Line $AB = 8$ cm
  b–d Check $AM = MB = 4$ cm and 90° angle
2 a Line length 64 mm
  b, c Check bisector 90° angle, bisects the line at 32 mm
3 Check perpendiculars
4 Check 90° angle passing through the point X
5 a Check angle bisector 27° **b** Check angle bisector 41°
  c Check angle bisector 51°
6 a Check angle bisector 35° **b** Check angle bisector 55°
  c Check angle bisector 45° **d** Check angle bisector 65°
  e Check angle bisector 25° **f** Check angle bisector 42.5°

11.3A

1 a Students construct a horizontal line of 8 cm. Students construct 8 cm arcs from end points of the line to form two triangles.
  b Students construct a horizontal line of 6 cm. Students construct 5 cm arcs from end points of the line to form two triangles.
2 a Construction arcs of 3 cm and 4 cm on the 9 cm line do not cross so not possible to construct the triangle.
  b Construction arcs meet at a point on the third line so not possible to construct the triangle.
3 a yes   **b** yes   **c** yes   **d** no
  e no   **f** no   **g** yes   **h** yes
4 a, b Check $BC = 12$ cm, $AC = 7$ cm and angle $B = 30°$.
  c SSA triangles are not unique.
5 Six equilateral triangles constructed inside a circle with radius 4 cm.
6 Check students' construction – perpendicular from point X to the line MN. Distance $= 130$ m

---

7 Yes, two SSS congruent triangles are constructed.
8 No, line will be perpendicular but will not bisect the line.
9 a i Line joining $(-4, -2)$ and $(2, 6)$.
  ii Check line is perpendicular.
  iii $(-1, 2)$
  b i Line passing through $(-6, -6)$ and $(6, 6)$.
  ii Line passing through $(-6, 4)$ and $(4, -6)$
  iii $(-1, -1)$

11.4S

1 **a** **b** **c**
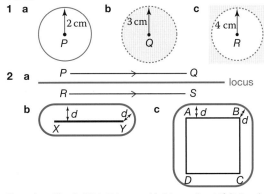

2 a

  b **c**

3 a, b Check AB is 7.8 cm and is bisected at 90° (3.9 cm).
4 a, b Check $\angle PQR$ is 76° and is bisected (38°).
5 a, b   **6 a, b, c**

7 a, b   **8**
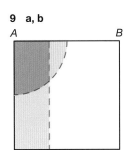

9 a, b   **10 a, b**
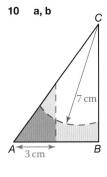

11 a, b, c, d   *****12 a, b**
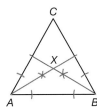
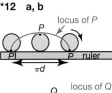

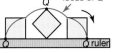

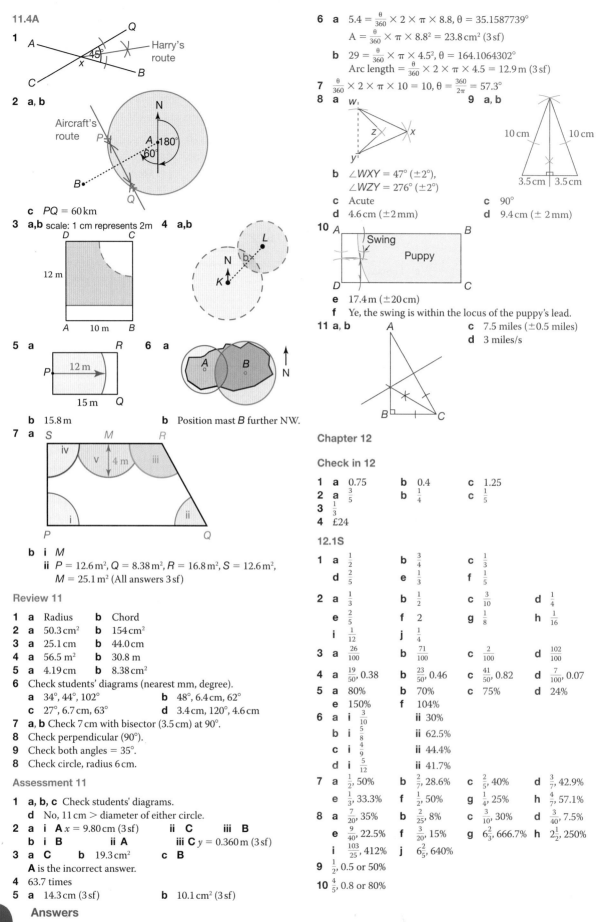

11.4A

1

A Harry's route, Q, 45°, x, B, C

2 a, b

Aircraft's route, N, P, A •180°, 60°, B•, Q

c $PQ = 60\,\text{km}$

3 a,b scale: 1 cm represents 2 m

D, C, 12 m, A, 10 m, B

4 a,b

L, N, b, K

5 a

R, P•, 12 m, 15 m, Q

b 15.8 m

6 a

A, B, N

b Position mast B further NW.

7 a

S, M, R, iv, v, 4 m, iii, i, ii, P, Q

b i M
ii $P = 12.6\,\text{m}^2$, $Q = 8.38\,\text{m}^2$, $R = 16.8\,\text{m}^2$, $S = 12.6\,\text{m}^2$, $M = 25.1\,\text{m}^2$ (All answers 3 sf)

Review 11

1 a Radius **b** Chord
2 a $50.3\,\text{cm}^2$ **b** $154\,\text{cm}^2$
3 a $25.1\,\text{cm}$ **b** $44.0\,\text{cm}$
4 a $56.5\,\text{m}^2$ **b** $30.8\,\text{m}$
5 a $4.19\,\text{cm}$ **b** $8.38\,\text{cm}^2$
6 Check students' diagrams (nearest mm, degree).
 a 34°, 44°, 102° **b** 48°, 6.4 cm, 62°
 c 27°, 6.7 cm, 63° **d** 3.4 cm, 120°, 4.6 cm
7 a, b Check 7 cm with bisector (3.5 cm) at 90°.
8 Check perpendicular (90°).
9 Check both angles = 35°.
8 Check circle, radius 6 cm.

Assessment 11

1 a, b, c Check students' diagrams.
 d No, 11 cm > diameter of either circle.
2 a i $\mathbf{A}\ x = 9.80\,\text{cm}$ (3 sf) **ii** C **iii** B
 b i B **ii** A **iii** C $y = 0.360\,\text{m}$ (3 sf)
3 a C **b** $19.3\,\text{cm}^2$ **c** B
 A is the incorrect answer.
4 63.7 times
5 a 14.3 cm (3 sf) **b** 10.1 cm² (3 sf)

6 a $5.4 = \frac{\theta}{360} \times 2 \times \pi \times 8.8$, $\theta = 35.1587739°$
 $A = \frac{\theta}{360} \times \pi \times 8.8^2 = 23.8\,\text{cm}^2$ (3 sf)
 b $29 = \frac{\theta}{360} \times \pi \times 4.5^2$, $\theta = 164.1064302°$
 Arc length $= \frac{\theta}{360} \times 2 \times \pi \times 4.5 = 12.9\,\text{m}$ (3 sf)
7 $\frac{\theta}{360} \times 2 \times \pi \times 10 = 10$, $\theta = \frac{360}{2\pi} = 57.3°$
8 a w, z, x, y
 b $\angle WXY = 47°\ (\pm 2°)$,
 $\angle WZY = 276°\ (\pm 2°)$
 c Acute
 d 4.6 cm (± 2 mm)

9 a, b

10 cm, 10 cm, 3.5 cm, 3.5 cm

 c 90°
 d 9.4 cm (± 2 mm)

10

A, Swing, B, Puppy, D, C

 e 17.4 m (± 20 cm)
 f Ye, the swing is within the locus of the puppy's lead.

11 a, b

A, B, C

 c 7.5 miles (± 0.5 miles)
 d 3 miles/s

Chapter 12

Check in 12

1 a 0.75 **b** 0.4 **c** 1.25
2 a $\frac{3}{5}$ **b** $\frac{1}{4}$ **c** $\frac{1}{5}$
3 $\frac{1}{3}$
4 £24

12.1S

1 a $\frac{1}{2}$ **b** $\frac{3}{4}$ **c** $\frac{1}{3}$
 d $\frac{2}{5}$ **e** $\frac{1}{3}$ **f** $\frac{1}{5}$
2 a $\frac{1}{3}$ **b** $\frac{1}{2}$ **c** $\frac{3}{10}$ **d** $\frac{1}{4}$
 e $\frac{2}{5}$ **f** 2 **g** $\frac{1}{8}$ **h** $\frac{1}{16}$
 i $\frac{1}{12}$ **j** $\frac{1}{4}$
3 a $\frac{26}{100}$ **b** $\frac{71}{100}$ **c** $\frac{2}{100}$ **d** $\frac{102}{100}$
4 a $\frac{19}{50}$, 0.38 **b** $\frac{23}{50}$, 0.46 **c** $\frac{41}{50}$, 0.82 **d** $\frac{7}{100}$, 0.07
5 a 80% **b** 70% **c** 75% **d** 24%
 e 150% **f** 104%
6 a i $\frac{3}{10}$ **ii** 30%
 b i $\frac{5}{8}$ **ii** 62.5%
 c i $\frac{4}{9}$ **ii** 44.4%
 d i $\frac{5}{12}$ **ii** 41.7%
7 a $\frac{1}{2}$, 50% **b** $\frac{2}{7}$, 28.6% **c** $\frac{2}{5}$, 40% **d** $\frac{3}{7}$, 42.9%
 e $\frac{1}{3}$, 33.3% **f** $\frac{1}{2}$, 50% **g** $\frac{1}{4}$, 25% **h** $\frac{4}{7}$, 57.1%
8 a $\frac{7}{20}$, 35% **b** $\frac{2}{25}$, 8% **c** $\frac{3}{10}$, 30% **d** $\frac{3}{40}$, 7.5%
 e $\frac{9}{40}$, 22.5% **f** $\frac{3}{20}$, 15% **g** $6\frac{2}{3}$, 666.7% **h** $2\frac{1}{2}$, 250%
 i $\frac{103}{25}$, 412% **j** $6\frac{2}{5}$, 640%
9 $\frac{1}{2}$, 0.5 or 50%
10 $\frac{4}{5}$, 0.8 or 80%

11 $\frac{5}{8}$, 0.625, 62.5%

12 A 30%   **B** 25.9%   **C** 23.3%   **D** 20.8%

12.1A

1 **a** 72%, 0.74, $\frac{3}{4}$ **b** 0.29, $\frac{3}{10}$, 31%

    c 0.78, 0.8, 83%, $\frac{7}{8}$ **d** 0.39, $\frac{2}{5}$, 41%, $\frac{3}{7}$

    e $\frac{3}{4}$, $\frac{8}{10}$, 83%, 0.84   **f** $\frac{3}{11}$, 28%, $\frac{7}{24}$, 0.3

2 **a** **i** Maths   **ii** English

    b Use a weighted average of her marks so far.

3 50 cars

4 Peter is correct. $\frac{165}{220} = \frac{3}{4}$

5 Jada is correct. Check students workings.

6 **a** 40, 50   **b** 50 employees.

7 Jenni wrong. The shapes are both half shaded.

8 $\frac{5}{16}$

9 **a** 64 cheeses   **b** 32%

12.2S

1 **a** 1:3   **b** 3:1   **c** 1:3   **d** 1:7

    e 1:10   **f** 5:1   **g** 3:1   **h** 1:15

    i 1:8

2 **a** 2:1   **b** 3:1   **c** 9:7   **d** 5:3

3 **a** 2:1   **b** 1:3   **c** 3:2   **d** 9:8

    e 3:4   **f** 4:3   **g** 10:9   **h** 4:5

4 **a** £27:£63   **b** 287 kg:82 kg

    c 64.5 tonnes:38.7 tonnes   **d** 19.5 litres:15.6 litres

    e £6:£12:£18

5 **a** £40, £35   **b** £350, £650   **c** 260, 104

    d 142.86 g, 357.14 g   **e** 242.29 m, 385.71 m

6 **a** 2:5   **b** 11:16   **c** 5:2   **d** 5:3

    e 5:3   **f** 8:5

7 **a** £100:£250:£150   **b** 180°:45°:135°

    c 900 m:400 m:700 m

8 **a** 1:3   **b** 1:4   **c** 1:2   **d** 1:5

    e 1:2   **f** 1:3   **g** 1:5   **h** 1:3

    i 1:2   **j** 1:5   **k** 1:2   **l** 1:6

9 **a** 1000 m   **b** 4000 m   **c** 5000 m

    d 250 m   **e** 7250 m   **f** 5 m

10 **a** 325 m   **b** 0.6 cm

11 **a** 50 m   **b** 4 cm

12 **a** 200 m   **b** 12 cm

13 **a** 116 m   **b** 180 cm

12.2A

1 **a** 66.67%   **b** 360 cm

2 **a** 120%   **b** 102 kg

3 **a** 72 g   **b** 115 g Copper, 69 g Aluminium

4 $\frac{11}{50}$

5 **a** 300 m   **b** 120 m

6 75 cm by 200 cm

7 **a** 6 m, 9 m, 18 m   **b** 162 m²

8 0.7

9 **a** £280   **b** £80, £200

10 **a** 3:6:2   **b** $20 ($60 compared to $40)

11 8 sides

12 P = 55, Q = 125

12.3S

1 **a** 0.5   **b** 0.6   **c** 0.25   **d** 0.085

    e 0.0015   **f** 0.0001

2 **a** £40   **b** 260 cm   **c** 3.2 kg   **d** 20 m

    e 190 p   **f** £35   **g** 3 kg   **h** £6.20

3 **a** 325.35 kg   **b** $120   **c** 10.35 kg   **d** 5.88 kg

    e £21.70   **f** 78.2 m

4 **a** £224   **b** £385.20   **c** 1458   **d** £77

    e €13.35   **f** £465.83

5 10% of £350 (£35 > £30)

6 8% of £28 (£2 < £2.24)

7 **a** £549.60   **b** £2519   **c** £842.72   **d** £1167.90

8 **a** £790   **b** £1109.25   **c** £54.60   **d** £132.43

9 **a** 1.2   **b** 1.3   **c** 1.45

10 **a** 0.6   **b** 0.4   **c** 0.65

11 **a** £495   **b** 672 kg   **c** £756   **d** 392 km

    e £658   **f** 256 m

12 **a** £275   **b** £2264   **c** £18 060   **d** £2520

    e £4.23   **f** £2000

13 **a** £385   **b** 70.3 kg   **c** £550.20   **d** 491.4 km

    e 1128 kg   **f** £216

14 **a** £397.80   **b** 524.9 kg   **c** £1758.96   **d** 599.555 km

    e $3423.55   **f** 2154.75 m

12.3A

1 **a** 60%   **b** 48%   **c** 60%

2 **a** 51.35%   **b** 28.57%   **c** 3.83%

3 **a** 20%   **b** 20%

4 **a** 12.5%   **b** 20%

5 9.23% per year

6 £56 is 80% of the original price, the original price is not 120% of £56. The original price is £70.

7 £5

8 **a** 50   **b** 40   **c** 80   **d** 110

9 **a** £6   **b** £80

10 **a** Francesca: £13.13, Frank £12.80. Francesca has the bigger pay increase.

    b £78.03

11 **a** £50   **b** £30.94

12 £8450

13 13 600

14 **a** £321.63   **b** £1749.60

Review 12

1 **a** $\frac{2}{5}$ **b** 60%

2 Physics

3 **a** 6:7   **b** 3:5   **c** 1:400

4 **a** £7, £28   **b** £40, £50

5 **a** 1:3   **b** $\frac{1}{4}$

6 **a** Butter 240 g, Flour 600 g   **b** 750 g   **c** $\frac{5}{7}$

7 **a** 63 cm   **b** 3 cm

8 **a** 10 km   **b** 0.75 cm

9 **a** 57   **b** 106.5

10 £4060

11 **a** 2.5%   **b** 30%

12 **a** £1.20   **b** £44

Assessment 12

1 **a** $\frac{1}{3}$ **b** $\frac{1}{6}$ **c** $\frac{1}{8}$ **d** $\frac{1}{2}$

    e $\frac{3}{8}$ **f** $\frac{3}{4}$

2 **a** 0.52 = 52%   **b** 117 000 rand

3 £13 110

4 **a** 12:5   **b** 2.4:1

5 55

6 **a** Chocolate 27, plain 36.   **b** 33

7 **a** 8:2:3   **b** Butter 45 g, cheese 67.5 g

8 **a** 6:5   **b** $\frac{5}{3}$:1

9 **a** Girls 32, boys 56

    b Girls 44, boys 77, 44:77 = 4:7.

10 **a** 200 m   **b** 1.17 km   **c** 20.75 cm

11 **a** 40.5 miles   **b** 25.6 in   **c** 236 miles

12 **a** 50p   **b** £728

13 **a** Gavin   **b** Gemma

    c Steven has calculated 30% of 90 kg. $90 \times 1.03 = 92.7$ kg

    d Shayda has calculated 45.4% of 550 ml. $550 \times (1 - 0.454) = 300.3$ ml

14 44.44%

15 30%

16 £2 712 500

17 90.82%

18 $(35 - 24) \div 24 = 45.83\% > 45\%$. They are correct.

19 a 230   **b** 217

20 £2200

21 £2480

22 £236

Revision 2

1 9

2 72 cm²

3 a

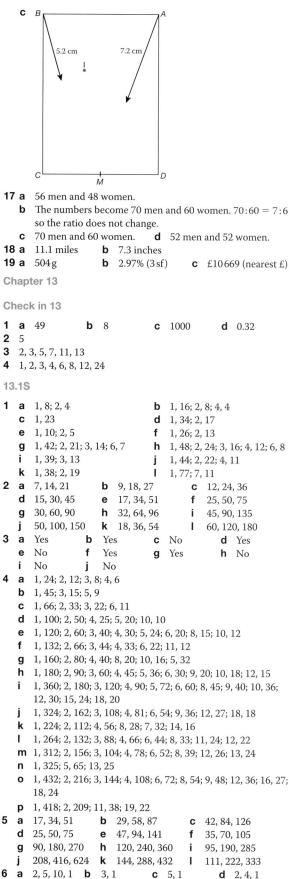

     b 90°

     c 30 km²

4 a 39 609 m²   **b** 3.96 Ha (3 sf)

5 a Enlargement, scale factor 3, centre (0, 0)

b Translation by vector $\binom{-1}{-2}$.

c

6 a i 360   **ii** 360   **iii** 366.3

     b 365.270 **is** 0.0104 from the exact figure.

7 a 120 cm   **b** 40 000 g or 40 kg

     c Length = 110.2 cm,   mass = 41 800 g or 41.8 kg.

     d Percentage error for length = 8.89%, percentage error for mass = −4.31%

     e Estimate for the mass.

8 a i Upper bound = 119.5 m², lower bound = 71.5 m².

     ii Upper bound = 86.625 m³, lower bound = 39.375 m³.

     b 3600 kg/m³

     c Upper bound = 311 850 kg, lower bound = 141 750 kg.

9 4 cm × 25 cm

10 $r < 100$

11 26 10p coins and 15 50p coins.

12 53 racquets

13 a 96   **b** 32

14 a 188 in (3 sf)   **b** 2 802 cm²

15 a 5 cm   **b** 19.6 cm² (3 sf)

     c 824 cm² (3 sf)   **d** 78.5% (3 sf)

16 a, b

17 a 56 men and 48 women.

     b The numbers become 70 men and 60 women. $70:60 = 7:6$ so the ratio does not change.

     c 70 men and 60 women.   **d** 52 men and 52 women.

18 a 11.1 miles   **b** 7.3 inches

19 a 504 g   **b** 2.97% (3 sf)   **c** £10 669 (nearest £)

Chapter 13

Check in 13

1 a 49   **b** 8   **c** 1000   **d** 0.32

2 5

3 2, 3, 5, 7, 11, 13

4 1, 2, 3, 4, 6, 8, 12, 24

13.1S

1 a 1, 8; 2, 4   **b** 1, 16; 2, 8; 4, 4

     c 1, 23   **d** 1, 34; 2, 17

     e 1, 10; 2, 5   **f** 1, 26; 2, 13

     g 1, 42; 2, 21; 3, 14; 6, 7   **h** 1, 48; 2, 24; 3, 16; 4, 12; 6, 8

     i 1, 39; 3, 13   **j** 1, 44; 2, 22; 4, 11

     k 1, 38; 2, 19   **l** 1, 77; 7, 11

2 a 7, 14, 21   **b** 9, 18, 27   **c** 12, 24, 36

     d 15, 30, 45   **e** 17, 34, 51   **f** 25, 50, 75

     g 30, 60, 90   **h** 32, 64, 96   **i** 45, 90, 135

     j 50, 100, 150   **k** 18, 36, 54   **l** 60, 120, 180

3 a Yes   **b** Yes   **c** No   **d** Yes

     e No   **f** Yes   **g** Yes   **h** No

     i No   **j** No

4 a 1, 24; 2, 12; 3, 8; 4, 6

     b 1, 45; 3, 15; 5, 9

     c 1, 66; 2, 33; 3, 22; 6, 11

     d 1, 100; 2, 50; 4, 25; 5, 20; 10, 10

     e 1, 120; 2, 60; 3, 40; 4, 30; 5, 24; 6, 20; 8, 15; 10, 12

     f 1, 132; 2, 66; 3, 44; 4, 33; 6, 22; 11, 12

     g 1, 160; 2, 80; 4, 40; 8, 20; 10, 16; 5, 32

     h 1, 180; 2, 90; 3, 60; 4, 45; 5, 36; 6, 30; 9, 20; 10, 18; 12, 15

     i 1, 360; 2, 180; 3, 120; 4, 90; 5, 72; 6, 60; 8, 45; 9, 40; 10, 36; 12, 30; 15, 24; 18, 20

     j 1, 324; 2, 162; 3, 108; 4, 81; 6, 54; 9, 36; 12, 27; 18, 18

     k 1, 224; 2, 112; 4, 56; 8, 28; 7, 32; 14, 16

     l 1, 264; 2, 132; 3, 88; 4, 66; 6, 44; 8, 33; 11, 24; 12, 22

     m 1, 312; 2, 156; 3, 104; 4, 78; 6, 52; 8, 39; 12, 26; 13, 24

     n 1, 325; 5, 65; 13, 25

     o 1, 432; 2, 216; 3, 144; 4, 108; 6, 72; 8, 54; 9, 48; 12, 36; 16, 27; 18, 24

     p 1, 418; 2, 209; 11, 38; 19, 22

5 a 17, 34, 51   **b** 29, 58, 87   **c** 42, 84, 126

     d 25, 50, 75   **e** 47, 94, 141   **f** 35, 70, 105

     g 90, 180, 270   **h** 120, 240, 360   **i** 95, 190, 285

     j 208, 416, 624   **k** 144, 288, 432   **l** 111, 222, 333

6 a 2, 5, 10, 1   **b** 3, 1   **c** 5, 1   **d** 2, 4, 1

     e 7, 1   **f** 2, 5, 10, 1   **g** 2, 4, 1   **h** 3, 9, 1

     i 2, 3, 6, 1

Scale 1 cm : 1 m

7 a 30, 60   **b** 36, 72   **c** 12, 24   **d** 30, 60
   e 42, 84   **f** 60, 120
8 a 2   **b** 5   **c** 6   **d** 8
   e 15   **f** 18   **g** 25   **h** 12
   i 15   **j** 16
9 a 12   **b** 40   **c** 36   **d** 75
   e 42   **f** 150
10 2, 3, 5, 7, 11, 13, 17, 19, 23, 29, 31, 37, 41, 43, 47, 53, 59, 61, 67, 71, 73, 79, 83, 89, 97

13.1A

1 a 220 (or $220 + 20k$ for any integer k)
   b 105 or 120 or 135   **c** 72 or 78 or 84 or 90 or 96
2 24, 54, 84
3 1, 2 (any value of n that does not have 3 as a factor)
4 5, 11 (any number of the form $6m - 1$ for integer m)
5 a 9, 25 (any product of two odd numbers)
   b 2   **c i** 5, 11, 17   **ii** 35
6 56 seconds
7 20 cm
8 12 arrangements
9 12 packs burgers, 5 packs buns
10 a Luca wins   **b** 72
11 a 80, 2, 40; 216, 6, 36; 54, 3, 18; 300, 5, 60; 375, 5, 75
   b HCF × LCM = product
   c LCM = product ÷ HCF
12 a $1 + 2 + 14 + 4 + 7 = 28$
   b $1 + 2 + 60 + 4 + 30 + 8 + 15 + 24 + 5 + 40 + 3 + 6 + 20 + 10 + 12 = 240$
   c $1 + 2 + 248 + 4 + 124 + 8 + 62 + 16 + 31 = 496$

13.2S

1 a 1, 18; 2, 9; 6, 3   **b** 1, 14; 2, 7
   c 1, 30; 2, 15; 3, 10; 5, 6   **d** 1, 48; 2, 24; 4, 12; 8, 6; 16, 3
2 a 2   **b** 6   **c** 2   **d** 7
   e 6   **f** 8   **g** 12   **h** 9
3 a 12   **b** 24   **c** 36   **d** 75
   e 84   **f** 78   **g** 200   **h** 75
4 a 12   **b** 8   **c** 4   **d** 6
5 a $77 = 7 \times 11$ **b** $51 = 3 \times 17$ **c** $65 = 5 \times 13$
   d $91 = 7 \times 13$ **e** $119 = 7 \times 17$ **f** $221 = 13 \times 17$
6 9, 3, 3; $18 = 2 \times 3 \times 3$
7 a 8   **b** 75   **c** 40   **d** 63
   e 180   **f** 441
8 a $18 = 2 \times 3 \times 3$ **b** $24 = 2 \times 2 \times 2 \times 3$
   c $40 = 2 \times 2 \times 2 \times 5$ **d** $39 = 3 \times 13$
   e $48 = 2 \times 2 \times 2 \times 2 \times 3$ **f** $82 = 2 \times 41$
   g $100 = 2 \times 2 \times 5 \times 5$ **h** $144 = 2 \times 2 \times 2 \times 2 \times 3 \times 3$
   i $180 = 2 \times 2 \times 3 \times 3 \times 5$ **j** $315 = 3 \times 3 \times 5 \times 7$
   k $444 = 2 \times 2 \times 3 \times 37$ **l** $1350 = 2 \times 3 \times 3 \times 3 \times 5 \times 5$
9 a $36 = 2^2 \times 3^2$ **b** $120 = 2^3 \times 3 \times 5$
   c $34 = 2 \times 17$ **d** $25 = 5^2$
   e $48 = 2^4 \times 3$ **f** $90 = 2 \times 3^2 \times 5$
   g $27 = 3^3$ **h** $60 = 2^2 \times 3 \times 5$
   i $72 = 2^3 \times 3^2$
10 a $2^2 \times 263$ **b** $2^9 \times 5$
   c $2 \times 3^2 \times 5 \times 7$ **d** $3 \times 5^2 \times 11$
   e $5 \times 11 \times 13$ **f** $7 \times 11 \times 13$
   g 3×73 **h** 17^2
   i $2^3 \times 5 \times 71$ **j** $5 \times 7^2 \times 11$
   k $7 \times 13 \times 19$ **l** $2 \times 3^2 \times 11 \times 17$
   m $2^2 \times 11 \times 13 \times 17$ **n** $2 \times 5 \times 7 \times 13^2$
   o $2^2 \times 23 \times 31$ **p** $3^3 \times 13 \times 29$
11 a 5   **b** 16   **c** 3   **d** 5
   e 14   **f** 15
12 a 48   **b** 800   **c** 66   **d** 416
   e 280   **f** 5040

13 a 1260, 60   **b** 8085, 7   **c** 1680, 48   **d** 9216, 2
   e 314706, 2   **f** 82944, 16

13.2A

1 a i Jack has forgotten to include the factor 7
     ii $126 = 2 \times 3^2 \times 7$
   b i $105 \div 3 = 35$ (Jack thought it was 21), then divide by 5: $35 \div 5 = 7$
     ii $210 = 2 \times 3 \times 5 \times 7$
2 a 2×3^2, $2 \times 3 \times 5^2$ **b** $2^3 \times 5$, $2 \times 3 \times 5^2$
   c $2 \times 3 \times 5^2$ **d** $2^3 \times 5$
3 a $30 = 2 \times 3 \times 5$, $105 = 3 \times 5 \times 7$, $45 = 3 \times 3 \times 5$ (many more solutions)
   b $210 = 2 \times 3 \times 5 \times 7$, $36 = 2 \times 2 \times 3 \times 3$, $81 = 3 \times 3 \times 3 \times 3$, $910 = 2 \times 5 \times 7 \times 13$, $46189 = 11 \times 13 \times 17 \times 19$ (many more solutions)
   c $108 = 2^2 \times 3^3$, $162 = 2 \times 3^4$, $243 = 3^5$, $200 = 2^3 \times 5^2$ (more solutions)
   d $64 = 2^6$ or $96 = 2^5 \times 3$
4 18 cm
5 a $(2 \times 3) \times 5 \times 7$; $(2 \times 5) \times 3 \times 7$; $(2 \times 7) \times 5 \times 3$; $(3 \times 5) \times 2 \times 7$; $(3 \times 7) \times 2 \times 5$; $(5 \times 7) \times 2 \times 3$
   b $6 \times 5 \times 7$; $10 \times 3 \times 7$; $14 \times 3 \times 7$; $15 \times 2 \times 7$; $21 \times 5 \times 2$; $35 \times 3 \times 2$
6 a $1815 = 3 \times 5 \times 11^2$
   b 15 cm × 11 cm × 11 cm; 55 cm × 3 cm × 11 cm; 33 cm × 5 cm × 11 cm; 121 cm × 3 cm × 5 cm
7 120 seconds
8 a $2 \times 6 = 12$; $2 \times 9 = 18$; $LCM(6, 9) = 2 \times 3 \times 3 = 18$; $LCM(12, 18) = 2 \times 2 \times 3 \times 3 = 32 = 2 \times 18$
   b 20 and 25 (many solutions)
9 45
10 12 factors

13.3S

1 a 16   **b** 64   **c** 400   **d** 125
   e 343   **f** 1000
2 a $16 = 8 + 8$ **b** $8 = 4 + 4$
3 a 36   **b** 121   **c** 196   **d** 529
   e 961   **f** 2209   **g** 64   **h** 216
   i 512   **j** 2197   **k** 5832   **l** 9261
4 a 13   **b** 16   **c** 28   **d** 80
   e 36   **f** 233   **g** 71   **h** 47
   i 509   **j** 1071
5 a 6.25   **b** 2401   **c** 32.77   **d** 23.04
   e 53.29   **f** 117.65   **g** 1.44   **h** 0.25
   i 98.01   **j** 970.30   **k** 25 cm²   **l** 64 m³
6 a ± 5 **b** ± 3 **c** ± 4 **d** ± 1
   e ± 2 **f** ± 8
7 a ± 6.32 **b** ± 7.81 **c** ± 13.42 **d** ± 15.78
   e ± 26 **f** ± 31.62
8 a 2   **b** 5   **c** -1 **d** 4
   e 10   **f** -3
9 a 3.42   **b** 8   **c** 15   **d** 4.64
   e 29   **f** -31
10 a 2   **b** 2   **c** 3   **d** 3
   e 3   **f** 3
11 a 81   **b** 1   **c** 128
   d 729   **e** 1 000 000   **f** 256
12 a 1728   **b** 46 656   **c** 9261
   d 1 048 576   **e** 2197
13 a 2   **b** 3   **c** 2   **d** 3
   e 5   **f** 6
14 a 3723.88   **b** 387.42   **c** 1667.99
   d 217678.23   **e** 2460.38   **f** -1338278.22
15 a 25   **b** 40   **c** 259947
   d 8000   **e** $\pm\frac{5}{256}$ **f** $\pm\frac{7}{6}$

1 2, 5, 8, 10, 13, 17, 18, 20, 25, 26, 29, 32, 34, 37, 40, 41, 45

2 **a,b** 1F, 2C, 3E, 4A, 5B, 6G, 7H, 8D

3 144 cm

4 17 cubes

5 **a** 12 and 13 **b** 17 and 18 **c** 8 and 9

6 2.645751 is a rounded number and not exactly equal to $\sqrt{7}$

7 56 and 57

8 15

9 **a** 27

b Yes, using clues 1 and 2 or 2 and 3, but we need to know how old a teacher might reasonably be (e.g. not 1, 8 or 729)

10 $a = 2$, $b = 3$ (many solutions)

11 $p = 3$, $q = 6$ (many solutions)

12 **a** $0.5^2 = 0.25$ and $0.25 < 0.5$; $0.7^2 = 0.49$ and $0.49 < 0.7$ (Any number <1 and >0)

b $\sqrt{0.64} = 0.8$ and $0.8 > 0.64$; $\sqrt{0.04} = 0.2$ and $0.2 > 0.04$ (Any number <1 and >0)

c When you square 0 or 1, you get the same number you start with. When you square a number between 0 and 1 the answer is smaller than the number; when you square a number greater than 1 or less than 0, the answer is larger than the number. The square root of 0 or 1 is the number you started with. The square root of a number between 0 and 1 is bigger than the number; the square root of a number greater than 1 is smaller than the number.

13 **a** $2^2 \times 19^2$ **b** $2 \times 19 = 38$

Review 13

1 **a** 1, 2, 3, 4, 6, 12 **b** 1, 19

c 1, 2, 3, 4, 6, 8, 12, 16, 24, 48

2 **a** 5, 10, 15, 20, 25 **b** 13, 26, 39, 52, 65

c 18, 36, 54, 72, 90

3 **a** No, 5 is also a factor. **b** Yes, exactly two factors

c No, 3 and 19 are factors

5 $72 = 2 \times 2 \times 2 \times 3 \times 3$

6 **a** $2 \times 3 \times 3$ **b** $2 \times 2 \times 7$

7 **a**

Prime factors of 60	Prime factors of 70
2 3	2 7
5	

b **i** 10 **ii** 420

8 **a** 4 **b** 27 **c** 1 **d** 1

9 **a** 12 **b** 24 **c** 99 **d** 143

10 **a** **i** 2×3^3 **ii** $2^3 \times 5^2$

b LCM $= 2^3 \times 3^3 \times 5^2$, HCF $= 2$

11 **a** 4 **b** 9 **c** 2 **d** 10

12 **a** 9 **b** 49 **c** 125 **d** 32

13 **a** 3 **b** 3

Assessment 13

1 No, 1×42, 2×21, 3×14, 6×7.

2 **a** 3 is not a factor.

b **i** 100, 102, 104, 106, 108 **ii** 102, 105, 108

iii 100, 104, 108 **iv** 100, 105

v 102, 108 **vi** 105

vii 104 **viii** 108 **ix** 100

3 **a** The last two digits are not divisible by 4. **b** 5, 10, 11

4 Students' workings, HCF $= 9$.

5 $36 = 2 \times 2 \times 3 \times 3$, $60 = 2 \times 2 \times 3 \times 5$, LCM $= 2 \times 2 \times 3 \times 3 \times 5 = 180$.

6 1st or 3rd

7 9 pm

8 $28 = 1 \times 2 \times 4 \times 7$, $1 + 2 + 4 + 7 + 14 = 28$.

9 **a** 4 **b** 6 **c** 84

10 Isa, $2^3 \times 3^2 \times 5^2 \times 11$.

11 **a** Yes, $1024 > 625$. **b** Yes, $1024 > 100$.

c No, $16 = 16$.

---

12 **a** No, $33 = 3 \times 11$, factors $= 1, 3, 11, 33$.

b No, all the numbers in column **A** are odd.

c No, 2 is the only even prime number and column **B** is even.

d 43 and 47 **e** No, all the numbers in **D** are multiples of 4.

13 **a** $25 = 2 + 23$ **b** $16 = 3 + 13 = 5 + 11$

c $36 = 5 + 31 = 7 + 29 = 13 + 23 = 17 + 19$

14 **a** Glen, 3^5 **b** Giorgia, 14^6

15 **a** George, no negative number can have a square root. Giorgia (partly correct), all positive numbers have 2 square roots.

b No, a positive number has a positive cube root, a negative number has a negative cube root.

16 23

17 **a** $41 - 0 + 0^2 = 41$, $41 - 3 + 3^2 = 47$, $41 - 6 + 6^2 = 71$

b 41, $41 - 41 + 41^2 = 41^2$

Chapter 14

Check in 14

1 **a** 8 **b** 5 **c** 8 **d** $3\frac{1}{2}$

2 **a** -1 **b** -5 **c** 1 **d** -7

3 Grid with points correctly plotted.

14.1S

1 **a** A(2, 4), B(0, 3), C(−3, 3), D(2, −1), E(−3, −2), F(−2, −3), G(−2, 1), H(3, 2)

b I(2, 3), J(2, 0), K(2, −3), L(0, −3), M(−4, −3), N(−4, 0), O(−4, 3), P(−1, 3), Q(0, 2)

2 **a i** and **b i**   **a ii** and **b ii**

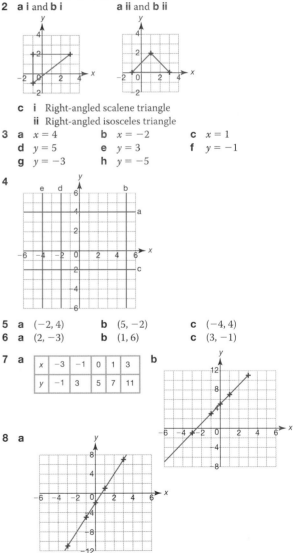

c **i** Right-angled scalene triangle

ii Right-angled isosceles triangle

3 **a** $x = 4$ **b** $x = -2$ **c** $x = 1$

d $y = 5$ **e** $y = 3$ **f** $y = -1$

g $y = -3$ **h** $y = -5$

4

5 **a** $(-2, 4)$ **b** $(5, -2)$ **c** $(-4, 4)$

6 **a** $(2, -3)$ **b** $(1, 6)$ **c** $(3, -1)$

7 **a**

x	−3	−1	0	1	3
y	−1	3	5	7	11

b

8 **a**

b

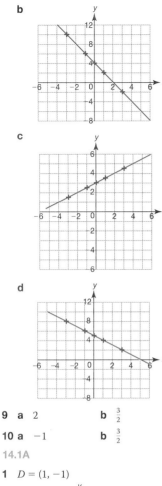

c

d

9 a 2   **b** $\frac{3}{2}$

10 a -1 **b** $\frac{3}{2}$

14.1A

1 $D = (1, -1)$

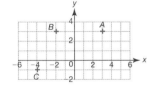

2 a $(1, 2)$

  b Below $y = 5$, above $y = 2$, above $y = 6 - x$ and above
      $y = 2x - 3$ (many solutions)

3 a No. If $x = 3$ then $y = (2 \times 3) + 1 = 7$, hence $(3, 8)$ does not
      lie on the line.

  b $(3, 7)$ (many solutions)

4 a

Pounds	0	1	10
NZ dollars	0	2.4	24

  b $(0, 0)$; $(1, 2.4)$; $(10, 24)$ (many solutions)

  c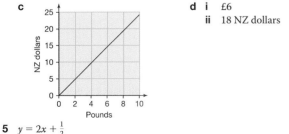

  d i £6
    ii 18 NZ dollars

5 $y = 2x + \frac{1}{2}$

  $2y - 4x = 1$ is the same as $y = \frac{1}{2} + 2x$

  $4x + 2y = 1$ is the same as $y = \frac{1}{2} - 2x$ (odd one out)

  $8x - 4y = -2$ is the same as $y = \frac{1}{2} + 2x$

6 a $y = 5 - x$ **b** $y = 3 + x$

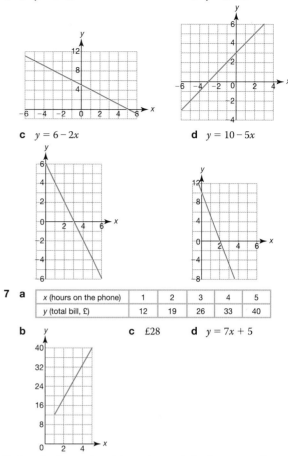

  c $y = 6 - 2x$ **d** $y = 10 - 5x$

7 a

x (hours on the phone)	1	2	3	4	5
y (total bill, £)	12	19	26	33	40

  b   **c** £28   **d** $y = 7x + 5$

8 Let x be the number of chores done and y be the amount of
    pocket money. Then $y = 0.2x + 3$ gives the most money if I do
    more than 40 chores, and $y = 0.15x + 5$ gives me the most money
    if I do less than 40 chores. If I do 40 chores then I get the same
    amount (£11) under both options (where the line graphs cross).

9 a

x (number of people)	1	2	3	4	5	6	7	8	9	10	11	12	13	14	15
y (cost of the campsite)	18	21	24	27	30	33	36	39	42	45	48	51	54	57	60

  b 7 people

  c £50 − £15 = £35, which is not a multiple of £3

  d $y = 3x + 15$ where x is the number of people and y is the
      cost of the campsite

  e £96

14.2S

1 a $y = 5$ **b** $x = -8$

2 a $3; -1$ **b** $2; 5$ **c** $4; -3$

  d $\frac{1}{2}; 2$ **e** $5; 1$ **f** $-3; 7$

3 a ii   **b** iii   **c** i

  d iv   **e** v

4 a i $1; 2$ **ii** $y = x + 2$

  b i $2; 3$ **ii** $y = 2x + 3$

  c i $3; 0$ **ii** $y = 3x$

  d i $-2; 3$ **ii** $y = -2x + 3$

5 $y = 3x + 5$; $y = 5x - 2$; $y = -2x + 7$; $y = \frac{1}{2}x + 9$; $y = -\frac{1}{4}x - 3$;
    $y = 4$; $y = x$

6 a $y = 5 - x$ **b** $y = 3 + x$ **c** $y = -2 - x$

  d $y = x - 3$ **e** $y = 6 - 2x$ **f** $y = 9 - 5x$

  g $y = -2 - 3x$ **h** $y = 5 + 2x$ **i** $y = 2 - \frac{1}{2}x$

j $y = 4 + \frac{1}{2}x$ **k** $y = 4 - \frac{1}{2}x$ **l** $y = 4 - 3x$

7 a b, d   **b** a, c   **c** g, l   **d** i, k

8 $y = 3x + 7$ (any solution of the form $y = 3x + c$)

9 $y = -4x + 3$ (any solution of the form $y = -4x + c$)

10 a $y = 7x + 5$ **b** $y = \frac{1}{2}x + 3$ **c** $y = 4x - 4$
   d $y = 3x - 5$ **e** $y = -2x + 5$ **f** $y = \frac{1}{4}x - 2$

14.2A

1 a $y = 4x + 1$ **b** $y = x + 2$ **c** $y = 8x - 6$

2 a 3   **b** $y = 3x + 5$ **c** $-2; y = -2x + 16$

3 a A, B, E   **b** C, F   **c** D
   d A   **e** C, F

4 a $y = -4x - 2$ **b** $y = \frac{3}{2}x - 2$

5 a True   **b** False   **c** False   **d** False

6 a $y = 6x + 50$. The gradient tells you that for every year older you get, you grow 6cm, and the intercept tells you that at birth you are 50cm tall.

   b $y = \frac{1}{2}x + 10$. The gradient tells you that for every year older you get you have to have another half a driving lesson, and it is not sensible to interpret the intercept because it says that a newborn baby would need 10 lessons to pass the driving test!

7 The gradient tells you that for every mark more scored on Paper 1 another mark was scored on Paper 2. It does not make sense to interpret the intercept because it says that if you scored nothing in Paper 1 you would score -10 in Paper 2, and this is impossible.

14.3S

1 a i 20 km   **ii** 40 km
   b i 45 mins   **ii** 2 hrs 15 mins

2 a 10:45 am   **b** She stops for 30 mins.

3 a 40 km   **b** 15 mins   **c** 45 mins
   d 4 pm   **e** 80 km/h
   f 14:00 to 14:30. This is the steepest section of the graph.
   g i 35 km   **ii** 55 km   **iii** 140 km

4 a 30 km/h   **b** 20 km/h   **c** 12.5 km/h

5 a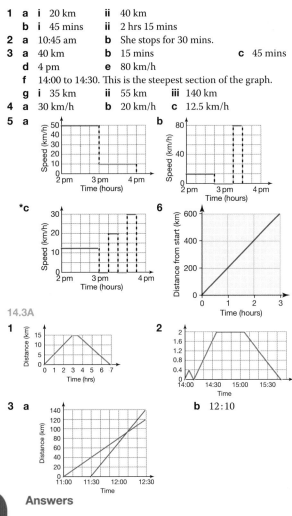

b

*c

6

14.3A

1   **2**

3 a   **b** 12:10

Answers

4 a Sian   **b** Asif overtook Sian.   **c** Asif

5 a The 2nd section goes backwards in time.
   b Distance is always positive.
   c The distance changes instantaneously in the 2nd section.
   d Time goes backwards in the 2nd and 4th section.

6 a C   **b** A   **c** B

7 a i 0.03 km/s²   **ii** 0.05 km/s²
   b 0.042 km/s² (2 sf)

Review 14

1 $A\,(2, 5),\ B\,(-3, 5),\ C\,(0, -4),\ D\,(3, 0),\ E\,(-2, 3)$

2 a $y = -6, -3, 0, 3, 6$ **b** $y = 3, 5, 7, 9, 11$

3 a 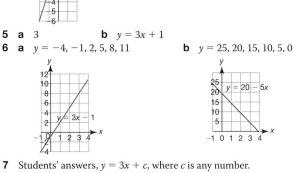   **b**   **4**

5 a 3   **b** $y = 3x + 1$

6 a $y = -4, -1, 2, 5, 8, 11$ **b** $y = 25, 20, 15, 10, 5, 0$

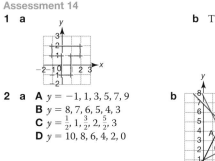

7 Students' answers, $y = 3x + c$, where c is any number.

8 $y = 7x - 4$

9 $y = 4x + 2$

10 a 4 km   **b** Speed   **c** 20 km/h

Assessment 14

1 a   **b** T

2 a **A** $y = -1, 1, 3, 5, 7, 9$ **b**
   B $y = 8, 7, 6, 5, 4, 3$
   C $y = \frac{1}{2}, 1, \frac{3}{2}, 2, \frac{5}{2}, 3$
   D $y = 10, 8, 6, 4, 2, 0$

   c i (2, 5)   **ii** (0, 1)   **iii** (4, 3)   **iv** (1, 6)

3 a–d $y = 6 - x$ **e** 4.5

4 No, point (0, 2) does not line on the line, $2 \times 0 + 3 \neq 2$.

5 a No   **b** $y = x - 1$

6 a 2, 7   **b** 4, 9   **c** 6, -11 **d** $-4, 12$
   e $\frac{4}{7}, -2$ **f** $\frac{-15}{14}, \frac{35}{14}$

7 a $y = x + 1$ **b** $y = -x + 1$
   c $y = 2x + 6$ **d** $y = -4x + 13$

8 a i A to B   **ii** E to F   **iii** C to D
   iv B to C   **v** D to E

b D to E, the gradient is steepest at this point.

9

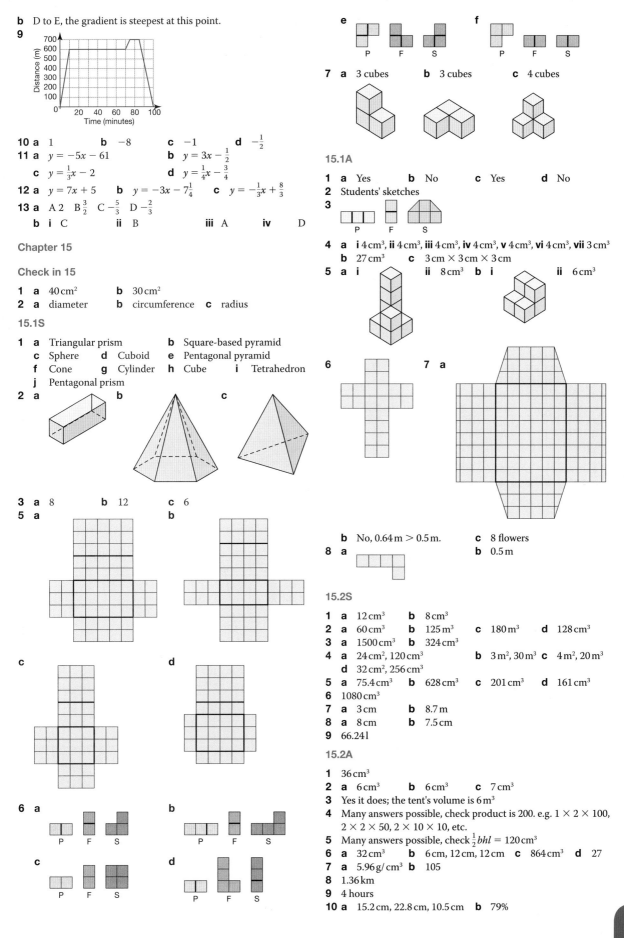

10 a 1   **b** -8 **c** -1 **d** $-\frac{1}{2}$

11 a $y = -5x - 61$ **b** $y = 3x - \frac{1}{2}$

     c $y = \frac{1}{3}x - 2$ **d** $y = \frac{1}{4}x - \frac{3}{4}$

12 a $y = 7x + 5$ **b** $y = -3x - 7\frac{1}{4}$ **c** $y = -\frac{1}{3}x + \frac{8}{3}$

13 a A 2   B $\frac{3}{2}$ C $-\frac{5}{3}$ D $-\frac{2}{3}$

     b i C   **ii** B   **iii** A   **iv** D

Chapter 15

Check in 15

1 a $40\,\text{cm}^2$ **b** $30\,\text{cm}^2$

2 a diameter   **b** circumference   **c** radius

15.1S

1 a Triangular prism   **b** Square-based pyramid

     c Sphere   **d** Cuboid   **e** Pentagonal pyramid

     f Cone   **g** Cylinder   **h** Cube   **i** Tetrahedron

     j Pentagonal prism

2 a   **b**   **c**

3 a 8   **b** 12   **c** 6

5 a   **b**

c   **d**

6 a   **b**

P   F   S   P   F   S

c   **d**

P   F   S   P   F   S

e   **f**

P   F   S   P   F   S

7 a 3 cubes   **b** 3 cubes   **c** 4 cubes

15.1A

1 a Yes   **b** No   **c** Yes   **d** No

2 Students' sketches

3

P   F   S

4 a **i** $4\,\text{cm}^3$, **ii** $4\,\text{cm}^3$, **iii** $4\,\text{cm}^3$, **iv** $4\,\text{cm}^3$, **v** $4\,\text{cm}^3$, **vi** $4\,\text{cm}^3$, **vii** $3\,\text{cm}^3$

     b $27\,\text{cm}^3$ **c** $3\,\text{cm} \times 3\,\text{cm} \times 3\,\text{cm}$

5 a i   **ii** $8\,\text{cm}^3$ **b i**   **ii** $6\,\text{cm}^3$

6   **7 a**

     b No, $0.64\,\text{m} > 0.5\,\text{m}$.   **c** 8 flowers

8 a   **b** $0.5\,\text{m}$

15.2S

1 a $12\,\text{cm}^3$ **b** $8\,\text{cm}^3$

2 a $60\,\text{cm}^3$ **b** $125\,\text{m}^3$ **c** $180\,\text{m}^3$ **d** $128\,\text{cm}^3$

3 a $1500\,\text{cm}^3$ **b** $324\,\text{cm}^3$

4 a $24\,\text{cm}^2$, $120\,\text{cm}^3$ **b** $3\,\text{m}^2$, $30\,\text{m}^3$ **c** $4\,\text{m}^2$, $20\,\text{m}^3$

     d $32\,\text{cm}^2$, $256\,\text{cm}^3$

5 a $75.4\,\text{cm}^3$ **b** $628\,\text{cm}^3$ **c** $201\,\text{cm}^3$ **d** $161\,\text{cm}^3$

6 $1080\,\text{cm}^3$

7 a 3 cm   **b** 8.7 m

8 a 8 cm   **b** 7.5 cm

9 66.24 l

15.2A

1 $36\,\text{cm}^3$

2 a $6\,\text{cm}^3$ **b** $6\,\text{cm}^3$ **c** $7\,\text{cm}^3$

3 Yes it does; the tent's volume is $6\,\text{m}^3$

4 Many answers possible, check product is 200. e.g. $1 \times 2 \times 100$, $2 \times 2 \times 50$, $2 \times 10 \times 10$, etc.

5 Many answers possible, check $\frac{1}{2}bhl = 120\,\text{cm}^3$

6 a $32\,\text{cm}^3$ **b** 6 cm, 12 cm, 12 cm   **c** $864\,\text{cm}^3$ **d** 27

7 a $5.96\,\text{g/cm}^3$ **b** 105

8 1.36 km

9 4 hours

10 a 15.2 cm, 22.8 cm, 10.5 cm   **b** 79%

15.3S1

1 **a** 4   **b** 10   **c** 20   **d** 48
2 **a** 28 cm²   **b** 22.26 cm²
3 **a** 7.5 cm²   **b** 11.76 cm²
4 15 cm²
5 12 cm²
6 **a** 154 cm²   **b** 78.5 cm²
7 **a** 48 cm³   **b** 75 m³   **c** 56 cm³   **d** 48 mm³
   e 1.28 cm³   **f** 1.68 m³
8 **a** 42 cm³   **b** 1.77 cm³   **c** 69.2 cm³   **d** 64 cm³
   e 263.9 cm³   **f** 87.1 cm³
9 **a** 12.6   **b** 78.5   **c** 37.7   **d** 205.3
   e 419   **f** 603.2   **g** 28.3   **h** 92.5
10 **a** 1437 cm³   **b** 57.9 cm³   **c** 9203 cm³   **d** 5575 cm³
   e 65450 cm³   **f** 124788 cm³ **g** 21688 cm³ **h** 2226 cm³
   i 9855 mm³   **j** 8.4 m³

15.3S2

1 **a** 96 cm²   **b** 12 m²
2 **a** 226.2 cm²   **b** 235.6 cm²   **c** 804.2 cm²   **d** 1846.0 cm²
   e 253.3 cm²   **f** 192.8 mm²
3 **a** 326.7 cm²   **b** 392.7 cm²   **c** 2412.7 cm² **d** 2648.3 cm²
   e 1467.2 cm² **f** 1254.6 mm²
4 **a** 101 cm²   **b** 408 cm²   **c** 201 cm²   **d** 165 cm²
5 **a** 128.8 cm²   **b** 56.5 cm²   **c** 42.4 cm²   **d** 23.6 cm²
   e 221.0 cm²   **f** 784.6 cm²
6 **a** 207.3 cm²   **b** 84.8 cm²   **c** 106.0 cm²   **d** 43.2 cm²
   e 362.0 cm²   **f** 1357.2 cm²
7 **a** 264 cm²   **b** 126 cm²
8 **a** 615.8 cm²   **b** 72.4 cm²   **c** 2123.7 cm² **d** 1520.5 cm²
   e 7854.0 cm² **f** 12076.3 cm²   **g** 3761.0 cm²
   h 824.5 cm² **i** 2222.9 mm²   **j** 20.0 m²

15.3A

1 **a** 189 cm²   **b** 580 cm²
2 **a** Y   **b** 134 cm³
3 69.2 cm²
4 **a** 443.4 mm² **b** 249.4 cm²
5 66.67%
6 **a** Cube   **b** Cube
7 **a** 96 cm², 64 cm³   **b** 384 cm², 512 cm³
   c **i** 4   **ii** 8
8 **a** Check students' diagrams. Surface area = 84 cm²
   b 9.9 cm × 9.9 cm   **c** 14.3%

Review 15

1 **a** 6, 12, 8   **b** 5, 8, 5
2 **a** cone   **b** (triangular) prism
3 **a** [rectangle]   **b** [circle]   **4** [3D cube arrangement image]

5 **a** 12 m³   **b** 34 m²   **c** 30 kg
6 **a** 7600 cm³   **b** 1380 cm²   **c** 2140 cm²
7 37.7 cm²
8 2140 cm³

Assessment 15

1 **b** Tetrahedron **c** Sphere   **e** Hexagonal-based prism
2 **a** Triangle-based pyramid   **b**
[net diagram: central 4 cm × 4 cm square with triangles of slant 5 cm on each side]

3 **a** A   **b** B

Answers

---

4 **a** 0.372   **b** 2.75   **c** 5
   d 8   **e** 9   **f** 6.25
5 **a** 512   **b** 64   **c** 8
   d **i** 18   **ii** 26
6 1 × 1 × 60, 1 × 2 × 30, 1 × 3 × 20, 1 × 4 × 15, 1 × 5 × 12,
   1 × 6 × 10, 2 × 2 × 15, 2 × 3 × 10, 2 × 5 × 6, 3 × 4 × 5
7 **a** Correct   **b** Incorrect, 1 000 000 cubic cm
   c Incorrect, 1 000 000 000 cubic mm
8 **a** 432
   b No, ratio of volumes = 432, ratio of surface areas = 51.2 (3 sf).
9 **a** 22 200 cm³ (3 sf)   **b** 0.378 m²
10 **a** 50.4 in²   **b** 13.5 in³
11 14.7 cm³
12 **a** 47666.7 cm³ (1 dp)   **b** 364
13 **a** 4.71 cm³ (3 sf)   **b** 12.3 g (3 sf)
14 **a** Yes, 13.5 × 0.3 × 0.3 = 6 × 0.45 × 0.45 = 1.215 cm³.
   b Chunky Dunkies = 11.205 cm² < 16.38 cm²
15 **a** 94.2 cm³ (3 sf)   **b** 104 cm² (3 sf)

Lifeskills 3

1 40
2 **a**

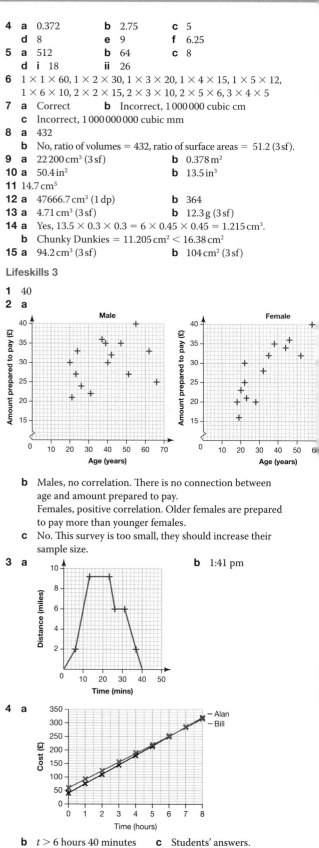

   b Males, no correlation. There is no connection between
     age and amount prepared to pay.
     Females, positive correlation. Older females are prepared
     to pay more than younger females.
   c No. This survey is too small, they should increase their
     sample size.
3 **a** [distance-time graph]   **b** 1:41 pm

4 **a** [cost-time line graph for Alan and Bill]
   b $t > 6$ hours 40 minutes   **c** Students' answers.
5 **a** 17240.7 cm³ (1 dp)   **b** 1 449 000 cm³ = 1.449 m³
   c 35.7% (1dp)   **d** 396.5 cm² (1 dp)
   e 43.0 cm² (1 dp)   **f** 10.6 cm (1 dp)
   g 18 cans   **h** Cartons, you can fit more in
     the cupboard.

6 a £866

b

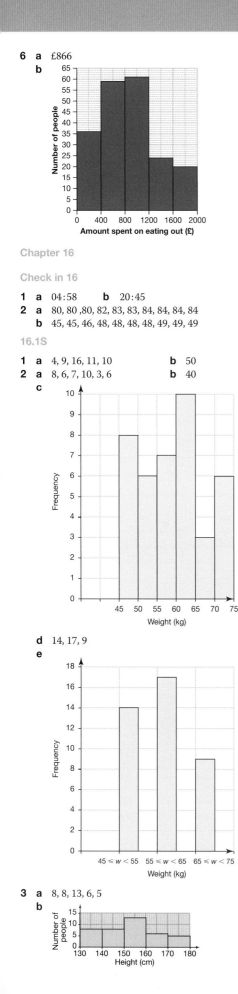

Chapter 16

Check in 16

1 a 04:58   **b** 20:45
2 a 80, 80 ,80, 82, 83, 83, 84, 84, 84, 84
   b 45, 45, 46, 48, 48, 48, 48, 49, 49, 49

16.1S

1 a 4, 9, 16, 11, 10   **b** 50
2 a 8, 6, 7, 10, 3, 6   **b** 40
   c

d 14, 17, 9
e

3 a 8, 8, 13, 6, 5
   b

4

5

1 a i 2   **ii** 1
   b 85–90 m   **c** 10 athletes
2 a i, ii Not possible
        iii £161 is possible. (Smallest possible range is £200.01 − £39.99 = £160.02).
   b 41   **c** $\frac{14}{41}$
3 $x = 5$, frequencies plotted at 5, 13 and 10.
4 7, 11, 9, 5, 4

16.2S

1 a 1, 6, 8, 3, 4, 3   **b** 70 to 74   **c** 70 to 74
2 a 3   **b** 13, 18, 23, 28
   c 13, 108, 46, 28. Mean = 19.5 mph
3 a i $10 < t \le 15$ **ii** $15 < t \le 20$ **iii** 16.1
   b i $10 < t \le 20$ **ii** $20 < t \le 30$ **iii** 23.5
   c i $5 < t \le 10$ **ii** $10 < t \le 15$ **iii** 14.7
   d i $15 < t \le 25$ **ii** $25 < t \le 35$ **iii** 30
4 a $165 \le h < 170$ **b** 164.3 cm   **c** $165 \le h < 170$

16.2A

1 a December = 50, January = 59.7
   b December: modal $40 < m \le 60$, median $40 < m \le 60$.
      January: modal $40 < m \le 60$, median $40 < m \le 60$
   c Less variation in miles travelled in January, less short journeys.
      The most common journey length does not change.
      The mean is greater for January than December.
2 a Teachers = 23.1 minutes, office workers = 38.25 minutes.
   b Teachers: $20 < t \le 30$, office workers: $30 < t \le 40$
   c On average, office workers take longer travelling home.
3 Maximum possible range is 40 g but mean is 53.125 g. Unlikely to be Granny Smith apples.
4 Modal and median class is $24.5 \le w < 25.5$. Estimated mean = 25.28 g which is close to 25 g. However as data is grouped the weight of the packets could be less than 25 g.
5 Machine A estimated mean = 0.3007 mm, machine B estimated mean = 0.2907 mm. The estimated mean indicates that machine A produces paper closest the required thickness. The modal and median classes both back up this conclusion.

16.3S

1 a Negative correlation   **b** No correlation
   c Positive correlation
2 a A: Poor exam mark, lots of revision; B: Very good exam mark, lots of revision;
      C: Very good exam mark, little revision; D: Poor exam mark, little revision;
      E: Average exam mark, average amount of revision
   b A: Not much pocket money, equal eldest; B: Lots of pocket money, equal eldest;
      C: Lots of pocket money, equal youngest; D: Not much pocket money, equal youngest;
      E: Average pocket money, middle age
   c A: Low fitness level, lots of hours in gym,
      B: Good fitness level, lots of hours in gym,
      C: Good fitness level, few hours in gym,

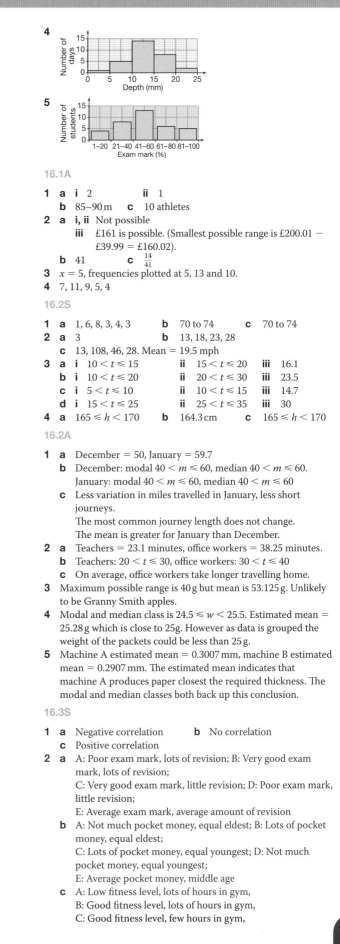

D: Low fitness level, few hours in gym,
E: Medium fitness level, medium hours in gym

3 a No correlation   **b** Positive correlation
  c Negative correlation

4 a
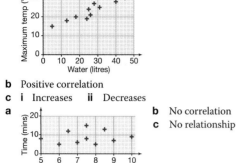

  b Positive correlation
  c i Increases   **ii** Decreases

5 a
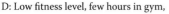
                           b No correlation
                           c No relationship

16.3A

1 a Positive correlation.
  b Paper 1 = 24, paper 2 = 80. Student performed poorly in paper 1 and well in paper 2.
  c 35
  d The line predicts that the student scores 110% which is impossible.

2 a 16   **b** 24
  c One student scored 20 marks on paper 1, but only 5 marks on paper 2

3 a, c Check students' diagrams.
  b Positive correlation.   **d** 20 years
  e, f 1 year old gives a negative diameter. Predictions outside the range of data values can be unreliable.

4 a, c Check students' diagrams.
  b Positive correlation, students who do well in maths tend to do well in statistics.
  d 68%
  e Predictions outside the range of data values can be unreliable.

16.4S

1
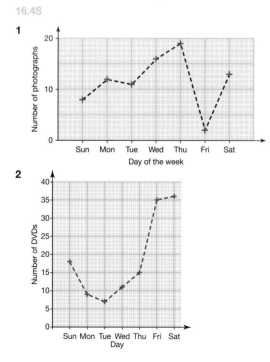

2

3
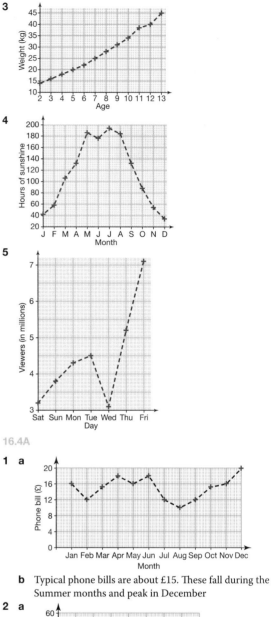

4

5

16.4A

1 a
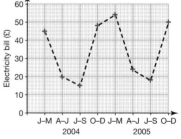

  b Typical phone bills are about £15. These fall during the Summer months and peak in December

2 a

  b Electricity bills are highest in the Winter months and lowest in the Summer months. This annual pattern repeats itself; there is a slight trend for bills to rise from year-to-year.

3 a
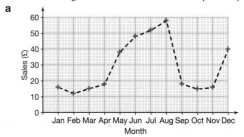

b Icecream sales grow steadily during Spring and Summer but drop sharply in the Autumn. Sales are low during Autumn and Winter except for a peak in December.

4 **a**

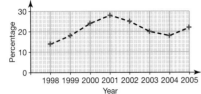

b The percentage of students using the library grows steadily from about 15% in 1998 to 28% in 2001. It has since fallen back to around 20%.

5 The average suggests Sell-a-lot, 116 is a valid range for both data sets.

Review 16

1 Frequency = 4, 5, 6, 1
2 **a** $12 \leqslant l < 14$ **b** $10 \leqslant l < 12$ **c** 11.5
3 **a** Negative
  b No, the correlation may be a coincidence or Y be affecting X.
4 **a, c**

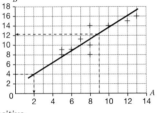

  b Positive
  d 12−13
  e **i** 2−3
    ii Unreliable estimate because there is no data in this range.

5

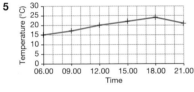

Assessment 16

1 **a**

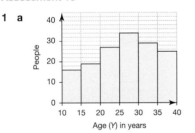

  b People who break the speed limit have been grouped with people who don't. The bottom class only contains people driving between 10 and 20 mph but has been recorded as 0 − 20.

3 **a** 20 − 30   **b** It is the highest bar.
  c 76   **d** 20 ± 30
4 **a**

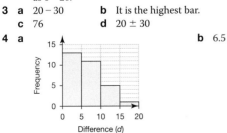

  b 6.5

---

c d is always recorded as positive, so 395 tea bags and 405 tea bags would both be recorded as a difference of 5.

5 **a**

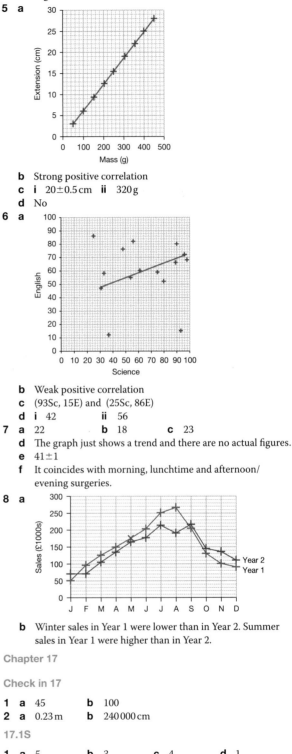

  b Strong positive correlation
  c **i** 20±0.5 cm **ii** 320 g
  d No

6 **a**

  b Weak positive correlation
  c (93Sc, 15E) and (25Sc, 86E)
  d **i** 42   **ii** 56
7 **a** 22   **b** 18   **c** 23
  d The graph just shows a trend and there are no actual figures.
  e 41±1
  f It coincides with morning, lunchtime and afternoon/ evening surgeries.

8 **a**

  b Winter sales in Year 1 were lower than in Year 2. Summer sales in Year 1 were higher than in Year 2.

Chapter 17

Check in 17

1 **a** 45   **b** 100
2 **a** 0.23 m   **b** 240 000 cm

17.1S

1 **a** 5   **b** 3   **c** 4   **d** 1
  e 2   **f** 6
2 **a** 6.32   **b** 7.81   **c** 13.42   **d** 15.78
  e 26   **f** 35.13
3 **a** 3   **b** 8   **c** 15   **d** 4.64
  e 29   **f** 25.27
4 **a** 25   **b** 8   **c** 27   **d** 64
  e 144   **f** 169
5 **a** 81   **b** 1   **c** 128   **d** 729
  e 1 000 000   **f** 256

6 a 1728 b 46656 c 9261 d 1048576
  e 2197 f 7529536
7 a 5 b 6 c 1 d 1
  e 1 f 41 g 1 h 0
8 a 7^3 b 3^3 c 5^3 d 6^4
  e 5^2 f 8^5 g 9^6 h 8^8
9 a 6^5 b 4^9 c 2^{13} d 11^7
  e 1 f 7^{12} g 3^{12} h 9^{10}
10 a 7^2 b 8^4 c 3 d 9^3
  e 4^6 f 1 g 12^2 h 1
11 a 2^6 b 4^{10} c 7^4 d 5^{15}
  e 3^{16} f 6^4 g 5^{21} h 10^{16}
12 a 3 b 5^{10} c 4^2 d 7^2
  e 8^{11} f 9^{10}
13 a 4^2 b 6^2 c 9^2 d 8
  e 5^3 f 6^5 g 8^2 h 1
14 a 4^4 b $\frac{5^5}{3}$ c 6^7 d 5^{12}
  e 2^{10} f 7^{12} g 3^3 h 9^5

17.1A

1 $3^2 + 3^2 + 3^2 = 3 \times 3^2 = 3^3 = 27$
2 a $6^2 \times 6^3 = 6^5$ b 1296
3 a 6 b 6 c 10 d 5
4 $n^3 = n \times n^2 = 1 \times n^3$
5 a $2^3 \times 2^5 = 2^8$ b Correct c Correct
   d $(7^3)^4 = 7^{12}$ e $4^5 \times 4^3 = 4^8$ or $4^5 \div 4^3 = 4^2$
   f Correct
6 $0^8, 1^7, 8^0, 7^1, 6^2, 2^6, 5^3, 3^5, 4^4$
7 a Yes b No c No d Yes
   e Many answers possible, e.g. 5 and 12, 6 and 8, etc.
8 a 1000 b 10000 c level 8
9 a i 4^3 and 8^2 ii 2^{10} and 2^9 iii 1000 and 10^2
      iv 125 and $5 \times 5 \times 5 \times 5$
      v 4^3 and 4^{-3} vi 4^3 and 4^{-3}
      vii 125 and $5 \times 5 \times 5 \times 5$ viii 125 and 0.2
   b i 0.2 ii $3^2 + 3^2$ iii 2^{10} iv 4^{-3}

17.2S

1 a $\frac{8}{15}$ b $\frac{29}{35}$ c $\frac{5}{56}$ d $\frac{44}{45}$
   e $4\frac{7}{8} = 4.875$ f $\frac{59}{60}$ g $\frac{19}{40} = 0.475$ h $9\frac{17}{36}$
2 a $\frac{1}{2} = 0.5$ b $\frac{5}{3}$ c $\frac{25}{16} = 1.5625$ d $\frac{8}{15}$
   e $\frac{14}{9}$ f $14\frac{3}{10} = 14.3$ g $2\frac{37}{55}$
   h $26\frac{4}{9}$
3 No exact truncated decimal form exists if the denominator has factors other than 2 or 5
4 a $\frac{17}{30}$ b $1\frac{67}{87}$ c $\frac{1}{3}$ d $\frac{207}{1715}$
   e $1\frac{133}{324}$ f $-\frac{8}{21}$
5 a 2 b $\frac{5}{4}$ c $\frac{5}{4}$ d $\frac{5}{4}$
6 a 2π b 3π c $2 + 2\pi$ d 10π
   e 18π f 16π g $\frac{21\pi}{2}$ h 12π
7 a 32π b $7 + 2\pi$ c $28 + 4\sqrt{2}$ d 62π
   e 6π f 4π
8 $64 - 4\pi$ cm
9 a $\frac{3}{4}x + 1$ b $2x + 5\pi$ c $8x + 5\pi$ d $\frac{3}{4}x - 1$
   e $\frac{1}{4}x + 1$ f $5x + 8\pi$
10 a 29 m b $17 + 6\pi$ m c 51 m² d 9π m²
   e $51 + 9\pi$ m²

17.2A

1 $\sqrt{100} - \sqrt{64} = 6^2 \div 18$
2 $12, 10 - \pi, 2 + \pi, 2 + \pi, 8 - 2\pi, 3\pi - 6$
3 Triangle, $3 + 3\pi > 3 + 2\pi$
4 $5.5 \times 0.2^2 \times \pi$ m
5 a i $2(2 + \pi)$ ii $12 + \pi$ iii 4π iv $4 + 3\pi$
   b Circle diameter = Square side length = 2 cm

6 a $\frac{9\pi}{4}$ in² b 3π in c $\frac{51\pi}{4}$ in² d $\frac{15\pi}{32}$ in³
   e $\frac{153\pi}{16}$ in³
7 a 1440π cm³ b 40.8%

17.3S

1 a 10^2 b 10^1 c 10^3 d 10^0
   e 10^4 f 10^6 g 10^5 h 10^8
2 a 10^{-2} b 10^{-1} c 10^{-3} d 10^{-5}
   e 10^{-4} f 10^{-7} g 10^{-6} h 10^0
3 a 1000 b 1000000 c 100000
   d 1000000000 e 10000
   f 10 g 100 h 10000000
4 a 1 b 0.01 c 0.00001 d 0.001
   e 0.0000001 f 0.1 g 0.0001 h 0.000001
5 a 10^5 b 10^9 c 10^8 d 10^3
   e 10^4 f 10^4
6 a 10^{-2} b 10^{-4} c 10^{-8} d 10^{-8}
   e 10^{-8} f 10^6
7 a 2×10^2 b 8×10^2 c 9×10^3 d 6.5×10^2
   e 6.5×10^3 f 9.52×10^2 g 2.358×10^1 h 2.5585×10^2
   i 3×10^{-1} j 4.7×10^{-3} k 7.8×10^{-5} l 4.485×10^{-1}
8 a 500 b 3000 c 100000 d 250
   e 4900 f 3800000 g 750000000000
   h 8100000000000000000
9 a 6×10^2 b 4.5×10^4 c 6.5×10^0 d 5×10^6
   e 2.8×10^{-1} f 4×10^{-2} g 1.35×10^{-3} h 1.2×10^{-7}
10 a 4×10^5 b 9×10^7 c 2.5×10^8 d 2.4×10^{13}
   e 5×10^{-1} f 9.2×10^{-8}
11 a 2×10^2 b 2×10^4 c 5×10^1 d 7.5×10^2
12 a 7.74×10^{-3} b 9.63×10^5
   c 4.38×10^{-5} d 2.55×10^2
   e 3.4×10^5 f 4.47×10^{-2}

17.3A

1 a 10^{-2} km b 2×10^{-3} g c 5×10^{-6} m d 1.1×10^{-2} l
2 a Correct b 3.276×10^{13}
   c 4.3×10^5 d Correct
3 1000
4 $333333.\dot{3} = \frac{1000000}{3}$
5 10^{48}
6 500
7 a 3.3×10^{-9} s b 9.47×10^{15} m
8 a 1.2×10^{-2} kg b 2×10^{-26} kg
9 1.79×10^6
10 4.3×10^4 km
11 a Jupiter b 2.668612×10^{27} kg

Review 17

1 a 5^{10} b 3^4 c 7^{12}
   d 2^0 e 3^3 f 4^4
2 a 128 b 16 c 5 d 64
   e 1 f 27 g 6 h 2
3 a $\frac{3}{5}$ b $\frac{3}{14}$ c $\frac{3}{5}$ d $\frac{13}{24}$
   e $\frac{31}{24}$ f $2\frac{1}{20}$ g $2\frac{4}{9}$ h $\frac{11}{14}$
   i $\frac{2}{7}$ j $\frac{1}{6}$
4 a $1\frac{1}{5}$ b $\frac{5}{8}$
5 a $2 + 3\pi$ b 5π c $\frac{5\pi}{12}$ d $\frac{9\pi}{2}$
6 a i 64π ii $\frac{25}{2}\pi$
   b i 16π ii $5\pi + 10$
7 a 1.37×10^9 b 5.46×10^7
   c 6.97×10^{-2} d 6.25×10^{-5}
8 a 350000 b 821000000
   c 0.0027 d 0.000000207
9 a 6×10^{10} b 7×10^4
   c 4×10^5 d 6×10^{-2}

Answers

Assessment 17

1 a Soraya $0^2 = 0 \times 0 = 0$ **b** Peter $49^{\frac{1}{2}} = \sqrt{49} = 7$
  c Soraya $(-3)^2 = -3 \times -3 = 9$

2 a 9.261   **b** 2687 (3 sf)   **c** 974 (3 sf)
  d 29   **e** 8.77 (3 sf)   **f** 8.68 (3 sf)

3 a No, 15^{12} **b** Yes, 3^{20} **c** Eliza, $(3^4)^0 = 3^0 = 1$
  d $7^{7+2-6} = 7^3$

4 12.167 cm³

5 a 216   **b** 5

6 a 5   **b** 4   **c** 4

7 Students' answers, for example, $\frac{1}{4} + \frac{2}{8} = \frac{1}{2}$, $\frac{1}{3} + \frac{2}{12} = \frac{1}{2}$, etc.

8 a $\frac{1}{5}$ **b** 192

9 a 40   **b** 15

10 a $\frac{14}{15}$ **b** $\frac{1}{15}$ **c** 54

11 a Juliet, £350   **b** £9030

12 $\frac{7}{10}$

13 a English Channel 29 000 mi², Baltic Sea 1 46 000 mi²,
  Bering Sea 876 000 mi², Caribbean Sea 1 060 000 mi²,
  Malay Sea 3 140 000 mi², Indian Ocean 28 400 000 mi²
  b 1×10^9, 1
  c i 4   **ii** 2   **iii** 6   **iv** -3
  d i 0.0008 joules   **ii** 10 763 km
  e 4×10^{-6} km   **f** 4 mm

14 a 8 710   **b** 199 times

15 a 1.332×10^{-6} **b** $\frac{1}{751\,000}$

16 a 226 km (3 sf)   **b** 2.26×10^2 km

Chapter 18

Check in 18

1 a 4   **b** 5.061   **c** -26.368

2

3 $y = 1, 3, 5, 7, 9$

18.1S

1 a

x	-3	-1	0	1	3
y	9	1	0	1	9

  b, c, d Check students' plots.

2 Translations upwards of $y = x^2$.

3 Translations downwards of $y = x^2$.

4 a iv   **b** i   **c** ii   **d** iii

5

Straight Line	Parabola
$y = 3x - 2$	$y = x^2 - 2$
$3x + 2y = 8$	$y = x^2 + 2x + 1$
$y = x$	$y = 10 + x^2$
e.g. $y = x + 1$	e.g. $y = x^2 + 1$

6 a, c Students' graphs
  b

x	-4	-3	-2	-1	0	1	2	3	4
x^2	16	9	4	1	0	1	4	9	16
$y = x^2 - 2$	14	7	2	-1	-2	-1	2	7	14

7 a i

x	-4	-3	-2	-1	0	1	2	3	4
$y = x^2 + 3$	19	12	7	4	3	4	7	12	19

  ii, iii Students' graphs   **iv** Minimum point is (0, 3).

b i

x	-4	-3	-2	-1	0	1	2	3	4
$y = 2x^2$	32	18	8	2	0	2	8	18	32

  ii, iii Students' graphs   **iv** Minimum point is (0, 0).

c i

x	-4	-3	-2	-1	0	1	2	3	4
$y = 3x^2 - 1$	47	26	11	2	-1	2	11	26	47

  ii, iii Students' graphs   **iv** Minimum point is (0, -1).

d i

x	-4	-3	-2	-1	0	1	2	3	4
$y = x^2 + x$	12	6	2	0	0	2	6	12	20

  ii, iii Students' graphs
  iv Minimum point is (-0.5, -0.25).

8 a

x	-3	-2	-1	0	1	2	3
x^2	9	4	1	0	1	4	9
y	7	3	1	1	3	7	13

  b Students' graphs
  c The approximate minimum is 0.75 and it occurs when $x = -0.5$.

9 B, C and D.

18.1A

1 a The parabola's y-intercept is 10. Its x-intercepts are -2 and 5. Its maximum point is at (1.5, 12.25).
  b The parabola's y-intercept is 9. Its minimum point is at (3, 0).
  c The parabola's y-intercept is -5. Its x-intercepts are -1 and 2.5. Its minimum point is at (0.75, -6.125).

2 a

Time (x)	0	1	2	3	4	5
$20x$	0	20	40	60	80	100
$4x^2$	0	4	16	36	64	100
Height (y)	0	16	24	24	16	0

  b Students' graphs
  c i The maximum height is 25 metres, which happens when $x = 2.5$.
    ii $x = 0.7$ and $x = 4.3$.
    iii $x = 0.92$ and $x = 4.08$ when $y = 15$. So the interval is 3.16 seconds.

3 a

t	0	10	20	30	40	50
h	100	480	660	640	420	0

  b Maximum height is 676 cm.
  c Length of the throw is 60 m.

4 a $x = -1.5$ or $x = 2.5$.   **b** $x = -0.5$ or $x = 1.5$.
  c $x = -1$ or $x = 3.0$.

5 a $x = -1.4$ or $x = 3.2$.   **b** $x = 0.0$ or $x = 2.0$.
  c $x = 1.0$.   **d** $x = -0.4$ or $x = 3.3$.
  e $x = -1.5$ or $x = 2.6$ **f** $x = -1.0$ or $x = 3.0$.

6 a Draw $y = x^2$ and $y = 7$. The solutions are -2.6 and 2.6.
  b Draw $y = x^3$ and $y = x - 1$. The solution is -1.3.
  c Draw $y = 2x^2$ and $y = x$. The solutions are 0 and 0.5.
  d Draw $y = x^3$ and $y = x^2 + 2$. The solution is 1.7.

18.2S

1 a Cubic: $y = x^3 - x - 6$, $y = x^3$
  b Reciprocal: $y = 3 - \frac{2}{x}$, $y = \frac{4}{x}$

2 a, c Students' graphs
  b

x	-3	-2	-1	0	1	2	3
y	-26	-7	0	1	2	9	28

  d i y is approximately 4.4   **ii** y is approximately 2.4

3 a

x	-2	-1	0	1	2	3
x^3	-8	-1	0	1	8	27
$x^3 - 4$	-12	-5	-4	-3	4	23

  b

x	-2	-1	1	2	3
$\frac{1}{x}$	$-\frac{1}{2}$	-1	1	$\frac{1}{2}$	$\frac{1}{3}$
y	-1	-2	2	1	$\frac{2}{3}$

c

x	−2	−1	0	1	2	3
x^3	−8	−1	0	1	8	27
$x+1$	−1	0	1	2	3	4
y	−9	−1	1	3	11	31

d

x	−2	−1	1	2	3
$\frac{3}{x}$	$-\frac{3}{2}$	−3	3	$\frac{3}{2}$	1
y	$-\frac{1}{2}$	−2	4	$\frac{5}{2}$	2

4 a Cubic passing through $(-3, -36)$, $(-2, -14)$, $(-1, -4)$, $(0, 0)$, $(1, 4)$, $(2, 14)$ and $(3, 36)$

  b Cubic passing through $(-3, -32)$, $(-2, -12)$, $(-1, -4)$, $(0, -2)$, $(1, 0)$, $(2, 8)$ and $(3, 28)$

  c Cubic passing through $(-3, -34)$, $(-2, -14)$, $(-1, -6)$, $(0, -4)$, $(1, -2)$, $(2, 6)$ and $(3, 26)$

  d Cubic passing through $(-3, -45)$, $(-2, -18)$, $(-1, -5)$, $(0, 0)$, $(1, 3)$, $(2, 10)$ and $(3, 27)$

5 a Cubic passing through $(-3, 36)$, $(-2, 14)$, $(-1, 4)$, $(0, 0)$, $(1, -4)$, $(2, -14)$ and $(3, -36)$

  b Cubic passing through $(-3, 32)$, $(-2, 12)$, $(-1, 4)$, $(0, 2)$, $(1, 0)$, $(2, -8)$ and $(3, -28)$

  c Cubic passing through $(-3, 34)$, $(-2, 14)$, $(-1, 6)$, $(0, 4)$, $(1, 2)$, $(2, -6)$ and $(3, -26)$

  d Cubic passing through $(-3, 45)$, $(-2, 18)$, $(-1, 5)$, $(0, 0)$, $(1, -3)$, $(2, -10)$ and $(3, -27)$

  e Reflections of question **4** in the x-axis.

6 a Cubic passing through $(-3, -44)$, $(-2, -14)$, $(-1, 0)$, $(0, 4)$, $(1, 4)$, $(2, 6)$, $(3, 16)$

  b i x is approximately -2.3.

  ii y is approximately 4.8.

7 a Reciprocal curve, with asymptote at $x = 2$, $y = 1$, passing through $(-2, -3)$, $(-1, -4)$, $(0, -6)$, $(1, -12)$, $(3, 12)$, $(4, 6)$, $(5, 4)$, $(6, 3)$

  b Reciprocal curve, with asymptote at $x = -4$, $y = 1$, passing through $(-3, -3)$, $(-2, -1)$, $(-1, -\frac{1}{3})$, $(0, 0)$, $(1, \frac{1}{5})$, $(2, \frac{1}{3})$, $(3, \frac{3}{7})$ and $(4, \frac{1}{2})$

  c Reciprocal curve, with asymptote at $x = 0$, $y = -2$, passing through $(-3, -3)$, $(-2, -3.5)$, $(-1, -5)$, $(1, 1)$, $(2, -0.5)$ and $(3, -1)$

  d Reciprocal curve, with asymptote at $x = 1$, $y = 2$, passing through $(-3, 2.25)$, $(-2, 2.333)$, $(-1, 2.5)$, $(0, 3)$ $(2, 1)$ and $(3, 1.5)$

18.2A

1 a Positive parabola passing through $(0, 0)$ and $(-8, 0)$

  b $y = x^2 + 8x$. $a = 1$, $b = 8$, $c = 0$.

2 a Negative parabola passing through $(1, 0)$ and $(5, 0)$

  b When $s = £1$ or $s = £5$.

  c When $s = £3$, the maximum profit is $4p$.

3 a Negative parabola passing through $(50, 0)$ and $(120, 0)$

  b When $s = £50$ or $s = £120$.

  c When $s = £85$, the maximum profit is £1225.

4 a Maximum profit is £506.25

  b The profit will be zero if $s = £15$ or $s = £60$.

5 a Negative parabola passing through $(39, 0)$ and $(-1, 0)$

  b Maximum height is 400 cm.

  c The length of the shot is 39 m.

6 a When $x = 0.5$ m.   **b** Maximum height is 0.625 m.

  c $y = -10x^2 + 5x$

7 a-A-3, b-B-2, c-D-1, d-C-4

8 S-shape passing through $(-2, 0)$, $(-1, 0)$ and $(0, 0)$.

9 S-shape passing through $(-4, 0)$, $(-1, 0)$ and $(2, 0)$.

10 b As a varies, the slope of the graph varies, but it always passes through $(-1, 0)$.

11 a The graph is $y = x^3$ translated vertically with y-intercept being a.

b i The graph is $y = x^3$ translated horizontally with x-intercept being $-a$.

  ii The graph looks similar to the reciprocal $y = \frac{1}{x}$ but with slope a.

  iii The graph is $y = \frac{1}{x}$ translated vertically with y-intercept being a.

18.3S

1 a 200   **b** 225   **c** 25

  d Points plotted at (O, 400), (N, 325) and (D, 400) joined with straight-lines.

  e The sales dropped in November and picked up in December, possibly due to sales season over Christmas holiday.

  f From January to September, the sales are quite flat and at low level; from October to December the sales increase a lot and peak in October and December.

2 a 0.7 m   **b** 0.5 m   **c** 1.9 m   **d** 0.1 m

  e Unlikely. Because its rate of growth is slowing down.

3 A-R, B-P, C-S, D-Q.

18.3A

1 a Quadratic function

2 A Quadratic

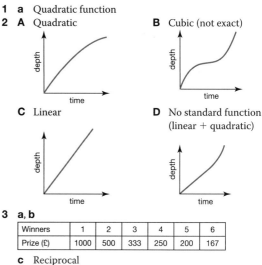

B Cubic (not exact)

C Linear

D No standard function (linear + quadratic)

3 a, b

Winners	1	2	3	4	5	6
Prize (£)	1000	500	333	250	200	167

  c Reciprocal

4 a Check coordinates.   **b** Reciprocal

  c Yes, because as the temperature goes down, the pulse size goes up. Moreover, there is a lower limit on pulse size, which is the feature of reciprocal functions.

5 c $a = \frac{1}{2}$, $b = -5$, $c = 12\frac{1}{2}$.   **d** $(10, 0)$

Review 18

1 a $(0, -5)$ **b** $(-1, 0)$, $(5, 0)$ **c** −1 and 5   **d** $(2, -9)$

2 a $y = 16, 9, 4, 1, 0, 1, 4, 9, 16$

  b

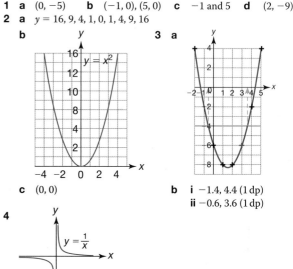

  c $(0, 0)$

3 a

  b i −1.4, 4.4 (1 dp)

  ii −0.6, 3.6 (1 dp)

4

Answers

5 a Reciprocal, $y = \frac{2}{x}$ **b** Linear, $y = 2x + 3$
c Quadratic, $y = x^2 + 3x$ **d** Cubic, $y = x^3 - 4x$
6 a 15 km/h **b** 20 km/h
8 a

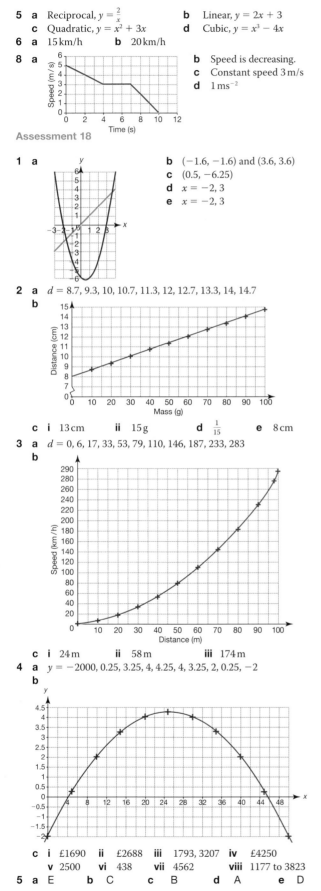

b Speed is decreasing.
c Constant speed 3 m/s
d $1\,\text{ms}^{-2}$

Assessment 18

1 a

b $(-1.6, -1.6)$ and $(3.6, 3.6)$
c $(0.5, -6.25)$
d $x = -2, 3$
e $x = -2, 3$

2 a $d = 8.7, 9.3, 10, 10.7, 11.3, 12, 12.7, 13.3, 14, 14.7$
b

c i 13 cm **ii** 15 g **d** $\frac{1}{15}$ **e** 8 cm
3 a $d = 0, 6, 17, 33, 53, 79, 110, 146, 187, 233, 283$
b

c i 24 m **ii** 58 m **iii** 174 m
4 a $y = -2000, 0.25, 3.25, 4, 4.25, 4, 3.25, 2, 0.25, -2$
b

c i £1690 **ii** £2688 **iii** 1793, 3207 **iv** £4250
 v 2500 **vi** 438 **vii** 4562 **viii** 1177 to 3823
5 a E **b** C **c** B **d** A **e** D

6 a, b

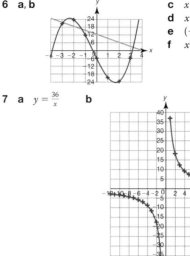

c $x = 4$
d $x = -4, 0, 4$
e $(-3, 21), (-1, 15), (4, 0)$
f $x^3 - 13x - 12 = 0$

7 a $y = \frac{36}{x}$ **b**

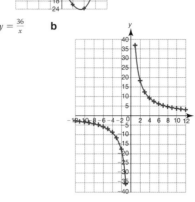

c 4 and 9, the two points of intersection between $y = 13 - x$ and $y = \frac{36}{x}$ are $(4, 9)$ and $(9, 4)$.

Revision 3

1 a 48 **b** HCF = 30, LCM = 1260
c $180 \times 5 = (2^2 \times 3^2 \times 5) \times 5 = 2^2 \times 3^2 \times 5^2$
2 a 121 **b** 6.76 **c** 21 **d** 1.5
e 1 **f** -27 **g** 32.768 **h** 3.6
i -8.4 **j** 6.1
3 a 8×10^3 **b** 7.5×10^5
c i 1.6^3 **ii** $4.096\,\text{cm}^3$
d $2^3 \times 2^{-3} = 2^{3 + -3} = 2^0 = 1$
4 a DE because the can empties rapidly
b FG because the level of water goes down slowly.
c EF because the graph is steepest in this section.
d BC because graph is horizontal meaning he has stopped filling the can.
5 a $3 \times -2 + 4 \times 7 = -6 + 28 = 22$
b i A **ii** C **iii** B **iv** D
c i $y = -4x + 3$ **ii** $y = -4x - 7$
iii $y = -4x - 245$
6 No, A forms an open box, B is not a net and C forms a cube.
7 a $232\,\text{cm}^2$ (3 sf) **b** $333\,\text{cm}^3$ (3 sf)
8 $1900\,\text{cm}^3$
9 $4000 = \frac{4}{3} \times \pi \times r^3$ so $r = \sqrt[3]{\frac{3 \times 400}{4\pi}} = 9.85\,\text{m}$ (3 sf)
10 a $9450\,\text{cm}^3$ **b** $3150\,\text{cm}^2$
11 a $2.1 \leqslant t < 2.6$ **b** $2.1 \leqslant t < 2.6$
c 19 **d** 2.4 mm (1 dp)
12 a

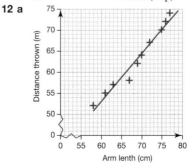

b Strong positive correlation
c i 58.5 m **ii** 71 cm

13 a

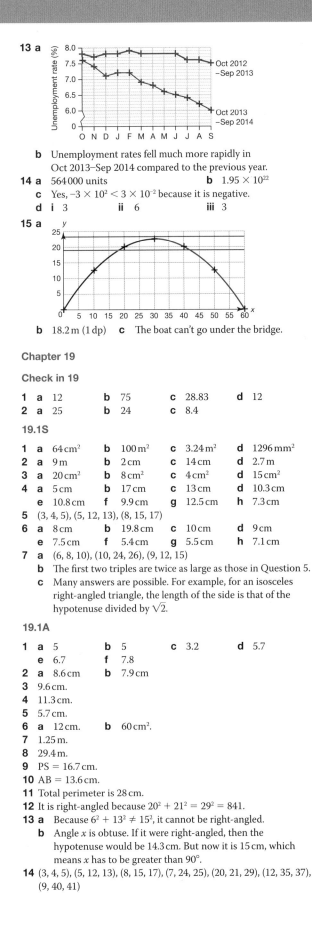

Oct 2012
–Sep 2013

Oct 2013
–Sep 2014

b Unemployment rates fell much more rapidly in Oct 2013–Sep 2014 compared to the previous year.

14 a 564 000 units   **b** 1.95×10^{22}

   c Yes, $-3 \times 10^2 < 3 \times 10^{-2}$ because it is negative.

   d i 3   **ii** 6   **iii** 3

15 a

b 18.2 m (1 dp)   **c** The boat can't go under the bridge.

Chapter 19

Check in 19

1 a 12   **b** 75   **c** 28.83   **d** 12

2 a 25   **b** 24   **c** 8.4

19.1S

1 a 64 cm²   **b** 100 m²   **c** 3.24 m²   **d** 1296 mm²

2 a 9 m   **b** 2 cm   **c** 14 cm   **d** 2.7 m

3 a 20 cm²   **b** 8 cm²   **c** 4 cm²   **d** 15 cm²

4 a 5 cm   **b** 17 cm   **c** 13 cm   **d** 10.3 cm

   e 10.8 cm   **f** 9.9 cm   **g** 12.5 cm   **h** 7.3 cm

5 (3, 4, 5), (5, 12, 13), (8, 15, 17)

6 a 8 cm   **b** 19.8 cm   **c** 10 cm   **d** 9 cm

   e 7.5 cm   **f** 5.4 cm   **g** 5.5 cm   **h** 7.1 cm

7 a (6, 8, 10), (10, 24, 26), (9, 12, 15)

   b The first two triples are twice as large as those in Question 5.

   c Many answers are possible. For example, for an isosceles right-angled triangle, the length of the side is that of the hypotenuse divided by $\sqrt{2}$.

19.1A

1 a 5   **b** 5   **c** 3.2   **d** 5.7

   e 6.7   **f** 7.8

2 a 8.6 cm   **b** 7.9 cm

3 9.6 cm.

4 11.3 cm.

5 5.7 cm.

6 a 12 cm.   **b** 60 cm².

7 1.25 m.

8 29.4 m.

9 PS = 16.7 cm.

10 AB = 13.6 cm.

11 Total perimeter is 28 cm.

12 It is right-angled because $20^2 + 21^2 = 29^2 = 841$.

13 a Because $6^2 + 13^2 \neq 15^2$, it cannot be right-angled.

   b Angle x is obtuse. If it were right-angled, then the hypotenuse would be 14.3 cm. But now it is 15 cm, which means x has to be greater than 90°.

14 (3, 4, 5), (5, 12, 13), (8, 15, 17), (7, 24, 25), (20, 21, 29), (12, 35, 37), (9, 40, 41)

19.2S

1 a, b, c 'hypotenuse' opposite the right angle, 'opposite' opposite θ, 'adjacent' on the third side.

2 a 0.53   **b** 0.62   **c** 0.42   **d** 2.36

   e 0.75   **f** 1.48

3 a i 1.50   **ii** 0.75   **iii** 0.53

     iv 0.43   **v** 0.63   **vi** 2.36

   b, c   **i** 56°   **ii** 37°   **iii** 28°

         iv 23°   **v** 32°   **vi** 67°

4 a 15.3°   **b** 17.4°   **c** 67.3°   **d** 40.5°

   e 26.0°   **f** 18.8°   **g** 51.9°   **h** 65.0°

   i 37.0°   **j** 25.8°   **k** 33.2°   **l** 23.6°

5 Check students' scale drawings.

19.2A

1 a 3.63 cm   **b** 11.3 cm   **c** 10.8 cm   **d** 3.02 cm

   e 74.5 cm   **f** 7.00 cm

2 a 6.38 cm   **b** 3.79 cm   **c** 30.5 cm   **d** 14.2 cm

   e 6.06 cm   **f** 12.3 cm

3 102.5 m.

4 a 167.8 m

   b Carl does not measure the angle from the ground (as he has some height).

5 Yes. 9.5 m + 1.6 m > 10 m

6 a For a right-angled isosceles, the two acute angles must be equal and add up to 90°, hence each of them must be 45°.

   b $\tan 45° = \frac{1}{1} = 1$.

7 a $h = \sqrt{3}$ cm.   **b** $\tan 60° = \frac{\text{opposite}}{\text{adjacent}} = \frac{\sqrt{(2^2 - 1^2)}}{1} = \sqrt{3}$.

   c $\tan 30° = \frac{\text{opposite}}{\text{adjacent}} = \frac{1}{\sqrt{(2^2 - 1^2)}} = \frac{1}{\sqrt{3}}$.

8 9.03 m.

19.3S

1 a 0.94   **b** 0.82   **c** 0.64   **d** 0.31

   e −0.17   **f** −0.77

2 a 20°   **b** 35°   **c** 50°   **d** 72°

   e 100°   **f** 140°

3 a 0.34   **b** 0.57   **c** 0.77   **d** 0.95

   e 0.98   **f** 0.64

4 a 20°   **b** 35°   **c** 50°   **d** 72°

   e 79°   **f** 40°

5 a $\cos 90° = 0$ **b** $\sin 90° = 1$

   c $\cos 60° = \frac{1}{2}$ **d** $\sin 60° = \frac{\sqrt{3}}{2}$

   e $\cos 45° = \frac{\sqrt{2}}{2}$ **f** $\sin 45° = \frac{\sqrt{2}}{2}$

   g $\cos 30° = \frac{\sqrt{3}}{2}$ **h** $\sin 30° = \frac{1}{2}$

   i $\cos 0° = 1$ **j** $\sin 0° = 0$

6 a $a = 67.7°$ **b** $b = 35.7°$ **c** $c = 73.0°$ **d** $d = 60.0°$

7 a $a = 27.7°$ **b** $b = 40.1°$ **c** $c = 41.0°$ **d** $d = 30.0°$

8 a $a = 26.4°$ **b** $b = 25.7°$ **c** $c = 56.2°$ **d** $d = 67.4°$

   e $e = 17.6°$ **f** $f = 23.6°$ **g** $g = 55.1°$ **h** $h = 19.8°$

19.3A

1 a 5.14 cm   **b** 10.4 cm   **c** 5.26 cm   **d** 4.75 cm

   e 9.50 cm   **f** 11.1 cm   **g** 8.20 cm   **h** 8.30 cm

   i 51.8 cm   **j** 10.9 cm

2 31.8 cm²

3 36.7 cm²

4 a Use Pythagoras and the hypotenuse is $\sqrt{1^2 + 1^2} = \sqrt{2}$ cm.

   b $\cos 45° = \text{adjacent} \div \text{hypotenuse} = \frac{1}{\sqrt{2}}$, and

       $\sin 45° = \text{opposite} \div \text{hypotenuse} = \frac{1}{\sqrt{2}}$.

5 a $\cos 60° = \text{adjacent} \div \text{hypotenuse} = \frac{1}{2}$, and

       $\sin 60° = \text{opposite} \div \text{hypotenuse} = \frac{\sqrt{3}}{2}$.

   b $\cos 30° = \text{adjacent} \div \text{hypotenuse} = \frac{\sqrt{3}}{2}$, and

       $\sin 30° = \text{opposite} \div \text{hypotenuse} = \frac{1}{2}$.

6 The height of the other end is $2.8 \times \sin 35° = 1.61$ m, so she is not correct.

7 The height above the ground is 0.53 km.

8 The bar has to be at least $2 \times 2 \times \sin 37° = 2.41$ m long. So 2 meter bar is not long enough.

19.4S

1 a 4 right, 3 up   **b** 2 right, 5 up
  c 1 left, 4 up   **d** 3 left, 3 down
  e 2 up   **f** 4 left

2 Equal vectors are: $\overrightarrow{AB} = \overrightarrow{BC} = \overrightarrow{FH}$, $\overrightarrow{AF} = \overrightarrow{DJ} = \overrightarrow{BH}$, $\overrightarrow{CH}$ $= \overrightarrow{EI} = \overrightarrow{BF}$, $\overrightarrow{FG} = \overrightarrow{GH} = \overrightarrow{DE} = \overrightarrow{IJ}$, $\overrightarrow{FI} = \overrightarrow{DG} = \overrightarrow{GJ} = \overrightarrow{EH}$ $= \overrightarrow{BE}$, $\overrightarrow{DF} = \overrightarrow{EG} = \overrightarrow{GI} = \overrightarrow{HJ} = \overrightarrow{BD}$

3 a $\overrightarrow{XA}, \overrightarrow{DX}, \overrightarrow{EF}, \overrightarrow{CB}$ **b** $\overrightarrow{AB}, \overrightarrow{FX}, \overrightarrow{XC}, \overrightarrow{ED}$
  c $\overrightarrow{AX}, \overrightarrow{XD}, \overrightarrow{FE}, \overrightarrow{BC}$ **d** $\overrightarrow{BA}, \overrightarrow{XF}, \overrightarrow{CX}, \overrightarrow{DE}$

4 a $\overrightarrow{OJ}, \overrightarrow{NK}$ **b** $\overrightarrow{OM}, \overrightarrow{QK}$ **c** $\overrightarrow{OP}, \overrightarrow{LK}$ **d** $\overrightarrow{JO}, \overrightarrow{KN}$
  e $\overrightarrow{MO}, \overrightarrow{KQ}$ **f** $\overrightarrow{PO}, \overrightarrow{KL}$

5 a 1 right, 1 down   **b** 6 right, 8 up
  c 5 right, 9 up   **d** 4 right, 3 up

6 a 2 right   **b** 1 right, 2 up
  c 1 right, 2 up   **d** 1 right, 2 down
  e 1 left, 2 up   **f** 1 left, 4 up

7 a   [diagram: g + g]   **b**   [diagram: g + h]
  c   [diagram: h – g]   **d**   [diagram: g – h – h]

8 a $\begin{pmatrix} -8 \\ 4 \end{pmatrix}$ **b** $\begin{pmatrix} 3 \\ -4 \end{pmatrix}$

9 a i $\begin{pmatrix} 1 \\ 1 \end{pmatrix}$ **ii** $\begin{pmatrix} -1 \\ 7 \end{pmatrix}$ **iii** $\begin{pmatrix} -5 \\ 7 \end{pmatrix}$ **iv** $\begin{pmatrix} 3 \\ 5 \end{pmatrix}$ **v** $\begin{pmatrix} 5 \\ 5 \end{pmatrix}$ **vi** $\begin{pmatrix} 8 \\ 0 \end{pmatrix}$
  b Check students' diagrams.

19.4A

1 a r   **b** p + r   **c** −p − r   **d** p − r
2 a l   **b** −j + l   **c** −j − l   **d** −j
3 a 3p   **b** 5p   **c** 5p + 2q   **d** 5p + 6q
  e 4q   **f** p + 3q   **g** p + 6q   **h** 3p + 3q
  i p + 2q   **j** p + 6q   **k** p + q   **l** −p + q
  m −4q   **n** −p − 4q   **o** −3p − 4q   **p** −p + 3q
  q p − 3q   **r** −p − 3q   **s** −4p − 4q   **t** −p − 4q
  u −3p + 4q
4 a 2 right   **b** 3 up
  c 2 right, 3 up   **d** 3 right, 1 down
  e 2 left, 1 up   **f** 1.5 right, 1.5 up
  g 2 right, 2 up   **h** 1 right, 1.5 down
  i 3 right, 4 up
5 a $\frac{3}{2}$j   **b** $-\frac{3}{2}$j
6 a d   **b** a   **c** a − d   **d** a − d
7 Check students' proofs.

Review 19

1 $a = 21.6$ cm, $b = 10.7$ cm
2 $x = 9$ cm, $y = 5$ cm
3 a i $\frac{5}{13}$ **ii** $\frac{12}{13}$ **iii** $\frac{5}{12}$
  b 22.6°
4 $a = 7.64$ cm, $b = 4.77$ cm, $c = 3.11$ cm

5 a $\frac{\sqrt{3}}{2}$ **b** 1   **c** 1   **d** $\frac{\sqrt{3}}{2}$
6 a 58.3 km (3sf)   **b** 124°
7 a $\begin{pmatrix} 5 \\ 9 \end{pmatrix}$ **b** $\begin{pmatrix} 6 \\ -6 \end{pmatrix}$ **c** $\begin{pmatrix} 15 \\ -5 \end{pmatrix}$ **d** $\begin{pmatrix} 1 \\ 0 \end{pmatrix}$
8 $\mathbf{u} = \begin{pmatrix} 2 \\ 4 \end{pmatrix}$, $\mathbf{v} = \begin{pmatrix} -3 \\ -2 \end{pmatrix}$

Assessment 19

1 a $a = \sqrt{181}$ m $= 13.45$ (2 dp)
  b $b \neq \sqrt{(14.2^2 + 6.1^2)}$, $b = \sqrt{(14.2^2 - 6.1^2)}$ m $= 12.82$ (2 dp).
2 a $2^2 + 3^2 = 13$, not 13^2.
  b $20^2 + 99^2 = 10\,201 = 101^2$, 101.
3 9.85 (3 sf)
4 a 1×28, 2×14, 4×7
  b 1×28, $d = 28.02$; 2×14, $d = 14.14$; 4×7, $d = 8.062$
5 a Yes, $9^2 + 12^2 = 225 = 15^2$
  b No, $9^2 + 14^2 = 277 \neq 17^2$
  c Yes, $1.6^2 + 3.0^2 = 11.56 = 3.4^2$
  d No, $11^2 + 19^2 = 482 \neq 22^2$
  e Yes, $3.6^2 + 7.7^2 = 72.25 = 8.5^2$
6 a $\angle HBN = 118°$ (IA), $x = 90°$ (AP)   **b** 4.38 km (3 sf)
  c 10.5 km (3 sf)   **d** 297°
7 a i Sine, 41.8°   **ii** 4.47 cm (3 sf)
  b i Cosine, 37.9°   **ii** 11.7 (3 sf)
  c i Tangent, 40.6°   **ii** 45.5 cm (3 sf)
8 a 8.09 m (3 sf)   **b** 7.89 m (3 sf)
9 a Check students' drawings.
  b i $\begin{pmatrix} 2 \\ 13 \end{pmatrix}$ **ii** $\begin{pmatrix} -6 \\ 1 \end{pmatrix}$ **iii** $\begin{pmatrix} 6 \\ -1 \end{pmatrix}$ **iv** $\begin{pmatrix} 8 \\ 12 \end{pmatrix}$
  v $\begin{pmatrix} 4 \\ -14 \end{pmatrix}$ **vi** $\begin{pmatrix} 14 \\ -9 \end{pmatrix}$ Check students' drawings
  c i $\sqrt{173}$ **ii** $\sqrt{37}$ **iii** $\sqrt{37}$ **iv** $\sqrt{208}$
  v $\sqrt{212}$ **vi** $\sqrt{277}$
  d The vectors are parallel, and the same length.
10 a Correct   **b** Incorrect, $-2\mathbf{x} + 7\mathbf{y}$
  c Incorrect, $-6\mathbf{y}$ **d** Correct
  e Incorrect, $2\mathbf{x} + 3\mathbf{y}$ **f** Correct
11 a $\begin{pmatrix} 35 \\ -10 \end{pmatrix}$ **b** $\begin{pmatrix} 16 \\ -10 \end{pmatrix}$ **c** $\begin{pmatrix} -1 \\ 3 \end{pmatrix}$ **d** $\begin{pmatrix} 3 \\ -3 \end{pmatrix}$
  e $\begin{pmatrix} 3 \\ 4 \end{pmatrix}$ **f** $\begin{pmatrix} -40 \\ 18 \end{pmatrix}$ **g** $\begin{pmatrix} 56 \\ -15 \end{pmatrix}$

Chapter 20

Check in 20

1 a $\frac{2}{3}$ **b** $\frac{1}{2}$ **c** $\frac{3}{4}$ **d** $\frac{1}{4}$
  e 1
2 a 1   **b** 1   **c** $\frac{3}{10}$ **d** $\frac{2}{5}$
3 a 0.9   **b** 0.4   **c** 0.85
4 a 2, 3, 5 or 7   **b** 1, 4 or 9   **c** 1, 3, 6 or 10
  d 4 or 8   **e** 1, 2, 5 or 10

20.1S

1 a P = {1, 4, 9, 16, 25, 36, 49, 64, 81, 100}
  b R = {Canada, Mexico, USA}
  c S = {2, 3, 5, 7, 11, 13, 17, 19, 23, 29}
  d T = {1, 2, 3, 4, 6, 9, 12, 18, 36}
2 a P∩T = {1, 4, 9, 36}   **b** S∩T = {2, 3}
  c P∩S = ∅
  d P∪S = {1, 2, 3, 4, 5, 7, 9, 11, 13, 16, 17, 19, 23, 25, 29, 36, 49, 64, 81, 100}
3 a Factors of 10.   **b** Even numbers
  c Vowels
  d All the possible outcomes when you toss two coins.
  e Types of coins that you can get in pounds sterling.
  f The first 10 multiples of 3.
4 a A = {1, 2, 3, 4, 5, 6, 7, 8, 9, 10}
  b {1, 3, 5, 7, 9}
  c {2, 3, 5, 7}
  d {1, 4, 9}

5 a $B \cap C = \{3, 5, 7\}$ b $B \cap D = \{1, 9\}$
c $B \cup D = \{1, 3, 4, 5, 7, 9\}$ d $C \cup D = \{1, 2, 3, 4, 5, 7, 9\}$
6 A contains all the elements of B, C and D.
7 a Yes, multiples of 10.   b Yes, 2.
c Yes, 1, 4, 9, 36.   d Yes, 9.   e No
f No   g No   h Yes, 3.   i No
8 a i 8   ii 13   iii 9   iv 21
b Students that don't play hockey or tennis.   c $\frac{13}{25}$
9 a i $\frac{29}{50}$ ii $P(H') = 1 - P(H) = 1 - \frac{31}{50} = \frac{19}{50}$
iii $\frac{17}{50}$ iv $\frac{43}{50}$ b 14

20.1A

1 a $P \cap Q = \{1, 3, 7, 21\}$ b 21
2 a 21   b 3   c 7
d A pupil who is not a girl and does not have dark hair.
3 a 58   b 110   c 5
d Takes Biology GCSE, walks to school and is an only child.
4 a Sketch of an isosceles right-angled triangle.
b An equilateral triangle cannot have a right angle.
5 a $18 + (8 - x) + (14 + x) + 2 = 42$
b i $\frac{11}{25}$ ii $\frac{2}{5}$ iii $P(A \cup B) = 1 - \frac{2}{50} = \frac{48}{50}$
iv $\frac{2}{25}$
c $x = 8$
d B is a subset of A; i.e. every element of B is contained in A.
6 5
7 3

20.2S

1 a $P(10) = \frac{3}{36} = \frac{1}{12}$ b $\frac{1}{6}$
c $\frac{7}{36}$ d $P(\text{less than 5}) = \frac{6}{36} = \frac{1}{6}$
e {2, 3, 4, 5, 6, 7, 8, 9, 10, 11, 12}
2 a $P(10) = \frac{2}{36} = \frac{1}{18}$ b $\frac{19}{36}$
c $P(\text{square}) = \frac{8}{36} = \frac{2}{9}$ d $P(\text{less than 5}) = \frac{8}{36} = \frac{2}{9}$
e {1, 2, 3, 4, 5, 6, 8, 9, 10, 12, 15, 16, 18, 20, 24, 25, 30, 36}
3 a

	1	2	3	4	5	6
1	0	1	2	3	4	5
2	1	0	1	2	3	4
3	2	1	0	1	2	3
4	3	2	1	0	1	2
5	4	3	2	1	0	1
6	5	4	3	2	1	0

b {0, 1, 2, 3, 4, 5}
c i $P(0) = \frac{6}{36} = \frac{1}{6}$
ii $P(3) = \frac{6}{36} = \frac{1}{6}$
iii 0
iv $\frac{4}{9}$
4 a {HHH, HHT, HTH, HTT, THH, THT, TTH, TTT}
b 3   c 2
5 a

	1	3	5
1	2	4	6
2	3	5	7
4	5	7	9

b {2, 3, 4, 5, 6, 7, 9}
c i $P(2) = \frac{1}{9}$
ii $P(3) = \frac{1}{9}$
iii $P(\text{even}) = \frac{3}{9} = \frac{1}{3}$
6 a

	1	3	5
1	1	3	5
2	2	6	10
4	4	12	20

b {1, 2, 3, 4, 5, 6, 10, 12, 20}
c i $P(2) = \frac{1}{9}$
ii $P(3) = \frac{1}{9}$
iii $P(\text{even}) = \frac{6}{9} = \frac{2}{3}$
7 Product
8 a ≈ 8 times   b ≈ 6 times
c More often.   d About the same.
9 a i $\frac{1}{6}$ ii $\frac{1}{6}$ iii $\frac{1}{12}$ iv $\frac{1}{12}$
b 4 and 5 and 7

20.2A

1 a $\frac{1}{5}$ b $\frac{4}{5}$ c 1
2 $\frac{1}{4}$

3 a

Spinner

Dice	B1	R2	Y2	G3	B2	R4
1	1	2	0	3	2	4
2	2	4	0	6	4	8
3	3	6	0	9	6	12
4	4	8	0	12	8	16
5	5	10	0	15	10	20
6	6	12	0	18	12	24

b i $\frac{1}{9}$ ii $\frac{5}{12}$

4 a $\frac{9}{100}$ b $\frac{9}{10}$ c $\frac{19}{100}$
5 a i $\frac{2}{15}$ ii $\frac{3}{5}$ b $\frac{19}{30}$
6 a $\frac{5}{12}$ b $\frac{1}{2}$
*7 a i $\frac{1}{36}$ ii $\frac{1}{6}$ iii $\frac{1}{9}$
b A and C, A and E, B and D, B and E.

20.3S

1 a, b

Gender / Pension?
$\frac{1}{3}$ (120) M
80% (96) Y $P(M \cap Y) = \frac{4}{15}$
20% (24) N $P(M \cap N) = \frac{1}{15}$
$\frac{2}{3}$ (240) F
70% (168) Y $P(F \cap Y) = \frac{7}{15}$
30% (72) N $P(F \cap N) = \frac{3}{15}$

2 a, b   c 24   d $\frac{4}{25}$

Travel / Late?
$\frac{2}{5}$ (60) Car
10% (6) Y $P(C \cap Y) = \frac{1}{25}$
90% (54) N $P(C \cap N) = \frac{9}{25}$
$\frac{3}{5}$ (90) Train
20% (18) Y $P(T \cap Y) = \frac{3}{25}$
80% (72) N $P(T \cap N) = \frac{12}{25}$

3 a, b   c $\frac{5}{13}$

Live / Debt worries?
$\frac{8}{13}$ (2000) Home
$\frac{1}{4}$ (500) Y $P(H \cap Y) = \frac{2}{13}$
$\frac{3}{4}$ (1500) N $P(H \cap N) = \frac{6}{13}$
$\frac{5}{13}$ (1250) Uni
$\frac{3}{5}$ (750) Y $P(U \cap Y) = \frac{3}{13}$
$\frac{2}{5}$ (500) N $P(U \cap N) = \frac{2}{13}$

4 a The red dice is even and the blue dice is a multiple of 3.
b $\frac{1}{6}$ c $\frac{1}{6}$
d Yes, $P(A \cap B) = P(A) \times P(B)$.

5 a

Brakes fail? / Lights fail?
10% (300) Y
40% (120) Y $P(Y \cap Y) = 4\% = 0.04$
60% (180) N $P(Y \cap N) = 6\% = 0.06$
90% (2700) N
20% (540) Y $P(N \cap Y) = 18\% = 0.18$
80% (2160) N $P(N \cap N) = 72\% = 0.72$
b 840
c Cars can fail for other reasons.

6 a   b 103   c 8

Taking drugs? / Test
20% (100) Home
$\frac{19}{20}$ (95) P $P(Y \cap P) = 0.19$
$\frac{1}{20}$ (5) N $P(Y \cap N) = 0.01$
80% (400) Uni
$\frac{1}{50}$ (8) P $P(N \cap P) = 0.016$
$\frac{49}{50}$ (392) N $P(N \cap N) = 0.784$

20.3A

1 a 0.84   b $P(\text{Loss}) = 1 - (0.3 + 0.6) = 0.1$
c Students' answers; e.g. losing one game may have a negative impact on the following game.

2 a 1st Attempt 2nd Attempt 3rd Attempt

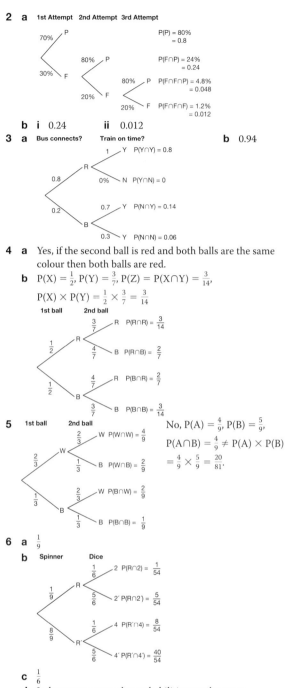

$P(P) = 80\% = 0.8$

$P(F\cap P) = 24\% = 0.24$

$P(F\cap F\cap P) = 4.8\% = 0.048$

$P(F\cap F\cap F) = 1.2\% = 0.012$

b i 0.24   **ii** 0.012

3 a Bus connects? Train on time? **b** 0.94

$P(Y\cap Y) = 0.8$
$P(Y\cap N) = 0$
$P(N\cap Y) = 0.14$
$P(N\cap N) = 0.06$

4 a Yes, if the second ball is red and both balls are the same colour then both balls are red.

b $P(X) = \frac{1}{2}$, $P(Y) = \frac{3}{7}$, $P(Z) = P(X\cap Y) = \frac{3}{14}$,

$P(X) \times P(Y) = \frac{1}{2} \times \frac{3}{7} = \frac{3}{14}$

1st ball 2nd ball

$P(R\cap R) = \frac{3}{14}$
$P(R\cap B) = \frac{2}{7}$
$P(B\cap R) = \frac{2}{7}$
$P(B\cap B) = \frac{3}{14}$

5 1st ball 2nd ball

$P(W\cap W) = \frac{4}{9}$
$P(W\cap B) = \frac{2}{9}$
$P(B\cap W) = \frac{2}{9}$
$P(B\cap B) = \frac{1}{9}$

No, $P(A) = \frac{4}{9}$, $P(B) = \frac{5}{9}$,

$P(A\cap B) = \frac{4}{9} \neq P(A) \times P(B)$

$= \frac{4}{9} \times \frac{5}{9} = \frac{20}{81}$.

6 a $\frac{1}{9}$

b Spinner Dice

$2\ P(R\cap 2) = \frac{1}{54}$
$2'\ P(R\cap 2') = \frac{5}{54}$
$4\ P(R'\cap 4) = \frac{8}{54}$
$4'\ P(R'\cap 4') = \frac{40}{54}$

c $\frac{1}{6}$

d It does not matter, the probabilities are the same.

e i Hindrance, you cannot score an odd number using the spinner.

ii Help, you cannot score an 8 unless you use the spinner.

Review 20

1 a

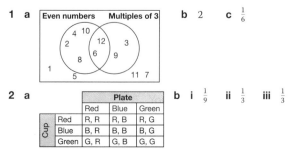

b 2   **c** $\frac{1}{6}$

2 a

		Plate	
		Red	Blue	Green
Cup	Red	R, R	R, B	R, G
	Blue	B, R	B, B	B, G
	Green	G, R	G, B	G, G

b i $\frac{1}{9}$ **ii** $\frac{1}{3}$ **iii** $\frac{1}{3}$

---

3 a First Set Second Set   **b i** 0.06   **ii** 0.56

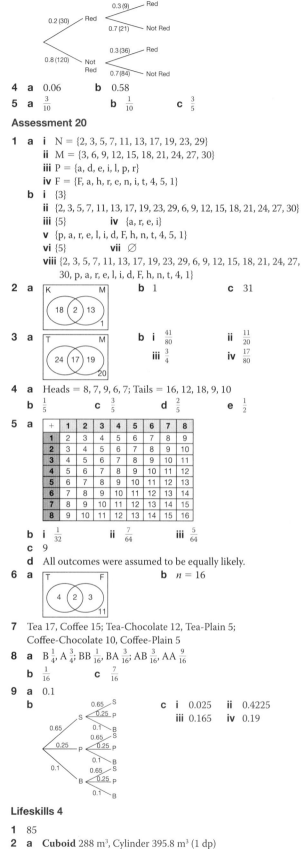

0.2 (30) Red
0.3 (9) Red
0.7 (21) Not Red
0.8 (120) Not Red
0.3 (36) Red
0.7 (84) Not Red

4 a 0.06   **b** 0.58

5 a $\frac{3}{10}$ **b** $\frac{1}{10}$ **c** $\frac{3}{5}$

Assessment 20

1 a i $N = \{2, 3, 5, 7, 11, 13, 17, 19, 23, 29\}$

ii $M = \{3, 6, 9, 12, 15, 18, 21, 24, 27, 30\}$

iii $P = \{a, d, e, i, l, p, r\}$

iv $F = \{F, a, h, r, e, n, i, t, 4, 5, 1\}$

b i $\{3\}$

ii $\{2, 3, 5, 7, 11, 13, 17, 19, 23, 29, 6, 9, 12, 15, 18, 21, 24, 27, 30\}$

iii $\{5\}$ **iv** $\{a, r, e, i\}$

v $\{p, a, r, e, l, i, d, F, h, n, t, 4, 5, 1\}$

vi $\{5\}$ **vii** $\varnothing$

viii $\{2, 3, 5, 7, 11, 13, 17, 19, 23, 29, 6, 9, 12, 15, 18, 21, 24, 27, 30, p, a, r, e, l, i, d, F, h, n, t, 4, 1\}$

2 a K M [Venn: 18, 2, 13, 1]   **b** 1   **c** 31

3 a T M [Venn: 24, 17, 19, 20]   **b i** $\frac{41}{80}$ **ii** $\frac{11}{20}$ **iii** $\frac{3}{4}$ **iv** $\frac{17}{80}$

4 a Heads = 8, 7, 9, 6, 7; Tails = 16, 12, 18, 9, 10

b $\frac{1}{5}$ **c** $\frac{3}{5}$ **d** $\frac{2}{5}$ **e** $\frac{1}{2}$

5 a

+	1	2	3	4	5	6	7	8
1	2	3	4	5	6	7	8	9
2	3	4	5	6	7	8	9	10
3	4	5	6	7	8	9	10	11
4	5	6	7	8	9	10	11	12
5	6	7	8	9	10	11	12	13
6	7	8	9	10	11	12	13	14
7	8	9	10	11	12	13	14	15
8	9	10	11	12	13	14	15	16

b i $\frac{1}{32}$ **ii** $\frac{7}{64}$ **iii** $\frac{5}{64}$

c 9

d All outcomes were assumed to be equally likely.

6 a T F [Venn: 4, 2, 3, 11]   **b** $n = 16$

7 Tea 17, Coffee 15; Tea-Chocolate 12, Tea-Plain 5; Coffee-Chocolate 10, Coffee-Plain 5

8 a B $\frac{1}{4}$, A $\frac{3}{4}$; BB $\frac{1}{16}$, BA $\frac{3}{16}$; AB $\frac{3}{16}$, AA $\frac{9}{16}$

b $\frac{1}{16}$ **c** $\frac{7}{16}$

9 a 0.1

b [tree diagram: 0.65 S, 0.25 P, 0.1 B, etc.]

c i 0.025   **ii** 0.4225   **iii** 0.165   **iv** 0.19

Lifeskills 4

1 85

2 a **Cuboid** 288 m³, **Cylinder** 395.8 m³ (1 dp)

b Cuboid 312 m², Cylinder 358.1 m² (1 dp)

c Cuboid   **d** Total hire cost ÷ Area of base

e

8 cm × 12 cm rectangle; 3 cm × 12 cm; 3 cm × 8 cm

3 a 33.7°

  b True. RT = 141m, BT = 206m, AT = 361m, TC = 541m.

  c 8 mins 6 secs

4 £12.80

5 810, 4761

Chapter 21

Check in 21

1 a 3   **b** 6   **c** 5   **d** 6

2 a 3   **b** 4   **c** 5   **d** 4

3 a 4, 8, 12, 16, 20, 24   **b** 3, 6, 9, 12, 15, 18

  c 5, 10, 15, 20, 25, 30   **d** 6, 12, 18, 24, 30, 36

21.1S

1 a 2, 8, 14, 20   **b** 3, 8, 13, 18

  c 20, 18, 16, 14   **d** $-1, -5, -9, -13$

  e 10, 50, 250, 1250   **f** 0.5, 1, 2, 4

  g 64, 32, 16, 8   **h** 2, 4, 8, 16

2 a 11, 13   **b** 14, 17   **c** 17, 21   **d** 12, 14

3 a 6, 1   **b** 16, 12   **c** 10, 7   **d** 46, 36

4 a 18, 21   **b** 7, 5   **c** 30, 37   **d** 18, 10

5 23 and 25. 72 is not in this sequence because it is not an odd number.

6 a Differences are 1, 2, 3, 4. The next two terms are 17, 23.

  b Differences are $-1, -2, -3$. The next two terms are 11, 6.

  c Differences are 1, 3, 5, 7. The next two terms are 29, 41.

  d Differences are $-5, -4, -3$. The next two terms are 6, 5.

7 a 8, 16   **b** 16, 8   **c** 40, 80   **d** 3, 1

8 a 6, 10, 14, 18, 22   **b** 26, 21, 16, 11, 6

  c 10, 7, 4, 1, -2 **d** $-6, -8, -10, -12, -14$

  e $-23, -30, -37, -44, -51$

9 a 2, 4, 6, 8, 10   **b** 17, 19, 21, 23, 25

  c 4, 8, 12, 16, 20   **d** 24, 30, 36, 42, 48

  e 7, 12, 17, 22, 27   **f** 1, 4, 9, 16, 25

  g 2, 5, 10, 17, 26   **h** 2, 4, 8, 16, 32

10 a 4, 7, 10, 13, 16   **b** 25, 19, 13, 7, 1

  c 4, 8, 12, 16, 20   **d** 30, 22, 14, 6, -2

  e $-16, -10, -4, 2, 8$

11 a 31, 38. Increase by 7 each time.

  b 25, 33. Increase by 8 each time.

  c $-7, -4$. Increase by 3 each time.

  d 11, 27. Increase by 4 each time.

  e 11, 19. Increase by 8 each time.

  f 1, -4. Decrease by 5 each time.

  g $-11, 1$. Increase by 3 each time.

  h 13, -1. Decrease by 7 each time.

21.1A

1 a 4 squares

  b 4 triangles

  c A cross with each branch consisting of 4 dots (including the centre)

  d 4 houses

2 a

Pattern number	1	2	3	4	5
Number of matches	4	7	10	13	16

  b

Pattern number	1	2	3	4	5
Number of pencils	3	5	7	9	11

  c

Pattern number	1	2	3	4	5
Number of dots	1	5	9	13	17

  d

Pattern number	1	2	3	4	5
Number of matches	5	9	13	17	21

3 a 31 matches   **b** 21 pencils

  c 37 dots   **d** 41 matches

4 a Each side has 5 dots.

  b Increase the number of dots by 4 each time. So pattern number 10 has 40 dots.

5 a Each horizontal side has 5 squares.

  b Pattern number 10 has 35 squares.

6 a 5, 13, 29, 61, 125   **b** 509

7 a No   **b** Yes   **c** No   **d** Yes

8 39   **10** 9

9 16   **11** Check students' research.

21.2S

1 a 6, 11, 16, and the 10th term is 51

  b 11, 14, 17, and the 10th term is 38

  c 4, 12, 20, and the 10th term is 76

  d $-2, 4, 10$, and the 10th term is 52

  e 26, 28, 30, and the 10th term is 44

  f 20, 25, 30, and the 10th term is 65

  g $-13, -6, 1$, and the 10th term is 50

  h $-2, 2, 6$, and the 10th term is 34

  i 11, 14, 17, and the 10th term is 38

  j $-9, -3, 3$, and the 10th term is 45

2 a 8, 6, 4, 2, 0   **b** 1, 0, $-1, -2, -3$

  c 14, 10, 6, 2, -2 **d** 13, 6, $-1, -8, -15$

  e 13, 10, 7, 4, 1   **f** 1, $-4, -9, -14, -19$

  g $-7, -9, -11, -13, -15$ **h** 25, 20, 15, 10, 5

  i 5, 2, $-1, -4, -7$ **j** 4, $-6, -16, -26, -36$

3 Linear: a, d, e, f, i

  Non-linear: b, c, g, h

4 a 4   **b** The nth term contains the term $4n$

  c

Sequence	5	9	13	17	21
$4n$	4	8	12	16	20

  d $4n + 1$

5 a $6n + 5$ **b** $9n - 8$ **c** $7n + 8$ **d** $4n - 14$

  e $-3n + 23$ **f** $-4n + 19$ **g** $-8n + 24$ **h** $-8n + 39$

6 a $4n + 3$ **b** $4n - 10$ **c** $-9n + 41$ **d** $-6n + 21$

7 a $-5n + 6$ **b** $1.5n + 0.5$ **c** $-1.5n + 9.5$ **d** $0.6n + 0.8$

  e $0.5n + 1.5$ **f** $-0.5n + 3.5$

8 a 47   **b** 67   **c** 407

9 a 38   **b** $4n - 2$

10 No. If it were in the sequence, then $75 + 3$ would have to be a multiple of 5.

21.2A

1 a 13, 17. Each time increase by 4.   **b** $4n - 3$

  c 197

2 a (i) 18, 22. Each time increase by 4. (ii) $4n + 2$. (iii) 202.

  b (i) 9, 11. Each time increase by 2. (ii) $2n + 1$. (iii) 101.

  c (i) 13, 16. Each time increase by 3. (ii) $3n + 1$. (iii) 151.

3 No. The nth term is $3n + 2$ and 52 is not in such sequence.

4 79

5 57

6 Many answers are possible. For example, the first pattern is a triangle made by 3 dots with a square of 4 dots next to it. And each time we increase the pattern by 4 dots (or a square).

7 a Each time we increase by 3 edges, and the first one has 4 edges.

  b Each time we increase the number of white and coloured tiles by the same amount, and the white tiles are always 4 more than the black ones.

  c Each time the length of the rectangle is 1 more than the width. And by analogy to the area of a rectangle, the total number of nails is just the product of width and length.

8 No. Because one cannot find an integer solution to $5n - 3 = 75$.

9 Yes. Because we start from 2, an even number, and increase by 4 each time, which is also an even number.

10 $60 - 12 \times 2 = 7 \times 5 + 1 = 36$ and $n = 2$ and 5 for the two sequences respectively.

11 a First sequence is $-7n + 147$ and the second is $5n - 64$. When $n = 17$, both become 21.
  b 126, 91, 56, 21, -14, -49

21.3S

1 a 21 **b** 36 **c** 216
2 Check students' drawings of triangles of size 1, 3, 6, 10 and 15
3 a arithmetic **b** arithmetic
  c Fibonacci-type **d** quadratic
  e geometric **f** arithmetic
  g Fibonacci-type **h** geometric
  i quadratic **j** arithmetic
  k geometric **l** geometric
4 a 6, 8, 10 **b** 8, 16, 32 **c** 6, 10, 16
  d Many solutions e.g. 8, 14, 22
5 a 15 **b** 20 **c** 15 **d** 17
6 Both are correct, the square numbers form a quadratic sequence.
7 Yes.
8 No. This is the cubic sequence.
9 Yes. The difference of the neighbouring terms form an arithmetic sequence.
10 a 3, 6, 12, 24 **b** 10, 50, 250, 1250
  c 3, 1.5, 0.75, 0.375 **d** 2, -6, 18, -54
  e $\frac{1}{2}, \frac{1}{4}, \frac{1}{8}, \frac{1}{16}$ **f** -3, 6, -12, 24
  g 4, $4\sqrt{3}$, 12, $12\sqrt{3}$
11 (a, b) could be $(1, 11)$ or $(2, 10)$ or $(3, 9)$ etc.
12 a $\frac{31}{4}$ **b** $\frac{1023}{128}$
  c The sum of the first n terms is $\frac{2^n - 1}{2^{n-3}}$

21.3A

1 1, 5, 9, 13, ...
  2, 6, 10, 14, ...
  3, 7, 11, 15, ...
  4, 8, 12, 16, ...
2 1, 2, 4, 8, ...
  3, 6, 12, 24, ...
  5, 10, 20, 40, ...
  7, 14, 28, 56, ...
3 a 64
  b It is the 8th in the square numbers and the 4th in the cubic numbers.
4 True, for successive values in the triangular sequence; i.e. $1 + 3 = 4, 3 + 6 = 9$.
5 Yes. Because she is increasing the difference by 1 each term, forming an arithmetic sequence for the differences, and hence a triangular sequence of numbers.
6 i 4, 5, 6, 7, 8,... By increasing by 1 each time.
  ii 4, 8, 16, 32, 64,... By multiplying by 2 each time.
  iii 4, 1, 5, 6, 11,... By fixing the 2nd term to be 1 and adding up two neighbouring terms to get the next one.
  iv 4, 5, 7, 10, 14,... By increasing the difference by 1 each time.
7, 8 Check students' research.
9 By adding the first term and the last term, and then the 2nd term and the last but one term, etc., each sum being 101. There are 50 such sums so the total sum is 5050.
10 Check students' research.
11 Hannah should choose Option 3 and Sam should choose Option 3.

12

$T(n) = n^2 - n + 1$	1, 3, 7, 13, ...	C	Quadratic
$T(n) = n^2$	1, 4, 9, 16, ...	D	Square
$T(n) = 4n - 3$	1, 5, 9, 13	B	Arithmetic
$T(n) = \frac{1}{2}n(n+1)$	1, 3, 6, 10, ...	A	Triangular

Review 21

1 a i 21, 25, 29 **ii** 4, -6, -16 **iii** 2.7, 3.3, 3.9
  b i $+4$ **ii** -10 **iii** $+0.6$
2 a 29 **b** 61 **c** 3 **d** $8\frac{1}{2}$

3 a 27 **b** 146 **c** 10 002
4 a $3n$ **b** $5n - 1$ **c** $6n + 2$ **d** $22 - 2n$
5 a
  b Number of matchsticks = 4, 7, 10, 13, 16, 19
  c $m = 3s + 1$
6 a 24, 48, 96 **b** Geometric
7 a i 125, 216 **ii** 15, 21
  b i Cubic **ii** Triangular
8 a Geometric **b** Fibonnaci-type
  c Arithmetic **d** Quadratic

Assessment 21

1 a i $-20, -30$ **ii** 30 **iii** ↓ **iv** -10 **v** -10
  b i 188, 210 **ii** 154 **iii** ↑ **iv** $+22$ **v** $+22$
  c i 3, 6 **ii** $-9, -3$ **iii** ↑ **iii** $+3$ **iv** $+3$
  d i $-10, -21$ **ii** 12 **iii** ↓ **iv** -11 **v** -11
  e i 49, 54 **ii** 34, 39 **iii** ↑ **iv** $+5$ **v** $+5$
  f i 3.1, 3.5 **ii** 1.9 **iii** ↑ **iv** $+0.4$ **v** $+0.4$
  g i 84, 96 **ii** 24, 60 **iii** ↑ **iv** $+12$ **v** $+12$
  h i 1.07, 1.06 **ii** 1.09 **iii** ↓ **iv** -0.01 **v** -0.01
  i i $-1.15, -1.20$ **i** $-1, -1.1$ **iii** ↓ **iv** -0.05 **v** -0.05
  j i 1 000 008, 1 000 011 **ii** 999 999, 1 000 002 **iii** ↑
  iv $+3$ **v** $+3$
  k i $-0.75, -1.5$ **ii** 0.75 **iii** ↓ **iv** -0.75 **v** -0.75
  l i $-38, -45$ **ii** -24 **iii** ↓ **iv** -7 **v** -7
  n i 6.625, 6.75 **ii** 6.125 **iii** ↑ **iv** $+0.125$
  v $+0.125$
2 a v **b** iv **c** vii **d** vi
  e i **f** iii **g** viii **h** ii
3 a $+10$ **b** $10n - 9$
4 a Jack
 b $2n - 1$ **c** 19
5 a 16, 49, 144, 196 **b** Square numbers $1^2, 2^2, 3^2, 4^2, 5^2, ...$
6 a 9 red, 10 blue **b** 49 red, 50 blue
  c 99 red, 100 blue **d** $n - 1$ red, n blue
7 a i $2n + 8$ **ii** $2n + 2$ **b** 54 **c** 42
8 a $3n - 1, 3 \times 1 - 1 = 2, 3 \times 2 - 1 = 5, 3 \times 3 - 1 = 8,$
  $3 \times 4 - 1 = 11.$
  b 3, -3, each term is 6 less than the previous.
  c $33 - 6n$ **d** -279
9 a Correct, $2 \times 10 + 7 = 27$.
  b Incorrect, $6 \times 1 - 5 = 1, 6 \times 2 - 5 = 7, 6 \times 3 - 5 = 13$.
  c Incorrect, $13 - 3 \times 100 = -287$.
  d Incorrect, $10^2 - 10 = 100 - 10 = 90$.
  e Correct. $15 - 3 \times 100^2 = 15 - 30 000 = -29 985$.
11 No, 1 : 2, 1 : 4, 1 : 6

Chapter 22

Check in 22

1 £28.00 **2** 1.5 **3** £68

22.1S

1 a 7.7 m/s **b** 7.1 m/s **c** 6.8 m/s **d** 5.1 m/s
2 a 160 km **b** 161 miles **c** 54 m **d** 288 miles
3 a 3 hours **b** 4 hours **c** 20 minutes $= \frac{1}{3}$ hours
  d 15 minutes $= \frac{1}{4}$ hours
4 1.425 g/cm³
5 a 9.4575 g **b** 6.15 litres
6 a 5704 kg **b** 3400 kg **c** 61 824 kg
  d 0.000125 m³ = 125 cm³ **e** 0.0000247 m³ = 24.7 cm³
  f 0.002 m³
7 a 1.29 N/m² **b** 19.3 N/cm² **c** 0.46875 m³ **d** 7424 N
  e 12.64 cm² **f** 68.88 N
8 £2.10/m
9 £11.95/hour

10 a 2.5 litres/s   **b** 1.6 litres/s
11 1200 litres
12 a 2.4 units/hour   **b** 57.6 units
13 a 500 km   **b** 20 litres   **c** 12.5 km/litre
14 a $1.50 per £ **b** $187.50   **c** £80

22.1A

1 89.25 mph
2 300 kg
3 8.5 g/cm³
4 32 mph
5 525 km, 2 hours 12 minutes 30 seconds, 37.3 kmph, 215 km, 37.1 kmph
6 a 5 g/cm³   **b** 87.88 g
7 a 5.96 g/cm³   **b** 105 blocks
8 a 10 500 kg/m³   **b** 4500 kg/m³   **c** 2700 kg/m³
9 a £89.25   **b** 15 hours
10 £42.24
11 4 hours 15.5 s
12 a 550 N/cm²   **b** 17.29 N/cm²

22.2S

1 a £8.25   **b** £12.38   **c** £18.84   **d** £75.76
2 a £17.25   **b** £11.90   **c** £7.76   **d** £16.73
3 12 kg
4 75 kg
5 a £0.60   **b** £1.80
6 a £0.40   **b** £2.80
7 a £0.03   **b** 0.52 g
8 a £1.88   **b** £1.48   **c** £1.73   **d** £2.04
  e £1.86
9 £3.64
10 a Check conversions   **b** 100 mm
11 a 0 lbs, 22 lbs, 11 lbs
  b Straight line through (0, 0) and (10, 22).
  c **i** 4.5 kg   **ii** 2.2–2.3 kg   **iii** 6.6 lbs   **iv** 5.5 lbs
12 a **i** 32 km   **ii** 97 miles
  b Charlie   **c** $y = 1.6x$

22.2A

1 Pack A: 1.2p per pin, Pack B: 1.15p per pin. Pack B is better value.
2 Regular is better value.
3 a 640 g   **b** 960 g   **c** 1120 g   **d** 2400 g
4 a 720 g   **b** 2040 g   **c** 2880 g   **d** 12 000 g
5 a 6.5 tonnes   **b** 51 : 14
6 No, his rate of pay for the four hours is £6/hour, the rate of pay for eight hours is £6.25/hour.
7 No, the formula is: number of radiator = days worked × 12
8 a Yes, the ratio is 1 : 1.4 for each size.
  b Straight line passing through (148, 210) and (841, 1189)

22.3S

1 a doubled   **3 a** 60 man hours
  b halved   **b** 10 builders
  c multiplied by 6   **c** 4 hours
  d divided by 10   **4 a** 1800 weeks
  e multiplied by 0.7   **b** 6 weeks
2 a halved   **c** 1.5 weeks
  b doubled   **5 a** 10 people
  c divided by 6   **b** 25 hours
  d multiplied by 10   **c** 40 hours
  d divided by 0.7
6

Inverse proportion	Direct proportion	Neither
$y = \frac{20}{x}$	$y = 4x$	$y = 5x + 1$
$y = \frac{13.5}{x}$	$y = \frac{x}{10}$	$y = \frac{x + 1}{10}$

7 a

x	$\frac{1}{4}$	1	2	4	8	16
y	64	16	8	4	2	1

  b Reciprocal curve passing through correct coordinates.
  c Yes
8 a 12
  b This equation models the situation in question **3**.
  c ? = 1800

22.3A

1 2 hours
2 a 5.3 minutes   **b** 3 minutes
3 a **i** 250 kg   **ii** 250 kg
  b Safeload = 3200 kg. Not enough to support 3500 kg.
4 10 left over.

x	0.5	1.5	2	3	4	4.8	7.5
y	48	16	12	8	6	5	3.2

5 a product always 48
  b

Length (cm)	Width (cm)	Perimeter (cm)
1	48	98
2	24	52
3	16	38
4	12	32
6	8	28
8	6	28
12	4	32
16	3	38
24	2	52
48	1	98

The perimeter and length do not have a constant product.

22.4S

1 a 1.3   **b** 0.7   **c** 1.03   **d** 0.97
  e 1.25   **f** 0.75   **g** 1.025   **h** 0.975
2 a

Minutes	Population
60	8
80	16
100	32
120	64
140	128
160	256
180	512

  b Check coordinates   **c** **i** 180   **ii** 360
  d Population has doubled.
3 a

No. of half-lives	Time (years)	Amount (grams)
2	60	250
3	90	125
4	120	62.5
5	150	31.25
6	180	15.625

  b Check coordinates   **c** 30 years
                               b 67.232%
4 a

Day	Price of boots
Tuesday	40
Wednesday	32
Thursday	25.6
Friday	20.48
Saturday	16.39

5 a

Year	Number of trout
0	800
1	680
2	578
3	491
4	417
5	355
6	302
7	256
8	218

  b Check coordinates
  c **i** After n years, population = 800 × 0.85 × ... × 0.85 = 800 × 0.85ⁿ

6 a After n hours, population = $800 \times 1.5 \times ... \times 1.5 = 800 \times 1.5^n$
  b 400×1.35^n c 6.5–7 years (in the 7th hour).
7 a

End of year	Amount in the account (£)
2	2163.20
3	2249.73
4	2339.72
5	2433.31
6	2530.64
7	2631.86
8	2737.14
9	2846.62
10	2960.49

  b 48%
  c i $(100 + r)\% = 1 + r/100$, so $A = P(1 + r/100)^n$
    ii £2960.49

22.4A

1 a i £1157.63   ii £157.63
  b i £316.33   ii £66.33
  c i £1075.27   ii £235.27
  d i £66053.11   ii £21 053.11
2 a Lily calculated the simple interest. She should multiply each amount by 1.06 to work out the amount at the end of each year.
  b £131.24
3 a £8352.10   b 9 years
4 a 56859   b 10 years
  c The population may not grow at a constant rate over a long time period.
5 a 350
  b The number of workers needed is expected to decrease by 4% each year.
  c Decreasing exponential curve starting at (0, 350)
6 a The number of vehicles per day when the road opens in 2400. The number of cars using the road increases by 8% each month.
  b Increasing exponential curve starting at (0, 2400)
  c The number of cars using the road may not grow at a constant rate over a long time period. There may not be 2400 cars on the first day.
7 a Account A (£7058.30 > £7057)
  b He may decide to withdraw his money before 3 years; account B is better for the first two years.
8 a C   b D   c B   d A
9 She is incorrect. The yearly interest on the half-yearly saver account is 1.0404%.

Review 22

1 a 0.14 m   b 16 min, 40 s
  c 3000 cm³   d 18 km/h
2 2.5 N/m²
3 £0.24 per 100 g
4 18 000 cm²
5 a i 64 cm³   ii 1728 cm³
  b i 96 cm³   ii 864 cm³
6 a $y = 4x$
  b i 32   ii 11
7 a $y = \frac{2}{x}$
  b i $\frac{2}{3}$ ii 9
8 a $P = 3N$ b 3   c Price per magazine
9 a A   b C
10 a i £2040   ii £2122.42
  b $V = 2000 \times 1.02^t$

Assessment 22

1 a 30 min   b 30 mph   c 39 mph
2 8 min 20 s
3 a 39.37 in   b £9.75   c 14 min 56 s
4 The wombat, Bolt's speed = 36.73 km/h.

5 a 7.28 g/cm³   b 23.81 cm³   c 1.03 kg
6 a 45 tins   b Yes, 100 ÷ (3 ÷ 2) = 67 days.
  c No, 30 ÷ (9 ÷ 2) = 7 days.
7 a 157.5 litres   b 379 cm
8 a 69.12 kg   b $x = 24$ cm, $y = 10$ cm
9 a $\tan(x) = 0, 0.0175, 0.0349, 0.0524, 0.0699, 0.0875, 0.1051, 0.1228, 0.1405, 0.1584, 0.1763$
  b
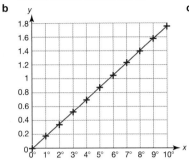
  c The graph is a straight line in this range.

10 a
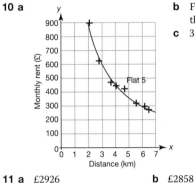
  b Flat 5 lies furthest from the curve.
  c 3.5 km

11 a £2926   b £2858
12 a £27 000, 5%   b £9546

Revision 4

1 6.71 km (3sf)
2 a 7.14 km   b 161 cm   c 8.6°
3 a The diagonal from the vertical angle to the base bisects the base.
  b 14.5 cm (3sf)
4 a 17.0   b $\theta = 17.9°$
5 $RQ = p - 2q$, $NM = \frac{p}{2} - q$, $NM = \frac{1}{2}RQ$
6 $\frac{24}{a} = \frac{b}{35}$
7 a $\begin{pmatrix} 15 \\ -30 \end{pmatrix}$ b $\begin{pmatrix} 6 \\ -18 \end{pmatrix}$ c $\begin{pmatrix} -9 \\ -9 \end{pmatrix}$ d $\begin{pmatrix} 6 \\ -15 \end{pmatrix}$ e $\begin{pmatrix} 6 \\ -15 \end{pmatrix}$
  f $\begin{pmatrix} 3 \\ 6 \end{pmatrix}$ g $\begin{pmatrix} 6 \\ -30 \end{pmatrix}$ h $\begin{pmatrix} -39 \\ 69 \end{pmatrix}$ i $\begin{pmatrix} -0.5 \\ 0 \end{pmatrix}$ j $\begin{pmatrix} 0 \\ 0 \end{pmatrix}$
8 a Yes. $1^2 + 1 = 2, 2^2 + 2 = 6, 3^2 + 3 = 12, 4^2 + 4 = 20$
  b 2550, 10 100   c $2n + 2$ d 102, 202
9 6
10 **Liddi 27p**, Addle 27.5p
11 a cm   b kg   c ml
  d mm   e l   f km
  g g   h m   i tonnes
12 a

3	4	5
6	10	15
3	6	10
9	16	25

  b i 1275   ii 5050   c r^2
  d $\frac{r(r-1)}{2}$ e $\frac{2(2-1)}{2} = 1$ f 1770
13 a 1300 ml   b 0.076 kg
14 a 40 mph   b 1h 26m   c 6 minutes
15 a 15.75 g/cm³   b 5.48 kg (3sf)   c 2.86 cm³ (3sf)
16 a i $A \propto h$ ii $S \propto r^2$ iii $m \propto \frac{1}{d^3}$ iv $T \propto \sqrt{l}$
  b i 75   ii 4.84

A

acceleration 304, 306, 308
accuracy, measurement and 196–9, 200
acute angle 48, 64, 144
addition 12–15
  algebraic terms 30
  compensation 12
  of fractions 104, 106
  mental methods 12, 14
  of negative numbers 12
  partitioning 12
  of probabilities 426, 428
  in standard form 366
  of vectors 408, 410
  written method 12, 14
adjacent side 400, 402, 404, 406, 412
algebraic convention 26, 30, 120, 130
alternate angle 48, 50, 64
angle(s) 48–51, 160
  acute 48, 64, 144
  alternate 48, 50, 64
  at a point 50
  at a point on a straight line 50
  bisector 244, 248, 252
  corresponding 48, 50, 56, 58, 64, 152
  exterior 60, 62, 64
  interior 60, 62, 64
  measuring 144–7
  obtuse 48, 64, 144
  in pie chart 82, 84
  in polygons 60–3
  in quadrilateral 52
  reflex 48, 64, 144
  right 48, 64
  on straight line 48, 50, 52, 62
  in triangle 52
  vertically opposite 50
anticlockwise 152
approximation 188–91, 200
arc 234, 236, 238, 252
arc length 238, 240
area 160
  of 2D shape 148–51
  of circle 234, 236
  of parallelogram 148, 150
  of rectangle 148, 150
  of trapezium 148, 150
  of triangle 148, 150, 322
arithmetic sequence 438, 446, 448, 450
ascending order 106, 112
averages 86–9, 340–3
axis (axes) 308

B

balance method 126, 206, 208, 210, 212, 222, 226
bar chart 78, 80, 90, 336, 338
bar-line chart 78, 90
base 34, 36, 42, 148, 150, 360, 370
bearing 50, 64, 144, 146
bias 70, 72, 170, 174, 176, 182
BIDMAS 16, 18, 30, 188, 206
bisection of angle 244, 248, 252
bracket keys (calculator) 192
brackets 26, 42
  multiplication of 38, 40
  expanding 38–41,132, 134, 210, 212

C

calculator methods 192–5
cancelling 100, 104, 112
capacity 196, 198, 200
category in pie chart 82
cause, variables and 344
census 70
centimetres 144, 146
centre 234
centre of enlargement 156, 158, 160
centre of rotation 152, 160
certainty 166, 168, 182

chord 234, 252
circle 234–41, 252
  area of 234, 236
  formulae for 364
circumference 234, 236, 252
class interval 336, 338, 340, 342
clockwise 152
compound units 456–9
coefficient 34, 42, 134, 220
column 70
common difference 438, 442, 444, 446, 450
common factor 38, 40, 100, 112, 278, 290
compasses 234, 242, 244
compensation 12, 20
complement of set 418, 430
compound interest 468, 472
compound measure 196, 198
cone 314, 328
  surface area 324, 326
  volume of 322
congruence 56–9, 64, 152
congruent trapezium 148
constant 134
constant of proportionality 464, 466, 472
construction 242–7, 252
construction lines 252
continuous data 196, 336, 338, 340
conversions 460
coordinate grid 308
coordinates 346
correlation 344–7, 352
corresponding angle 48, 50, 56, 64, 152
cosine ratio (cos) 404, 406, 412
counter-example 130
cross-section 318
cube 328
cube number 286, 288, 290, 446, 448, 450
cube root 286, 288, 290, 358, 370
cubic function, graph of 380, 382, 388
cuboid 328
  volume of 318, 320
  see also three-dimensional shape
cylinder 314, 328
  formula for 364
  surface area of 324, 326
  volume of 318, 320

D

data 70
  organising 74–7
  representing 78–83
data collection sheet 70, 72, 90
decagon 60
decay 468–71
decimal equivalent 112
decimal places (dp) 8, 20
decimal points, lining up 12, 14
decimal system 4, 20
decimals 96–9, 112
  addition 12, 14
  conversion between fractions and 96, 98
  conversion of fraction to 260
  conversion to fraction 108, 110
  conversion to percentage 260
  equivalent 96, 98
  proportion and 258, 260
  recurring 100, 108, 110, 112, 362, 370
  subtraction 12, 14
  terminating 108, 112, 362, 370
decreasing quantity 384
decreasing sequence 438
degrees 48, 144
denominator 96, 98, 104, 106, 108, 112
  common 110, 112
density 196, 198, 200, 456, 458
dependent event 430
descending order 106, 112
diagonal line 296, 298, 300
diameter 234, 236, r 252

Difference Of Two Squares (DOTS) 133
digit 4, 6, 20
direct proportion 460–3, 472
discrete data 336, 338
distance-time graph 304, 306, 308
division 16–19
  algebraic terms 30
  of fractions 104, 106
  of indices 34, 36
  of inequality 224
  by negative number 16
  of powers 358, 360
  by prime numbers 282
  in standard form 366
  by 10 4, 6

E

edge 314, 316, 328
elements of set 418, 430
elevation 314, 316, 328
elimination 218, 220
empty set 418, 420, 430
enlargement 56, 146, 156, 158, 160
equal 112
equal likelihood 166, 174, 176, 182
equal to 4, 6
equality signs 4, 6
equation 26, 42, 128–31
  linear 206–13
  of straight line 300–3
equidistance 152, 250
equilateral triangle 52, 58, 246
equivalent fraction 96, 98
estimated mean 340, 342, 352
estimated probability 166, 168, 174
estimation 12, 14, 16, 18, 188–91, 200, 340
even chance 166, 168, 182
event 166, 174, 182
exact calculations 362–5, 370
exchange rate 460
exhaustive events 178
expansion 38–41, 132–5, 136
  of brackets 210, 212
expected frequency 170, 172, 182
expected outcome 170–3
experimental probability 182
explicit function 298
exponential decay 468
exponential growth 468
expression 26–9, 42, 128, 130
  simplifying 30–3
exterior angle 60, 62, 64
extrapolation 346

F

face 314, 316, 328
factor pair 282
factor tree 282, 284
factorisation 38–41
factorising 132–5, 136, 214, 216, 226
factors 278–81, 290
Fibonacci-type sequence 446, 448, 450
FOIL acronym 132, 134
force 456
formula (formulae) 26, 42, 128, 130
  standard 124–7
  substituting into 120–3, 126
formula triangle 456, 458
fractions 96–105, 112
  addition of 104, 106
  calculations with 104–7
  conversation of percentage to 108, 110
  conversion between decimals and 96, 98
  conversion of decimal to 108, 110, 260
  decimal equivalent and 362
  division of 104, 106
  equations involving 210
  equivalent 96, 98
  improper 96, 104, 106, 112

multiplication and 100, 102
multiplication of 104, 106
probability as 426
proportion and 258, 260
reciprocal of 104, 106
simplification 96
subtraction of 104, 106
frequency 84, 90
frequency diagram 336–9
frequency table 74, 76, 90
frequency tree 426, 430
function 128, 130, 136
function machine 124, 136

G
general term (nth term) 442, 444, 450
geometric sequence 446, 448, 450
gradient 300, 302, 306, 308, 462
graph
    cubic function 380, 382
    distance–time 304, 306, 308
    of horizontal line 384, 386
    kinematic 304–5
    line 348, 350, 352
    linear equations 296
    quadratic function 376–9, 382
    real-life 384–7
    reciprocal 380, 382, 464, 466
    S-shaped 380
    scatter 344–7, 352
    of simultaneous equations 218, 220
    speed–time 306
    straight-line 296–9, 300, 384, 386
    straight line, passing through origin 460, 462
    time series 348, 352
greater than (>) 4, 6, 222, 224
greater than or equal to (≥) 4, 6, 222, 224
grouped data 336, 338
grouped frequency table 336, 340
growth 468–71

H
hemisphere 326
heptagon 60
hexagon 60, 62
    regular 60
highest common factor (HCF) 38, 40, 42, 278, 280, 282, 284, 290
histogram 336
horizontal axis 348
horizontal line 296, 298
hypotenuse 246, 396, 398, 400, 404, 406, 412

I
identically equal to 128
identity 128, 130, 136
image 160
implicit function 298
implied accuracy 196, 200
impossibility 166, 168, 182
improper fraction 96, 104, 106, 112
increasing quantity 384
increasing sequence 438
independent events 428, 430
index (indices) 34–7, 42, 286–9, 370
    calculating with 358–61
index laws 34, 42, 370
indices *see* index (indices)
inequalities 222–5, 226
inequality signs 4, 6, 224
input 124, 136
integer 222
interest 266
interior angle 60, 62, 64
interpolation 346
intersection 48, 218, 220, 282, 284, 418, 420, 430
invariant 160
inverse operation 124, 126, 136, 206
inverse proportion 464–7, 472

irregular pyramid 322
isosceles trapezium 52
isosceles triangle 52, 58

K
key 74
kilometres 144
kinematic graph 304–7
kite 52

L
length 160, 196, 198
    measuring 144–7
less than (<) 4, 6, 222, 224
less than or equal to (≤) 4, 6, 222, 224
life skills
    business plan 116–17
    getting ready 332–3
    launch party 434–5
    starting the business 230–1
like terms 30, 32, 38, 120, 136
likelihood 166, 168, 182
line 48–51
    angles on 48, 50, 52, 62
    parallel 48
    perpendicular 48
line graph 348, 350, 352
line of best fit 344, 346, 352
line of symmetry 60
line segment 398
linear equations, graphs of 290
linear function, graph of 382
linear relationship 344, 352
linear sequence 438, 442, 446, 450
liquids, measurement of 318
locus (loci) 248–51, 252
lower bound 196, 198
lowest common denominator 112
lowest common multiple (LCM) 278, 280, 282, 284, 290

M
mapping 152
mass 196, 198, 200, 320
maximum point 376, 378, 382
mean 86, 88, 90
    estimated 340, 342
measurement
    accuracy and 196–9
    of angles 144–7
    of lengths 144–7
    of liquids 318
median 86, 88, 90, 340, 342
member of set 418, 430
mental methods 12
    percentages 100
metre 144, 146
metric measurements 196, 198
midpoint 244
millimetre 144
minimum point 376, 378
mirror image 60
mirror line 152, 160
mixed number 96, 112
    conversion to improper fraction 104, 106
modal class 340, 342, 352
mode 86, 88, 90
multiple 278–81, 410, 412
multiplication 16–19
    algebraic terms 30
    brackets and 38, 40, 132, 134
    fractions and 100, 102
    of fractions 104, 106
    index notation 34
    of indices 34, 36
    of inequality 224
    by negative number 16, 132
    of powers 358, 360
    of probabilities 426, 428

    in standard form 366
    by 10 4, 6
    of vectors 408
multiplication signs 26
multiplier 468
multiple 290
mutually exclusive events 178–81

N
negative number 2, 4, 6
    addition 12
    division by 16
    multiplication by 16, 132
    subtraction 12
net 314, 316, 328
nonagon 70
not equal to 4, 6
nth term 442, 444, 450
numerator 96, 98, 104, 106, 108, 112

F
object 160
obtuse angle 48, 64, 144
octagon 60
operation 124, 136
opposite side 400, 402, 404, 406, 412
order of operations 16, 18, 188, 192, 194
ordering numbers 4, 74
origin 308
outcome 174, 176, 182, 424, 426, 428
outlier 86, 90, 344, 346
output 124, 136

P
parallel lines 48, 300
    vectors and 408, 410
parallel vectors 410
parallelogram 52, 54
    area of 148, 150
partitioning 12, 20
pentagon 60
percentage change 266–9
percentage decrease 266, 268, 270
percentage increase 266, 268
percentage interest 270
percentages 100–3, 270
    conversion to fraction 108, 110
    mental methods 100
    proportion and 258, 260
perimeter 36, 148, 150, 160
    of semicircle 236, 238, 240
perpendicular bisector 244, 246, 248, 250, 252
perpendicular distance 244, 246
perpendicular height 148, 150
perpendicular lines 48
pi (π) 234, 236, 362, 364, 370
pictogram 78, 80, 90
pie chart 82, 84
place value 4–7, 16, 18, 20
plan 314, 316, 328
polygon 64
    angles in 60–3
    regular 60, 62
population 70, 72, 90
position 450
position-to-term rule 442, 450
positive number 286, 288
possibility space 422–5, 430
power 34, 42, 370
    raising to 358, 360
    of zero 358, 360
    see also index
power of a power 34, 36
pressure 456, 458
prime factor decomposition 282–5, 290
prime factors 282, 284, 290
prime number 278, 280, 290
    division by 282

prism 314, 316, 328
  square-based 314
  triangular 314
  volume of 318–21, 326
probability 424, 426, 428
  estimated 166, 168, 174, 182
  as fraction 426
  theoretical 174–7, 182
probability experiments 166–9
product 278, t 464
proof 130
proportion 82, 258–61, 270, 472
proportional quantities 472
protractor 82, 84, 144, 146
pyramid 314, 322, 328
  irregular 322
  square-based 314, 322
  triangular-based 314
  volume of 322, 326
Pythagoras' theorem 326, 396–9, 412

Q

quadratic equations 214–17
quadratic expression 134, 136, 226
quadratic function 388
  graph of 382
  properties of 376–9
  sketching 380–3
quadratic sequence 446, 448, 450
quadrilateral 52–5, 60, 64
  angles in 52, 54
quarter-circle 239

R

radius 234, 244, 252
random sample 70, 72
range 86, 88, 90, 342
rate 456, 458, 472
rate of flow 456
rate of pay 456
ratio 146, 262–5, 270, 460, 462
real-life graphs 384–7
rearranging equation/formula 124, 136
reciprocal 104, 106
reciprocal curve 464
reciprocal equation 380
  graph of 380, 382, 466
reciprocal function 388
rectangle 52, 54
  area of 148, 150
recurring decimal 108, 110, 112, 362, 370
reduction 146
reflection 152, 154, 160
reflex angle 48, 64, 144
regular polygon 60
regular shape 60
regular tetrahedron 322
relative frequency 166, 168, 176, 182, 426
representation of data 78–83
resultant vector 408, 412
reverse percentage problem 268, 270
rhombus 52, 54
right angle 48, 64
right-angled triangle 52, 246, 396, 398, 400,
    402, 404
root 286–9, 388
  calculating with 358–61
rotation 152, 154, 160
rounding 8–11, 20, 188
row 70
rule 438
ruler 144, 146

S

S-shaped graph 38
sample 90
  biased 70, 72
  random 70, 72
sample space 422, 430
sampling 70–3

scale 50, 262, 264, 270
scale diagram 264
scale drawing 146, 270
scale factor 56, 146, 156, 158, 160, 262
scalene triangle 52
scatter graph 344–7, 352
sector 234, 252
  in pie chart 82, 84
sector area 238, 240
segment 234, 252
self-similarity 58
semicircle
  area of 236, 238
  perimeter of 236, 238, 240
sequence 438, 440, 442, 444, 450
  special 446–9
sequence rules 438–41
set 418–21, 430
significant figure (sf) 10, 188, 190, 200
  first 8, 20
similarity 56–9, 64
simple interest 266, 270, 468, 472
simplest form 262, 426
simplification 28, 42
  of fractions 96
simplify (ratio) 262, 270
simultaneous equations 218–21, 226
sine ratio (sin) 404, 406, 412
sketching functions 380–3
solution 214, 218, 226, 388
speed 196, 198, 200, 304, 306, 308, 456, 458
  average 304
sphere 322, 328
  surface area 324, 326
  volume of 326
spread 86–9, 340–3
square 52, 54
square number 286, 288, 290, 446, 448
square root (√) 286, 290, 358, 370
square-based prism 314
square-based pyramid 314, 322
SSS rule 218
standard form 366–9, 370
standard formulae 124–7
stem-and-leaf diagram 74, 76
straight line 52
  angles on 48, 50, 52, 62
  equation of 300–3, 308
  graph 296–9
  plotting 460, 462
subject of formula 124, 126, 136
substitution 28, 42, 120, 206, 212
subtraction 12–15
  algebraic terms 30
  compensation 12
  of fractions 104, 106
  mental methods 12, 14
  of negative numbers 12
  partitioning 12
  in standard form 366
  of vectors 408, 410
  written method 12, 14
surface area 322–7, 328
  of 3D shape 316
  of cone 324, 326
  of cylinder 324, 326
  of sphere 324
survey 70, 72, 90

T

tally chart 70, 90
tangent 252
tangent ratio (tan) 400, 402, 404, 406, 412
term 26–9, 30, 438, 440, 450
  algebraic 128, 130
term-to-term rule 438, 440, 450
terminating decimal 108, 112, 362, 370
tetrahedron, regular 314, 322
theoretical (expected) outcome 170–3

theoretical probability 174–7, 182
theta 238
three-dimensional shapes 314–17
  surface area of 316
  volume of 316, 320
three-figure bearing 50, 64
transformation 152–9, 160
translation 152, 154, 160
trapezium 52
  area of 148, 150
  congruent 148
  isosceles 52
tree diagram 426–9, 430
trend 348, 350, 352
trial 166, 170, 174, 182
triangle 52–5, 60, 64
  angles in 52, 54
  area of 148, 150, 322
  congruent 56, 58
  construction of 243–7
  equilateral 52, 58, 246
  isosceles 52, 58
  right-angled 52, 246, 396, 398, 400, 402, 404
  scalene 52
triangular number 446, 448, 450
triangular prism 314
triangular-based pyramid 314
trigonometry 400–7
turn 152
turning point 382, 388
two-dimensional shape 314
  area of 148–51
two-way table 70, 76

U

union of sets 418, 420, 430
universal set 418, 420, 430
unknown 28, 42, 210, 212
unlikelihood 166, 168, 182
upper bound 196, 198

V

variable 120, 136, 344
varies 460, 472
vector 408–11, 412
Venn diagram 282, 284, 418, 420, 430
vertex (vertices) 60, 314, 316, 328
vertical axis 78
vertical line 296, 298
vinculum 98
volume 200, 322–7, 328
  of cone 322
  of cuboid 320
  of cylinder 318, 320
  of prism 318–21, 326
  of pyramid 322
  of 3D shape 316

W

weight 200
written method 12

X

x-coordinate 296

Y

y-coordinate 296
y-intercept 300, 302, 308, 388